Introduction to
Environmental Science

Introduction to Environmental Science

Joseph M. Moran
Michael D. Morgan
James H. Wiersma
University of Wisconsin, Green Bay

W. H. Freeman and Company
San Francisco

Sponsoring Editor: *Gunder Hefta*
Developmental Editor: *Linda Chaput*
Project Editor: *Betsy Dilernia*
Manuscript Editor: *Suzanne Lipsett*
Designer: *Perry Smith*
Production Coordinator: *Linda Jupiter*
Illustration Coordinator: *Cheryl Nufer*
Artists: *Cyndie Jo Clark; Eric G. Hieber and Associates*
Compositor: *York Graphic Services, Inc.*
Printer and Binder: *Kingsport Press*

Cover photograph by Steven C. Wilson/ENTHEOS.

Library of Congress Cataloging in Publication Data

Moran, Joseph M
   Introduction to environmental science.

   Includes bibliographies and index.
   1. Human ecology.   2. Environmental protection.
3. Environmental policy.   I. Morgan, Michael D.,
joint author.   II. Wiersma, James H., joint
author.   III. Title.
GF41.M67   301.31   79-19007
ISBN 0-7167-1020-X

Printed in the United States of America

2 3 4 5 6 7 8 9

Contents

Preface

Much of the excitement that characterized concern about the environment in the late 1960s had waned by the middle of the 1970s, and in its place the problem-oriented, interdisciplinary field of study known as environmental science had begun to evolve. This relatively new field, now an established academic program, is marked neither by emotionalism nor by fatalism; rather, it represents an attempt to assess the environmental problems of our technological society objectively and rationally. In this context, problems and their possible solutions are studied carefully, and decisions more often involve compromise and cooperation than finger-pointing and reproof.

As the title suggests, *Introduction to Environmental Science* is meant to introduce students to this field of study. The book is intended for students who do not plan to major in physical or biological sciences, and we presume no prior experience with science. We examine environmental problems in the framework of well-founded physical and biological principles. However, the discussion of each of these principles is developed only to the extent that it directly contributes to the students' understanding of the problem at hand. The product, we hope, is an uncluttered, systematic exposition that will appeal to reason rather than to emotion.

One of our fundamental objectives is to show students how environmental problems and controversies affect them. We try to accomplish this by briefly reviewing the normal functioning of the relevant part of the environment; examining the problem in its historical perspective; identifying the attitudes and activities that contribute to it; and pinpointing (when possible) its political, legal, economic, and social aspects.

In coming to recognize the personal sacrifices and trade-offs necessary to improve environmental quality or to contribute to resource conservation, readers will gain an appreciation of the difficulties of resolving environmental problems on a local, national, or worldwide scale. At the same time, they will learn to determine what each of them can do to help solve such problems.

We do not presume to tell students what course of action they should adopt. Instead, we present a number of alternatives—including inaction—and we discuss their consequences. It is our hope that the holistic view of problems presented here will help readers to make informed, realistic choices. They should come to understand that different situations dictate different responses, and that blind adherence to the extreme of environmental purity or that of unbridled exploitation may result in a situation to which there can be only one outcome: severe hardship.

Organization

Beyond the introductory chapter, the book is divided into three major parts: Part I, Concepts of Ecology (Chapters 2-5); Part II, Environmental Quality and Management (Chapters 6-15); and Part III, Fundamental Problems: Population, Food, and Energy (Chapters 16-20).

Part I surveys the fundamental principles that govern the functioning of the environment. What is the natural flow of energy and materials through the environment? How do organisms and ecosystems respond to change? Why and how do populations grow? Readers will gain an understanding of these and other points that they can apply to their comprehension of more specific issues later in the book.

Part II explores dominant issues of environmental quality and management: water and air pollution; exploitation of the earth's rock, mineral, and fuel resources; waste disposal; endangered species; and conflicts in land use.

Part III focuses on problems at the core of most environmental issues: growing human population and shortages in food and energy resources.

This book is designed for one-semester courses on environmental science. Recognizing that topic coverage in such courses varies from college to college, we have included a wide variety of topics, organizing the coverage in such a fashion that chapters or parts of chapters—especially in Part II—can be omitted without loss of continuity.

Features

We have included several features in our book that make it an effective teaching and learning text.

Pedagogical Aids

Each section begins with a statement of objectives, and each chapter ends with conclusions, summary statements, review questions, suggestions for group and individual projects, and an annotated bibliography. Both metric and British units of measure appear throughout the text and in many tables and illustrations. There are appendixes of scientific conversion factors, geologic time, and powers of ten notation.

Boxes

Where a deeper scientific explanation of certain topics—for example, net energy analysis and the nature of nuclear power plants—seems desirable, we have included complementary information set off from the text in boxes. Where specific examples will illuminate the social, political, economic, or personal aspects of such environmental conflicts as the channelization of the Kissimmee River and the poisoning of the Love Canal, brief case studies are included in box form.

Illustrations

To bring more realism to the book we have included an unusually large number of high-quality photographs and line drawings. More than 400 illustrations—photographs, drawings, maps, and graphs—illustrate and clarify important points.

Glossary

All important terms are italicized and defined at first use in the text. They appear again in the glossary, which, with more than 350 entries, functions as a minidictionary of environmental topics and terms.

Supplementary Materials

The text is accompanied by an Instructor's Manual that contains learning objectives for each chapter, test questions (each with a parenthetical reference to the page in the book on which it is answered), topics of current concern for discussion or research, and a list of recommended slides and films that instructors might wish to obtain for classroom use.

Acknowledgments

We are especially indebted to Professor John F. Reed for his continuous interest and encouragement. We are also grateful to other colleagues at the University of Wisconsin-Green Bay, particularly to Professors Charles A. Ihrke, Robert W. Lanz, Charles R. Rhyner, Dorothea B. Sager, Paul E. Sager, Leander J. Schwartz, Ronald H. Starkey, Ronald D. Stieglitz, Richard B. Stiehl, and Joanne Westphal, for their valuable contributions. To the students of Environmental Science 102 we are thankful for critiques of course materials. We extend a special thanks to Jeanne Broeren, Nancy Lambert, Janice Mastin, and Joy Phillips for their enthusiastic, prompt, and accurate typing of the manuscript's innumerable drafts. We also appreciate the extra efforts of photographer Michael L. Brisson and documents librarian Kathy Pletcher.

We also thank reviewers who criticized the manuscript at one time or another during its development. They are: Professor Richard Anthes, The Pennsylvania State University; Professor Gary Barrett, Miami University; Professor Russell F. Christman, University of North Carolina, Chapel Hill; Dean Earl Cook, Texas A&M University; Professor Eckhard Dersch, Michigan State University; Dr. John M. Fowler, National Science Teachers Association; Professor Garrett Hardin, University of California, Santa Barbara; Dr. John P. Harley, Eastern Kentucky University; Professor Marilyn Houck, The Pennsylvania State University; Professor Richard E. Pieper, University of Southern California; Professor David Pimentel, Cornell University; Professor Clayton H. Reitan, Northern Illinois University; Professor Wayne M. Wendland, University of Illinois at Urbana-Champaign; Professor Susan Uhl Wilson, Miami-Dade Community College.

Few authors have had the opportunity to work with as talented, dedicated, and cooperative a group of professionals as the editorial and production staff of W. H. Freeman and Company. Their patience, organization, and imagination transformed a formidable task into a rewarding experience. Their ability to translate ideas into reality and their tireless dedication to the project served as continual inspiration to us. In this regard we are especially grateful to Gunder Hefta, Linda Chaput, Betsy Dilernia, Perry Smith, and Ruth Allen.

One of the most difficult challenges of writing an introductory science text for nonscience majors is communicating sophisticated concepts in terms that are readily understood. Suzanne Lipsett, our manuscript editor, and John Hendry, the editor of boxed material, met this challenge with skill and precision.

September 1979

Joseph M. Moran
Michael D. Morgan
James H. Wiersma

Introduction to Environmental Science

Satellite view of North and South America and adjacent portions of the Atlantic and Pacific Oceans. The photograph was taken from an orbit more than 35,000 kilometers (22,000 miles) above the earth's surface. (NASA.)

Chapter **1**

People and Nature in Conflict

Aldrin, anchovies, and asbestos: what do these seemingly unrelated things have in common? Each in its own way will somehow affect the well-being of many people. Aldrin is a pesticide that contaminates food, anchovies are a major protein source now threatened by overfishing, and asbestos is a valuable mineral that, as an air pollutant, causes lung cancer. These are only three in a perplexing array of serious environmental problems that plague our modern society. With ever increasing frequency, it seems, the media report on some insidious poison that somehow wound up in the wrong place, on a bitter confrontation between conservationists and resource developers, or on a municipality's costly battle to maintain an outmoded waste disposal system. Just where did these problems come from? To find their roots, we must go back many years.

The early settlers of North America found themselves in a land of plenty. Resources were so abundant and the population so small that the government literally gave parcels of land to anyone willing to exploit the timber, fuel, minerals, and running water. And exploit they did, with enthusiasm and thoroughness. They saw no reason not to; the supply of resources seemed endless.

The abundance of natural resources spurred the growth of industry and technology, and gradually a simple, agrarian society became a complex, industrial one. Smokestacks belching noxious fumes were viewed as beacons of economic prosperity. And the ability of rivers and streams to wash away pollutants was considered more than equal to even the most offensive industrial and municipal wastes.

The growth of industry and prosperity encouraged a steady increase in population, and each succeeding generation demanded more and better products and convenience items and a wider variety of services. Agriculture and industry responded by drawing on increasingly sophisticated technology, which in turn used fuel and materials at an ever accelerated rate. Consumption of energy and mineral resources soared. More water was required to extract resources and produce goods, and in these processes more waste was generated. The nation prospered, but at the expense of environmental quality. Today we enjoy the fruits of our nation's progressive past, but we are also faced with the undesirable by-products of its rapid technological advance.

Many readers, undoubtedly burdened with enough worries of their own, might dare to ask, "But why should *I* worry about it?" The answer is that we all must concern ourselves with the contemporary problems of overpopulation, resource exploitation, and pollution simply because these problems affect every one of us. For example, as our population continues to grow, competition for available jobs gets stiffer. Well-qualified job applicants find, to their dismay, that their preferred professions and occupations are closed to them simply because the number of people seeking entry exceeds the number of jobs. Also, as our population grows, our demands for certain resources (petroleum, for example) are sometimes not met. Aside from immediate effects (cold homes, stilled cars) such a shortage has serious long-term economic ramifications: as resources are used up, production tails off and with it jobs in manufacturing, sales, and transportation. Resource shortages also contribute to the spiraling cost of living. When resource supply fails to keep pace with demand, scarcity-induced inflation further elevates the prices of goods and services.

Pollution, too, imposes an economic burden on each of us. As the effects of air and water pollution have become better understood, pollution abatement has become more urgent, and governments have responded with strict legislation forcing industries to install expensive air and water quality control devices. Invariably, the cost of these measures is passed on to consumers. To meet water quality regulations, many municipalities must build new sewage treatment facilities. Some construction costs for these plants are covered by federal grants funded by taxes; the rest of the construction costs and all of the operating expenses must be met locally by higher fees for water and sewage treatment or increased property taxes.

But, before we back away from pollution control because of the financial burdens it imposes on us, we must remember the main reason we are concerned about pollution: it directly threatens our personal health. And what price can be put on good health? Although we are making some progress in cleaning up air and water, pollutant health hazards will probably be with us for a long time to come. Most of us may never suffer a debilitating illness as a result of exposure to pollution, but all of us bear an increased risk of becoming ill. Unless we begin to reduce that risk by ridding ourselves of the cause, our legacy to future generations could be one of chronic illness and fear.

When we try to escape our polluted cities by vacationing in the mountains or at the seashore, we meet our neighbors there by the thousands. Our population is so large and mobile that we simply cannot avoid one another. Campgrounds are overcrowded, rural highways are jammed, and even wilderness areas are overrun by people seeking solitude.

Like it or not, then, we are all involved in the problems of population, resource availability, and environmental quality. In facing these issues, we will do well to remember that environmental concern is not unique to our generation. Even in the early days of our nation's industrial growth, some people recognized the connection between health and pollution, and realized that many common exploitive practices were self-defeating.

Awareness eventually spurred some individuals into action. At the turn of this century, a conservation ethic emerged. Efforts were made to protect wildlife prized for hunting, such as deer and waterfowl. Some farmers began to see the merits of protecting cropland from wind and water erosion. Certain timberlands were managed for sustained production instead of being logged over and abandoned. Lands were set aside to preserve their natural beauty and to provide space for recreation. Still, although the conservation movement goes back about a century, major battles against pollution began only recently. In the late 1960s, concern over deteriorating air and water quality gave birth to the environmental movement. Public demonstrations, television programs, and such popular books as Rachel Carson's *Silent Spring* heightened public awareness of the impact of human activity on the environment.

Soon it became politically expedient for legislators to formulate stricter air and water quality laws. The National Environmental Policy Act (NEPA), which requires that an environmental impact assessment be prepared prior to any federally financed project, was passed in 1969. Simultaneously, changes were taking place on a more personal level. Some far-thinking people realized that their lifestyles conflicted with their insights about environmental quality, and they began to make voluntary adjustments. They saw the advantages of limiting their family size, for instance, and of recycling certain kinds of waste.

In the mid-1970s the Arab oil embargo demonstrated our country's strong dependence on foreign petroleum. Cold homes, long lines at service stations, and closed factories triggered an intense search for alternatives to natural gas and petroleum. And, again, some people began to alter their living habits. They began to trade in their large cars for ones that consumed less fuel. Some two-car families decided that perhaps they could make do with one car, after all.

Today public interest in ecology and the environment has declined. For the most part people seem unconcerned about our energy supplies. One reason for this is that a single issue seldom holds public attention for very long. Another is that significant progress has been made in many areas. The laws we noted here are just a few of the many that have been passed to improve environmental quality.

Yet, despite the progress, many problems remain and new concerns continue to emerge. We have halted the decline in the quality of our air and water, but we have taken only a small step toward returning these priceless resources to their original condition. Nonrenewable resources, including minerals and fossil fuels, continue to dwindle, and we are still using land as if it were in endless supply. Moreover, the fundamental source of our environmental problems remains: our population continues to grow, and so do our individual demands and expectations. Believing that we are still living in an era of unlimited resources, most of us continue to pursue the American dream.

Thus, a great deal remains to be done. What we do today will affect not only our own prospects but also those of future generations. What can we do?

To take constructive action, we must first identify the conflicts. That done, we must consider every alternative and learn enough to analyze the consequences of each. Although many environmental problems are formidable, solutions do exist. For example, it is true that petroleum supplies are falling behind demand, but vast supplies of coal and alternative energy sources (solar, wind, and geothermal) have

yet to be tapped. Also, we have a vast storehouse of knowledge from which to develop solutions; and history warns us against underestimating human ingenuity.

On the other hand, we must recognize that we are working under severe constraints. For example, using more coal will accentuate air pollution, and more air pollution will increase the health hazards. Also, developing new technologies is expensive, and the amount of money we can afford to spend on air pollution control is limited.

Everyone is affected by environmental problems, and therefore everyone has a stake in (if not a personal responsibility for) solving them. But our collective success depends on how much we know and how creatively we apply our knowledge. In forthcoming chapters, we will explore topics that will provide the reader with a rational basis for deciding our environmental future. First we will examine how nature works and the many ways in which it serves us. Then we will consider how our activities affect the proper functioning of these natural processes, which in turn influence our own well-being. After examining the roots of our problems with population, resources, and the environment, we will explore current attempts to correct them, analyzing the results to date, both the benefits and the costs. Finally, we will consider what the future may hold for us.

It will be clear from these discussions that for every problem we identify many responses are possible. Some readers may decide that these issues are unimportant and may choose to do nothing. Others, believing the problems to be critical, may decide to devote their lives to solving them. Most of us will probably take a position somewhere between these extremes. The choice we make will hinge on our knowledge, our personal prejudices, and our ideals.

Challenging days are ahead. Many futurists predict that major changes will take place, not all of them welcome. But only by making the proper decisions can we hope to direct our own destinies. We are at the crossroads now, and a sense of purpose anchored in understanding has never been more important.

Part I

Concepts of Ecology

To find practical answers to environmental problems, we must first understand how the environment works. In this part of the book, we consider some of the fundamental principles that govern nature's activities. In Chapter 2, we look at the flow of energy through food webs and the movement of materials through the environment. An understanding of these flows is essential to the solution of such problems as human hunger and the decline in the quality of our air and water. Because the environment is constantly changing, either naturally or as a consequence of human activities, we consider in Chapter 3 the response of organisms and ecosystems to such factors as water and air pollution, fire, weather, and agriculture. In Chapter 4, we look at how populations grow and at what controls their growth. Only by understanding population growth can we control the pests that attack our crops and livestock, thwart some of the organisms that transmit human diseases, save certain endangered species, and limit human population growth. In Chapter 5 we consider the characteristics of the diverse ecosystems of the earth and the points at which these ecosystems are vulnerable to human activity.

In succeeding sections of this book, our understanding of the basic principles that govern the working of the environment will help us to better understand the environmental consequences of human activities. As our increasing numbers and activities continue to have an impact on the earth, a better understanding of ecological principles becomes ever more important to our well-being.

This fox, carrying its prey, vividly illustrates one step in the flow of energy through an ecosystem. (Wilford L. Miller, from National Audubon Society.)

Chapter 2

Ecosystems: The Flow of Energy and Materials

Opinion may differ as to which of our environmental problems are the most critical, but most of us probably share many of the same concerns. We worry about the pollution of our air and water. We fear that we will run short of fuel to drive our cars and heat our homes. Some of us are angered about plans to build a new highway through farmlands or to develop a waste disposal site in our neighborhood. Basically, such concerns are rooted in the fact that most of our resources are limited—when demand exceeds supply someone suffers from the shortages.

On a larger scale too, all of our environmental problems are economically and ecologically interrelated: each one has resulted from the inability of the environment to maintain its integrity in the face of our unrelenting demands for more material goods. Each of our environmental problems involves the movement of energy and materials through the environment, and this movement is governed by certain inviolable principles that we all too often forget to take into account. In our disregard of these principles, we simply try to make nature do more than it can do.

Components of Ecosystems

To understand the laws that govern the movement of materials and energy in the environment, we will study these flows in the framework of *ecosystems*. An ecosystem is a functional unit of the environment that includes all organisms and physical features within a given area. An ecosystem thus consists of both living, or *biotic,* and nonliving, or *abiotic,* components. The biotic community is made up of *producers* and *consumers,* which are distinguished by their major functions. Producers, mainly green plants, manufacture their food from water and carbon dioxide, using sunlight as a source of energy. (This process is called *photosynthesis,* and it is discussed further in subsequent paragraphs.) In contrast, consumers are incapable of producing their own food, and must consume other organisms for energy and nutrition.

Consumers fall into one of four classes on the basis of their food source. A consumer that eats only plants is a *herbivore,* whereas a consumer that eats only animals is a *carnivore.* An *omnivore* eats both plants and animals. Thus, a pheasant feeding on corn is a herbivore, a hawk making a meal of the pheasant is a carnivore, and a human being eating both corn and pheasant is an omnivore. The fourth group of consumers feeds upon *detritus*—the freshly dead or partially decomposed remains of plants and animals. This class, called the detritusfeeders, or *decomposers,* includes bacteria, fungi (for example, molds and mushrooms), and such animals as termites and maggots. In feeding upon detritus, decomposers acquire energy and nutrients. In addition, they return simpler decomposition products, such as their own wastes, to the soil and air, where these materials can again be taken up by green plants and recycled.

The physical environment comprises the abiotic parts of an ecosystem. These components include chemical substances, which can be subdivided into two groups—*inorganic* and *organic.* Examples of inorganic chemicals are water, oxygen, carbon dioxide, and essential minerals. Most organic substances are produced by organisms. Examples of organic substances are carbohydrates, fats, proteins, and vitamins. Besides chemicals, other abiotic components include such physical factors as temperature, light, and wind, all of which are manifestations of energy.

We commonly picture ecosystems as being relatively untouched areas—deserts, forests, mountain ranges, and prairies—such as those shown in Figure 2.1. But areas that have been greatly changed by human activity—for example, a cornfield—also

Figure 2.1 *Above:* A desert ecosystem in Saguaro National Monument. (U.S. Department of the Interior, National Park Service photo by Fred Mang, Jr.) *Opposite, top:* A mangrove ecosystem along a waterway in the Florida Everglades. (U.S. Department of the Interior, National Park Service photo by Jack Boucher.) *Middle:* An evergreen forest ecosystem on the Olympic Peninsula, Washington. (U.S. Department of the Interior, National Park Service photo by Jack Boucher.) *Bottom:* Grasslands in the Wichita Mountains Wildlife Refuge, Oklahoma. (U.S. Department of the Interior, Fish and Wildlife Service, photo by Rex Gary Schmidt.)

function as ecosystems. Although a farmer may intend to grow only corn, other producers, called weeds in this context, spring up as well. Herbivorous insects eat portions of the corn plants, and birds and carnivorous insects feed upon the corn-eating insects. Bacteria, earthworms, and other decomposers live in the soil. Even an urban area can be considered an ecosystem, albeit a highly modified one. Apart from pigeons, rats, and pets, people are the major consumers in cities. Unlike other ecosystems, urban ecosystems have few producers and decomposers. Food must be brought into an urban ecosystem, and wastes (garbage and sewage) must be processed by special treatment plants or hauled out of the city to a dumping site, where they are broken down by decomposers.

Movement of Energy

The components of an ecosystem exhibit varying degrees of activity, and all activity is the result of the movement of energy. An example from the physical world is the energy that drives the wind. An example from the biological world is the food energy that enables us to carry out our daily activities. In this section, we focus primarily on energy flow involving organisms. In later chapters, we will explore how energy sustains the water cycle, the weather, and landscape evolution. For a general description of energy and how nature regulates its flow, see Box 2.1 (pp. 14–15).

Food Webs

Most of us eat our food without thinking very much about where it comes from or how it functions to keep us alive. We eat to obtain the energy and nutrients we need for sustenance and growth, and then go about our business. But with each meal we form a link with other organisms, thereby playing a role in the flow of energy through ecosystems.

All organisms, dead or alive, are potential sources of food energy for other organisms. This energy always moves in only one direction—from producers to consumers. Consider one simple example, illustrated in Figure 2.2: food energy is transferred from a clover plant to a rabbit to a fox, as the clover is eaten by a rabbit that subsequently falls prey to a fox. This pathway is called a *food chain*. Natural ecosystems consist of complicated networks of many interconnected food chains called *food webs*. Figure 2.3 is a stylized food web; the arrows indicate the flow of energy in the form of food from one organism to another. Though it may seem complicated, the diagram is simplistic. Hundreds more species would have to be added to portray the actual complexity of the food webs that exist in nature.

Each group of organisms along an energy pathway occupies a *trophic*, or feeding, level. All green plants—that is, the producers—in an ecosystem belong to the first trophic level. Herbivores compose the second trophic level, carnivores that eat herbivores the third trophic level, carnivores that consume other carnivores the fourth trophic level, and so on. Omnivores function at more than one level: when a bear eats a pawful of berries, he is functioning at a lower trophic level than when he gulps down a fish.

The ultimate source of the energy used by organisms is the sun. A small portion of the sun's energy enters food webs through the photosynthesis that takes place in green plants. For example, by the process of photosynthesis plants in a clover field use light energy from the sun to combine low-energy substances from their environment (carbon dioxide from the air, and minerals and water from the soil) to produce food with a higher energy content. The food produced by photosynthesis is sugar, a form of carbohydrate. Clover changes this sugar into other types of food—proteins, fats, and starch (another form of carbohydrate). These foods not only serve as the basic building blocks to construct and repair living *cells* (the basic structural unit of organisms), but they are also used by the clover as the energy source for performing the work of maintenance, growth, and reproduction.

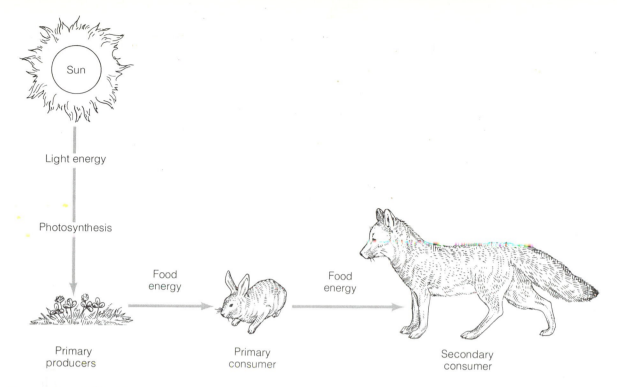

Figure 2.2 A grazing food chain.

The other product of photosynthesis is oxygen. Some of the oxygen is used by the clover in various metabolic processes, and the remainder escapes into the surroundings. Figure 2.4 illustrates the sources of raw materials and the distribution of the products of photosynthesis within a plant. The process of photosynthesis is summarized by the following word equation:

$$\text{Carbon dioxide} + \text{water} + \text{light energy} \rightarrow$$
$$\text{sugar} + \text{oxygen}$$

In a clover field, rabbits, mice, and other herbivores feed upon plants to obtain needed energy and nutrients. In turn, carnivores, such as foxes, weasels, and hawks feed on the herbivores, gaining from them the raw materials for their movement, maintenance, growth, and reproduction.

Our clover field is an example of a *grazing food web*. In grazing food webs, carnivores feed directly upon herbivores, which feed directly upon living plants. Living plants, therefore, form the base of grazing food webs. Eventually, however, all plants and animals die, and their dead matter form the base of *detritus food webs*, often equally important but generally less conspicuous than grazing food webs. We can get a sense of how a detritus food web works by examining energy flow within a marsh at the edge of a lake, as shown in Figure 2.5. When wetland plants and animals die and settle, their remains are broken down into smaller and smaller fragments by decomposers. Both the detritus and the decomposers themselves are eaten by scavenging animals such as crayfish, shrimp, and snails. These animals then serve as food for small fish (perch) that are subsequently eaten by larger game fish (northern pike and walleyed pike) and fish-eating birds (eagles and herons).

Detritus pathways usually predominate in terrestrial ecosystems and in aquatic ecosystems such as streams, rivers, and marshlands. In fact, as little as 10

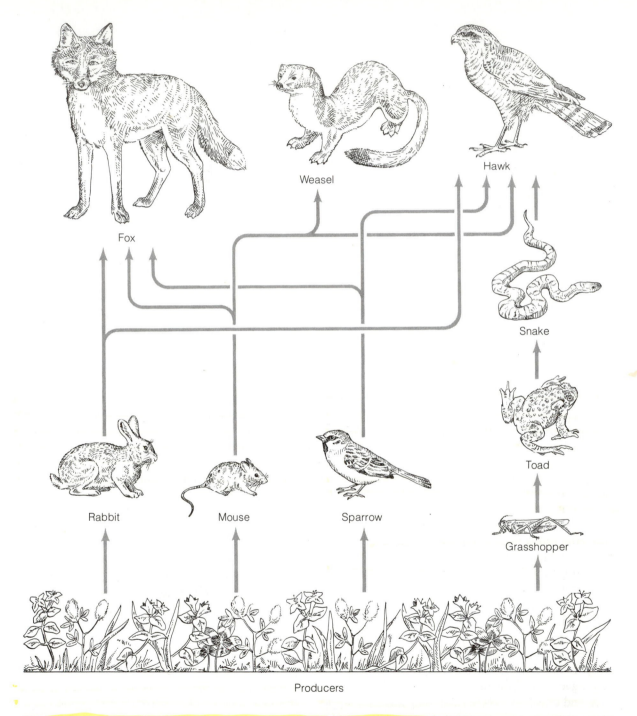

Figure 2.3 A very simplified food web, made up of a network of interconnected food chains.

Weasel

Fox

Hawk

Snake

Toad

Rabbit

Mouse

Sparrow

Grasshopper

Producers

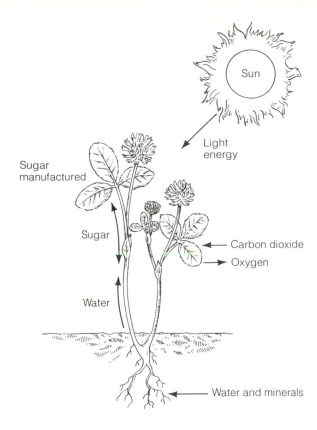

Figure 2.4 Sources of raw materials and the distribution of the products of photosynthesis in a clover plant. Sugars can move upward through the stem to be used for growth or stored in seeds or fruits, or they can move downward to be used for growth or stored in roots.

the relative significance of the type of pathway hinges on the type of ecosystem within which it exists.

Human beings function primarily as herbivores in grazing food webs. Plant materials such as cereals, vegetables, and fruit account for about 64 percent of the food energy in the American diet. (Worldwide, the figure is 89 percent.) Furthermore, even when we function as carnivores, most of our meat comes from herbivores—beef cattle, chicken, and hogs. When people eat fish and shellfish (for example crabs, clams, and oysters) or mushrooms they tap food energy from detritus pathways, but these foods make up only a very small portion of the human diet (less than 1 percent in the United States).

It is important to remember that energy moves through food webs in only one direction—it never moves from carnivores to herbivores to green plants. Organisms at each trophic level depend on those at lower trophic levels for the energy with which they sustain themselves and reproduce. For example, although human beings function as omnivores, we cannot convert energy from the sun into food energy; we depend on green plants to make the transformation for us. Thus, all consumers are ultimately dependent on green plants for their food energy.

The Efficiency of Food Webs

Food webs are the sources of food energy, but nature limits the amount of food energy accessible to the organisms within those webs. Not all food energy is transferred from one trophic level to the next. If we were to envision the amount of energy present in all organisms at each trophic level, we would see a pyramid in which the producers form a broad base and each succeeding level contains considerably less energy than the one below it. Figure 2.6 is an example of such an ecological pyramid. The percentage of energy transferred from one level to the next is referred to as *ecological efficiency,* and it can be determined by dividing consumer production by prey production. *Production* refers to the accumulation of

percent of the living plant material in a forest is eaten by herbivores; the rest dies and is funneled through detritus pathways. Decomposers in a forest include bacteria, fungi, millipedes, and certain insect larvae. These organisms, in turn, fall prey to carnivores such as centipedes, beetles, spiders, and rodents, which subsequently may be eaten by snakes and owls. In other types of ecosystems, however, energy moves predominantly through grazing food webs rather than detritus food webs. For example, in open water where microscopic, free-floating plants (*phytoplankton*) are the chief producers, as much as 90 percent of the algae is eaten by microscopic animals (*zooplankton*) and thus enters the grazing web, leaving only 10 percent to die and enter detritus pathways. Hence,

Box 2.1

The Flow of Energy

Energy is a deceptively familiar word. We all speak of food energy, atomic energy, abundant energy, cheap energy, and so on. Moreover, we all know that our highly industrialized society would grind to a standstill without great quantities of energy. But energy is more than simply the force that turns the wheels of civilization. It pervades, indeed powers, the whole universe—from the radioactive decay of atomic nuclei to the awesome power of hurricanes to the movement of planets, stars, and galaxies. Despite energy's pervasiveness and its seeming familiarity, it is an abstract and elusive concept. It cannot be seen, touched, tasted, or smelled. What, then is it?

Physicists, who study this mysterious something that is responsible for everything that happens everywhere, usually define energy as *the ability to do work or produce change*. Work is done—change occurs—energy is used—whenever a leaf flutters, a baby cries, a truck climbs a hill, a lightbulb glows. A hummingbird can flit about only because the food energy (sugar) in the nectar it sips is transformed biochemically into mechanical energy in the form of a blurred pair of beating wings. We can cook a meal on an electric stove only because somewhere a turbine is being driven by falling water or pressurized steam to produce electricity.

Energy exists, then, in a variety of forms. These include heat, light, chemical (food or fuel), kinetic, electrical, and nuclear energy. Every form of energy can be transformed into another. Scientists observing these transformations always find that no new energy is created in the process and that no energy is ever destroyed. This observation underlies the *first law of thermodynamics,* also known as the *law of the conservation of energy: Energy can neither be created nor destroyed*. This law holds for all systems, living and nonliving.

Let us restate the conservation law as it relates to living organisms: *No organism can create its own food supply*. Every organism must fulfill its energy needs by relying upon energy transformations within its ecosystem or adjacent ones. Thus plants must depend on light energy, and animals must rely upon plants or other animals for energy.

A familiar example of an energy transformation takes place in an automobile engine. As gasoline burns in the cylinders, much of its chemical energy is transformed into motion (kinetic energy). A great deal of that chemical energy, however, is not converted into kinetic energy. Instead, it escapes through the engine walls or goes out the exhaust pipe in the form of heated combustion products: carbon dioxide, water vapor, and partially burned fuel. Because the engine obeys the first law of thermodynamics, no energy is created or destroyed in the process; energy input in fuel exactly equals energy output. In other words, only energy *transformations* occur (although some of them are undesirable, or, in other words, do no work).

Efficiency is a measure of the fraction or percentage of the total energy input of a process that is transformed into work or some other usable form of energy. It can be determined by the following formula:

$$\text{Percent efficiency} = \frac{\text{energy output as work} \times 100}{\text{total energy input}}$$

A standard gasoline-powered auto engine typically

has an efficiency of 20–25 percent. The efficiency of an incandescent light is much lower, about 5 percent. This means that only 5 percent of the electricity input is transformed into light energy; all the rest is transformed into heat. (This is why a "burning" light bulb is hot to the touch.) Muscles, internal combustion engines, and other systems that convert chemical energy into work all operate at efficiencies below 50 percent. Devices that are designed solely to produce heat are generally more efficient. Home oil furnaces are around 65 percent efficient. Electric space heaters—which need no heat-wasting chimneys or vents—are 100 percent efficient.

It would be convenient if the biological and physical systems we exploit approached efficiencies of 100 percent. But this is rarely possible, because energy's natural and unavoidable tendency is to spread out (become disorganized or disordered). The degree of disorder in a system can be measured and expressed mathematically as *entropy*. The higher the entropy value, the more disorder in a system, and so the less work the system's energy can do. The chemical energy that goes into an auto engine, for example, is more highly ordered energy than the mechanical and heat energy that comes out.

Another example should help clarify the concept of disordered energy, or energy that can do little work. When we heat a bathtubful of water with an oil-fired heater, we convert the chemical energy in a small volume of fuel to heat energy and transfer it into a very much greater volume of water. The heat energy is now quite disorganized. It can't drive a turbine; it can't even heat a can of soup above the temperature of the bathwater itself.

Nor is there any way to transform this heat energy in the bathwater back into a more concentrated (ordered) form. For all practical purposes, then, it is forever lost. This is the inevitable consequence of the *second law of thermodynamics,* which may be stated as follows: *In every energy transformation, some energy is always lost in the form of heat that is thereafter unavailable to do further work.* In other words, in all systems energy "flows downhill" from more ordered to less ordered forms. And the most disordered form of all—the form every bit of energy ultimately takes—is heat.

One consequence of entropy, or the downhill flow of energy, is that organisms—which are highly organized systems—require a continual input of energy to maintain their order. For example, because our red blood cells break down at the rate of about 5 million per second, we have a never-ending need for energy to manufacture (organize) new cells.

Entropy has another important consequence for organisms and the ecosystems in which they live. As we shall see later in this chapter, much of the earth's materials are recycled again and again through biological and physical systems. But energy cannot be recycled through a system once it reaches the bottom of the "energy hill" in the form of heat. Fortunately for us and for all living systems, the energy forever lost as heat is continually being replaced by the inflow of solar energy, which is more ordered than heat energy and which can be transformed by plants into chemical energy through the process of photosynthesis.

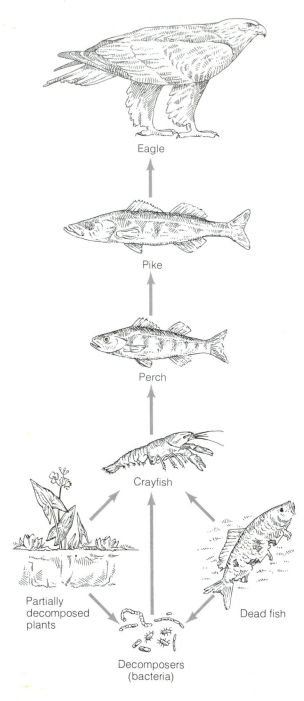

Eagle

Pike

Perch

Crayfish

Partially
decomposed
plants

Decomposers
(bacteria)

Dead fish

Figure 2.5 A simplified detritus food web.

energy at a given trophic level by means of growth and reproduction, and is often determined by measuring a change in the *biomass* (the total weight or mass of organisms in an area) in that trophic level.

The efficiency of energy transfer between trophic levels generally ranges from 5 to 20 percent, depending on the types of organisms and environmental conditions. But an ecological efficiency of 10 percent (the 10 percent rule) is used frequently in general discussions because calculating 10 percent of production at each succeeding trophic level is convenient. For example, if we apply the 10 percent rule to the grain-beef-people food chain, we can predict that 100 kilograms (220 pounds) of grain produces 10 kilograms (22 pounds) of beef that in turn produces only 1 kilogram (2.2 pounds) of people. (See Appendix I for conversion factors between the English and metric systems.) Hence, we see that only a small amount of food energy is transferred from one trophic level to the next.

Causes of Low Efficiencies

It is important that we know why ecological efficiencies are low—that we are able to account for the apparent loss of energy between trophic levels—if we are to meet the challenges of feeding a growing human population and managing the earth's plant and animal life.

To determine where the apparent energy losses occur, we must begin by realizing that many steps lie between an organism's attempts to obtain food and the utilization of food for energy by that organism's cells. First, the predator must capture and eat its prey. But all organisms possess characteristics that reduce their vulnerability to being eaten. Although plants cannot pull up their roots and run away, they do have many mechanisms to ward off attack by herbivores. Needles and thorns on thistles and cacti certainly discourage hungry herbivores, and some plant leaves or stems contain poisons or release noxious vapors that repel them. Animals also possess characteristics that help them to escape predators.

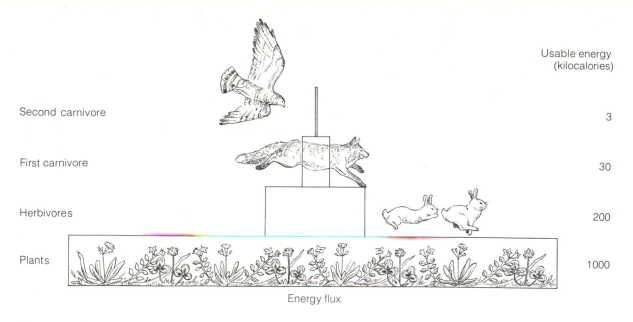

	Usable energy (kilocalories)
Second carnivore	3
First carnivore	30
Herbivores	200
Plants	1000

Energy flux

Figure 2.6 A pyramid of energy illustrating the low efficiency of energy transfer between trophic levels.

Prairie dogs find refuge in their burrows, skunks emit a pungent odor, fleet-footed antelope outrun their predators, and, as shown in Figure 2.7, walking sticks blend unnoticed into their surroundings. Although predators in turn have capabilities to overcome the defense mechanisms of their prey, their ability to find and capture prey is limited. Hence, often no more than 10–20 percent of the material present at one trophic level is actually harvested by the organisms in the next trophic level. Those organisms that escape predators eventually die, and their energy enters detritus pathways.

Once the prey is harvested and ingested, not all of it is digested. Most animals cannot digest hair, feathers, skeletons, and plant tissue containing fibers composed of cellulose. Therefore, energy contained within these materials is unavailable to most animals. Undigested wastes are excreted and subsequently used by detritus feeders.

Further energy losses occur when digested food is "burned" in living cells. This "burning" process, called *respiration*, releases from digested food the energy for movement, maintenance, growth, and reproduction, and is performed by all living things. Sugars are usually the most important raw material for respiration. Respiration is an inefficient process; that is, less than half of the energy present in sugars is converted into work. Most of the remaining energy in sugars is transformed into heat energy that cannot be used to perform work, and is eventually lost to the environment. Two other products of respiration are carbon dioxide and water, which are usually released into the surroundings. The process of respiration is summarized by the following word equation:

Sugar + oxygen ⟶

energy + heat + carbon + water
for   energy   dioxide
work

As you can see, several factors account for the 10 percent rule. Often, only a small percentage of the production at one trophic level is harvested by the organisms in the next trophic level; the remainder eventually dies and enters detritus food webs. Sec-

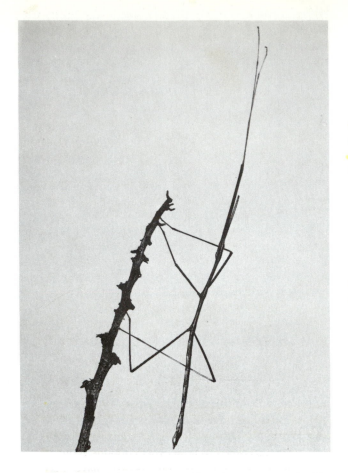

Figure 2.7 A walking stick, which often goes unnoticed by potential predators because of is resemblance to its surroundings. (Des Bartlett, Photo Researchers.)

Consequences of Low Efficiencies

In view of the 10 percent rule and the fact that each organism has a minimum food requirement for survival, it is clear that there is a limit to the number of organisms that can survive at a particular trophic level. This fact applies to all organisms, including human beings: the earth can support only so many consumers. Given the size of our species' world population and its growth rate during the last century, we cannot afford to lose sight of this natural limit in our struggle to eliminate hunger.

Through an understanding of food web dynamics, human beings have developed the means for greatly improving the amount of food energy available to

ond, only some of the ingested food is actually digested; the remainder is lost as waste, which also enters detritus pathways. And, third, much of the energy in the digested food is lost as heat when it is released through respiration for performing the essential functions of organisms.

These inefficiencies occur not only in grazing food webs but also in detritus food webs. Hence, as illustrated in Figure 2.8, all the solar energy that enters the food webs of an ecosystem eventually is lost from the ecosystem as heat. Because of the continual loss of energy as heat, an ecosystem requires a continual influx of solar energy.

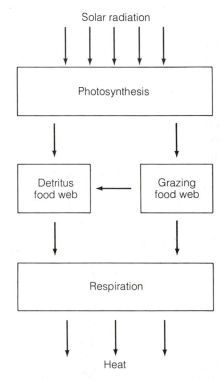

Figure 2.8 Basic pathways of energy flow through an ecosystem.

them—thus, it might be added, contributing to the unprecedented growth of the human population. One means of applying this knowledge has to do with the fact, noted earlier, that the presence of large amounts of unpalatable or undigestible plant materials such as wood can lower ecological efficiencies. Accordingly, if we clear away forests and replace them with crops, we channel a greater percentage of photosynthetic energy into seeds and fruits, materials that are digestible by us and domestic livestock. During the last two-hundred years, timberland in the United States has been reduced by some 20 percent, an area about the size of Texas. Many of these forests were cleared to provide land for agriculture.

This means of increasing food production has its natural limits, however. Most forested land suitable for agriculture has already been converted. Also, existing forests must be maintained for recreational purposes, for sustaining wildlife habitats, and for fulfilling our needs for wood products. Therefore, as is always the case with environmental issues, human inventiveness is met with constraints from nature and our own conflicting demands on resources.

In fact, limited success at best seems to be the rule rather than the exception for human attempts at increasing ecological efficiency. Animal and plant breeders, for example, have expended considerable effort to increase the efficiency of energy use by livestock and crops. Dairy farmers point with pride to dramatic increases in milk production per cow. Fifty years ago the average cow in the United States produced 2000 liters (2100 quarts) of milk per year. Whereas the average cow now gives over 4900 liters (5200 quarts) per year. However, most of the increased milk production is the result not of improved efficiency, but rather of better feed and breeding practices that have selected for larger animals. Today's farmer raises more productive cows, but the cows are larger and therefore require more feed. The net effect is about the same milk production per cow for the same food energy input.

A similar situation exists in crop breeding. New varieties of wheat and rice have been developed that

partition more of the sun's energy into the grain head (the prime human food source) than into the plant stalk. As a result, grain production has improved severalfold in some areas of the world. But the attainment of these goals is costly: it requires the production and application of more fertilizer and pesticides and better management of water resources. These agricultural innovations use considerable amounts of petroleum. (For an analysis of recent on-the-farm changes in energy consumption, see Box 2.2.) Not only do these changes represent increased energy expenditures, but they add to our concerns over health-threatening pollution and wasteful land management as well.

If we ate less meat and more fruits, vegetables, and cereals, the amount of food energy available to us would be increased automatically. A gram of grain yields about the same amount of energy as a gram of meat. So, for example, according to the 10 percent rule, approximately ten times more food energy would be available to us if instead of eating beef, we ate the grain we now feed to beef cattle. A strict vegetarian diet is unappealing to many Americans, who are accustomed to a diet that contains about 40 percent animal products, but people have been forced to shorten their food chains in many of the overcrowded areas of the world. The diet of Southeast Asians, for example, consists almost entirely of rice or wheat; occasional meals of fish for protein supplement amount to about 10 percent of their diet. Most of these people simply cannot afford the energy loss—about 90 percent, according to the 10 percent rule—that occurs between the trophic levels of herbivore and carnivore.

Another way we increase the food energy available to us is by reducing or eliminating populations of other organisms that occupy our trophic levels. We apply pesticides to kill insects that compete with us for food, thus increasing the share of available energy that we receive. If our competitors get the bigger share—which they do, for example, during locust plagues (see Figure 2.9)—then fewer people can be fed. Famine and death may result unless food

Box 2.2

Net Energy Analysis: The Energy Costs of What We Eat

How much energy does it take to produce your food? If you lived in one of the world's few remaining hunting and gathering societies, it would take relatively little—just what you expended in searching out the berries, seeds, roots, and other plant products you ate and in catching and cooking your game or fish. If you lived in an unmechanized agricultural society, the amount of energy expended in food production would still be modest.

But you live in an industrialized society, which means that it takes a lot of energy to produce the food you consume each day. In 1974, American agriculture consumed 44 billion liters (12 billion gallons) of petroleum.

The graph illustrates the historical changes in energy costs of food production. An *energy subsidy* is the amount of energy, in one form or another, that we must add to some process—in this case food production—to get the results we want. Energy subsidies for food production are commonly measured in kilocalories of energy added per "food calorie," or kilocalorie, of food produced.

Our century has witnessed great changes in the growing of crops and livestock. The development of cheap and abundant petroleum supplies fostered a change from *labor-intensive* farming practices to *energy-intensive* ones. In crop production, horses have been replaced by tractors and trucks. The supply of organic manures has been insufficient to meet farmers' demands for use in increasing grain production. Hence, American farmers now use more than 44 million metric tons of commercial inorganic fertilizers, whose manufacture consumes vast amounts of petroleum and electricity. Petroleum-based herbicides, insecticides, and fungicides are now used to increase crop yield and prevent spoilage. Directly or indirectly, petroleum is also used to run irrigation pumps, grain dryers, and other stationary farm machinery.

Animal husbandry, too, has undergone an energy-based revolution. Hens once lived off the land; the only energy subsidies to egg production were a little grain to supplement the hens' diets and the labor and materials needed to build their unheated shelters. But yesterday's henhouses have been replaced by today's egg factories—climate-controlled buildings where hens are confined in individual cages, nutrient-enriched feed is delivered to the cages by conveyor belts, and eggs are carried off by other conveyors for sorting and packing. The chicken farmer produces more eggs this way. But as the graph shows, the energy subsidy has increased to the point where more calories of fuel energy go into egg production than calories of food energy come out. This same shift from a net gain to a net loss has occurred in beef production.

As long as energy remained inexpensive, farmers could afford energy-intensive agriculture. But the picture began to change with the Arab oil embargo in 1973. Since then, petroleum prices have nearly tripled, and rising energy costs have forced many farmers to reexamine their production methods.

As energy costs have soared, the concept of *net energy analysis* has gained favor among economists and farmers. Simply put, it asks: what is the return in food-energy output for each dollar spent on food-producing energy input? A rancher might put the question this way: will cattle prices be high enough this season for me to afford the energy subsidy required to fatten my herd in feedlots? Beef processors must ask: if we can make a profit, will the prices of our beef be too high for most families to afford?

Net energy analysis can also be applied to other

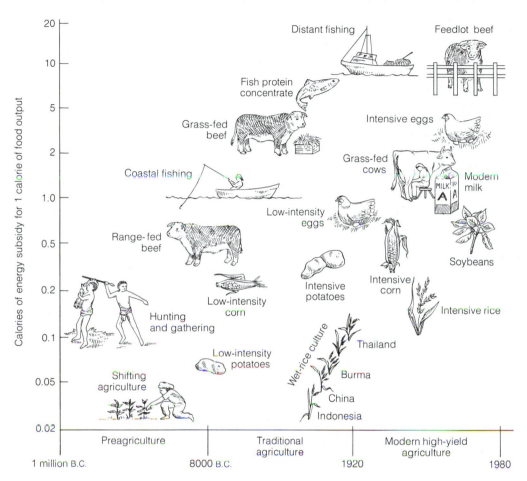

Energy subsidies for various food crops. After *Energy: Sources, Use and Role in Human Affairs* by Carol Steinhart and John Steinhart. © 1974 by Wadsworth Publishing Co., Inc., Belmont, Ca. 94002. Reprinted by permission of the publisher, Duxbury Press.

sectors of the economy. For example, how much petroleum energy must be expended to get more petroleum? Most of the known easy-to-reach resources have been tapped; now oil companies must go to the northern limits of Alaska and to offshore drilling sites for new supplies. But petroleum energy must be expended to find, extract, and transport this far-off oil to refineries. By one estimate, the energy consumed in constructing the Alaska pipeline was equivalent to approximately one-third of the known petroleum reserves in Northern Alaska when the pipeline was begun!

Figure 2.9 An invasion of locusts, which have been significant competitors with people for food over the ages. (FAO photo.)

can be imported from regions with surpluses. However, although insecticides play an important role in increasing the amount of food available to us, some of them also have a harmful impact on valuable organisms such as honey bees, which, by their pollination activities, play a major role in food production. Some pesticides also pose a threat to human health. Furthermore, the production and application of insecticides consume significant amounts of petroleum energy—in the United States, the energy equivalent of about 4 billion liters (1 billion gallons) of oil annually.

In all our efforts to increase available food, we risk the hazards of overharvesting our food sources. As mentioned earlier, many natural safeguards prevent predators from overharvesting their prey; consequently, predators rarely bring about the extinction of prey populations. Even roving bands of our prehistoric ancestors were limited in their ability to harvest their prey. And for all the advances of modern civilization, we modern human beings share the same limitations. Compared with many predators of equal size, we are generally rather slow afoot; we are rather poor at climbing up to, pouncing upon, or digging out our prey, and we do not possess exceptional strength. By learning to use tools and then making other technological advances, we have been able to overcome our physical limitations and increase our harvest of food energy. In attaining this success, however, we have, in some instances, overharvested and caused the demise of our prey. For example, overhunting has already contributed to the extinction of the passenger pigeon and great auk, and overfishing has severely depleted the numbers of many commercially important types of fish, including the Atlantic salmon, cod, haddock, and herring.

Unless we acknowledge and restrain our technological capabilities for overharvesting our food webs, it is probable that we will eradicate more species, thereby losing valuable food sources.

In spite of our ingenuity in overcoming ecological inefficiencies, then, nature continues to restrict our ability to exploit available food energy resources. There is, after all, no such thing as a free lunch. In our efforts to make more food available to people, we must try to foresee all the possible consequences of our methods. Then we can decide if, in the long run, we will be able to afford the losses that might follow from the gains.

The Stability of Food Webs

Although our understanding of what makes an ecosystem stable is still considerably limited, it is generally accepted that under some conditions increased complexity in an ecosystem leads to increased stability. For instance, most predators can consume several types of prey. Hence, if one prey species becomes scarce, a predator can switch to a more abundant prey population. As a consequence, energy flow from one trophic level to the next remains roughly the same. In turn, the ecosystem remains stable despite internal changes, assuming that many prey species exist and that not all of them become rare at the same time. In contrast, in a food web consisting of only a few food chains, energy flow may fluctuate tremendously, depending upon the well-being of a few prey species. We can find parallels to this principle in commerce. A city or a country with a variety of natural resources, manufacturing processes, and export markets is more economically stable in the long run than a city or country whose economy is based on only one or a few products.

The significance of food web stability for our food energy sources was demonstrated by the Irish potato famine of the 1840s. Ireland's soil and cool, moist climate are ill-suited for most crops. From the time of its introduction at the end of the sixteenth century, the potato was the main source of Ireland's food energy because it grew well and yielded more calories per hectare than other food crops. Potato production flourished, and the human population of Ireland doubled between 1780 and 1845, from roughly 4 million to 8.5 million. In that year, a fungus that causes a disease called potato blight entered Ireland from Europe. The potato plants were susceptible to the fungus and most of the crops destroyed during the five years of blight. Because the Irish had no major substitute food source, roughly one million people died from starvation or disease between 1845 and 1850 and another million emigrated. In a period of five years the population declined 25 percent, and the country's economic and social structure suffered a staggering blow—all largely the result of oversimplifying the food web.

A look at our own modern agricultural practices reveals that we have not yet learned history's lesson. Because of economic necessity, most farmers in the United States have turned to *monoculture*, that is, the planting of only one or two crops over large areas. In the central Midwest, most farms now grow only two crops—corn and soybeans—while on the Great Plains, wheat is grown almost exclusively. By growing only one or two crops, a farmer minimizes investments in expensive implements. Furthermore, by planting more acreage per crop, a farmer can buy larger implements, which in the long run cost less per acre to operate. But a one- or two-crop farm is itself a simple ecosystem, with the same vulnerability as Ireland's single-crop economy. An infestation of a single pest species or poor weather during a growing season can severely damage the entire enterprise. A farm with a fair array of crops, on the other hand, has a better chance of sustaining the loss of a single crop.

The inherent dangers of a much less obvious type of simplification—genetic simplification—were well illustrated in 1970, when millions of acres of corn in the United States were damaged by the southern corn leaf blight. Despite warnings by some crop scientists, the genetic makeup of the corn hybrids grown in the United States was becoming increasingly uniform. By 1970, 80 percent of the nation's corn was of a single type, one that gave a high yield, but was very susceptible to corn leaf blight.

Although the fungus that causes the blight had been present in the southern United States for many years, it had reduced corn yield only slightly (by less than 1 percent). But in 1970, a genetic change (a mutation) produced a new strain of the blight fungus that was more lethal than the former strain. Unfortunately, the more lethal fungus did not remain in the South, but was carried by winds into the corn belt states of Indiana, Illinois, and Iowa. Warm, moist weather and the susceptibility of corn plants enabled the fungus to spread rapidly throughout the corn belt.

Applications of fungicides to control the spread of the fungus would not have been economical because the cost of buying and applying fungicides would have exceeded the value of the crop. The blight, which was the result of genetic change and was exacerbated by the weather (both factors beyond human control), resulted in a deficit of almost 710 million bushels below the predicted 1970 level of consumption—a loss of 12 percent of the expected yield. Existing reserves were not used heavily, however, because as the reserves were tapped, corn prices rose and people turned to less expensive corn substitutes.

The following year, 1971, blight was nearly everywhere in the Midwest, but damage was slight. Planting weather was excellent, and good weather conditions prevailed through the growing season. During the winter of 1970-71, hybrid seed corn companies had grown blight-resistant varieties in Hawaii and South America to sell to farmers for planting the following spring. Consequently, a larger percentage of the corn grown in the Midwest in 1971 was genetically resistant to the blight. In addition, because of the blight damage in 1970, changes were initiated in federal feed grain programs that led to the planting of 280 additional square kilometers (7 million acres) of corn. The weather, the blight-resistant seeds, and the government progams led to a bumper corn crop in 1971. However, one critical factor in the increased corn yield, the weather, was still beyond human control. Nevertheless, since 1971, planting of resistant varieties of corn has greatly reduced yield losses from southern corn leaf blight.

These examples demonstrate our need to be aware of the consequences of trends toward monoculture. Not only are farmers planting one or two crops over large areas, but also the genetic makeup within each crop species is becoming uniform. As a result of these trends, modern agricultural ecosystems are growing more vulnerable to pest outbreaks and extremes in weather, and thus are potentially less reliable in their ability to produce an adequate food supply from year to year. Resolution of this problem is not easy. It is to the farmer's economic interest to plant the single most productive strain of crop each year and hope that the weather will be favorable and that pests can be controlled.

The United States is fortunate to have an abundance of good farmland spread over a large geographical area. This land has a high food-production capacity that can partially compensate for regional crop losses resulting from disease, insect infestations, or unfavorable weather. Unfortunately, few other countries have such advantages. For example, grain reserves are traditionally found in only a few grain-exporting countries—Canada, Argentina, Australia, and the United States. Thus, the tendency toward monoculture in a world where hunger is already a widespread reality and reserves against famine are usually quite limited appears to be a step in the wrong direction. Given even our limited insights into how the stability of food webs is maintained, should we be willing to risk long-term food web stability for short-term economic gains?

Movement of Materials

All organisms need certain materials to survive. Carbon dioxide and water are essential raw materials for photosynthesis—the process by which all organisms ultimately derive food energy. Nitrogen is a component of the proteins that sustain plants and animals. Phosphorus and calcium are essential for strong bones and teeth.

Where do these materials come from? Although the earth has a continuous source of energy—the sun—it has no comparable source of materials. For all practical purposes, their quantity on earth is fixed and finite. But for millions of years, these life-sustaining materials have been continually cycled and recycled within and among the earth's ecosystems. In this manner, organisms not only receive essential nutrients, but also play a major role in recycling them.

Unlike the one-way flow of energy through food webs, the movement of materials in ecosystems is accomplished through a complex set of earthbound, interlocking subcycles. In this section we consider as illustration four of the many cycles essential for all life.

The Carbon Cycle

The carbon cycle is composed of several subcycles, one of which is the exchange of carbon between organisms and the atmosphere. Carbon, an essential component of all living things, is cycled into food webs from the atmosphere, as shown in Figure 2.10. Green plants obtain carbon dioxide from air and through photosynthesis incorporate carbon into food substances. Both producers and consumers transform a portion of the carbon in food back into carbon dioxide as a by-product of respiration. This carbon dioxide is subsequently released to the atmosphere. Carbon that is tied up in dead plants and animals is also eventually returned to the atmosphere through the respiration of detritus feeders. The organism-atmosphere subcycle tends to remain in equilibrium; that is, about as much carbon is incorporated into plants every year as is released by the respiration of all plants and animals.

Some dead plant and animal materials are buried in sediments before they can be broken down completely by decomposers. This process has been going on since life began, over 600 million years ago. It was particularly important during the Carboniferous Period, 280–345 million years ago (see Appendix II), when exceptionally large amounts of materials were buried below the earth's surface. Heat and compression through subsequent eons transformed some plants and animal remains into coal, oil, and natural gas. Figure 2.11 shows the luxuriant swamp forests that grew millions of years ago and have been transformed into the coal reserves of today. When these materials, known as *fossil fuels,* are burned, the energy that is released can be channeled to perform work, and the stored carbon combines with oxygen in the air to form carbon dioxide, which enters the atmosphere. The subcycle involving fossil fuels interests us particularly at this point in history, when the amount of these fuels remaining on the planet is of critical importance. But our involvement with fossil fuels has lasted barely a moment in geologic time. Although the Chinese are said to have used coal 2100 years ago, it was not until the twelfth to thirteenth century that it was used commercially in smelters in China and in blacksmith shops in Europe.

The exchange of carbon dioxide between atmosphere and ocean makes up another important subcycle of the carbon cycle. This exchange occurs at the interface between air and water, and is enhanced by the activity of wind and waves. Here the flow of carbon dioxide occurs in two directions, with, on the average, about as much carbon dioxide moving from the surface water to the lower atmosphere as flows in the opposite direction. An increase in atmospheric carbon dioxide tends to be balanced by an increased absorption of carbon dioxide by the oceans, which helps to maintain an equilibrium in the carbon dioxide content of the atmosphere. The opposite is also true; that is, a decrease in atmospheric carbon dioxide is compensated for by the release of more carbon dioxide from the oceans to the atmosphere.

A third subcycle of the carbon cycle is the formation of limestone, dolomite, and carbonaceous shale (see Figure 2.12). By several rock-forming proc-

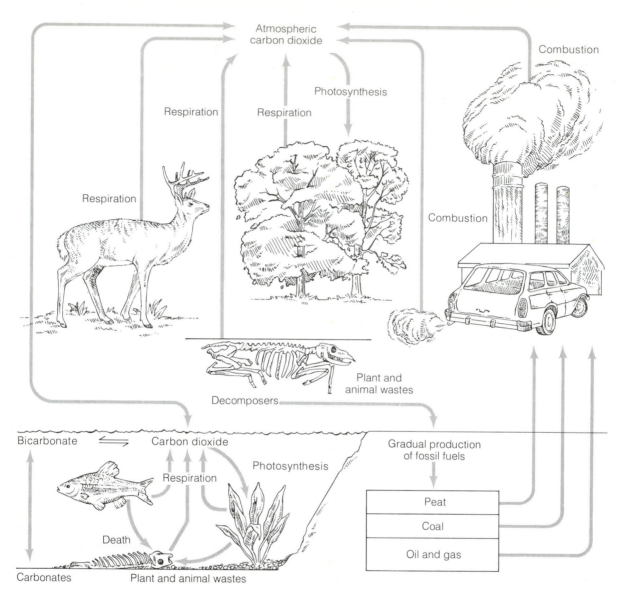

Figure 2.10 The carbon cycle.

esses, carbon is incorporated into these components of bedrock. Subsequently, chemical and physical processes (weathering) may break down these substances, releasing carbon dioxide into the atmosphere. Although they have little significance for us in the immediate future, these processes of formation and degradation (called rock cycling) of carbonaceous rock are important in balancing the amounts of carbon over periods of millions of years. For a more detailed discussion of rock cycling, see Chapter 12.

Figure 2.11 A diorama showing a carboniferous swamp forest like those that existed 280 to 345 million years ago. (Courtesy, Field Museum of Natural History, Chicago.)

Figure 2.12 The spectacular White Cliffs of Dover, England, the result of the massive accumulation of carbonaceous shells over a period of millions of years. (The British Tourist Authority.)

The Phosphorus Cycle

Whereas the direct source of carbon for plants is the atmosphere, the major direct source of phosphorus is the soil, which also supplies other major nutrients such as calcium, magnesium, potassium, and sulfur. These mineral nutrients are essential for the growth of both plants and animals. Phosphorus is taken up from the soil mainly in the form of phosphate (PO_4^{3-}) through a plant's root system (see Box 2.3), and is transported to its growing parts. Figure 2.13 illustrates the central role of the soil in the earth's phosphorus cycle. Within the growing parts of plants, phosphorus is incorporated into a variety of organic compounds, including fats. And, as you may recall, in animals it becomes a major component of bones and teeth as well as fat. When plants and animals die and decompose, the phosphorus is returned to the soil, where it can be taken up again by plant roots.

Within undisturbed terrestrial ecosystems, the subcycle between soil and organisms tends to be closed; that is, losses of phosphorus and other nutrients are minor. Plants serve as a protective covering for the soil; this and the plants' extensive root systems help to retard soil erosion and retain the supply of nutrients within the ecosystem. Such a means of retaining nutrients within an ecosystem is important for organisms, because the amount of available phosphorus in the soil is relatively small. Phosphorus is added to the soil through the weathering of phosphorus-containing rocks, but this process takes place very slowly. Thus, the most effective means of conserving the limited phosphorus supply in an undisturbed terrestrial ecosystem is through recycling by organisms.

Still, no undisturbed terrestrial ecosystem is completely leakproof, and small amounts of phosphorus are carried off to rivers and lakes, and eventually they may reach the oceans. In these aquatic ecosystems, phosphorus is taken up by algae and rooted aquatic plants and again passed along food webs including detritus pathways. The large congregation

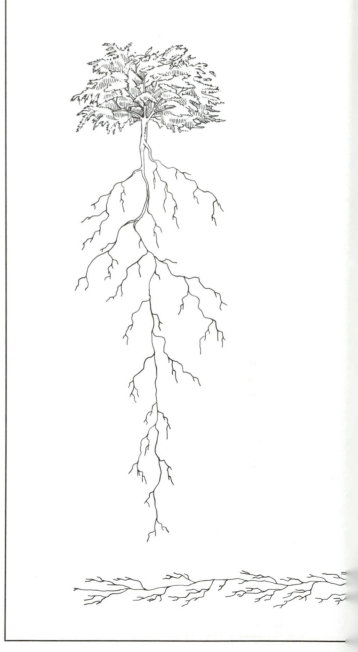

Box 2.3

Roots

All living things require many different elements to survive. Sixteen are found in all organisms; we humans require about twenty, including sodium, potassium, phosphorus, calcium, magnesium, and iron. Most of these life-supporting minerals are found in the soil, but their concentrations are very low—typically one part per million or less. Even if we could eat soil, we would have to consume tremendous quantities of it to extract the minimum amounts of minerals needed to sustain ourselves. Fortunately for our digestive tracts, plants extract our minerals for us.

A plant faces two problems in obtaining minerals from the soil. First, it must penetrate the ground in such a way as to gain the greatest possible contact with the soil (or, more precisely, with soil water in which minerals are dissolved). Second, the plant must be able to concentrate the extremely diluted minerals it has taken up from the soil.

The plant solves these problems by means of its root system. As the root system grows, it branches and branches again. The roots, in turn, extend millions of tiny, fingerlike *root hairs* into the soil. Roots and root hairs combined put an astoundingly large surface area in contact with the soil moisture and the minerals dissolved in it. For example, the root system of a single four-month-old rye plant is nearly 11,000 kilometers (7000 miles) long and has a surface area of 630 square meters (7000 square feet)!

Plant roots also have remarkable mineral-concentrating capabilities. Experiments have shown that the concentration of certain minerals within roots is as much as 10,000 times greater than in the surrounding soil. Once taken up, minerals are transported to growing parts of the plant, where they are incorporated into proteins, fats, vitamins, and many other organic compounds. Plants, then, are a concentrated source of essential minerals for consumers in both grazing and detritus food webs.

Plants "mine" staggering quantities of minerals from the soil—as much as 6 billion metric tons every year, according to one estimate. This is more than 6 times the amount of iron, copper, lead, and zinc ore produced by all worldwide mining operations in 1973.

We see, then, that plant roots are vital to our survival and the survival of other terrestrial animals. Without roots we could not obtain sufficient calcium to build bones and teeth, sodium to maintain blood pressure, and iron to help carry oxygen in the blood, as well as the many other minerals our bodies need to function properly.

From Emmanual Epstein, "Roots," *Scientific American*, May 1973, 48-58.

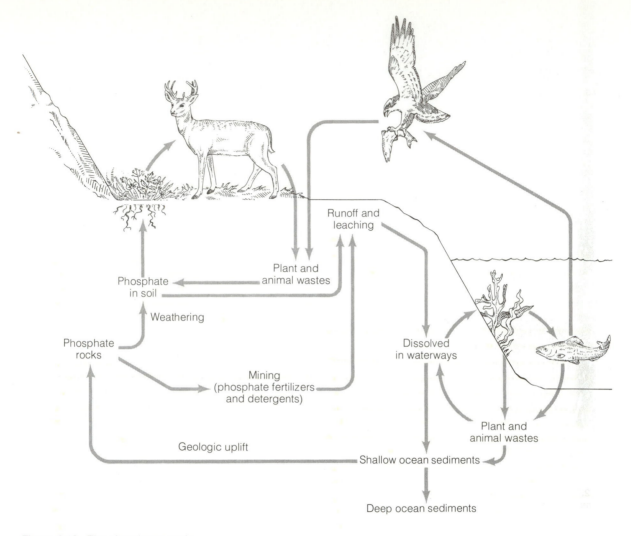

Figure 2.13 The phosphorus cycle.

of birds pictured in Figure 2.14, for instance, consumes huge amounts of fish. ==The wastes (guano) produced by these birds contain phosphorus and other nutrients. This material can be used as fertilizer to return these nutrients to the soil.== Also, the phosphorus that enters waterways may sink to the bottom. If the body of water is shallow, wave action can stir up bottom sediments, making the phosphorus again available for uptake by algae. Phosphorus deposited in sediments in deep portions of water bodies, however, is removed from circulation until geologic processes reexpose them. Reappearance may require hundreds of thousands or even millions of years.

The Nitrogen Cycle

==As a component of protein, nitrogen is essential to all forms of life.== Although 79 percent of the atmosphere is nitrogen gas, air cannot be used directly by plants

Figure 2.14 A large congregation of birds, whose droppings (guano) contain phosphorus and forms of nitrogen that can be used for fertilizer. (FAO photo.)

as a nitrogen source. In nature, two major processes change nitrogen gas into forms that plants can use. They are an essential part of the nitrogen cycle, illustrated in Figure 2.15. One of these processes is *atmospheric fixation,* which occurs during electrical storms. The electrical energy in lightning, shown in Figure 2.16, causes nitrogen gas to react with oxygen gas in the atmosphere to form nitrate (NO_3^-). This form of nitrogen is then captured by falling raindrops and carried to the soil, where it can be taken up by plants.

The other natural process that makes nitrogen available to plants is *biological fixation.* Within soil, special types of microorganisms combine nitrogen gas with hydrogen to form ammonia (NH_3). Some of these microbes, such as nitrogen-fixing blue-green algae, are free living. Others, certain specialized bacteria, live in nodules on roots of such leguminous plants as peas, beans, alfalfa, and clover. (Bacteria-containing nodules on soybean plants are shown in Figure 2.17.) The ammonia produced by biological fixation is acted upon by other specialized bacteria and is eventually changed to nitrate.

Nitrates are readily taken up by plant roots. Within plants, nitrates are incorporated into a variety of organic compounds, including proteins, and in

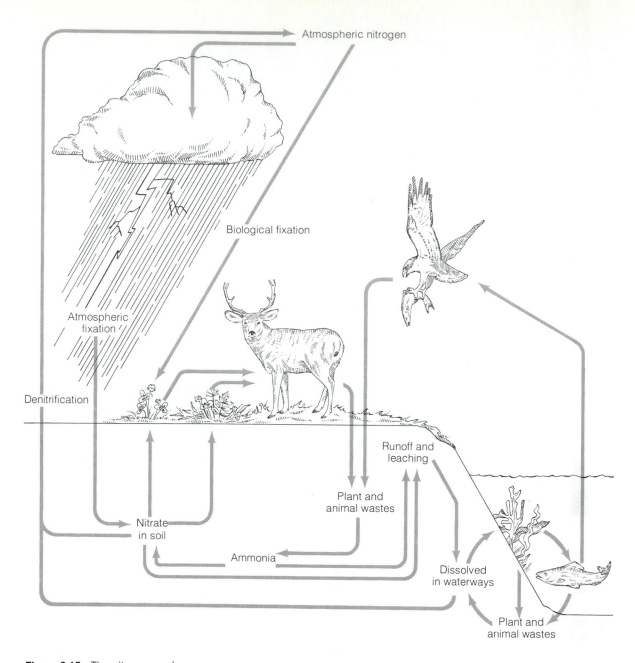

Figure 2.15 The nitrogen cycle.

these forms, nitrogen travels through food webs. When plants and animals die and decompose, nitrogen again is converted into ammonia, in which form it returns to the soil. Animal wastes, rich in urea, are another major recycling route. Organic nitrogen compounds such as proteins and urea are converted into ammonia by yet another specialized group of bacteria. Once the ammonia returns to the soil, it is

Figure 2.16 (*Left*) Lightning, a significant factor in the nitrogen cycle. (National Oceanic and Atmospheric Administration.)

Figure 2.17 Root nodules on soybean plants containing bacteria that transform nitrogen gas into ammonia. (U.S. Department of Agriculture.)

changed into nitrate by microbial action and can be taken up again by plants and recycled through food webs.

Ammonia and nitrate are lost from terrestrial ecosystems through two major pathways: denitrification and transport by water. *Denitrification,* a process carried out by a particular group of bacteria, is the conversion of nitrates back into nitrogen gas, which escapes from the soil into the atmosphere. Transport by water is a simple mechanical process: since both nitrate and ammonia are highly soluble in water, they are readily carried away by surface runoff or groundwater. But the nitrates and ammonia trans-ported in this way are not necessarily lost. In lakes, rivers, and oceans, they normally remain dissolved and can be taken up by aquatic plants and subsequently cycled through aquatic food webs. Ammonia does not move through the soil as readily as nitrate, since ammonia tends to adhere to soil particles.

The Oxygen Cycle

Oxygen is found almost everywhere on earth. As a gas (O_2), oxygen is an important component of air; it is also found dissolved in surface waters and in the pore spaces of soils and sediments. And oxygen combines chemically with a multitude of other elements to form important substances. These oxygen-containing substances include water (H_2O); such gases as carbon monoxide (CO), carbon dioxide (CO_2) and sulfur dioxide (SO_2); plant nutrients, including nitrate (NO_3^-) and phosphate (PO_4^{3-}); organic substances, such as sugars, starches, and cellulose; and virtually all rocks and minerals, including limestone ($CaCO_3$) and iron ore (Fe_2O_3). Clearly, in order to react with such a wide variety of materials, the oxygen cycle must interrelate with many other cycles of materials, including the carbon, phosphorus, and nitrogen cycles. The oxygen cycle is so intricate and complex that we can outline only a few of the more important subcycles here. Figure 2.18 illustrates some of these pathways in simplified form.

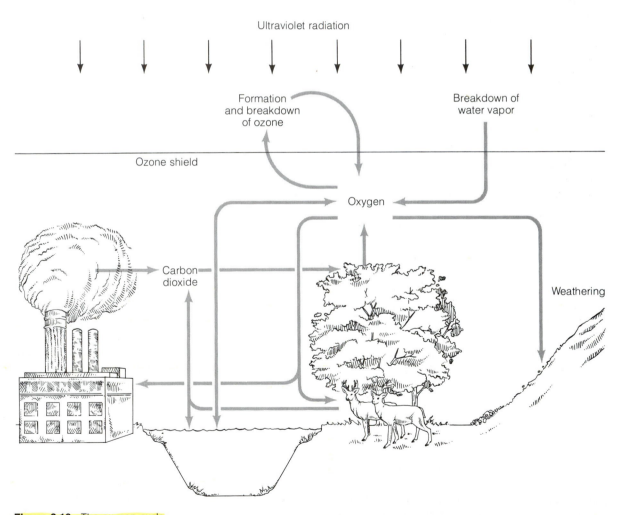

Figure 2.18 The oxygen cycle.

One oxygen subcycle consists of the interrelationship of photosynthesis and respiration. As you will recall, oxygen gas is combined with sugar through respiration, and one of the products is carbon dioxide. (In fact, organic materials such as wood, coal, or manure will burn only in the presence of oxygen, with carbon dioxide as an end-product.) On the other hand, in photosynthesis, carbon dioxide is combined with water in the presence of sunlight to produce sugar and oxygen. The equations for photosynthesis and respiration seem to suggest that each process is the reverse of the other. But in reality each process consists of a complex series of intermediate reactions, many of which occur only in photosynthesis or respiration exclusively.

Exchanges of oxygen between the atmosphere and the ocean make up another important subcycle of the oxygen cycle. As with carbon dioxide in the carbon cycle, the exchange of oxygen between the ocean and the atmosphere is balanced, so that the oxygen content of the atmosphere remains essentially constant. Ocean currents and waves carry dissolved oxygen into deeper waters to sustain aquatic life at great depths.

We have already seen that oxygen enters the nitrogen cycle through atmospheric fixation, whereby oxygen gas is combined with nitrogen gas to form nitrate. By means of denitrification, nitrate is transformed back into nitrogen gas, which escape back to the atmosphere.

Oxygen plays an important role in the natural chemical disintegration of rock. For a more detailed discussion of the role of oxygen in weathering, see Chapter 12.

Transfer Rates and Human Activity

Over time, the rates of transfer of energy and materials among various components of the environment have become relatively stable. But in recent years human activity has begun to disturb these transfer rates significantly. For example, many people are aware that the consumption of huge amounts of fossil fuels by developed nations has greatly increased the transfer of carbon to the atmosphere. But only very recently have scientists determined that human destruction of native forests and prairies has also accelerated carbon transfer to the atmosphere. Primeval forests with their massive trees were huge reservoirs of carbon. With the clearing and burning of these forests for agriculture and urban development, large amounts of carbon dioxide were released. Significant increases in the rate of transfer of carbon to the atmosphere have also resulted from the decay of vast amounts of *humus*—organic materials that are generally resistant to decay. The natural rate of humus decay has been accelerated by several factors—the harvest of forests, the spread of agriculture onto soils containing large amounts of organic matter (such as the prairie soils of Illinois and Iowa), and the destruction of natural wetlands. Hence, whereas world vegetation once served to store carbon by taking up more carbon dioxide through photosynthesis than was released through respiration, our destruction of virgin vegetation has actually resulted in a net release of carbon dioxide.

What has happened to the excess carbon dioxide in the atmosphere? Answers vary, because we still have much to learn about the carbon cycle. Some of the excess has been taken up by the oceans, yet much remains in the atmosphere. The graph in Figure 2.19 indicates that during the last hundred years—a period that has seen birth and development of industrial societies and the destruction of vast expanses of virgin vegetation—the carbon dioxide content of the atmosphere has increased by about 12 percent. If our consumption of fossil fuels and our demand for food continue to accelerate, the carbon dioxide content of the atmosphere will rise about 20 percent by the year 2000, and it could double within sixty years. This increase will occur despite compensating transfers into oceans. The added atmospheric carbon dioxide may alter the earth's climate, but unfortunately scientists do not know enough at present to fully com-

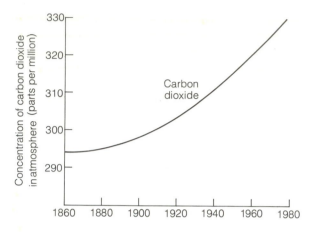

Figure 2.19 Changes in atmospheric carbon dioxide levels due to increased consumption of fossil fuels and increased burning of vegetation. (After C.F. Baes, H. Goeller, J.S. Olson, R.M. Rotty, "Carbon Dioxide and Climate: The Uncontrolled Experiment," *American Scientist* 65:310–320, May/June 1977.)

prehend and assess the consequences that such a disturbance could have.

The disruptive effects of human activities on cycling rates can also be seen in the phosphorus cycle. Normally, the rate of phosphorus loss from an undisturbed ecosystem is low. But if the protective vegetative cover is removed—for example, when all the trees in a forest are cut down—the loss of nutrients in runoff increases severalfold. The phosphorus cycle is also disturbed when we mine deposits of phosphorus as illustrated in Figure 2.20, and incorporate this mineral into fertilizers and detergents. Runoff from agricultural lands and inefficient phosphorus removal by municipal sewage treatment plants causes large amounts of phosphorus to be transported into lakes and rivers.

Human activities have also disrupted the nitrogen cycle. About thirty years ago, an industrial process was developed (the Haber process) that utilizes natural gas as an energy source to synthesize ammonia from atmospheric nitrogen. Today, large amounts of ammonia synthesized by the Haber process are added to the soil as fertilizer. Because ammonia and nitrate readily dissolve in water, runoff from agricultural land is carrying increased amounts of these chemicals into lakes and rivers.

Unfortunately, nitrogen and phosphorus are entering waterways faster than they are being cycled out. In effect, the result is a nutrient logjam in rivers and lakes—that is, a higher than normal concentration of phosphorus and nitrogen. These high levels of phosphorus and nitrogen act as fertilizers to accelerate the growth of algae and rooted aquatic plants. The accelerated plant growth in turn lowers water quality and is indirectly responsible for elimination of some species of fish in certain lakes.

Atmospheric oxygen levels do not seem to be endangered by human activity so far, but the corrosive effect of oxygen is visible on many of the products of civilization. Oxygen plays an important role in the chemical disintegration of our buildings and sculptures. On a less monumental level, oxygen rusts our automobiles and farm equipment, reminding us that even they are not exempt from interaction with the planet's natural cycles.

Transfer Rates and Pollution

As we have seen, materials and energy in their myriad forms are continually transferred among the components of ecosystems and among ecosystems themselves. Generally, materials and energy reach certain equilibrium concentrations within the various components of the environment—soil, air, waterways, and organisms. As you may recall from the preceding section, however, rates of transfer can be altered so that the amounts of energy and/or materials in a particular component are also changed. Figure 2.21 shows three possible effects of changes in transfer rates. But another change can influence these energy-material flows: a material may be introduced into a component where it was not formerly found. If the subsequent changes in concentration

Figure 2.20 Strip mining for phosphate ore in Florida. (Kit and Max Hun, Photo Researchers.)

adversely affect the well-being of organisms (including people), then the overall process is called *pollution,* and the type of material or form of energy involved is termed a *pollutant.* Hence, the increased flow of available phosphorus into waterways that we mentioned in preceding paragraphs is a form of pollution.

Note that, by definition, pollution need not have a *direct* effect on organisms. For example, human activities have greatly increased the amount of carbon transferred from the earth to the atmosphere. Some scientists believe that a continued increase in atmospheric carbon dioxide will result in a warming of the climate worldwide; this change would, in turn, have detrimental effects on many species of plants and animals, including human beings. In this example, the carbon dioxide would be an atmospheric pollutant because it would indirectly affect the well-being of organisms through its role in bringing about climatic change.

This definition of pollution encompasses the introduction of new materials into the environment by technology—obvious examples are pesticides and radioactive wastes. It also takes in changes in rates of transfer that occur independently of human activity. In 1815, for example, the volcano Tambora in Indonesia spewed millions of metric tons of volcanic ash into the air. Some scientists believe that this increased atmospheric dustiness triggered the unusual weather experienced in some parts of the world during the following summer. For example, in New England, 1816 is known as "the year without a summer." New England had snows in June, light frost in July and August, and severe frost in September. Thus, the volcanic ash that triggered the weather patterns causing these unusual summer events—if, indeed, it did—qualifies as a pollutant, albeit a natural one. However, it is important to note that the effects of sporadic, nonhuman pollution are usually short term, lasting a matter of days or perhaps a few years, whereas the continual impact of human activity may well have effects of extended duration.

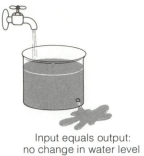

Input equals output:
no change in water level

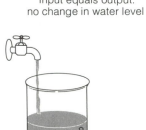

Output greater than input:
water level drops

Input greater than output:
water overflows

One more distinction remains to be made. This definition of pollution is objective; if organisms are harmed as a result of abnormal transfer rates of some form of material or energy, then pollution has occured. What is left for us to decide is whether a pollutant affects our well-being seriously enough to require corrective measures.

Conclusions

In this chapter, we confront for the first time two related concepts that recur continually in environmental science: interrelationship and balance. We find that we cannot define an ecosystem without describing the interactions among organisms that compose it. We cannot describe energy or material flow except by showing how each cycle affects and interacts with those around it. And we cannot really define environmental stability except by making reference to the balance maintained by these interactions—the dependency of each component or mechanism on the smooth working of all the other interrelated parts.

These concepts are particularly important when we try to measure the effect of a human activity on the environment. We can be sure that if we throw a rock into a pond the effects of that disturbance will be felt in ever-widening circles. One reason for studying environmental science is to make ourselves more sensitive to the effects of these ripples on ecosystem function and thus on our own personal well-being.

Summary Statements

Many of our environmental problems are caused by disruptions of the flow of energy and materials through the environment.

Ecosystems contain both living and nonliving components. Organisms are distinguished by their major functions. The main categories are producers (green plants) and consumers (animals and certain types of fungi and bacteria). The physical environment comprises the nonliving portion of an ecosystem.

Within ecosystems, food energy moves only in one direction, from producers to consumers, in a complex network of pathways called a food web. There are two important types of food webs: grazing and detritus. The relative significance of each depends on the type of ecosystem in question.

For many reasons, only a small percentage of food energy is transferred from one trophic, or feeding, level to the next. Because of these inefficiencies and the minimum food requirements for organisms, a geographical area can feed only a limited number of organisms.

The quantity of food available to people can be increased by (1) reducing the number of organisms that compete for the same food, (2) converting forests and rangeland into cropland, and (3) increasing the efficiency of energy use by livestock and crops. All these efforts are limited by the energy inefficiencies inherent in food webs. Such efforts have also resulted in significant environmental degradation.

Complex food webs are probably more stable than networks composed of only a few food chains. Modern-day agricultural ecosystems are less stable in the sense that they are more vulnerable to pest outbreaks and extremes in weather. Hence, they are potentially less reliable in providing an adequate food supply from year to year.

Movement of materials within and among ecosystems involves a complex set of interlocking subcycles. Though energy is supplied continually by the sun, the quantity of materials on earth is essentially fixed and finite. The carbon, phosphorus, nitrogen and oxygen cycles illustrate many of the pathways whereby materials are recycled.

Over past millennia, the rates of energy and material transfer among various components of the environment have become relatively stable. In recent years, human activity has begun to disturb these rates significantly. Some of the consequences include a decline in air and water quality and a possible change in climate.

Pollution is any change in transfer rates of materials or energy that either directly or indirectly affects adversely the well-being of organisms. Pollution can be either human-induced or natural.

Questions and Projects

1. In your own words, write out a definition for each of the terms italicized in this chapter. Compare your definitions with those in the text.

2. List the biotic and abiotic components of an ecosystem. Describe how these components interact to govern the flow of energy through a food web.

3. What is a detritus food web? How does it differ from a grazing food web?

4. Visit several types of ecosystems in your region. Describe how the different organisms present in each ecosystem constitute a food web.

5. List the foods you ate in your most recent meal. Trace the energy in each food back to its ultimate source by the most plausible food chain.

6. Chickens are considered to be energy efficient because they have energy transfer efficiencies approaching 20 percent. That is, 100 kilograms (220 pounds) of corn produce 20 kilograms (44 pounds) of chicken. But this is not the only energy transfer process involved in raising chickens and getting them to market. Describe other energy inefficiencies (both in food energy and fossil fuel energy) that are not being considered.

7. Food energy is used by livestock for movement, maintenance, growth, and reproduction. Describe several ways that livestock growers could manage their herds to increase the amount of food energy used for growth.

8. Describe why attempts to increase the efficiency of energy use of crops and livestock have met with only limited success.

9. The percentage of digestable food in the diet helps to determine the amount of energy transferred to the next trophic level. Per kilogram of food intake, which herbivore would obtain the most usable energy—a herbivore that eats seeds, wood, young foliage, or mature foliage? Explain your answer.

10. What is the "10 percent rule"? Why should we take this rule into consideration as we formulate policies to increase the amount of food available to human beings?

11. Americans obtain only two-thirds of their food energy from plant materials, while most of the world's people obtain nearly all of their energy from plants. How is it that Americans can afford to obtain so much of their energy from livestock products?

12. List several means whereby human beings can increase the amount of food available to them.

13. Describe the advantages that accrue to an animal that can occupy more than one trophic level.

14. For the southern corn leaf blight episode in 1971, what were the factors that were beyond human control? What were the factors that humans could, and did, control?

15. Describe the roles of respiration and photosynthesis in the carbon cycle and the oxygen cycle.

16. Describe how biological fixation of nitrogen differs from atmospheric fixation.

17. What does the nitrogen cycle tell us about the significance of inconspicuous microorganisms for the survival of life on earth?

18. Millions of metric tons of nitrogen fertilizer are produced annually by the Haber process. What effects might this transfer of nitrogen from the atmosphere to the soil have on the environment?

19. Describe some of the interactions between the flow of food energy and the movement of materials such as carbon and phosphorus in an ecosystem.

20. Cite examples, if any, of instances in your region in which human activity has significantly changed the transfer rates of some type of material or energy. What have been the consequences?

21. Cite examples of "natural pollution."

22. Make a list of the materials that human beings need for survival. Briefly describe the role of each material and its source. Make another list of materials not necessary for human subsistence, but used to maintain a particular standard of living.

Selected Readings

Annual Reviews, Inc. *Annual Review of Ecology and Systematics.* Published yearly. Palo Alto, California: Annual Reviews, Inc. Each volume contains articles that review current research in energy and materials flow in the environment.

Bell, R. H. V. 1971. "A Grazing Ecosystem in the Serengeti," *Scientific American 226:*86–93 (July). Describes how migrations of herds of zebra, wildebeest, and Thomson's gazelle are synchronized with the availability of specific food sources.

Brill, W. J. 1977. "Biological Nitrogen Fixation." *Scientific American 236:*68–81 (March). Examines how a few types of bacteria and blue-green algae are major contributors to earth's nitrogen cycle and how they can be used to improve soil fertility.

Collier, B. D., G. W. Cox, A. W. Johnson, and P. C. Miller. 1973. *Dynamic Ecology.* Englewood Cliffs, New Jersey: Prentice-Hall. A good ecology text covering energy flow and nutrient cycling.

Friden, E. 1972. "The Chemical Elements of Life," *Scientific American 227:*52–60 (July). A good introduction to the role of chemical elements in organisms.

Gosz, J. R., R. T. Holmes, G. E. Likens, and F. H. Bormann. 1978. "The Flow of Energy in a Forest Ecosystem." *Scientific American 238:*93–102 (March). A comprehensive examination of energy flow through food webs in a forest and adjacent stream.

Odum, E. P. 1971. *Fundamentals of Ecology.* Philadelphia: W. B. Saunders. The most recent edition of a classic ecology text. Contains chapters devoted to energy flow and nutrient cycling.

Ricklefs, R. E. 1976. *The Economy of Nature.* Portland, Oregon: Chiron Press. A well-written book containing several chapters on energy flow and nutrient cycling.

Scientific American. 1970. *The Biosphere 225* (September). A special issue providing basic references on the flow of energy and materials in the environment.

Woodwell, G. M. 1978. "The Carbon Dioxide Question," *Scientific American 238:*34–43 (January). A provocative look at our disruption of the carbon cycle and the potential outcome.

A minute change in elevation resulting in a major change in vegetation, in Everglades National Park. This palm hummock is only a few feet higher than the watery sawgrass prairie that surrounds it. (U.S. Department of the Interior, National Park Service photo by M. W. Williams.)

Chapter 3

Ecosystems and Environmental Change

Change is one of the most basic characteristics of our environment. Rivers and lakes rise and fall, weather differs from day to day, soil erodes, and forests and grasslands go up in flames. The very face of the planet changes shape as volcanoes erupt and mountains are sculpted by the weather. Animals die and are replaced by their offspring; populations grow and diminish. Human beings, too, work significant changes; they dam rivers, clear forests, fill in marshes, alter the rates of nutrient cycling, and spill out into the countryside from crowded cities.

To understand the impact of environmental change on plants and animals, we must first appreciate the fact that each organism is a unique and vastly complex entity whose survival depends on several fundamental processes. First, all living organisms obtain food and water from their surroundings. They then process, or metabolize, the nutrients they derive from food for self-maintenance, growth, and reproduction. They excrete the waste by-products of metabolism back into the environment. Finally, all organisms receive and respond to stimuli in their surroundings. For an organism to remain alive, each of these intricate processes must occur and must interact properly with the others. An environmental change can influence the functioning of any one of these processes, which may in turn affect the others. If even one process malfunctions, the well-being and perhaps survival of the organism is imperiled.

To varying degrees, organisms are able to adjust to changes in their environment. Changes in the season, for instance, initiate various physical and behavioral responses in the animals and plants that experience them. Foxes and wolves adjust to the coming of winter by growing thicker fur; deer congregate in small, sheltered areas and greatly reduce their movement and foraging activities; and chipmunks and groundhogs go into hibernation. Analogously, human beings adjust to colder weather by wearing insulated clothing, staying indoors more, and heating their homes. As we shall see in our discussion of natural selection, the ability of a species to make protective responses, or adaptations, accounts in part for its survival in an environment where changes occur.

Some environmental changes may be so great or may occur so rapidly, however, that some organisms are unable to adjust to them. The decline in numbers of bald eagles as a result of pesticide accumulation, and the increase in human respiratory ailments during air pollution episodes attest to the limitations on the ability of organisms to adjust to environmental change.

What factors limit the ability of organisms to respond to and survive environmental change? Why are some organisms able to adjust to certain changes while others are not? How do ecosystems respond to change? The answers to these questions will aid us in understanding the impact of pollution and the problems of resource management to be considered in subsequent chapters.

Limiting Factors

In the 1960s, many people began to notice a major change in water quality: many lakes and streams were turning green because of population explosions of algae. Figure 3.1 illustrates the rampant growth of other aquatic plants that also occurred at this time. Although the algal "blooms" had occurred before, their increased frequency and intensity made them more apparent. During this time, scientists detected another change in the water: abnormally high concentrations of phosphorus and nitrogen. The obvious question was whether these two observations were related.

We can come up with a partial answer ourselves by performing a simple experiment. If we place algae in an aquarium devoid of phosphorus, the algae will die; they, like all plants and animals, require a certain amount of phosphorus to survive. If we add more phosphorus than this minimum amount, the algal population will begin to grow. And if we continue to add phosphorus, a maximum population will eventually be reached. Above a certain concentration, however, the phosphorus will become toxic to some algae, and at a higher level still it will be lethal to the entire population.

To relate our experiment to nature, we can start by noting that pristine mountain lakes generally contain

Figure 3.1 Excessive growth of aquatic weeds in Lake Mendota, Wisconsin. (Photo by Don Chandler, compliments of Arthur D. Hasler, Laboratory of Limnology, University of Wisconsin-Madison.)

few algae and low concentrations of phosphorus. In marked contrast, lakes with massive algal growth contain high concentrations of phosphorus (as well as other plant nutrients). But rarely, if ever, does phosphorus in natural waterways reach a concentration that is toxic to algae. Of course, other factors—including water temperature, light intensity, and concentration of other nutrients such as nitrogen—influence algal growth as well. But we are safe in concluding that in most instances the excessive algal growth that appeared in the 1960s was indeed related to the increased amounts of phosphorus in the water.

Tolerance Limits

The response we have described of algal populations to changes in phosphorus concentration conforms with the *law of limiting factors*. This principle states that for each physical factor in the environment, there are a minimum limit and a maximum limit—both called *tolerance limits*—beyond which no members of a particular species can survive. This principle is illustrated by the limiting factor curve in Figure 3.2. A limiting factor curve has three critical points: the lower tolerance limit, the optimum concentration, and the upper tolerance limit. For the phosphorus-algae example, the lower tolerance limit represents the minimum concentration of phosphorus needed to sustain a marginal algal population. The optimum is the concentration of phosphorus that produces maximum population size. The upper tolerance limit is the concentration beyond which no algae can survive. The tolerance limits and optimum for each factor are determined by the *genetic makeup*—information contained within the genes—of the individual organism.

In some respects, the law of limiting factors supports the adage of moderation in all things. Algae require a minimum quantity of phosphorus to func-

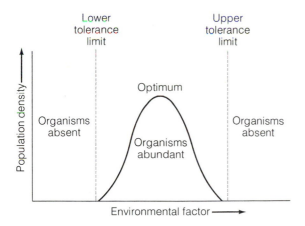

Figure 3.2 The law of limiting factors.

tion properly, as do all organisms, yet excessive amounts can be fatal to them. In a similar way, human beings require many vitamins and minerals to maintain good health, but improper amounts of these nutrients in the diet can lead to serious health problems. For example, a lack of vitamin A leads to drying of the skin, abnormal bone formation, and night blindness, whereas an excess of this vitamin produces gastrointestinal upset, dermatitis, loss of hair, and pain in the bones. Either too little or too much of any required factor (such as food energy, vitamins, minerals, water, air temperature, or oxygen) can seriously jeopardize the survival of individuals and perhaps entire species.

Toxic and Hazardous Materials

The environmental changes that human beings effect very often involve tolerance levels never even tested before. Besides changing the cycling rates of essential substances such as phosphorus and nitrogen, and of nonessential, naturally occurring substances such as lead and mercury, human beings have introduced numerous foreign chemicals into the environment. In our increasingly complex technological world, approximately seventy thousand chemicals are used in industry, with about a thousand new chemicals being introduced each year. Equally significant as new inputs into the environment are the waste products resulting from industrial processes. These chemicals are complex and extremely varied; they include, for example, pesticides, radioactive materials, petroleum products, plastics, and acids. Some of these substances eventually make their way into the water we drink and the air we breathe; some continually bombard wildlife and vegetation. Many have already been incorporated into the plants and animals that form the base of our food webs. What effects do these materials have on us and our fellow inhabitants on earth? Are our tolerance limits to these substances being exceeded?

The tragedy at Hopewell, Virginia, demonstrated some of the hazards of manufacturing and mishandling these chemicals. In the mid-1970s, nearly half of the 150 employees of Life Science Products, Incorporated, a small chemical company in Hopewell, developed symptoms of poisoning—blurred vision, tremors, loss of memory, liver damage, localized pain in the joints and chest, and, among males, sterility. Nearly thirty employees were hospitalized with these ailments, some of which defied treatment. The plant, which was a converted gas station, manufactured a very poisonous ant and roach killer named Kepone. As is apparent in Figure 3.3, the plant was poorly operated—Kepone dust filled the air, covered equipment, and contaminated pools of water on the floor.

Continual exposure to the poison under these illegal working conditions eventually led to poisoning of the employees—and to lawsuits for millions of dollars in damages, filed on behalf of them and their families.

This direct assault on human health was only a small part of what had been happening in Hopewell, however. For some time, Life Science Products, with the consent of the city of Hopewell, had illegally been dumping tons of Kepone-laden wastes into the municipal sewer system. The Kepone disrupted the normal bacterial processes of sewage treatment, resulting in the discharge of untreated sewage into the

Figure 3.3 Hazardous working conditions at Life Science Products, Inc. (Richmond Newspapers, Inc.)

James River, which flows through the city. Along with the raw sewage, Kepone itself was discharged into the James River, thereby contaminating aquatic life as far as 100 kilometers (62 miles) downstream. Kepone, a nerve poison, inhibits the growth and reproduction of fish and other wildlife. Because Kepone accumulated in fish, eating contaminated fish posed a public health problem. In response to this public health hazard, Virginia's governor closed 160 kilometers (99 miles) of the river to commercial fishing. Because thousands of kilograms of Kepone are now mixed with the river's bottom sediments and because Kepone breaks down very slowly, this ban is likely to remain in effect for many years to come.

The Hopewell incident is only one of innumerable cases in which hazardous chemicals have been released, either accidentally or intentionally, into the environment. If released in large enough amounts, these substances can weaken an organism and render it more vulnerable to other environmental stresses. If the amount of the substance exceeds an organism's tolerance limit, it is fatal. Unfortunately, even minute amounts of many substances are toxic—that is, the tolerance limit of most organisms is very low for these chemicals.

The release of such poisons or *toxic substances* as Kepone into the environment is worrisome in itself, but it is only one of many ways in which we are contaminating our habitat and harming ourselves. For example, concern is now growing about our increased exposure to *mutagens, carcinogens,* and *teratogens,* all of which can have catastrophic health effects. Mutagens are substances that cause hereditary changes (mutations); this means that if a person who has a mutation reproduces, the mutation can be passed on to succeeding generations. For this reason mutations pose the greatest danger to the human species. On the other hand, carcinogens, substances that cause cancer, pose the greatest threat to the survival of individual human beings. The incidence of cancer has climbed sharply in recent years. Today, it is second only to heart disease as a cause of death (see Figure 3.4); more than 350,000 Americans die

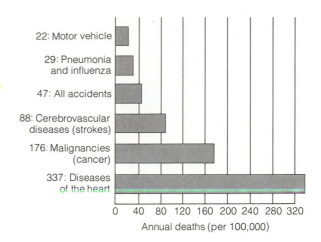

Figure 3.4 Leading causes of death among Americans. (After National Center for Health Statistics, U.S. Department of Health, Education and Welfare, 1978.)

from it each year. Teratogens are substances that cause monstrous deformities in developing fetuses. The thalidomide drug catastrophe that occurred in Europe in the early 1960s clearly illustrates the potential dangers of teratogens. Before this sedative was taken off the market, about eight-thousand pregnant women who took it had given birth to seriously deformed infants within a span of two years. Figure 3.5 shows a victim of the thalidomide disaster.

Unfortunately, we know very little about the potential health hazards of long-term exposure to toxic and hazardous substances at the levels at which they appear in the environment. Recognizing the complications of exposure to harmful substances, government agencies usually allow for a margin of error when setting "safe" exposure levels. Generally, the agency sets the exposure level ten to one hundred times lower than might be inferred from the latest experimental results. Such a policy raises considerable controversy. Some people feel that the standards should be even more restrictive, to provide an even

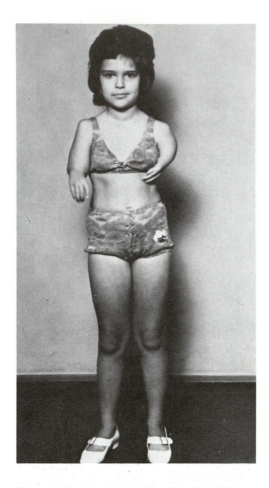

Figure 3.5 A victim of thalidomide. (D. Mühlenstedt.)

ing the safety margin is unjustified. Hence, we face the difficult task of setting standards that will simultaneously minimize our health risks and save us from bearing an unjustified economic burden, since the cost of pollution control is eventually passed on to the consumer of the manufactured product.

The critical problems caused by the release of hazardous materials into the environment are discussed in several subsequent chapters. For a more detailed consideration of the nature of hazardous materials and the results of exposure to them, see Box 3.1.

Determining Limiting Factors

Determining tolerance limits of a given organism for a particular substance can be exremely difficult. One complicating factor is that an organism usually has the genetic capacity to adjust to change in its physical surroundings. Thus, for example, visitors to high mountains experience headaches, dizziness, and shortness of breath because of the low concentration of oxygen in the air. These distressing symptoms indicate that the minimum tolerance level for oxygen is being approached. However, after one or two weeks of exposure to high altitudes these symptoms usually disappear. We say that people who adjust in this way have become *acclimatized* to the altitude. Their adjustment to low oxygen concentrations is the result of an increase in the number of red blood cells (carriers of oxygen) in the bloodstream. Some human beings are better able to adjust than others because of differences in their individual genetic makeup. Still, a minimum limit does exist, and below it no human being can adjust; thus, even native guides must be equipped with an auxiliary oxygen supply on Mount Everest.

Age is another factor that influences an organism's tolerance limits. In all species, the very young usually suffer the most harm from a stressful environment. A young child who is fed a diet deficient in protein might suffer permanent brain damage, whereas an adult on a similar diet probably would

greater margin of safety, and some past experiences support this position. For example, allowable levels of radiation exposure in uranium mines have had to be revised downward several times as a consequence of the high incidence of lung cancer discovered among mine workers. In contrast, other people feel that less restrictive standards should be established. They argue that the experimental evidence is insufficient to justify the setting of such low exposure levels. They also complain that the cost of maintain-

not suffer this effect because his brain is already fully developed. (The degree of retardation in the child would be a function of the severity and duration of diet deficiency.)

Despite these complications, determining tolerance limits is still relatively simple if we isolate one species and change only one factor at a time. But such isolation is only possible in a laboratory; in a natural, uncontrolled environment many factors are always present and these usually interact, often with unforeseen consequences. Some interactions create an effect called *synergism*. The interaction of two factors is said to be synergistic if the total effect is greater than the sum of the two effects evaluated independently. In other words, the whole may be greater than the sum of its parts. For example, sulfur dioxide fumes, a common air pollutant, attack the lungs and in high concentrations can be fatal. However, when laboratory animals are exposed to air that contains both sulfur dioxide and particulates, much greater damage is sustained by the lungs than would be expected from exposure to each pollutant separately in sequence. Thus, the interaction of sulfur dioxide with particulates is a synergistic interaction.

In another type of interaction, two factors are said to be *antagonistic* if the total effect is less than the sum of the two effects evaluated independently. Nitrogen oxide, for example, is an air pollutant whose effects are similar to those of sulfur dioxide. However, when nitrogen oxide and particulates interact, they counteract each other in some way, and the deleterious impact on the lungs is much less severe than would be expected. Thus, the interaction of these two factors is antagonistic. These examples show that in order to predict confidently the effect a specific substance will have on an organism in a natural environment, we would have to be able to identify all the other substances with which the first will interact and then determine the kinds of interactions they will have.

In the larger context of an ecosystem, investigations of limiting factors become extremely difficult. An ecosystem is composed of hundreds of species, each influenced both by the physical environment and by other species. In addition, the number and kind of species present and the physical factors such as air temperature and soil moisture fluctuate daily and seasonally.

You can begin to appreciate now why so many predictions about the impact of environmental change must be qualified; few simple cause-and-effect relations exist in nature. So many interactions are possible that scientists have difficulty identifying them, let alone predicting their outcome. Clearly, this kind of complexity can lead to disagreement about the causes, not to mention the solutions, of an environmental problem. In some cases, scientists are not even certain that a problem exists. Unfortunately, however, many people use the lack of complete understanding to justify apathy and inaction. They seem reluctant to acknowledge that, in our complex world, no such thing as one perfect solution exists. Every problem demands intelligent action based upon careful consideration of the options and their potential consequences. In some cases, we may find that the procedures in current use are the best solution under the prevailing circumstances. But the important point is that careful, intelligent consideration must be given to problems based upon the best information available.

The Consequences of Limiting Factors

We have seen that, if an environmental factor is beyond the tolerance limits of an organism, then that organism cannot survive. Now, in a larger context, we consider how an ecosystem as a whole responds when an environmental factor exceeds the tolerance limits of some organisms within it. One well-documented study of this question, on-going since 1961, concerns an oak-pine forest in the Brookhaven National Laboratory on Long Island, New York, that has been intentionally exposed to radiation. A source of radioactive rays was placed within the forest, and for twenty hours each day the forest plants and animals were exposed. After two years of

Box 3.1

Warning: The Environment

 Human beings and all other organisms are continually exposed to materials that are potentially harmful. Some of these substances, such as mercury and lead, have always been with us in trace amounts. But now industrial processes concentrate them to dangerous levels and release them into the environment. Some manufactured materials, including pesticides and industrial chemicals found in air, water, soil, and food, are lethal in small quantities. Many of them are also carcinogenic, mutagenic, and teratogenic.

Toxicity is the capability of a chemical substance to induce discomfort, illness, or death. One measure of toxicity is LD_{50}, the dosage of a substance that will kill (*Lethal Dose*) 50 percent of a test population. Toxicity is measured in units of poisonous substance per kilogram of body weight. For example, the very deadly chemical that causes botulism, a form of food poisoning, has an LD_{50} in adult human males of 0.0014 milligrams per kilogram. This means that if a dosage of only 0.14 milligrams—roughly equivalent to a few grains of table salt—is consumed by each of 100 adult human males each weighing 100 kilograms (220 pounds), approximately 50 of them will die.

Lethal doses are not, however, the only danger from toxic substances. Today industry and government are showing much concern over minimum harmful dosages, or *threshold dosages,* of poisons, as well as their sublethal effects.

The relative toxicity of a hazardous chemical depends in part on the age, sex, and general health of the exposed subject; a dose that is lethal to one individual may produce only mild symptoms in others. Toxicity further depends upon how the substance enters the body. In general, chemicals enter more slowly through the skin than by way of the digestive tract or the lungs. Hence ill effects are less likely to occur when a toxin is spilled on the skin than when it is ingested or inhaled.

Length of exposure further complicates the determination of toxicity values. *Acute exposure* refers to a single exposure lasting from a few seconds to a few days. *Chronic exposure* refers to continuous or repeated exposure for several days, weeks, months, or even years. Acute exposure usually is the result of a sudden accident—for example, the release of chlorine gas from a ruptured railroad tank car—that kills or permanently disables people immediately. Acute exposures often make disaster headlines in newspapers, but chronic exposure to sublethal quantities of toxic materials presents a much greater hazard to public health. For example, millions of city dwellers are continually exposed to low levels of such air pollutants as the sulfur dioxide produced by burning coal or fuel oil. Many deaths attributed to emphysema or cardiac arrest, may actually be brought on by a lifetime exposure to sublethal amounts of air pollutants.

Using toxicity values to determine safe levels of human exposure presents still another difficulty: few actual measurements have been made. Scientists must rely on data from accidental exposure of humans or, more frequently, on data from tests on other animals. But health experts cannot confidently apply the results obtained with laboratory animals to human beings. So it is difficult to know where to set the safe exposure level to any chemical.

As we have noted, direct poisoning is not the

May Be Dangerous to Your Health

only danger posed by toxic substances; many of them are also carcinogenic, teratogenic, or mutagenic. A disturbingly large number of materials that can be found in air, water, and food are known to cause cancer, at least in test animals. They include asbestos fibers, metals (cadmium, nickel, and chromium compounds), vinyl chloride and other ingredients of plastics, and "tar" in cigarette smoke. Radiation from radioactive substances and devices that emit x-rays and ultraviolet light can also induce cancer. Even sunshine may be carcinogenic: ultraviolet light from the sun is suspected to cause skin cancer. A 1978 study by the National Cancer Institute found that on-the-job exposure may cause at least 20 percent of all cancer deaths. Asbestos, a fire-resistant mineral much used in the construction industries and in the manufacture of such products as auto brake linings, is apparently responsible for more deaths than any other carcinogen encountered in the workplace.

Some industries are, of course, more hazardous to their employees' health than others. According to a 1977 study prepared for the National Institute for Occupational Safety and Health (NIOSH), these are the eight most hazardous industries and some of the carcinogens they use:

Industrial and scientific instruments (solder, asbestos, thallium).

Fabricated metal products (nickel, lead, solvents, chromic acid, asbestos).

Electrical equipment and supplies (lead, mercury, solvents, chlorohydrocarbons, solders).

Machinery other than electrical (cutting oils, quench oils, lubrication oils).

Transportation equipment (constituents of polymers or plastics, including formaldehyde, phenol, amines).

Petroleum and its products (benzene, naphalene, and others).

Leather products (chrome salts and other salts used in tanning).

Pipeline transportation (petroleum derivatives, metals used in welding).

Some environmental pollutants are not only carcinogenic but also teratogenic. A developing fetus is particularly vulnerable to harmful substances, and congenital defects are the third most common cause of death in the newborn. Known teratogens include methyl mercury, x-rays, and radiation emitted by radioactive materials. Smoking by a pregnant mother can also cause birth defects.

Researchers have discovered that many carcinogenic and teratogenic substances are also mutagenic. Mercury and lead compounds, ultraviolet light, x-rays, and radiation emitted by radioactive materials are among the mutagenic substances and agents that can cause spontaneous abortion and stillbirth, mental retardation, heart disease, bone and joint deformities, and other defects.

Efforts to determine safe exposure levels for carcinogens, teratogens, and mutagens are hampered by the same problems that plague attempts to establish LD_{50} levels. Test results vary with the subject's age, sex, general health, and length of exposure, as well as with pathway of entry into the body. A particularly difficult problem in the study of potential carcinogens is that cancer generally does not appear until ten to twenty years after exposure to the implicated substance.

Figure 3.6 An aerial view of radiation damage in an oak-pine forest in the Brookhaven National Laboratory, Long Island, New York. (Brookhaven National Laboratory.)

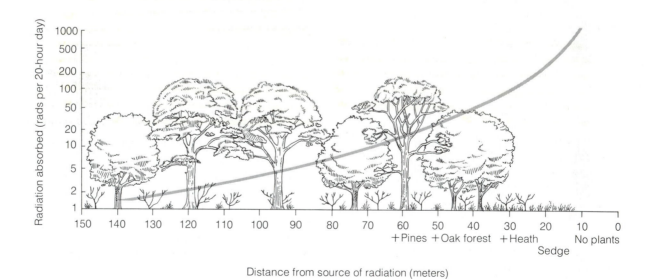

Distance from source of radiation (meters)

Figure 3.7 Affects of gamma radiation on the oak-pine forest. (Brookhaven National Laboratory.)

this exposure, striking results were noted. Figure 3.6 is an aerial view of the area taken at that time.

A concentric pattern of damage was evident around the radiation source. Close to the source, only mosses and lichens survived; the tolerance limits of the other plants to radiation had been exceeded. Farther from the source was a circle of low-growing sedges followed by a band of shrubs including blueberries and huckleberries. Surrounding the shrubs was a zone of oak trees, and farther out the original oak-pine forest, which remained intact; its growth had been inhibited, but no plants had been killed directly by radiation. Apparently, the pines were least able to tolerate radiation, whereas the lichens and mosses were most tolerant of it. Figure 3.7 shows the relation of growth inhibition to the level of radiation.

Other pollutants have been shown to have similar effects. For example, high concentrations of sulfur dioxide released by iron and copper smelters kill vegetation, as Figure 3.8 illustrates. A pattern similar to that at Brookhaven is found in the vicinity of an iron ore processing plant in Ontario, Canada. Close to the plant, no vegetation save a few decrepit ferns are growing. Zones downwind from the plant contain, successively, only herbaceous plants, then shrubs, and finally white spruce and balsam fir trees. Adjacent to the plant, where all vegetation has been killed, soil erosion and nutrient loss are severe.

Figure 3.8 Damage to vegetation produced by sulfur dioxide emissions from an iron-ore processing plant in Wawa, Ontario. (National Film Board of Canada, photo by George Hunter.)

Ecological Succession

The examples cited in the preceding section illustrate a pattern of response to *environmental stress*—that is, external change with adverse effects on the organisms involved—at a specific location over a short period of time. In such situations, if the stress is removed the damaged vegetation zones may eventually regain their original structure and composition. This would be accomplished through a series of successive changes taking place within each zone. The pattern of succession would be based on the composition of the zone at the start of recovery. For example, at Brookhaven the lichen-moss zone would be expected to change in time, first into a community of sedges, then shrubs, and finally an oak-pine forest. In contrast, the shrub community would probably change directly into an oak-pine forest. This sequence of changes is called *ecological succession.* Ecological succession is generally a predictable and orderly series of changes in which one ecosystem is replaced with another until an ecosystem is established that is best adapted to that environment. Ecological succession can be seen as a series of recovery stages that an ecosystem goes through after the stress is no longer present.

Whenever a new habitat is created (the formation of sand dunes along a lake, for example, or the bare area left by a melting glacier) or an existing habitat is damaged by stress (agriculture, forestry, fire, wind, or pollution), the resultant habitat serves as an open site for invading plants and animals. Successful invaders are those organisms that are well adapted to the stressful conditions of bare soil or sand. These *pioneer* organisms are gradually replaced by other types of colonizers. The process of replacement continues until a stage is reached where the assemblage of organisms is able to maintain itself on the site barring further significant environmental change. This final stage in ecological succession is called the *climax* ecosystem.

The type of climax vegetation that emerges depends on interactions of a large variety of factors, including soil type, amount of precipitation, temperature, drainage conditions, and altitude. For example, the natural climax ecosystem of the Great Plains was grasslands (now wheat fields grow there) primarily because grasses are more tolerant than trees of the drier environment and because thunderstorms, frequent in that area, ignited fires that killed woody vegetation and promoted the growth of grasses.

There are two basic types of succession, primary and secondary. Which one takes place depends on the soil conditions at the beginning of the process. *Primary succession* occurs on such surfaces as bare rock or sand dunes where no soil exists. Most plants cannot enter these barren areas until soil forms in which they can grow. Mosses and lichens, such as those shown in Figure 3.9, slowly invade cracks and

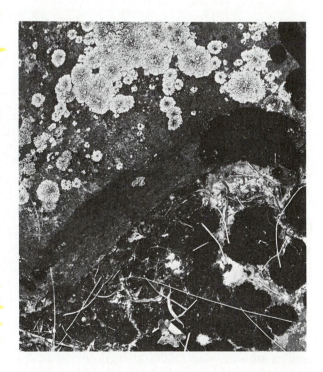

Figure 3.9 Lichens and mosses contributing to the slow process of rock disintegration and soil buildup. (M. L. Brisson.)

crevices of the bare rock and begin the slow process of soil building. Thousands of years may be needed for soil to form; hence, the initial stages of primary succession are extremely slow. Once soil has been formed, the subsequent stages of primary succession are usually similar to those for secondary succession in comparable environments. *Secondary succession* occurs where soil is present but stresses such as agriculture or fire have removed the natural vegetation. Because the soil usually remains after the stress is passed, revegetation normally occurs within a few weeks.

Secondary Succession

The succession that occurs in abandoned agricultural fields, known as old-field succession, serves to illustrate some of the processes of secondary succession. Let us examine what happened to abandoned farmland in the Piedmont region of North Carolina. European settlers cleared away the native oak-hickory forests there and farmed the land. Improper farming practices depleted the soil of its nutrients, thereby reducing soil fertility, and eventually farmers had to vacate their fields and migrate westward. We can reconstruct the successional events that occurred on these uncultivated farmlands by examining fields that have been abandoned for different lengths of time.

Abandoned fields do not immediately revert to the original oak-hickory forest, even if oak and hickory trees nearby act as a source of seeds. Rather, a series of other biotic communities successively occupies a field before a oak-hickory forest is reestablished. Figure 3.10 shows three stages in old-field succession. In the year following abandonment, a field is colonized by a mixture of pioneer species that includes crabgrass, pigweed, ragweed, and horseweed. The next year these plants are replaced by asters, which are succeeded the following year by a nearly pure stand of a grass commonly called broom sedge. The pace of succession then slows; the broom sedge persists for as long as twenty years, and during that time the field is gradually invaded by pine seedlings. Eventually, the pines mature and shade out the broom sedge. Later, oak and hickory seedlings slowly establish themselves beneath the pine trees, and as the pines grow old and die they are replaced by the hardwoods, the oaks and hickories. Eventually (approximately 150–200 years after abandonment), the climax oak-hickory forest is reestablished. As vegetation changes, the type of shelter and food available for animals changes, too. Consequently, concomitant alterations in the number and types of animals present take place as succession proceeds. Figure 3.11 illustrates the changes in wildlife that occur in old-field succession to an oak-hickory forest.

The physical environment also changes with succession. Trees shade the ground and slow the wind, and the climate beneath the tree canopy thus differs greatly from conditions outside the forest. Fluctuations in air temperature are reduced and more moisture exists in the soil and the air beneath the canopy than outside it.

As succession proceeds, the soil is enriched by the activities of plants and animals. Plant roots take up nutrients from the soil and incorporate them into their structures. When plants and animals die, decomposers act upon them, releasing nutrients into the soil. Burrowing animals such as earthworms and ants work the remaining humus deeper into the soil. More activity by decomposers enriches the fertile zone below the surface. Thus, plant and animal activities mix humus and nutrients into the soil and increase the soil's capacity to hold air and water.

In part, one plant community is succeeded by another because, ironically, the existing community alters its own environment to the point where it can no longer perpetuate itself. For example, pine seedlings can successfully invade an open old field during the broom sedge stage. But as pine seedlings mature into trees, they modify the local climatic and soil conditions, making the physical environment more favorable for successful invasion by oaks and hickories. These new conditions do not favor the continuation of the pine forest because pine seedlings

cannot survive in the shade of mature pine trees. Hence, as pine trees die, they tend to be replaced by oak and hickory saplings already present beneath the pine canopy.

We have examined one example of succession towards a climax community. But you should remember that the nature of the climax community varies from one geographical locality to another depending on climate and other environmental variables. It is as a consequence of these interactions that the landscape varies from the majestic rainforests of the Pacific Northwest to stands of scrub oak and pine along the coastal plain of the Southeast.

The Effects of Human Activities on Succession

As noted earlier, many types of stress can destroy a climax ecosystem. Natural stresses include fire, flooding, drought, hurricane-force winds, and insect infestations, and agriculture is probably the most significant of the human-caused stresses. In the United States alone, millions of hectares of climax forest and prairie have been cleared and plowed under for cultivation. Basically they have been replaced by a few grasses (corn, wheat, oats, rye, and barley) and some legumes (soybeans and alfalfa). Human beings play a similar role when they clear away forests and replace them with tree plantations. In the southeastern United States, row after row of pine trees now stand where climax oak-hickory forests once grew. In tropical areas, complex climax rain forests are being chopped down and replaced by ecosystems that are greatly simplified by comparison—for example, banana, rubber, and cocoa planta-

tions. Other major human-related sources of stress include air and water pollution.

Although ecosystems tend in their succession toward the climax suitable to their particular environment, succession can be halted at any stage before climax is reached. Succession can be interrupted when a developmental stage is of value to its human overseers. Cultivation of farmland, for example, prevents old-field succession by continually disrupting the soil. And in the southeastern United States, controlled burning is used to prevent oaks and hickories from invading pine plantations. When these climax species are allowed to invade a pine plantation, pine tree production is reduced, subsequent pine regeneration becomes difficult, and the presence of oak and hickory increases the fuel available for a destructive wildfire. To overcome these difficulties, forest managers periodically set controlled ground fires (such as the one shown in Figure 3.12) to eliminate the fire-sensitive invading climax species. Pines are fire-resistant and usually remain undamaged.

As succession takes place, each succeeding ecosystem exhibits specific trends with respect to the flow of energy and materials; thus, in certain fundamental ways climax stages may differ considerably from early successional stages (as Table 3.1 indicates). Given these differences, what are the consequences of large-scale human activities that replace climax stages with pioneer or developmental stages? Probably the most significant effect is a reduction in the diversity of the organisms present in the area. Pioneer ecosystems consist of relatively few species forming a relatively simple food web. A climax ecosystem consists of many species, however, and therefore has a more complex food web. As noted earlier, greater complexity generally means greater stability.

Figure 3.10 *Opposite:* Various stages of old-field succession. *Top:* A marginal farm such as this one is often abandoned after a few years of cultivation. *Middle:* This field is in the broom sedge stage, while the adjacent land is in the pine stage. The broom sedge is being burned back. *Bottom:* The climax stage of old-field succession in this area is an oak-hickory forest. (Tennessee Valley Authority.)

We can gain an appreciation of the stabilizing effect of diversity by comparing the small, diversified farms that predominated in the past with today's more specialized agricultural enterprises. Even though both types of farms are pioneer communities, those of the past were more self-sufficient than the monocultures are today by virtue of their diversity.

On a small, diversified farm horses or oxen provided power for cultivation and transportation, and a windmill pumped water. Farmers raised a few cows or goats for milk and meat and a few chickens for eggs. Someone in the vicinity usually maintained a small herd of sheep to provide wool for clothing. Diversified farmers usually had a large vegetable garden and a variety of crops for feeding livestock.

Figure 3.11 The changes in wildlife that take place as succession proceeds.

The planting of a variety of crops generally ensured that a period of inclement weather would not destroy all the food sources for the household and livestock. A nearby woodlot provided timber for building, and crop and animal wastes were recycled onto fields as fertilizers. Although small diversified farms were not completely self-sufficient, they usually had a large enough variety of food and materials to see the household through the tough times.

In marked contrast, today's large modern farms grow only one or two crops, and these usually are marketed. Many farmers no longer raise livestock or poultry. Power is now provided by electricity and gasoline or diesel fuel, all of which must be brought to the farm from an outside source. In fact, most of the daily needs (including food) of the modern farm family are met now from sources outside the farm. These characteristics result in instability for several

Figure 3.12 Controlled burning of a pine stand, preventing the invasion of fire-sensitive climax species and the build-up on the soil of plant litter that could fuel a destructive wildfire. (U.S. Department of Agriculture, Soil Conservation Service.)

reasons. As we noted in Chapter 2 the tendency toward monoculture makes the modern farm, and thus the nation's food resources, more susceptible to such stresses as pest infestations and weather extremes. When these stresses do occur, farmers must apply extra energy—in the form of fertilizers and pesticides, erosion control measures, and irrigation—to maintain stability.

The loss in self-sufficiency is another cause of the modern farm's characteristic instability. Growers have to depend upon outside sources for fuel, seeds, commercial fertilizer, and pesticides, but in today's complex political and economic world, they have no assurance that these needs will be met, at least not at prices they can afford. Without a dependable supply of these resources, food production itself becomes less dependable. Thus, modern agricultural enterprises are actually less reliable in meeting world food needs than were the smaller, independent efforts.

Another consequence of reversing succession, as indicated in Table 3.1, is related to the fact that pioneer communities are more susceptible to erosion and loss of soil nutrients than are climax ecosystems; that is, nutrient cycles in developmental stages are said to be open, whereas those in the climax stage are closed. This significant difference in nutrient cycling has been well documented by investigations carried out at Hubbard Brook Experimental Forest in New Hampshire since the mid-1960s. In one study, all vegetation in a small watershed was cut and subsequent regrowth artificially inhibited for two years by periodic application of herbicides. Nutrient losses were substantial. In the second year following clear-cutting, stream water concentrations of nitrate,

Table 3.1 Changes that occur during ecological succession.

	Developmental stages	Mature stages
Species diversity	Low	High
Stability (resistance to external change)	Poor	Good
Mineral cycles	Open	Closed
Net production (yield)	High	Low
Food chains	Linear	Weblike

averaged 56 times greater than before deforestation, exceeding levels recommended for safe drinking water. Clearly, this important plant nutrient was being lost by the soil through runoff. During the same period, substantial increases were also measured in the average stream water concentrations of other plant nutrients: for example, 16 times as much potassium as had been there before, 4 times as much calcium and magnesium, and 2 times as much sodium.

These findings are not necessarily applicable to other places, since particular characteristics in each watershed influence runoff there. Furthermore, herbicide use in this experiment was extreme. Some vegetation usually regrows during the first year after disturbance whether by natural regeneration or through the planting of trees or crops. Nonetheless, these data illustrate the potential magnitude of nutrient loss in a disturbed ecosystem. Losses may even be sufficient to inhibit an ecosystem from completely restoring itself to the original climax condition. However, where ecological succession is allowed to proceed to climax naturally, the nutrient value of the soil is increased. In contrast to pioneer stages, most climax stages, by virtue of extensive ground cover and root systems, tend to recycle nutrients efficiently within the ecosystem and experience relatively small losses. Also, a greater proportion of nutrients are tied up in the plants, not in the soil, and hence are less subject to runoff. Thus, a climax ecosystem con-

serves more nutrients than the early developmental stages. For other characteristics of climax vegetation, particularly forests, see Box 3.2.

Human disruptions of ecological succession often encourage the proliferation of organisms that damage the crops that were the original motivation for interrupting the process. Thus, though a farmer intends to prepare the way for a profitable crop, he may be opening the way for invasions by large numbers of animals (mostly rodents and insects) and "weedy" plants that are well adapted to the rigors of life in pioneer ecosystems. We derive little in the way of food, clothing, or shelter from these plants and animals, but we spend substantial amounts of time and energy trying to eradicate them. Ironically, by persisting in such practices as improperly abandoning strip mines and failing to maintain soil fertility, we are making much more land available to these invaders.

In contrast, while pioneer organisms are increasing, those species that require climax conditions for survival are disappearing. In dozens of cases of extinction or near extinction of plant and animal species, habitat destruction is the most significant factor. The fate of Attwater's prairie chicken, shown in Figure 3.13, illustrates the problem. This species once thrived in the lush tall-grass prairie along the gulf coast in southwest Louisiana and southeast Texas. But the plowshare changed the prairie into cropland, thus reducing the prairie chicken's range by 90 percent and its population by 99 percent. Today Attwater's prairie chicken is found in only a few isolated areas in southeastern Texas. Many other species are threatened by habitat loss as well, for example, the San Joaquin kit fox, the Key deer, the Florida panther, Kirtland's warbler, the ivory-billed woodpecker, the black-footed ferret, trailing arbutus, and several species of orchids. Though such species are not significant today from an economic point of view, we should think of them as irreplaceable resources that might be lost to the planet forever. Not only do they enrich our aesthetic enjoyment of the environment, but they contribute to the complexity,

Box 3.2

Nature's Free Pollution Removal Service

Although they seem peaceful, forests are places of intense activity. Not only are countless plants, animals, and microorganisms growing and reproducing, but in the process they are also filtering the air and water, regulating stream flow, and reducing soil erosion. Recent ecological studies have taught us much about the important role forests play in maintaining environmental quality. The findings reported here are from research conducted by F. H. Bormann, G. E. Likens, and their coworkers at Hubbard Brook Experimental Forest. These findings have been largely verified by studies conducted at other sites.

Forest soils act as filters between polluted rainfall and the streams and groundwater the rainfall enters. For example, the burning of fossil fuels is causing rainfall in the northeastern United States to become more acidic. As these fuels burn they create nitrogen and sulfur oxides that enter the atmosphere. The oxides combine with the water in rain (and snow) to produce highly toxic nitric and sulfuric acids. The resulting "acid rains," which find their way into streams and lakes, are deadly to fish. Data from Hubbard Brook indicate that forest ecosystems remove much of the acid from rainwater and melting snow. As a result, stream water draining from forests is about thirty times less acid than the rainwater that falls upon those forests.

Precipitation may also contain other pollutants, including copper, nickel, lead, and other heavy metals, pesticides such as DDT—and even radioactive materials released by nuclear-bomb tests. Much of these pollutants, too, are removed from rainwater as it percolates through forest soil.

An undisturbed forest also enhances streamwater quality by reducing soil erosion. The forest's extensive root system and the rich layer of litter and humus at the soil surface soak up the rainwater, thereby reducing its runoff into streams. Slow-moving water dislodges and transports less soil than swiftly flowing water, and so the water entering streams in forests is often virtually sediment-free. Reduced runoff also reduces the severity of downstream flooding.

Forest vegetation and soil also purify the air. Very fine particles of some air pollutants stick to leaves and branches; many gaseous pollutants enter leaves and are chemically detoxified there. Eventually most air pollutants that enter a forest reach the ground, where they are either detoxified by soil microorganisms, bound up in the soil, or absorbed by plant roots. Thus, air is considerably cleaner when it leaves a forest than when it enters.

This natural cleanup of water and air is accomplished at no expense to us; the forest's natural filtration and detoxification processes are powered by free solar energy. All too often, however, the useful work performed by forests and other natural ecosystems is undervalued or ignored. If a region becomes overdeveloped, the pollution-removing functions of natural ecosystems must be replaced or supplemented by costly, fuel-consuming technologies. Water treatment and sewage disposal plants must be larger and must employ more sophisticated pollution-control devices. Channels must be dredged, streams diverted or piped, and other flood control measures adopted. This adds up to a large—and often unnecessary—expense for the taxpayers.

Those who argue for the preservation of natural areas usually cite the need for parks and wildlife refuges. But the free antipollution services provided by natural areas offer two even stronger reasons for maintaining areas of natural vegetation within developed regions: such areas promote the health and the economic well-being of people.

Figure 3.13 Attwater's prairie chicken, in danger of extinction. (U.S. Department of the Interior, Fish and Wildlife Service, photo by Luther C. Goldman.)

and thus the stability, of their ecosystems. In Chapter 14, we will consider further the pros and cons of preserving endangered species.

In assessing the impact of our destruction of climax ecosystems, we must not disregard the many benefits we have reaped from these efforts, at least in the short run. As Table 3.1 indicates, pioneer stages produce a greater net yield than a climax stage. Furthermore, we have greatly increased the amount of available food by clearing the land and sowing crops that direct more of their production into seeds and fruits that are edible either directly by us or by our livestock. By encouraging the tendency toward monoculture, we have made possible the use of large, efficient machinery, permitting fewer farmers to produce more food. Between 1960 and 1977 on-farm population declined by 50 percent, whereas farm production increased by about 30 percent. We have increased nonagricultural production, too. For example, many successional trees are more productive and some are more desirable for wood products than are climax species. Examples of such desirable species include pines in the Southeast and Douglas fir in

the Northwest. In addition, many species of game—white-tailed deer, elk, moose, bobwhite quail, and pheasants—do better in successional habitats than in climax conditions. In fact, many more deer are present in the United States today than when the continent was first colonized.

In the contexts of agriculture, forestry, and wildlife management, then, the benefits we have derived from reversing succession are considerable. Still, as we have noted before, short-term economic gains may result in serious long-term losses of fundamental resources. Whether in the long run the benefits will outweigh the costs remains an open question.

Limitations on Succession

Can nature heal all wounds? That is, given enough time, can succession mend all disturbances, reestablishing the original climax community in every instance? Not always; there are limits to healing processes. A disturbance may be so severe and the changes in the environment so extreme that succession takes a new course, never reestablishing the original climax community. Such a tendency appears to be expressing itself throughout the tropical areas of the world, where thousands of square kilometers of climax rain forest are being cut down for timber and to provide additional land to raise crops and livestock. If these areas were suddenly left alone, the climax rain forest would be restored, although the process would probably take centuries. But, because of unrelenting pressures from the human population, natural regeneration has had little chance to take hold. Just a very few years after the initial cutting of timber, demand for food, lumber, and firewood has forced the reclearing of the land again. As a result of this overuse, soil erosion is severe, and coarse, low-nutritive grasses and bamboo thickets have quickly invaded the land. This vegetation is highly flammable during the dry season and is regularly burned to encourage new growth. Such continual human disturbance postpones reestablishment of the original rain forest, perhaps forever.

Nature's healing process takes a substantial amount of time. Even under normal circumstances, 150–200 years are necessary for a climax oak-hickory forest to become established on abandoned farmland. Even more time may be required for restoration after a severe disturbance. Following cultivation, timber cutting, or fire, soil erosion may be so intensive that the land is incapable of supporting vegetation. In areas where the land has been mined, freshly exposed waste heaps are usually quite low in nutrients. Thus, the extremely slow process of soil generation through primary succession must occur before the land can again be productive.

In fact, time is the most significant single factor influencing our efforts to manage many of our natural resources. For all practical purposes, natural succession cannot be considered a viable means of restoring much of the earth's disturbed land to its original state within our lifetime. To speed up restoration, we need to employ such management practices as adding fertilizer to the land and planting hybrid plants that do well under the poor growing conditions characteristic of many disturbed areas. But even these techniques meet with natural limitations: soil building remains a slow process and most trees take much longer than a few years to reach maturity. Slow as it is, ecological succession is the natural recovery response of ecosystems to environmental stress. We may discover that in some cases our only hope for restoring the environment lies in lowering our expectations and allowing this process to occur at its own rate. For a more detailed discussion of problems and progress in land restoration, see Chapter 12.

Adaptation

Individual organisms do not necessarily die when their environments change. As you may recall from our earlier mention of it, some organisms are able to adjust to changes in the environment, whereas others are not. The adjustments that organisms make to environmental conditions are called *adaptations*. Some organisms adapt to changes in seasonal climate by avoiding the stressful conditions. Birds fly south to avoid the cold weather, wind, and lack of food that characterize northern winters. Chipmunks, ground squirrels, and woodchucks descend into their dens and *hibernate* in winter. Their body temperatures drop to just a few degrees above freezing and their heart and breathing rates slow down. This sharp decrease in body activity diminishes their energy needs, so that by relying upon stored fat, they can live through the winter without ever having to leave their dens to search for food. Similarly, many trees and shrubs lose their leaves and become dormant as winter approaches. These responses are adaptations to changes that take place slowly—lowering temperatures, decreasing soil moisture, or decreasing day length. Organisms could not tolerate such changes if they occurred within minutes or hours.

Some individual organisms are unable to adjust to environmental change. These more vulnerable organisms die and are usually replaced by others whose particular characteristics are more suited to the altered environment. Thus, adaptation increases an organism's chances of surviving and reproducing within the environment in which it resides. Adaptations may be structural, functional, or behavioral. In the following sections, we explore the process of adaptation and its significance and limitations for the survival of organisms.

Natural Selection

The ability of an organism to adapt to a particular environment depends upon its genetic makeup. If a particular genetically determined characteristic makes an organism better suited for its environment, then the organism has a better chance of surviving and producing offspring than organisms without the characteristic. It is likely, therefore, that this organism will pass the beneficial characteristic on to its offspring and that these offspring also will have a greater probability of being favored by the environment. Eventually, the population will consist of many members that possess this beneficial characteristic. This sequence of events, in which a certain characteristic is "selected" by the environment, is called *natural selection*. The result of natural selection is that the population contains a greater number of individuals that are better adapted to a given environment.

Natural selection is illustrated by the classic example of the peppered moth that lives in the forests of England. This moth is active only at night; during the day it rests on tree trunks. Before the mid-1800s, most peppered moths were light colored, but occasionally a black peppered moth was observed. With the onset of the Industrial Revolution, however, the black variants of the moth became the predominant form. In Manchester, England, the first black specimen was caught in 1849, but, by 1895, the black form constituted 98 percent of the total population. Afer extensive experimentation, the following conclusions were reached as to what caused this dramatic shift in the predominant color of the peppered moth. While the moths were resting on tree trunks during the day, they were subject to predation by birds. Under natural conditions, the light-colored form is difficult to distinguish against a light-colored tree trunk. By contrast, as you can see in Figure 3.14, the darker variant is more visible to birds and thus has a

Figure 3.14 Light and dark forms of peppered moths at rest on a tree trunk. *Left*: The light form, difficult to see on the lichen-covered tree. *Right*: The light form, much easier to see on a soot-covered tree. (From the experiments of H. B. D. Kettlewell, University of Oxford, England.)

greater chance of being eaten (selected against). With the growth of industry, the large amounts of soot and ash belched into the air by factories gradually darkened nearby tree trunks. Under the changed conditions, as shown in the figure, the black form blended in with the dark tree trunk while the light form was more clearly visible to birds. Hence, the black forms were selected for and soon became the more abundant members of the peppered moth population. It is interesting to note that with the recent implementation of air pollution controls, tree trunks are becoming lighter, and in turn the light form of the moth is beginning to predominate again.

Several important points in this example need emphasis. First, environmental change did not produce the black form of the moth. The black variant was already present in the population. Natural selection, in response to environmental changes, can only work on the genetic variations already present in the individual members of the population. It favors only those members that possess a trait that makes them better suited to the changing environment.

Also significant is the short time period required for an almost complete shift in the predominating color of the population. The change occurred quickly because moths, like most insects, produce large numbers of young several times a year. (Table 3.2 presents some reproductive rates for comparison.) The black forms were able to quickly increase in numbers when birds were no longer an important limiting factor. Since generation times vary considerably among species, this factor has a significant effect on the length of time required for an adaptation to become established. Some bacteria can double their population size in as short a period as twenty minutes. Hence, if some members of a bacteria population are resistant to a particular antibiotic, the bacterial population can adapt to the presence of the drug within hours or a few days. In marked contrast, long-lived, slow-reproducing organisms such as the human being may require hundreds or thousands of years to undergo an adaptation of comparable magnitude.

Table 3.2 Comparison of some measures of reproductive capabilities among different species.

	Mean length of generation*	Time required to double population
Rice weevil	6.2 weeks	0.9 weeks
Brown rat	31.1 weeks	6.8 weeks
Human beings (1977 world rate)	15–20 years	39.8 years

* Mean length of generation is the average time between birth of individual members of a species and birth of their first offspring.
Source: After E. P. Odum, *Fundamentals of Ecology* (Philadelphia: Saunders), 1971.

The Significance of Adaptations

Adaptations can increase an organism's chances for survival in many different ways. To illustrate, we consider here how adaptations affect the interaction of organisms with each other through predation, and how they influence an organism's response to one component of the physical environment, the climate.

Many adaptations reduce an organism's vulnerability to predators. We have already seen how differences in wing color make peppered moths more or less vulnerable to predation by birds. Numerous other examples show how natural selection favors protective coloration, or camouflage, in certain populations in which a highly visible individual is in danger of being eaten. Many desert animals, particularly lizards, are light colored and blend with the light-colored background of desert soils. Some animals change colors with the seasons. The Arctic hare is white during the winter snow season, but in the spring it sheds its white fur and becomes brown, thus blending in better with the tundra vegetation. During wartime, people, too, have employed camouflage techniques to blend into their surroundings and thereby avoid detection by the enemy. Following the example of animals, soldiers wear all-white clothing in the snow and mottled green and brown clothing in a jungle. In each case, the protective coloring represents an adaptation to an environment in which low visibility is an advantage for survival.

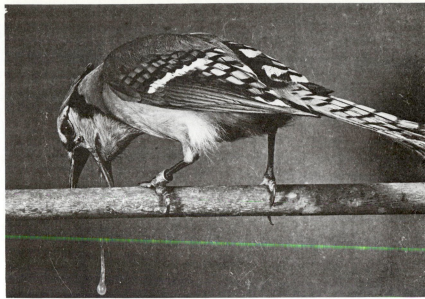

Figure 3.15 A freshly eaten monarch butterfly inducing vomiting in a blue jay. A viceroy butterfly, which is similar in appearance, would not be eaten by the jay, as a result of the jay's experience with the monarch. (Lincoln Brower.)

Many of the adaptations prevalent in insects are actually forms of mimicry. Walking sticks, for instance, look so much like sticks, and leafhoppers so much like leaves, that predators often are unable to tell the difference, and thus pass up many a nutritious meal. A particularly interesting type of mimicry is that expressed in species that look very similar to other species. The monarch butterfly and the viceroy butterfly, for example, are almost identical in appearance, but, when eaten by birds, the monarch induces vomiting, as shown in Figure 3.15. Although the viceroy is a tasty morsel, it is not often eaten by birds that have had bad experiences with its look-alike, the monarch. Again, these look-alike adaptations resulted from the action of natural selection on individuals that just happened to exhibit these characteristics in an environment that favored them. An insect that looked exactly like a stick had a good chance of living to reproduce, whereas one that looked to be what it was—a palatable source of food—had a good chance of getting eaten before it could breed. Thus, the former, favored by natural selection, proliferated, and its life-saving characteristic, its resemblance to a stick, became a populationwide adaptation. In contrast, the latter was selected against, and the characteristic was eliminated from the population.

The second sort of adaptation we consider here is that made in response to changes in climate. Some of these adjustments may occur in just a few minutes. Shivering, for example, is the body's normal response to exposure to cold. Shivering consists of enhanced muscle activity that is maintained by energy from the process of respiration. The by-product of this process is heat energy, which helps to maintain a normal body temperature. Thus, the individual who happens to be able to shiver would be better adapted to a sudden drop in temperature than one who did not possess this or some other temperature-compensating characteristic.

Some adjustments to climatic change are more gradual, requiring several days or even weeks to express themselves. For example, when we first turn down our thermostats in winter to 18°C (65°F), many of us are uncomfortable. But after exposure to several weeks of cool temperatures, our bodily functions adjust and we feel more comfortable. Earlier in this section, we saw the gradual adjustments of plants and animals to the onset of winter. Because such adjustments take place over time, the rate of environmental change is often as important as the magnitude of the change. An extended outbreak of severe cold temperatures is much more detrimental to mammals in midautumn, before they have adapted to winter by growing thicker coats and gaining fat, than it is in midwinter, when these animals are acclimatized.

Adaptation by Learning

Learning is another means animals have of adapting to change. An organism that can learn to adjust its behavior to changes in its environment has a great advantage over one that simply repeats the same behavior patterns regardless of environmental conditions. The two most important learning processes in the more complex animals are *trial and error learning* and *insight learning*. Trial and error learning often takes place as an animal develops its feeding habits. For example, a young squirrel just beginning to feed upon the seeds present in hazelnuts may spend a considerable amount of time gnawing and attempting to crack the nut. With more feeding on hazelnuts, the squirrel randomly finds the method that efficiently cracks it open. Subsequently, the squirrel still gnaws randomly on the hazelnut initially, but soon it begins to use the more efficient means to crack the nut. Hence, by trial and error, the squirrel learns to open the nut on its first attempt.

Insight learning is the ability to respond correctly the first time to a new situation. In effect, insight learning allows an animal to apply previous learning experiences without the process of trial and error. For example, if we place a hungry chimpanzee in a cage where boxes are scattered about the floor and bananas are hanging from the ceiling, the chimp will initially survey the situation and then collect the boxes, stack them beneath the bananas, and climb up on them to reach the bananas. This chimp, using past experiences to solve a problem the first time without resorting to trial and error, is experiencing insight learning.

Both types of learning can be passed on to other members of the population by some means of communication, such as sounds or visual displays. Communication is important because it allows a learned trait to spread rapidly throughout a population. In contrast, a genetic trait is only passed from parents to offspring. Hence, an adaptation by learning can be acquired by a large segment of the population more rapidly than can a genetic adaptation. Adaptation by learning is particularly important for slowly reproducing organisms such as human beings.

But it is important to remember that learning itself is not independent of genetic makeup. In fact, an animal's ability to learn is ultimately determined by its genetic makeup. Figure 3.16 illustrates a racoon's inability to solve a problem through insight, a genetically determined limitation. And we can cite an example of our own limitations: our genetic makeup enables us to speak and understand many human languages, but we apparently do not have the ability to learn to understand or communicate with most other animals.

The relative importance of adaptations by learning varies with the species in question. In fairly simple animals, such as worms and oysters, genetic adaptations have greater importance than adaptations through learning. As the brain becomes more complex, the significance of adaptations by learning grows. Our own species represents the current peak in brain development on this planet, and adaptations by learning are therefore extremely important for us.

Some Human Adaptations

Human beings are generalists. Although we are genetically adapted to perform a wide variety of activities, we excel at few. We can run and swim, but we are neither the fastest nor the slowest at these activi-

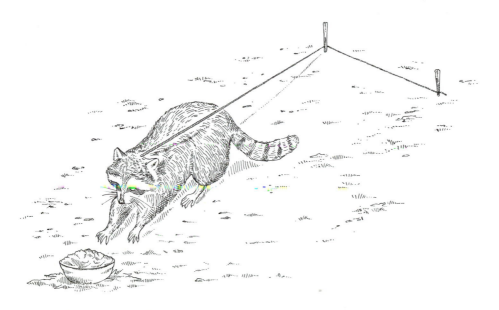

Figure 3.16 Lack of insight learning in a raccoon. Because its leash is looped around a stake, the raccoon cannot reach the food dish. Unlike a chimpanzee or person, who would immediately go back around the stake and then go to the dish, the raccoon must find the solution by trial and error.

ties. Many animals far exceed us in physical prowess and agility. Nonetheless, we do in fact possess several specialized adaptations, which, in combination with our general abilities, set us apart from other forms of life.

One of our few specialized adaptations is the thumb. Because the thumb moves in opposition to the fingers, we have the ability to grasp objects. This adaptation has played an essential role in our dominion over the earth's resources, for it has allowed us to employ tools to fulfill many of our basic needs and to avoid environmental stress.

The most significant of our specialized adaptations is our enlarged forebrain (cerebrum), the location of insight learning. As you can see in Figure 3.17, the forebrain is much larger in humans relative to other parts of the brain than in other animals. The development of the cerebrum, a genetic adaptation, has enabled us to add to our adapations through learning.

Although we were never the fastest or the strongest animal, by using the problem-solving ability centered in the forebrain we learned to domesticate other animals to supplement our speed and power. Later we learned to build machines that were faster and stronger than any animal. We were able to use first clubs, spears, and bow and arrows, and later firearms, in killing prey. These weapons more than adequately substitute for a carnivore's fangs. Although we never evolved armor, as the armadillo did, that would protect us from aggressors, through our technology we provided ourselves with metal shields, helmets, breastplates, and more recently, armored tanks. Another skill related to insight learning, our ability to contemplate the future, enabled us to give direction to our activities, both socially and technologically. Thus, not only did we develop the technological means to protect ourselves and enhance our culture, but we could visualize ways of applying

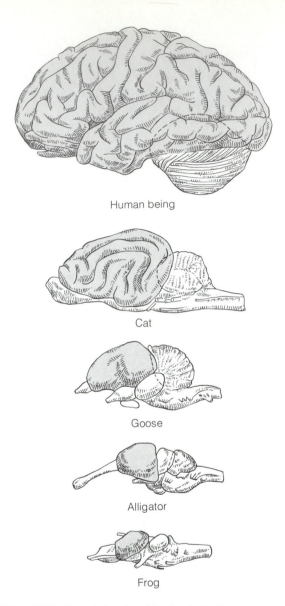

Figure 3.17 The relative size of the forebrain (the location of insight learning) in human beings as compared with four other animals.

Human being

Cat

Goose

Alligator

Frog

our predators, to significantly control many of our major competitors, to live in almost all habitats on earth, and even to visit the moon.

Limits to Adaptations

It is tempting to believe that physical and technological adaptations will solve all our environmental problems. Perhaps our lungs will become more resistant to air pollutants or we will develop a control technology to cleanse our air of all contaminants. And perhaps, if worldwide climatic changes take place, the planet's plants and animals will merely adapt to the wetter or drier, colder or warmer conditions. We must realize, however, that nature imposes strict limits on adaptations to which all organisms are subject. We have had occasion to mention the major natural limits in our discussion of adaptation, but we do well to reconsider them here. The following are three such natural constraints on the ability to adapt. First, each organism's capability to adjust to a particular stress is determined by its genetic makeup; hence, an organism cannot adjust to all stresses. Second, a changing environment selects only for characteristics that are already present in the population. And, third, even if an adaptive trait is present, the rate at which a population as a whole can adopt it is limited by the population's reproductive capabilities—both number of young per generation and length of generation.

Technological adaptations are limited as well. The development and large-scale implementation of most technological advances require considerable time and expense. And often, as we will point out many times in this book, our advances in technology produce as many problems as they solve.

them and imagine, in a limited way at least, the consequences of our actions.

It is to our capacity for learning, more than any other adaptation, that we owe our success as a species. This ability has enabled us to eliminate most of

Conclusions

In this chapter we have examined the way organisms and ecosystems respond to changes in the environment. We have seen that environmental change brings about a continual selection process, determining the characteristics of populations and ecosystems by favoring the continuing existence of some organisms over others. The net effect of environmental change, then, is change in the type of organisms present in an ecosystem.

In response to human activities, however, the environment itself is also changing. Some observers foresee a period of greater environmental change than we have yet experienced, as world population soars and human beings continue to strive for a higher standard of living. So far, our technological and cultural adaptations—the collective expression of our capacity for insight learning—have made us the most successful species in the history of the earth. Yet with these same adaptations we have fouled the planet and gained the ability to destroy ourselves and most life on earth. We do have the ability to contemplate the future and imagine the consequences of our actions. But the question still remains—how responsibly, and how successfully, will we use this ability in years to come?

Summary Statements

The environment is continually changing. Some organisms are able to adjust to changes while others succumb. Limiting factor curves aid us in understanding the influence of environmental change on the well-being of plants and animals, including people.

The variables of age, ability to adjust to change, synergism, and antagonistic interactions all make determining an organism's tolerance limits to environmental change difficult. Because so many interactions are possible among organisms and their environment, few simple cause-effect relations exist in nature.

One kind of change of special interest to modern society is the introduction of toxic and hazardous substances into the environment. These materials pose a significant threat to vegetation, wildlife, and people. Efforts to establish standards for safe exposure levels of poisons, carcinogens, mutagens, and teratogens encounter the same problems as those for determining an organism's tolerance limits for any environmental factor.

If an essential factor in the environment lies outside the tolerance limits of an organism, then that organism must either be able to avoid the stressing conditions or it will die. When many plants and animals are killed by a stress, the functioning of an ecosystem is severely disrupted.

If the stress is removed, a damaged ecosystem may eventually return to its original structure and composition via a sequence of changes called ecological succession. As populations of plants and animals succeed each other, the physical environment also changes. Many types of stress can destroy a climax ecosystem, and succession can be halted at any stage before climax is reached. Agriculture and forest management are important examples of the reversal of natural succession by human beings.

The stages preceding climax in ecological succession are called pioneer or developmental stages. The major consequence of maintaining pioneer ecosystems on a large scale is a reduced diversity in the kinds of organisms present. This relative lack of diversity makes large modern farms and tree plantations unstable in several ways. Another consequence of the loss of climax ecosystems is the imminent extinction of many species that require climax habitats. On the other hand, the advantages of maintaining pioneer ecosystems include increasing the land's capacity for food production, improving timber production, and enhancing the population size of many species of game animals.

Succession to climax often requires many decades or centuries. Even then, a disturbance may be so severe that succession takes a new course and the original climax ecosystem may never be established.

Organisms within an ecosystem adjust to change by means of adaptations, characteristics that make them better suited to their environment. Organisms that can adapt have a better chance of surviving and producing offspring than organisms that cannot. Because organisms that survive can pass on their beneficial traits to their offspring, eventually populations come to consist of many members that possess traits that are beneficial for survival in that particular environment.

Each organism's capability to adjust to environmental change is determined by its genetic makeup. A changing environment selects for (favors) only those characteristics already present in the population. If a trait is already present, the rate at which a population can adapt is limited by its reproductive capabilities.

An especially important means of adaptation for people is insight learning. Insight learning has enabled us to become the most successful species on earth. Yet this capability has also given us the potential to destroy ourselves and all other life on the planet. Our success at adapting through insight learning in the future is at this point an open question.

Questions and Projects

1. Write a definition in your own words for each of the italicized terms in this chapter. Compare your definitions with those in the text or glossary.

2. What characteristics distinguish a living thing from a nonliving thing?

3. Describe the law of limiting factors by using an example of an organism and an environmental factor from your area.

4. Use the concept of limiting factors to explain why certain plants and animals live in one type of habitat and not another.

5. What are the differences in the effects of carcinogens, teratogens, and mutagens on

the human body and for future generations? Why does the threat of potential exposure to these substances often produce a loud public outcry?

6. What is acclimatization? How does it affect our understanding of the concept of limiting factors?

7. Define a synergistic interaction; an antagonistic interaction. What is the importance of synergism and antagonism in assessing the potential effects of pollutants on the well-being of plants and animals?

8. What is the difference between primary and secondary succession?

9. For your area, list plants and animals that are associated with pioneer communities and those that are usually found in climax communities. Is one group valued more than another? If so, for what reasons?

10. We can often find within a limited geographical region small areas illustrating the different stages of succession. What natural forces may be responsible for the presence of a variety of successional stages? Why is it important for resource managers to assist nature in maintaining this diversity?

11. Contact the local office of your state's Department of Natural Resources or Conservation to find out where ecological succession is being used as a vegetation or wildlife management tool. If possible, visit several of these sites and describe what you see.

12. Although, for several reasons, the small, diversified farm is more stable than the large modern farm, the small family farm is disappearing from the American landscape. Explain this apparent paradox.

13. What do studies such as those at Brookhaven National Laboratory and Hubbard Brook indicate about the effects of stress on ecosystem stability?

14. In the process of natural ecological succession, decades and often centuries pass before a climax ecosystem is fully reestablished. List some of the management techniques that humans can utilize to speed up succession. What might be the negative consequences of these management practices?

15. Are weeds growing in a crack in a sidewalk an example of primary or secondary succession?

16. Give some examples of succession in a large, metropolitan area.

17. Describe the process of natural selection. How have human activities such as livestock and crop breeding and the pollution of air and water affected the process of natural selection?

18. Which group is better able to adapt to environmental change, human beings or insects such as flies and mosquitos? Defend your choice.

19. What is the difference between trial and error learning and insight learning? Which type is more important for human beings?

20. List some specialized adaptations in human beings. Describe how these adaptations have contributed to the success of the species.

21. List some ways in which human beings have used technology to modify their immediate environment and thereby reduce exposure to stresses such as an inhospitable climate, air and water pollution, and disease-bearing organisms.

22. What are the natural limitations of an organism's or a population's ability to adapt to environmental changes? How does the rate of environmental change influence an organism's or a population's ability to adapt?

23. What is the evidence, if any, to support the argument that greater environmental change has occurred in recent years than earlier in this century.

24. List some of the natural environmental changes that occur in your area daily, bimonthly, seasonally, and annually. How have the local plants and animals, including people, adapted to survive these changes?

Selected Readings

Bishop, J. A., and L. M. Cook. 1975. "Moths, Melanism and Clean Air," *Scientific American 232:*90-99 (January). Examines the processes that shaped the selection for wing color of the peppered moth in England.

Clarke, B. 1975. "The Causes of Biological Diversity," *Scientific American 233:*50-60 (August). Explores the role of natural selection in maintaining a diversity of genetic traits.

Collier, B. D., G. W. Cox, A. W. Johnson, and P. C. Miller. 1973. *Dynamic Ecology.* Englewood Cliffs, New Jersey: Prentice-Hall. A good ecology text containing several basic chapters on adaptations of organisms to stress. Also has a chapter on succession.

Copper, C. F. 1961. "The Ecology of Fire," *Scientific American 204:*150-160 (April). Examines the role of fire in ecological succession in grasslands and forests. Considers how fires are used as a resource management tool.

Folk, G. E. 1974. *Textbook of Environmental Physiology,* 2nd ed. Philadelphia: Lea and Febiger. A look at the response of animals, including people, to such stresses as cold and hot temperatures and high and low pressure environments.

Odum, E. P. 1971. *Fundamentals of Ecology.* Philadelphia: W. B. Saunders. The most recent edition of a classic ecology text. Contains several chapters devoted to limiting factors, succession, and natural selection.

Ricklefs, R. E. 1973. *Ecology.* Portland, Oregon: Chiron Press. An extraordinary ecology text containing many well-written chapters devoted to natural selection, adaptation, and environmental change. Also has one chapter on succession and stability.

U.S. Environmental Protection Agency. 1976. *Pollution and Your Health.* Washington, D.C.: Environmental Protection Agency. An overview of the many types of pollutants and what they are doing to our health.

Vitousek, P. M., J. R. Gosz, C. C. Grier, J. M. Melillo, W. A. Reiners, and R. L. Todd. 1979. "Nitrate Losses from Disturbed Ecosystems," *Science 204:*469–474 (May 4). A study of mechanisms of responses of forest ecosystems to disturbance.

Waldbott, G. L. 1978. *Health Effects of Environmental Pollutants,* 2nd ed. St. Louis: C. V. Mosby. A well-illustrated account of the basic effects of toxic and hazardous substances on human health. Plants and domestic livestock are considered to a lesser degree. Focus is primarily on air pollutants.

Wildebeests, migrating by the thousands across the plains of Africa. The population size of wildebeest herds is determined by many factors whose relative importance varies from place to place and even from time to time in a particular area. (Leonard Lee Rue, from National Audubon Society.)

Chapter 4

Population: Growth and Regulation

We share this planet with more than 1.6 million species of plants and animals, most of which exist in countless *populations*—groups of individuals of the same species occupying the same geographical area. Though we sometimes behave as if ours were the only significant species on earth, our well-being rests on a delicate balance among the other populations of living things.

Populations that become too large can interfere with our well-being. Overgrowth of algae, for instance, significantly reduces water quality, and great swarms of locusts periodically devastate crops. Large flocks of blackbirds and starlings endanger airplanes, do considerable damage to agriculture, and carry a fungus that causes a disease of the human respiratory tract. A large rat population in an urban area poses a threat to human health because it greatly increases the possibility that human beings will be exposed to the fleas rats carry. These fleas transmit typhus and bubonic plague, both deadly diseases, to people.

On the other hand, a reduction in certain populations can have an equally disruptive, and sometimes formidably complex, impact on other populations in an ecosystem. Obviously, the reduction in, say, the beef population in this country or the wheat population in Russia can have a noticeable effect on the diets of these countries' citizens. But the complex reverberations of population losses in nature are often less direct and less easily enumerated. For example, a decline in the numbers of such predators as foxes, owls, and hawks often allows rabbit, mice, and ground squirrel populations to grow virtually unchecked. In their increased numbers, these animals, in turn, can damage crop and grazing lands. The effects of such a shift are by no means limited to the species whose populations change initially, but are felt, by way of food webs, throughout the entire ecosystem. As we noted in the last chapter, an ecosystem consists of interrelationships too numerous even to identify, and all are sensitive to changes, small and large, throughout the system.

Because our interrelationships with other populations affect us so deeply, we often attempt to regulate their size and quality. Cattle and sheep raising for food are obvious examples. Another is our control of game animals (deer and rabbits, for instance) and waterfowl (geese and ducks) in the name of wildlife management. One of the goals of such efforts is to keep game populations matched to the land's capacity for supporting them. A deer population too large for a particular habitat, for instance, can overbrowse and damage vegetation, and these effects in turn can accelerate soil erosion and the loss of nutrients necessary for further vegetative growth. Sadly, however, too many hunters have little regard for the land's capacity to sustain wildlife over time. They put pressure on wildlife managers to allow deer herds to proliferate so that their chances of a kill will be improved. Members of the tourist industry, too, favor the uncontrolled growth of deer herds, since deer in the wild attract visitors who spend their money in nearby resort areas. Thus, not only must wildlife managers grapple with difficult wildlife population problems, but they must contend as well with the public's disregard for the total ecosystem in favor of short-run goals.

Our concerns about populations are not limited to fluctuations in other species, but focus also on the recent population explosion our own species has experienced. The growing numbers of human beings on the planet together with soaring demands of affluent and developing societies are putting stress on the capacity of the earth's ecosystems to sustain all forms of life. In response to our unrelenting demands for food and timber, to name but two of our myriad requirements, we have developed vast agricultural monocultures, creating conditions that favor new population explosions—of insects, rats, and other pests. Conversely, while we continue to create these artificial environments and disturb the natural balance among species through widespread pollution, overhunting, and destruction of habitat, we are pushing hundreds of species to the brink of extinction.

Population Size

To begin solving the problems associated with population size, we must understand how populations grow and what factors promote or inhibit their growth. Although many factors have their effect, the size of a population at any given moment can basically be seen as the net difference between inputs and outputs. A population's inputs are its *natality* (the production of new individuals by birth, hatching, or germination) and immigration (migration into the group). Outputs are *mortality* (deaths) and emigration (migration out of the group). If inputs exceed outputs, the population grows; conversely, if outputs exceed inputs, the population shrinks.

A population's natality rate, or birth rate, is sensitive to environmental change. If conditions are ideal, natality approaches a genetically determined maximum rate. But environmental conditions are rarely ideal, and a population therefore rarely reproduces at its maximum rate. Low food availability, for instance, reduces natality; in other words, the hungrier a population is, the fewer offspring it bears. Also, natality varies considerably from species to species. Fish commonly produce thousands, and frogs hundreds, of eggs a year, though not all the eggs result in new fish or frogs. Birds, however, lay from one (the albatross) to fifteen (the quail) eggs per clutch, with a higher proportion maturing into living young. Although most bird species lay one clutch a year, some birds, such as robins and bluebirds, may produce two or three. Small mammals, such as mice, may have a litter of four to six young four times a year, but larger mammals characteristically have only one or two offspring each year. In general, the number of offspring is small for those species whose young require the most care.

The mortality rate, or death rate, varies with environmental conditions as well. If the environment is optimal, organisms die of "old age," or senescence. Commonly, however, an organism's lifespan is shortened by a predator, parasites, or some other stress such as a lack of food or severe weather. Ecologists usually consider mortality to be more sensitive to environmental change than natality; hence, the death rate is the more important determinant of population growth rates.

When a population enters a new region containing

abundant resources, its growth may follow the pattern illustrated in Figure 4.1. At first, it grows slowly, but once it becomes established, the population increases rapidly and approaches its maximum reproductive capabilities. Eventually, some factor in the environment becomes limiting and the population size levels off. This growth pattern is called the *sigmoid (S-shaped) growth curve*.

The curve in Figure 4.1 illustrates the growth pattern of a population whose numbers have leveled off below the *carrying capacity* of the habitat—that is, the population size the habitat's total resources can support. But usually a population does not follow as simplistic a curve. Rather, population size tends to overshoot the carrying capacity. Then, when some factor becomes limiting, instead of leveling off, the population comes crashing down below the carrying capacity and then may oscillate about it. Two of the many possible oscillation patterns are shown in Figure 4.2. One explanation for this oscillation is that a population on the upswing includes a large percentage of young members who are not yet making full demands upon their environment. When the young mature and begin to require an adult share of the resources, the population exceeds the carrying capacity. In response, a portion of a population either migrates or dies, resulting in a return to a population level below the habitat's carrying capacity. This process is then repeated, and the overall effect on the curve is the oscillation around the carrying capacity.

Generally, the actual size of the fluctuations of a population in response to habitat changes depends on the particular species in question. Large, long-lived, slowly reproducing species—for instance, the sheep represented in Figure 4.3—are less influenced by changes in their surroundings than are small, short-lived, rapidly reproducing species, such as the phytoplankton whose population size is plotted in Figure 4.4. For large animals, daily variations in the environment have only a minor impact on population size. Even if climate or resources such as food, water, and shelter become limiting and diminish natality

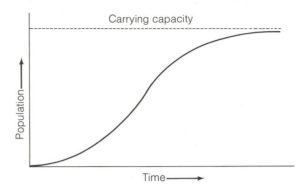

Figure 4.1 A sigmoid growth curve illustrating a population leveling off below the carrying capacity.

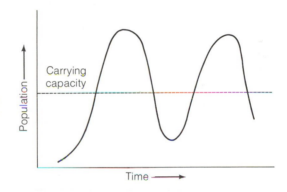

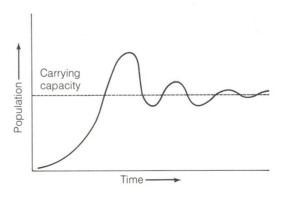

Figure 4.2 Two of the many possible patterns of oscillation of a population around the carrying capacity.

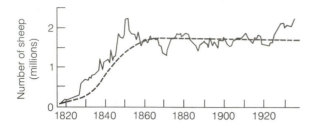

Figure 4.3 Growth of a sheep population in the more than one-hundred years after their introduction onto the island of Tasmania. Because sheep are large, long-lived, and reproduce slowly, the sheep population is relatively insensitive to environmental change and thus shows little year-to-year variation in number. The dotted line is intended to represent the overall trend in growth of the population. Although the line has the same shape as the sigmoid growth curve, you can see from the plotted points that actual population growth may not conform to the idealized sigmoid growth curve. (From J. Davidson, ''On the Growth of Sheep in Tasmania,'' *Transactions of the Royal Society of South Australia* 62:342–346, 1938.)

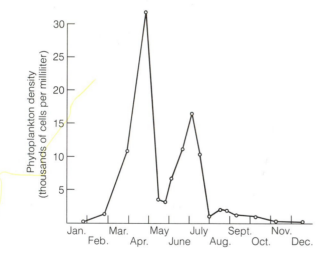

Figure 4.4 Variations in number of algae in water. Because they are small, short-lived, and reproduce rapidly, algae are quite sensitive to environmental change, and their numbers vary from very small to quite large (a bloom) in just one year. Data are from samples taken from the bay of Green Bay, Wisconsin.

one year, many adults survive to reproduce the following year. For example, the reproductive period of sheep extends from the age of one until death at ten years (90 percent of the life-span), and normally sheep produce one to two offspring each spring. Such reproductive characteristics in large animals tend to reduce the degree of variation in their population size in the face of habitat changes.

Small, short-lived, quickly reproducing species, however, are more sensitive to habitat change and exhibit greater fluctuations in population. Microscopic algae, for example, can complete their life cycle within a few days. If water conditions suddenly become ideal, their numbers explode. However, because of their short life span, a massive die-off follows. Hence, as we saw in Chapter 3, a seemingly clear lake may become pea-green with algae in only a day or so, and the bloom can disappear almost as quickly. Other small animals, such as insects and rodents (lemmings and mice are noted examples), also frequently exhibit dramatic fluctuations in numbers. Despite this characteristic fluctuation, however, these species seem to have an inherent ability to persist. Even if a local population is wiped out, a well-developed ability to disperse to other regions (particularly in insects) usually results eventually in a successful reinvasion of the locale.

Population Regulation

Having given an overview of how populations grow or decline, we turn now to a consideration of specific environmental factors that regulate population size. Both biological and physical factors play a role in the regulatory process. Biological factors consist of the interactions of the population in question with other populations in the habitat through predation, parasitism, and competition. Physical factors include such types of climatic stress as flooding, drought, and extremes in weather. You should remember that although we isolate these topics for purposes of discussion, several of these factors usually interact to regulate a population, and the relative significance of a particular factor depends on the nature of the specific population and its physical surroundings.

Biological Factors in Population Regulation

Predation. *Predation* is an interaction in which one organism (the predator) derives its sustenance by killing and eating another (its prey). The role of predation in keeping the predator species alive is clear. A more difficult question is the effect of predation on the prey population. In fact, the role of predation in regulating the size of a prey population remains a highly controversial and emotional issue. Because of our involvement with them, wild game and pests have been the focus of most studies on the regulation effects of predation. Results of these studies indicate that the significance of predators in controlling their prey depends on the type of predator, the type of prey, and the habitat conditions.

For small animals, such as insects, predation is one of the major agents in limiting prey population size. In fact, *biological control,* the regulation of pests by natural predators, is an important alternative to pesticides. A good example of pest control by a predator involves the romanticized ladybug beetles and the cottony-cushion scale, an insect that infests citrus plants, as shown in Figure 4.5. In the early 1870s, cottony-cushion scale was introduced accidentally into California from Australia. It spread rapidly, and within a few years the infestation of this insect seriously jeopardized the citrus industry. A resourceful government scientist journeyed to Australia to search out the natural predators of the scale. He turned up a species of ladybug beetle, which, when introduced into California citrus orchards, quickly brought the scale under control. The graph in Figure 4.6 illustrates the ladybug's effect on the scale population. Overall, the ladybug has been successful in keeping the scale population at a level where economic loss is insignificant. Ironically, during the late 1940s, the

Figure 4.5 The use of ladybugs to control the cottony-cushion scale insect on citrus trees, an example of biological control. (Florida Department of Agriculture, Division of Plant Industry, Gainesville, Florida.)

scale population exploded when DDT applications decimated the ladybugs.

Today one can cite nearly a hundred examples from around the globe in which insect pests have been at least partly controlled by predators. In the United States, other insect pests that are regulated by biological control include bark beetle and sawflies on ponderosa pine, spotted aphid on alfalfa, and codling moth on apple trees.

While predation is important in controlling insect populations, the role of predators in regulating game

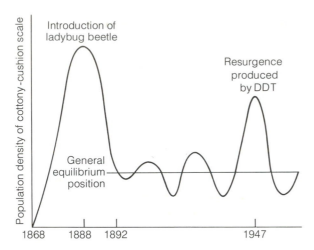

Figure 4.6 The effect of ladybug beetles on population size of cottony-cushion scale. (After V. Stern et al., "The Integrated Control Concept," *Hilgardia* 29:81–101, 1959.)

populations is much less well defined. Some studies of large herbivores and carnivores in Africa suggest that large predators may not be as important in limiting prey populations as was once thought. Studies in the early 1960s of the wildebeest herds on the Serengeti Plains in Tanzania (see the frontispiece to this chapter) indicated that their numbers changed little from year to year. The most obvious population control factor was predation by lions, cheetahs, and hyenas. But subsequent investigations have suggested that the take by these larger predators amounts to only a small percentage of the number of animals that must die annually if the population is to remain stable. Although many wildebeest die from disease, these losses are also inadequate to account for the number of necessary annual deaths. Apparently, the major reason for mortality is that young calves become separated from their mothers in the large migrating herds and subsequently die from starvation.

In contrast, some more recent studies indicate that at least under certain conditions, large predators may have a marked impact on prey population size. Although lions may have little influence on wild game herds on the Serengeti Plains, they do influence significantly the prey populations in Kruger National Park, South Africa. Several factors contribute to this difference. In the Serengeti, there are about one-thousand prey animals per lion, while in Kruger the ratio is only 110 to 1. All other things being equal, we would expect predation by lions to have a greater impact on prey population size in Kruger than in the Serengeti. Another contributing factor has been the weather. Because of a five-year period of unusually high rainfall, an abundance of natural and artificial water supplies has facilitated predation by lions. The large aggregations of wildebeest and zebra common during drier years have broken up and become dispersed in small groups over a large area. This distribution pattern allows young lions, which have been forced out of their prides, to hunt without contact and competition with adults. Hence, the total lion population has increased by about 60 percent in 2.5 years, thereby putting greater hunting pressure on wild game herds.

Closer to home, the significance of large predators in population control has been documented in Superior National Forest, Minnesota, where intensive wolf predation caused drastic declines in deer populations during unfavorable periods with several consecutive years of inclement weather and deteriorating habitat conditions. Hence, we see that changing habitat conditions may influence the importance of predation, and that the level of predation may vary not only from place to place but also from time to time within a particular area.

Although the role of large predators in controlling the numbers of prey requires further study, there is little question that predators do contribute in other ways to the stability of prey populations. Large predators tend to keep prey populations healthy by killing off the weak, diseased, and old members. These individuals are normally the first to be overtaken when their herd is migrating or being pursued; they tend to lag behind and lose the protection of the herd. Thus, some species of large predators may act more as scavengers than actual attackers. They contribute to the general well-being of the herd by eating animals suffering from lethal diseases that would have died anyway. Had these diseased animals lived longer, they would have increased the threat of disease to the remaining members of the herd.

In general, as we have seen, many prey species possess adaptations that allow them to escape predation, and these animals are usually victimized only when they are young, sick, or feeble—when their defense and escape mechanisms are undeveloped or weakened. But not all species are so equipped. Many insect pests, for example, do not run away and hide from predators. Aphids merely sit on leaves sucking up plant juices, as shown in Figure 4.7; hence, they are easy prey. But such insect pests compensate for their inherent vulnerability by producing offspring by the hundreds and thousands. Thus, though predation alone would have a significant effect in reducing population in such species, their inherent high natality rate maintains their numbers. This balancing effect demonstrates a fundamental concept that population size results from the interaction of many

Figure 4.7 Aphids, "easy pickings" for the predatory lacewing insect. (Grace Thompson, from National Audubon Society.)

factors, both inherent to the population and environmental.

Parasitism. In predation an organism almost always kills its prey for nourishment, but in *parasitism* one organism (a parasite) obtains its nutrients by living within or upon another living organism (the host). In obtaining their nourishment parasites sometimes kill their hosts. The means of death depends on the type of parasite. Some parasites attack the host cells and tissues directly, as in the case of the malaria organisms, which destroy large numbers of the red blood cells in which they live and reproduce. Others produce and release specific poisons that are lethal to their hosts. These parasites include the gas gangrene organism, the diphtheria organism, and the scarlet fever organism. Still others weaken their hosts by depriving them of essential nutrients and fluids. Adult hookworms, for example, cause anemia by anchoring to the inner surface of the host's intestines and sucking blood from blood vessels in the intestinal wall.

Like predation, parasitism is a *density-dependent factor*; that is, its controlling influence becomes stronger as population density (the number of individuals per unit area) increases. Also, when a host population exceeds the carrying capacity of the habitat, some of its members may be forced by intense competition into marginal areas. There, weakened by the relative lack of resources, they may be particularly susceptible to parasites.

The death of a host often means death for a parasite, however. A parasite population that fully infests and weakens a host population to the point of extinction loses its own habitat in the process. Therefore, natural selection, acting upon both parasites and hosts, often produces parasite–host interactions in which the parasite is less virulent (disease causing) and the host more immune (resistant to disease). This process is known as *coevolution*, or coadaptation.

An interesting example of coevolution occurred in Australia after the introduction of the European rabbit. When Europeans migrated to the far corners of the earth during the 1800s, they often became homesick and yearned for familiar surroundings. As a small compensation, they imported some components of their own culture, including familiar plants and animals. Thus, in 1859, Thomas Austin released thirteen English rabbits on his estate in Victoria, Australia. The number of rabbits quickly exploded and six years later Austin reported that he had killed twenty-thousand rabbits on his ranch and that at least ten thousand more remained. The rabbits rapidly spread, and by 1928 they had successfully invaded nearly two-thirds of the Australian continent, destroying the valuable grasslands that served as the base for Australia's main industry, sheep raising. Poisons, predators, and fences proved to be ineffective against them; the answer appeared to be a virus that causes myxomatosis, a disease related to small-pox. The virus was a parasite discovered in South American rabbits, which were immune to it. European rabbits, however, quickly died of myxomatosis when infected.

When the virus was introduced in Australia in 1951, hopes soared as myxomatosis quickly spread over much of southeastern Australia, killing hundreds of thousands of rabbits. The rapid dispersal of the virus was accomplished by mosquitos, which picked up the virus when they bit infected rabbits and subsequently spread the virus to healthy rabbits by biting them also. But enthusiasm waned during succeeding years as fewer and fewer of the remaining rabbits succumbed to the virus. Today, rabbits are again a problem.

Studies revealed that through natural selection the virus had lost its punch, and that simultaneously the rabbits had become less susceptible to myxomatosis. Rabbits infected by the most virulent strains died almost immediately, and this response among the rabbits had two effects. It inhibited the spread of the most virulent forms of the virus, and it stopped reproduction by the most susceptible rabbits. To understand why the spread of the most virulent forms was greatly diminished, we must realize that these mosquitos are not attracted to dead animals. Hence, if the virus quickly killed its host, there was much less time available for a mosquito to bite a host infected with the virulent strain of virus and then to transmit the strain to another rabbit. Meanwhile, rabbits infected with the less virulent strains in the virus population lived longer; thus, the chances of a mosquito transmitting the less virulent strains increased. Because they were more readily transmitted, these new forms became the dominant strains in the viral population. At the same time, natural selection continued to favor rabbits that had some immunity to the disease. Since both rabbits and viruses reproduce relatively rapidly, within a few years selection had occurred for both a less virulent strain of virus and rabbits with greater immunity. The result was a stable interaction between the relatively nonvirulent parasite and the effectively immune host. Once this balance was reached, the virus no longer acted as a significant limiting factor to the rabbit population, which again began to increase virtually unchecked.

Coevolution between parasite and host is by no means inevitable, however, particularly if the host is a long-lived species. Consider what happened when chestnut blight was introduced into the United States. Before the turn of the century, the American chestnut tree was a dominant member of the Appalachian forests. In 1904, Asiatic chestnut trees were brought to New York, carrying with them a parasitic fungus to which they were resistant. American chestnuts, however, had no resistance to the new parasite. The fungus quickly spread, and by the early 1950s the American chestnut had been virtually eliminated from the Appalachian forest ecosystem. In a few places, the roots of destroyed trees are still alive and they continue to send up new shoots, but these will eventually be killed by the fungus. Dutch elm disease, whose effects are illustrated in Figure 4.8, also demonstrates the destructive effect of a new parasite that becomes firmly established in a nonresistant long-lived host. Partly because of the long generation time of trees (it usually takes twenty years or more for a tree to first produce seed), the possibil-ity of coevolution between a very susceptible tree species and a new, virulent parasite is considerably lessened. Hence, instead of coevolution, extinction might well be the result.

In many instances, parasites are transmitted from one host to the next by a third organism, called a *vector*. Common vectors for parasites of human beings and animals include lice, fleas, ticks, and mosquitos, as you will recall from the example of myxomatosis. To understand the potential effects of parasites we must fully account for the ecology of their vectors. A clear demonstration of the parasite-vector-host interaction is the situation that resulted in the devastation of many species of birds on the Hawaiian Islands. For thousands of years, shorebirds and ducks migrated there from North America and Siberia. Although these birds undoubtedly carried disease-causing parasites, the native Hawaiian birds never became infected, because there were no vectors on the islands to transmit parasites to them from the migrating birds.

In 1826, however, a watering party from the ship *Wellington* accidentally introduced the tropical form of the night-flying mosquito onto the island of Maui. Because coastal areas of the islands are tropical, the mosquito quickly became established and successfully invaded the lowlands of the major islands. With the introduction of this vector, the populations of nonresistant native birds were soon greatly reduced by such diseases as avian malaria and birdpox. Interestingly, bird populations residing on mountainsides above an altitude of 600 meters (2000 feet) were seldom affected by these diseases, because the tropical mosquito is restricted to the warmer climate found below 600 meters. Hence, the higher mountainsides now serve as sanctuaries for the birds that remain. Unfortunately, from time to time inclement weather in the highlands drives birds to lower altitudes, where they again are subject to attack by the mosquito.

In this instance, then, we can identify a factor limiting the native bird population of the islands (the parasites brought in by migrating birds), a factor

Figure 4.8 The effects of Dutch elm disease, an example of the introduction of a new parasite. *Top*: Elm-lined Gillett Avenue in Waukegan, Illinois, as it appeared in the summer of 1962. *Bottom*: Gillett Avenue in 1969, after the elms were destroyed by Dutch elm disease. (Elm Research Institute, Harrisville, New Hampshire.)

limiting the parasite population (the vector), and a factor limiting the vector population (the colder climate at higher altitudes). Thus, as you can see, a single population is affected either directly or indirectly by many physical and biological factors. It is in this way that the influence of change is felt throughout an ecosystem.

The effects of parasites are not limited to plants and animals. Human beings have always been af-fected as well. In fact, human history has been shaped in part by the impact of parasitism. For an example, see Box 4.1.

Competition. Besides predation and parasitism, a third biological factor regulates population size: *competition*. The resources necessary for life (food, water, and shelter, for example) are finite, and some-times the demand for them exceeds the supply. The degree of subsequent competition among individuals for the limited resources depends on both the limits to the natural resources and the size of the populations dependent on the resources. Thus, in discussing competition we are once again dealing with the complex interactions between organisms and their environment. In this context, available resources affect population size, which in turn affects the resource supply. The more the demand exceeds the supply, the more significant competition becomes in population regulation; that is, competition is also density dependent.

Ecologists recognize two types of competition: intraspecific and interspecific. *Intraspecific competition* occurs among individuals within a single species, whereas *interspecific competition* takes place between populations of different species.

Competition Among Animals. Intraspecific competition among animals often occurs as social interactions that regulate population size. Many animals—songbirds, hawks, muskrats, and Alaskan fur seals, to name a few—establish territories that they defend from intruders of their own species. For instance, during the breeding season, male robins establish and defend the territory within which they and their mates build nests. The male robin uses his song to inform other robins of his territorial limits. As illustrated in Figure 4.9, if another male robin enters his established territory, he will attempt to drive the intruder out. This sort of aggressive defense, an instance of *territorial behavior,* serves to separate animals and tends to ensure to individuals within each territory an adequate supply of resources. Thus, territorial behavior favors population growth by allowing individuals to stay alive and reproduce. However, the same behavior pattern also tends to limit population, because the number of breeding pairs cannot exceed the number of available territories. Although the size of each breeding territory may shrink as population increases, every species has a minimum territorial size. Once the minimum is

Figure 4.9 Two male robins fighting over territory. (Phillip Strobridge, from National Audubon Society.)

reached, extra animals are forced to emigrate to marginal areas. Since the areas outside of prime territories have limited resources, many of these outcasts die. Some succumb because of inadequate food, water, or shelter, while others, in their weakened condition, easily fall victim to predators and disease. Poor habitat conditions also contribute to the absence of breeding among the outcasts. Occasionally, an outcast replaces a member of the breeding population that dies, but usually no outcast can force its way into a fully occupied habitat. Territorial behavior, therefore, appears to regulate population size by limiting the number of individuals that can breed in a favorable habitat.

The formation of *social hierarchies,* or peck orders, is another behavior pattern related to competition that affects population size. In some species, such as baboons, wolves, ring-necked pheasants, and chickens, encounters between individuals result in a system of dominant-submissive relations that eventually involves the entire population. For example, one individual may dominate all other animals in the local population, whereas another, though submissive to the first, dominates the remaining members, and so on. Of course, this example is very much oversimplified; in nature, patterns of dominant-submissive relations are far more complex.

Once a peck order has been established, greater stability and order prevail within the population. Because everyone "knows" his place, less fighting occurs; therefore, individuals do not injure one another, and the population as a whole remains healthier than it would if members were continually struggling to dominate. Thus, social hierarchies foster population growth by maintaining overall health but function to limit it as well, since individuals of low status often fail to breed. During times of stress, these submissive individuals are the first to be deprived of adequate food, water, and shelter. Even in the best of times, low-status individuals may be so harrassed by dominant individuals that they are driven out into marginal habitats, where they die.

As population densities increase, the frequency of

Tsetse Flies and Trypanosomes: Human Scourges or Guardians of African Ecosystems?

sleeping sickness was found in relatively few areas because tribal warfare and lack of roads restricted travel, thus preventing the disease from spreading. When Africa was opened up by the European colonial powers, sleeping sickness rapidly spread throughout the tsetse belt.

Trypanosomes cause about seven-thousand human deaths each year. But they also have a devastating impact on domestic livestock. Several forms infect cattle, causing Ngana, a disease similar to sleeping sickness in humans. Like the Rhodesian form that attacks people, these trypanosomes have another host: wild game animals. The game are unaffected by them, however, and so they are an ever-present source of trypanosomes that can be transmitted to cattle by tsetse flies. As a consequence, few cattle graze in a region that could otherwise support 125 million cattle for the protein-starved people of Africa. Herdsmen must concentrate their livestock on the limited amount of fly-free pasturage in or near the Sahel region just south of the Sahara. But when cattle must be driven to market through fly-infested country, as much as 25 percent of the herd perishes en route.

One way to combat Ngana would be to raise trypanosome-resistant cattle. The breed most favored by African herdsmen is the zebu, a large, humpbacked longhorn. It is well adapted to semi-arid conditions, and it produces a high yield of meat and milk, but it is susceptible to trypanosomes. Some cattle breeds are resistant to the parasites, but because their small size makes them less valuable they have not been well accepted. Unfortunately, crossbreeding the small, resistant cattle with the zebu or European breeds does not produce offspring that are immune to the trypanosomes.

Many measures have been taken to combat sleeping sickness in humans. They include insecticides, large-scale deforestation to destroy the tsetse's habitat, and the destruction of the large game on which both tsetse and trypanosome depend. Probably the most controversial of these measures was the large-game destruction program. Devised in the 1950s to open land in east Africa to human settlement, the plan was to eliminate both the tsetse fly and the Rhodesian form of trypanosome by eliminating the game. Although its logic seemed faultless, the scheme did not work. After the campaign, enough small animals remained to support the tsetse. Furthermore, as herdsmen moved their cattle into the cleared areas, the tsetse began feeding on both the cattle and the herdsmen. Thus the transmission of animal and human trypanosomes not only continued but intensified.

Despite about seventy years of research and eradication efforts, the tsetse and the trypanosome still reign over a large region of Africa. Most remedies are too severe to be practical. Most are also too costly in economic and technical resources for the emerging African nations to afford.

If the African nations do some day bring sleeping sickness under control, the cure may well prove worse than the disease. For as areas become free of trypanosomes, Africa's burgeoning human population, along with its cattle, will quickly invade these regions. Without proper planning and strong laws, the regions would soon suffer overgrazing, soil erosion, and extinction of wild game.

Perhaps it is best that the tsetse fly remain the guardian of many African ecosystems until humankind better understands how nature works—and until it is willing to control its actions for the benefit of all Africa's residents, both human and animal. What do you think?

encounters between individuals also increases. If these encounters are stressful, then increased crowding may eventually lead to a level of stress at which reproduction is halted, thereby limiting population growth until the necessary population-space ratio is reestablished. The classic controlled experiments of J. B. Calhoun and J. J. Christian conducted in the late 1950s suggest the potential social stress has for limiting populations. In these experiments, rats were confined to a system of interconnected boxes where they were provided excess water, food, and nesting sites—more than they needed to stay alive and reproduce. As the population increased, no living rats were removed; therefore, conditions became increasingly crowded. Soon the rats were falling over each other and, apparently as a result of social stresses resulting from overcrowding, their behavior changed radically. Female rats failed to build proper nests for their young, and often they either abandoned their offspring or ate them. Males either fought frequently or became recluses. Some even exhibited homosexual tendencies. The increased deviant behavior brought successful reproduction almost to a standstill, and in some cases individuals were under so much stress that they died. Collectively, these responses to social stress are referred to as *shock disease.*

The hypothesis that overcrowding and increased competition results in deviant behavior and reduced reproduction is intriguing and popular, but little evidence exists to show that shock disease regulates populations under actual field conditions. The high densities that produce shock disease in the laboratory are rarely found in nature, since weaker members in natural populations are usually forced to emigrate as densities increase.

Let us turn our attention now to interspecific (between species) competition among animals. On this subject, the conclusions of laboratory experiments and of field observations usually diverge significantly. Under laboratory conditions, one species always outcompetes the other and brings about its demise. These laboratory results led experimenters to formulate the *competitive exclusion principle:* two species having the same resource requirements cannot coexist indefinitely in the same habitat, because as both populations grow, resources eventually become limiting. Yet casual field observations indicate that many similar species seemingly requiring the same resources do coexist in the same habitat. Five different species of herons in Florida, for example, occupy the same coastal marshes and prefer similar fish for their diet. And in Maine, five species of warblers apparently forage for the same insects in the same spruce trees. These birds coexist in apparent violation of the principle of competitive exclusion, but nevertheless they thrive.

In a classic study conducted in the mid-1950s, the noted ecologist R. H. MacArthur spent hundreds of hours in the field studying the feeding behavior of the five species of Maine warblers. He observed that although all five species feed on basically the same type of insects found in spruce trees, competition is greatly diminished by differences in feeding habits. As shown in Figure 4.10, the hunting activities of each species is confined to a specific part of a tree. Although feeding zones overlap to some extent, other behavioral patterns further reduce competition—for example, one kind of warbler captures insects on top of spruce needles while another hunts the insects hidden under needles. Also, because the five species all have different nesting times, their times of greatest food need differ, too. Hence, these five species are able to coexist with a minimum of competition, since they have evolved through natural selection to exploit slightly different resources.

Many such mechanisms serve to reduce competition among species. A common example is the difference in the hunting periods of hawks and owls. These species feed on similar types of animals. But, because hawks capture their prey during the day whereas owls are nocturnal, they usually do not compete for the same prey species. Other feeding habits disperse species throughout an entire habitat, thereby reducing competition. For example, some species of birds feed on the ground, while others seek food in trees and shrubs, and a few

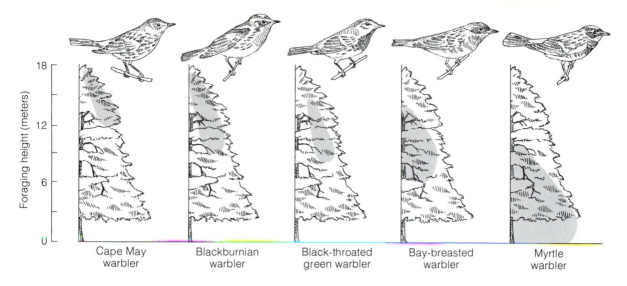

Figure 4.10 The foraging areas used by five species of warblers in spruce forests in Maine. (After R. MacArthur, "Population Ecology of Some Warblers in Northeastern Coniferous Forests," *Ecology* 36:533–536, 1958.)

species even capture their prey in the air. Structural adaptations also serve to isolate one species from another. The downy woodpecker and hairy woodpecker occupy similar habitats in North America. As you can see in Figure 4.11, these species are similar in appearance, but the hairy woodpecker is larger and has a proportionately larger bill. These structural differences allow the hairy woodpecker to eat larger insects and to reach insects hidden deeper within tree trunks. With respect to nesting sites, we find again that birds have evolved to use different existing resources. Some birds nest on the ground, while others nest on a tree branch or in the tree trunk. A few species even use a burrow in the ground or a hole in the side of a cliff.

Thus, adaptations that reduce competition are probably more common in nature than competitive exclusion. Though the result of competition, popularly referred to as "survival of the fittest," conjures up images of animals fighting tooth and nail for survival, such violent encounters are rare in nature. When competition is reduced through adaptations that reduce confrontation, more energy can be directed to growth, maintenance of the individual,

reproduction, and the rearing of young. The term "fittest," therefore, applies best to those organisms that avoid competition.

As with all other factors influencing population, the two types of competition—intraspecific and interspecific—are interrelated. As Figure 4.12 shows, their interaction partially determines the geographical distribution of an animal species. Interspecific competition restricts a species to regions where it is the best competitor, usually areas where resource conditions are nearly optimal for its survival. As the population density in the occupied habitat increases, however, intraspecific competition forces weaker members into marginal areas, thereby expanding the population's territory. With the encroachment of the population into habitats occupied by other populations, the influence of interspecific competition grows more significant once again. Hence, a species' territorial boundaries expand or contract with changes in its population density and that of its competitors.

Competition Among Plants. As it does with animal populations, intraspecific competition has a signifi-

Figure 4.11 *Left:* A hairy woodpecker. (Karl H. Maslowski, Photo Researchers, Inc.) *Right:* A downy woodpecker. (Henry C. Johnson, from National Audubon Society.)

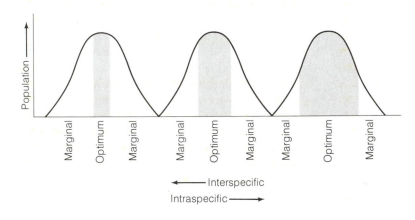

Figure 4.12 The effect of competition on population distribution within a habitat. When interspecific competition is greater, the population tends to be restricted to areas where habitat conditions are near optimum. When intraspecific competition is greater, the population tends to spread out and occupy marginal areas. (After E. Odum, *Fundamentals of Ecology*, 3rd ed. Philadelphia: Saunders, 1971.)

cant effect in controlling the size of plant populations. Generally, most seeds fall close to the parent plant, and, when seeds germinate, a dense growth of seedlings results. Since, unlike animals, plants are unable to make compensating adjustments in spacing to reduce competition, the whole mass of seedlings begins to compete for sunlight, water, and nutrients. As competition becomes more severe, plant mortality increases and a self-thinning process occurs. The graph in Figure 4.13 shows how the number of

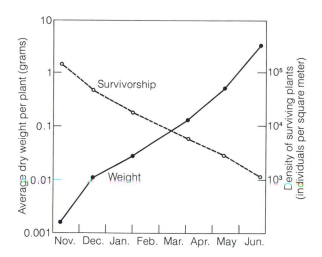

Figure 4.13 A self-thinning curve for an experimental planting of horseweed. The horseweed was sown at a density of 100,000 seeds per square meter. (After J. L. Harper, "A Darwinian Approach to Plant Ecology," *Journal of Ecology* 55:247–270, 1967.)

individual plants in a population of horseweed decreases over time, while the total plant weight increases. Self-thinning curves illustrate the carrying capacity of an area for total plant growth. At low population densities, most seedlings grow vigorously, but at high densities only a few plants reach a large size, while most survivors are stunted.

Let us now consider interspecific competition among plants. In direct contrast to our evidence on animals, many field examples of competitive exclusion exist in the plant kingdom. One such example concerns the competitive interactions of bluebunch wheatgrass and cheatgrass on the rangelands of the Columbia River Basin in Washington. Before the advent of domestic grazing, this rangeland was dominated by bluebunch wheatgrass along with a rich diversity of other plants. But heavy overgrazing destroyed the vegetative cover, creating an open habitat for invasion. The most successful invader was cheatgrass, a native of Europe, which soon dominated several million hectares previously occupied

by bluebunch wheatgrass. Ranchers much prefer to have their cattle grazing on wheatgrass, but attempts to reestablish wheatgrass have failed often and continue to do so, even where grazing has been stopped temporarily and wheatgrass seeds have been sown.

Field and laboratory experiments indicate that the cheatgrass is able to competitively exclude wheatgrass because it is a better competitor for soil moisture. Although seeds of both species germinate at approximately the same time in moist autumn conditions, cheatgrass continues growing during winter while wheatgrass becomes dormant. As a result, the extensive root system of cheatgrass extends deeper and gains control of the site before wheatgrass seedlings can become adequately established. Cheatgrass also has the advantage because it matures four to six weeks earlier than wheatgrass, thereby withdrawing much of the water from the soil before the wheatgrass needs it. During the summer dry season, wheatgrass seedlings usually succumb because the cheatgrass has already used up available soil moisture. Hence, in places where summer rainfall is low, cheatgrass excludes wheatgrass by out-competing it for soil moisture. In areas with more plentiful summer rains, however, cheatgrass is an unimportant competitor, present only on sites of recent disturbance.

Competitive exclusion also occurs during ecological succession. The most obvious means of competition is by shading out. In our discussion of succession, we saw that during old-field succession pine trees in the Piedmont region of the southeastern United States eventually shade out broom sedge grass. Nevertheless, broom sedge is a successful species. It invades farmland within three years after abandonment, quickly replaces the pioneer weeds already present, and subsequently maintains itself for fifteen to twenty years, often in almost pure stands. How does broom sedge play such a dominant role in early stages of succession? Investigations by E. Rice and his coworkers have shown that its roots and shoots contain chemicals that inhibit the growth of other plant seedlings. Once broom sedge enters a field, its chemicals retard the growth of pioneer weeds, and, after the broom sedge has become established, its decaying parts prevent significant competition from other potential invaders, thereby slowing the rate of secondary succession.

Although competitive exclusion does sometimes result from interspecific competition in plants, abundant observational data suggest that plants like animals, have evolved distinct strategies of resource use that reduce competition within a habitat. If we visit a forest in early spring, we will find the forest floor carpeted with a spectacular array of spring wildflowers. These spring ephemerals renew growth, flower, and set seed within three to five weeks before the leafing out of trees. Then they die back into underground dormant structures. This rapid growth cycle allows the plants access to sunlight before the tree canopy closes.

Growth to differing heights also reduces competition for space among plants. In forests we can observe a vertical stratification of vegetation, as shown in 4.14. For example, in mature forests in the southern Appalachians tall trees such as beech, sugar maple, basswood, and tulip tree make up the upper canopy. Beneath this layer are smaller trees, including dogwood, magnolia, and ironwood, that seldom or never attain upper-canopy position. Next come the shrubs, such as spice bush, witch hazel, and papaw. And beneath the shrubs are wildflowers and a ground layer of mosses, lichens, and blue-green algae. Vertical stratification is most fully developed in tropical rain forests, which have three to five overlapping tree strata.

Another competition-reducing adaptation is found in conditions of heavy rainfall and uniform temperature—the *epiphytes* or "air plants." These plants grow on other plants, usually trees, but they are not parasitic and have no structural connection with the plants on which they grow. By growing on tree limbs rather than on the ground, they receive more sunlight than they would on the forest floor. Epiphytic plants include ferns, mosses, orchids, and members of the cactus and pineapple families. Perhaps the most

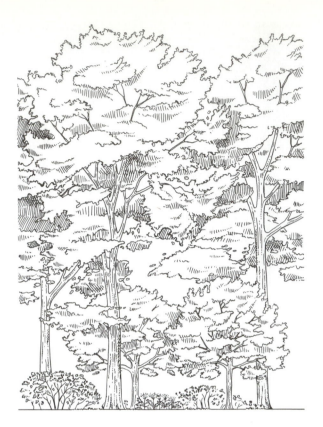

Figure 4.14 Vertical stratification in a forest. Large trees are about 30 meters tall; small trees, 5–10 meters; shrubs, 1–3 meters; mosses, 2–5 centimeters.

familiar epiphyte is Spanish moss—shown in Figure 4.15—which is not really a moss, but a rootless member of the pineapple family.

Competition, Adaptation, and the Complexity of Ecosystems

The notion that adaptations reduce competition sheds some light on the complexity of natural ecosystems. Species are not necessarily eradicated by the brute force of stronger competitors; rather, through natural selection, they evolve ways of filling their needs in a slightly different place, at a slightly different time, or by a slightly different means. Interestingly, as a species evolves, it may become a potential

Figure 4.15 Spanish moss, a common epiphyte ("air plant") in the southern United States. (U.S. Department of the Interior, National Park Service photo by George Grant.)

new resource for another species, and, by its presence, may set up conditions that allow the second species to evolve in a particular direction. It is the wide variety of resources, including organisms, in an ecosystem that makes adaptations possible. The diversity of species contributes in turn to the stability of the whole system by improving the efficiency of energy and nutrient transfer.

The Effects of Underpopulation

Most population problems stem from overpopulation, but for some species underpopulation also poses dangers. Consider what can happen to bobwhite quail when their population becomes abnormally small. Except when feeding, a covey of bobwhite quail usually forms a compact circle, as shown in Figure 4.16. During cold weather, the quail draw together more tightly in their circle, thereby decreasing heat loss by exposing less body surface to the cold air. Field observations show that members of large coveys survive severe winter temperatures, while single birds or small coveys often succumb. Also, note that the birds pictured face outward from the circle. A covey, in effect, watches in all directions for the approach of predators. If a predator comes near, the alarmed birds produce a phenomenon known as the *confusion effect*—they fly up suddenly in all directions, thereby disorienting the predator so that he fails to capture even a single bird. A covey

Figure 4.16 Gathering in a circle, quail protect themselves from cold and predators. (Jack A. Stanford, Missouri Conservation Commission.)

with too few birds to form a circle loses this advantage.

This use of circling as a protective strategy is exhibited in other species as well. For example, both pronghorn antelope and musk oxen form circles and stand their ground against predators, with adult males forming a protective ring around the females and young. Also, following examples of this behavior in animals, pioneers of the West "circled up" their wagons, as Figure 4.17 illustrates, as a protective measure. In each case, a group too small to make a tight circle would not gain the protective effect. In fact, if a pronghorn antelope herd numbers less than fifteen, the antelopes will attempt to run away from a predator rather than form a circle and stand their ground.

For predators that hunt as social groups, such as wolves, lions, and hyenas, underpopulation can impair their ability to stalk, attack, and kill their prey. For example, moose and caribou can repel an attack by one or two wolves, but wolves operating as a pack attack a moose from all sides, thereby enhancing their chances of bringing it down. Lionesses have a less obvious means of cooperating in a hunt: they spread out, and each one stalks toward the prey from a different direction, often growling. Upon detecting the presence of a lioness, the prey animal may bound off and come within striking range of another that it had not seen. Clearly, when cooperative hunting suffers due to a loss in population, the survival of the remaining individuals may be seriously threatened.

A more general effect of undercrowding is a possible drop in reproduction rate. The smaller the population the smaller are the chances of females and

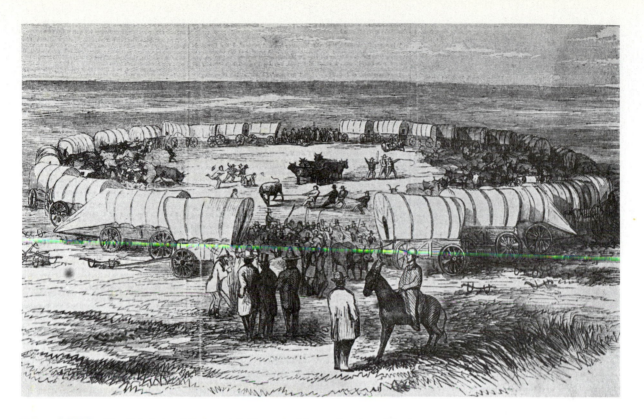

Figure 4.17 Pioneers "circled up" their wagons to protect themselves from intruders. (The Bettmann Archive, Inc.)

males of reproductive age getting together. For some species, particularly colonial birds such as penguins, a certain minimum number of individuals must be present in order for hormonal changes essential for mating to be initiated. In populations smaller than this minimum size, reproduction is severely curtailed.

When the reproduction rate drops in an already reduced population, the group may be unable to compensate for normal losses sustained from predation, disease, accidents, or natural disaster. Such a population is doomed to extinction unless natural immigration or the artificial planting of individuals from another population adds to its numbers.

Another possible consequence of the drop in breeding-age individuals in small populations is inbreeding (breeding between close relatives). Inbreeding sometimes reduces a population's vigor by restricting the input of beneficial characteristics and allowing the buildup of deleterious characteristics within the population. A weakened population has less resistance to predation, parasitism, competition, and climatic extremes. Another consequence of inbreeding is sterility, as observed among the young of the Mississippi sandhill crane, shown in Figure 4.18. For more information on the nature of inbreeding and its consequences, see Box 4.2.

As we have said, the effects of a population loss are felt throughout an ecosystem, not just by the population in question. A significant drop in a prey population can lead predators to feed more heavily on other prey species or, if insufficient alternative prey are present, to starve or migrate. A drop in a predator population can result in the unchecked growth in

Figure 4.18 The Mississippi sandhill crane, whose greatly reduced population is being threatened further by inbreeding. (Harry Engels, from National Audubon Society.)

the populations of its prey. Each of these responses in turn has an effect on the other populations and the physical environment with which the altered population interacts.

Physical Factors in Population Regulation

Short-term climatic shifts to more severe conditions may temporarily reduce population size, but populations usually recover from these effects. For example, if seasonal drought lowers the water level in marshes, exposing the entrances to muskrat lodges, muskrats become more vulnerable to attack by foxes and mink. But when water levels return to normal, the muskrat population soon returns to its original size. In another example, when bobwhite quail cannot scratch for food because of deep snow cover, their populations suffer dramatic losses. But when spring populations are small, the quail's reproductive success is often unusually high during the summer, and by fall populations are restored to near normal numbers.

Even in environments where seasonal climatic shifts are unusually harsh physical factors do not have a significant effect on the long-term regulation of population size. It was long believed that the severe climatic conditions in Arctic regions produced greater population fluctuations. But recent studies and reassessments of previous investigations suggest that fluctuations of an Arctic population are no more pronounced than those fluctuations of a population of the same species residing further south.

The population size of plant species too can be severely influenced in the short run by sudden stresses such as fires, hurricanes, or frost, but, through a variety of mechanisms, most plant species avoid extinction. For instance, some plant species exist primarily as seed populations. Seeds are dormant structures that can remain viable for many years despite such stresses as extreme heat or drought. Other plant species die back yearly into underground structures (the familiar tulip bulb and underground stems of many grasses are examples), thus avoiding exposure during seasons when stressful conditions occur. Some plant species are able to withstand the stressful conditions in an area for several years or more. Such populations persist in an area because they are well adapted to the major recurrent short-term stresses there. Consequently, such hazards do not play a significant role in determining population size.

Although short-term variations in physical factors are generally insignificant as population control measures, long-term changes in the physical environment usually are quite important. As we saw in Chapter 3, natural selection favors individuals that are best able to adapt to such changes. The physical environment may change in such a way that a species is less effective as a competitor, a parasite, or a predator. Or the species may become more vulnerable to predators or parasites. In either case, if the

Box 4.2

The Risks of Inbreeding, or
Why You Shouldn't Marry Your Cousin

In many human societies today, it is unlawful to marry one's first cousin or a closer blood relative. In our own country, at least 30 states forbid first-cousin marriages. A major reason for these laws is that relatively rare hereditary defects are likely to occur more frequently in the offspring of related individuals than in children of unrelated persons. For example, although only about one in a thousand marriages in the United States are between first cousins, about one of every twelve albino children are from first-cousin marriages. Other hereditary disorders that occur more often among offspring of closely related parents include clubfoot, harelip, and more serious effects such as some forms of mental retardation, and hemophilia, a very dangerous condition in which the blood flowing from a wound will not clot. Moreover, still births and infant deaths occur more frequently in offspring of first-cousin marriages.

Inbred populations can be found in many small religious communities. One of these is the Old Order Amish community in Lancaster County, Pennsylvania, whose founders immigrated from Europe before the American Revolution. The extent of inbreeding among the Old Order Amish is indicated by the fact that only eight different surnames are found among its eight-thousand or so members. One result of this frequent inbreeding is an increased incidence of the rare defect known as six-finger dwarfism. Afflicted individuals have six fingers on each hand and are small in stature. Only about a hundred cases of the defect have been recorded since it was first described in 1860, but at least fifty-five of them occurred in this small Amish population.

The consequences of inbreeding are not always negative. For example, brother–sister marriages were common in ancient Egypt, and were in fact the rule among the pharaohs. By the beginning of the Eighteenth Dynasty (around 1580 B.C.), brother–sister marriages had long been practiced in the royal line—yet there occurred a succession of the most brilliant rulers Egypt had ever known. And today the inbred populations of numerous small islands exhibit no signs of hereditary defects.

Such examples show that it is not inbreeding *per se* that is responsible for hereditary defects. Whatever effect inbreeding may have is a result of the inheritance received. If that inheritance is bad, the effect will be bad. If the inheritance is good, the effect will be good. Nevertheless, most individuals do carry harmful genes, and the more closely related two individuals are the greater the chance that they have the same harmful genes. So children of closely related parents are more likely to inherit the same harmful gene from both parents. If they do, they will be afflicted by a genetic defect.

changed environmental conditions persist, the species may significantly decline in numbers, perhaps to extinction. On the other hand, a physical environmental change may allow a species to be more successful in its interactions with other species and it may flourish. Hence, we should note that although in the short run population interactions may act as more significant regulators than changes in physical factors, it is the long-term changes in the physical environment that set the stage for these interactions.

Conclusions

The well-being of any population, including our own, depends on its physical environment and its interactions with other populations. In the long run, these factors regulate the population's size and tend to keep it in balance with the habitat. Although the fluctuations in size may be quite large, population growth rarely ranges out of control for long, nor do most populations become extinct.

Environmental changes, however, can upset this balance. In recent decades, human intrusion into ecosystems has dramatically altered population-regulating mechanisms. Ecosystems have been greatly simplified. In many cases, new predators, parasites, and competitors have been introduced, while in other instances a controlling agent such as a predator or competitor has been removed. The result has been the explosion of populations of some species (we call them pests) and the extinction or near extinction of many other species. Because our well-being rests on a delicate balance among other populations, we need to ask ourselves how much longer can humans afford to thoughtlessly disturb the "balance of nature."

Summary Statements

Population size at any given time is the difference between inputs (natality and immigration) and outputs (mortality and emigration). Both natality and mortality vary with environmental conditions and the type of organism.

Population growth may follow a sigmoid growth curve, with the population size eventually fluctuating about the habitat's carrying capacity for that species. The amplitude of fluctuations depends upon the nature of the species and habitat conditions.

Both biological and physical factors play a role in regulating population size. Physical factors include climatic stress, and biological factors include predation, parasitism, and competition, which are density-dependent factors. Density-dependent interactions become more important in limiting growth as a population approaches the land's carrying capacity.

For populations that can escape predation, predators often appear to play a minor role in controlling population size. For animals such as insects that do not run away and hide, predation often is more important in controlling their population numbers.

Establishment of a new parasite-host relation often triggers a devastating impact on the host population, particularly if the host is a long-lived species. If both parasite and host reproduce relatively rapidly, a stable interaction may soon result from a selection for a less virulent parasite and a more immune host. Such an interaction is known as a coevolution.

Territorial behavior and peck orders are examples of intraspecific competition (competition among members of the same species) that may regulate population size. Many similar animal species can coexist in the same habitat. Factors that reduce interspecific competition (competition between species) include differences in timing of hunting activities, location of feeding activities, and types of food.

The term "survival of the fittest" is best applied to those organisms that avoid competition. The relative severity of interspecific and intraspecific competition partially determines the geographical distribution of animal populations.

Self-thinning curves illustrate the effects of intraspecific competition of plants. Differences in timing of growth and vertical stratification reduce competition among plant species.

Selection for adaptations that reduce competition is probably more common than competitive exclusion. The variety of species that results helps to optimize the efficiency of energy and nutrient transfer within an ecosystem.

Inability to ward off predators, inbreeding, and reduced reproductive success are some of the dangers posed by underpopulation.

Short-term physical stresses may temporarily reduce population size, but they generally do not play a significant role in the long-term regulation of populations. The density-dependent population interactions exert the primary influence on population growth, but long-term changes in the physical environment establish the conditions within which population interactions occur.

Questions and Projects

1. In your own words, write a definition for each of the terms italicized in this chapter. Compare your definitions with those in the text.

2. Identify the environmental factors that influence the rate of natality and those that influence the rate of mortality. In the long run, which influences have the greatest effect on population size?

3. Examine the sigmoid growth curve shown in Figure 4.1. Determine the points on the curve where the rate of increase is smallest and where it is the largest.

4. Why do some populations overshoot the capacity of their habitat to support them? What might be the impact on the habitat during the time of overpopulation?

5. Describe the characteristics of a species and its habitat that would produce large population fluctuations. Contrast these characteristics with those that would promote small population fluctuations.

6. Describe why predation appears to be quite important in controlling the size of certain prey populations whereas it has relatively little effect on the numbers of other prey populations.

7. Although some predators may not be the primary agents in controlling the size of prey populations, they do contribute to the stability of these populations. Explain.

8. Compare and contrast predation and parasitism as population control mechanisms.

9. For parasites that require a vector, why is it important to take into account interactions between the vector and its environment to properly assess the potential impact of the parasite on a host population?

10. Describe how territorial behavior and peck orders function in population regulation. Identify animals in your area that demonstrate these social interactions.

11. Compare the complexity of a laboratory set-up to test competition with the complexity found in nature. How might these differences explain why competitive exclusion may occur in laboratory experiments, though little evidence for this outcome exists in nature?

12. How would you define the word "fittest" in the phrase "survival of the fittest"? Would you define the term any differently if it were applied exclusively to human beings?

13. Visit a park or natural area in your region and observe the many species of plants and animals in the ecosystems. Describe the adaptations that reduce competition for resources among these species.

14. How does the relative severity of intraspecific and interspecific competition help to determine the geographical distribution of a species?

15. Some scientists claim that the valiant efforts being made to save certain nearly extinct species are doomed because the populations involved are too small. What is the basis for this viewpoint?

16. Why are short-term physical stresses of relatively little significance in the long-term regulation of population size?

17. Describe how physical and biological factors may interact to control the population size of a species. Cite examples from your area.

18. How does the population size of a particular plant or animal influence the well-being of people? Give specific examples from your region.

19. How have human activities affected the population size of plant and animal species? Give specific examples from your region.

20. The size of some populations fluctuates widely. Can these large fluctuations be reconciled with the concept of "the balance of nature"?

21. Check with your local office of the State Department of Natural Resources, Department of Conservation, or Department of Agriculture. Learn whether they are involved in projects dealing with pest control or overpopulation of wildlife, and, if so, determine which population control measures they are using.

Selected Readings

Annual Reviews, Inc. *Annual Review of Ecology and Systematics.* Published yearly. Palo Alto, California: Annual Reviews, Inc. Each volume contains up-to-date reviews of research areas in population ecology.

Bertram, B. C. R. 1975. "The Social System of Lions," *Scientific American 232:*54–61 (May). An account of how behavior plays a role in the population ecology of lions.

Boughey, A. S. 1973. *Ecology of Populations,* 2nd ed. New York: Macmillan. An examination of the basic aspects of population ecology.

Krebs, C. J. 1972. *Ecology: The Experimental Analysis of Distribution and Abundance.* New York: Harper & Row. A basic ecology text containing many examples of population studies, including biological control.

Myers, J. H., and C. J. Krebs. 1974. "Population Cycles in Rodents," *Scientific American 230:*38–46 (June). An examination of the causes for fluctuations in field populations of small rodents.

Ricklefs, R. E. 1976. *The Economy of Nature.* Portland, Oregon: Chiron Press. A well-written book that contains several chapters on population ecology.

Smuts, G. L. 1978. "Interrelations Between Predators, Prey, and Their Environment," *Bioscience 28:*316–320.

Wilson, E. O., and W. H. Bossert. 1971. *A Primer of Population Biology.* Sunderland, Massachusetts: Sinauer. A concise coverage of the basic concepts of population ecology.

Wading birds. These birds, which feed in the water and nest in trees, illustrate some of the linkages between terrestrial and aquatic ecosystems. (© Robert Hermes, from National Audubon Society.)

Chapter 5

The Earth's Major Ecosystems

We may walk through majestic redwood forests, trudge across tundra, slosh through streams, raft down roaring rapids, or float lazily down a quiet river. Indeed, our small planet is covered with an astounding variety of ecosystems. Each ecosystem owes its uniqueness to a singular combination of myriad organisms and physical conditions. This means that each ecosystem houses a different set of resources, whose exploitation generates a different group of problems. Because each ecosystem is unique, the solutions to these problems are also necessarily unique.

Every ecosystem on earth has by now been modified to some extent by human activity. The most severely modified ecosystems are those that have been most heavily populated or purposefully altered to meet human needs. The once great forests of eastern North America, Europe, and China, for instance, were originally cleared for agriculture and villages, and now these areas house a significant percentage of the world's urban and industrial centers. Much of the world's grasslands, including the Great Plains of North America, the pampas of Argentina, and the steppes of Russia, have been greatly modified by grazing and cultivation.

Today, as populations continue to soar, societies must travel farther from home to meet their resource demands. Hence, ecosystems that were once considered remote are now bearing the brunt of human exploitive activities. The North American tundra and continental shelves are being altered as we seek to meet our demands for oil, natural gas, and minerals. Tropical rain forests of Central and South America are being cleared to meet the human demands for wood and agriculture. In the meantime, we continue to disturb those ecosystems that already have been significantly changed by human activity.

Terrestrial Ecosystems

Because we spend much of our lives enclosed in climate-controlled cars, homes, schools, and even shopping centers, many of us forget that the terrestrial environment is a severe, often hostile place, whose two most common limiting factors are moisture and temperature. Air has a drying influence, and all terrestrial organisms face the threat of dehydration. To retard dehydration, animals are covered with hair, scales, or feathers, and they conserve internal water by excreting concentrated urine and relatively dry feces. Bark protects tree trunks and branches from drying out, while leaves are covered with a waxy layer that is relatively impervious to water. However, the continual movement of air provides at least one benefit to terrestrial life: an atmosphere in which the gases most important to life—oxygen and carbon dioxide—are fairly uniform from place to place.

Air temperatures may fluctuate both diurnally and annually in terrestrial environments. In the continental interior, far from the moderating influence of oceans, temperatures are particularly variable. In the Midwestern United States, for example, the temperature may change by as much as 10–15°C (18–27°F) in a day; between summer and winter it may vary by about 70°C (126°F). To survive seasonal extremes, many animals either migrate or hibernate and most plants become dormant.

Soil is vital for the proper functioning of terrestrial ecosystems. As you will recall from Chapter 2, soil is the major source of essential mineral nutrients and water for plants and is the home of most terrestrial decomposers. Soil also contains humus and broken-down rock, usually in the form of sand, silt, and clay. The pore spaces in soil contain varying amounts of air and water. The quantity of available mineral nutrients, the water-holding capacity, and the aeration of the soil play major roles in determining soil fertility, and these factors in turn govern plant productivity and the number of consumers a terrestrial ecosystem can support. These soil properties are determined by five factors: composition of parent bedrock, climate (prevailing moisture and temperature), topography (quality of drainage), types of plants and animals, and time.

A vertical profile of most soils reveals a sequence of distinct layers, called soil *horizons,* produced by the gradual upward transition from undisturbed

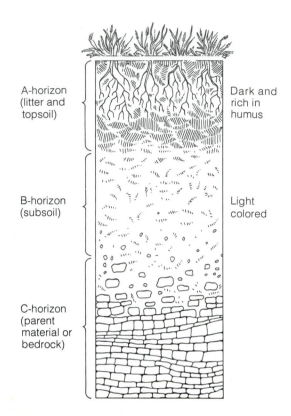

Figure 5.1 A soil profile.

A-horizon (litter and topsoil) — Dark and rich in humus

B-horizon (subsoil) — Light colored

C-horizon (parent material or bedrock)

bedrock to surface vegetation. Figure 5.1 illustrates a soil profile. Horizons are classified on the basis of physical and chemical characteristics, and these in turn are determined for each soil type by the degree of *leaching* to which each layer is subjected. Leaching is the process by which seeping water dissolves, transports, and redeposits soluble soil constituents.

Three horizons can be identified in most soil profiles. Surface soil, or topsoil, called the *A-horizon,* is the major source of mineral nutrients and therefore usually contains most of the plant roots. The A-horizon also has a dark surface layer of humus. Some of the soluble soil constituents of the A-horizon (for example, calcium carbonate) are transported from it by leaching to the subsoil, called the *B-horizon*. Because the B-horizon contains less *organic matter*, that is, materials of plant and animal origin, it is less fertile than the A-horizon. The B-horizon may also be a zone of clay accumulation. The *C-horizon* is composed of the broken or partially decomposed underlying *bedrock*. Because of its low fertility, it is penetrated by few plant roots.

Biomes of North America

The North American continent, like all of the continents on earth, is covered by large communities called *biomes*, which consist of plants that are similar in structure. Figure 5.2 shows the distribution of the major biomes of the world. For example, most of the eastern United States is covered by a deciduous forest whose trees lose their leaves in the winter whereas the midsection of the continent is occupied by vast grasslands. Although the presence of a particular biome is determined mainly by the regional climate, there is considerable variation in vegetation within a biome, since the interaction of such factors as soil, local climate, altitude, and drainage produces differing climax vegetation in localized areas. As a result of disturbances, both human and nonhuman, several stages of succession exist simultaneously within a biome. Hence, the map in Figure 5.2 is an oversimplification of the complex mosaic of ecosystems. In the following sections we will discuss these biomes one by one.

Tundra. Near the top of the continent lie the treeless plains of the *tundra*, shown in Figure 5.3. Low growing-season temperatures and a short growing season (sixty days or less) make the tundra one of the least productive biomes in the world (see Table 5.1). The tundra biome is relatively simple; that is, it has a low diversity of organisms. Year-round inhabitants include caribou, Arctic hare, lemmings, musk oxen, wolves, and Arctic foxes. In the summer, large num-

Table 5.1 Estimates of plant production for various ecosystems. (Numbers indicate grams of carbon per square meter per year.)

Type of ecosystem	Mean plant productivity
Desert scrub	32
Tundra	65
Temperate grassland	225
Savanna	315
Boreal forest	360
Temperate deciduous forest	540
Tropical rain forest	900

Source: (Data after R. Whittaker and G. Likens, "Carbon in the Biota," in G. Woodwell and E. Pecan, eds. *Carbon and the Biosphere* (Washington, D.C.: Technical Information Center, USAEC, 1973).

bers of migratory birds, mostly geese and ducks, are present. The vegetation consists of lichens ("reindeer moss"), grasses, sedges, and dwarf woody plants such as willows and birches. Because low temperatures slow the rates of chemical and biological changes, tundra soils are poorly developed. That is, the A- and B-horizons are thin and poorly differentiated from each other.

Low species diversity, poorly developed soils, and a severe climate make the tundra biome extremely fragile. Given the low diversity, few species are available to replace key species that are eliminated. The poor development of the soils contributes to natural erosion and slow rates of plant growth. The latter, in turn, results in a slow rate of recovery from disturbance.

For hundreds of years, the tundra biome was largely ignored, but that era passed with the discovery of vast deposits of petroleum on the North Slope of Alaska. Today, plans are being made to exploit coal deposits (almost two-thirds of North America's estimated coal reserves) and the vast deposits of copper, lead, zinc, and other metals that lie beneath the tundra. Access to these minerals could be gained by improving the trucking road that runs parallel to the Alaska pipeline, shown in Figure 5.4. Plans for facil-

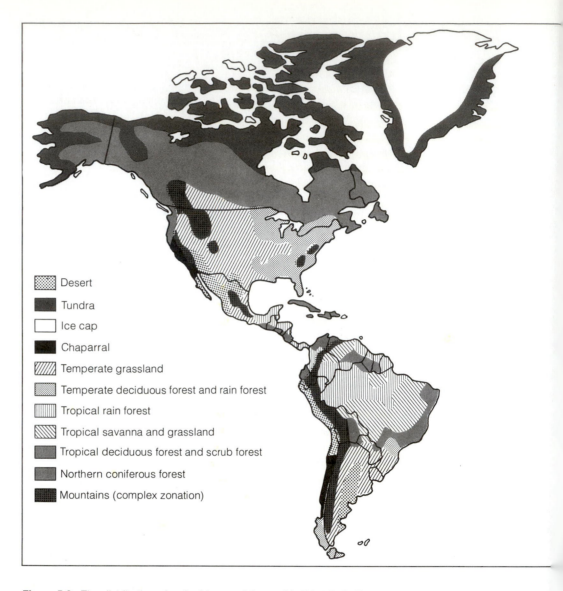

Figure 5.2 The distribution of major biomes of the world. (After P. R. Ehrlich, A. H. Ehrlich, and J. R. Holdren, *Ecoscience: Population, Resources, and Environment.* San Francisco: W. H. Freeman and Company, 1977.)

ities to process the minerals, petrochemical complexes, and hydroelectric projects are slated for the future.

We have already said that the tundra ground cover is very fragile and the soil vulnerable to erosion. The danger that human activity will disturb this fragile biome is compounded by the fact that vast areas of the tundra are underlain by *permafrost,* or ground that is frozen the year round. Permafrost thaws when its vegetative cover is disturbed. To protect it, all roadbuilding must be carried out only during the winter months. (For more about permafrost and the

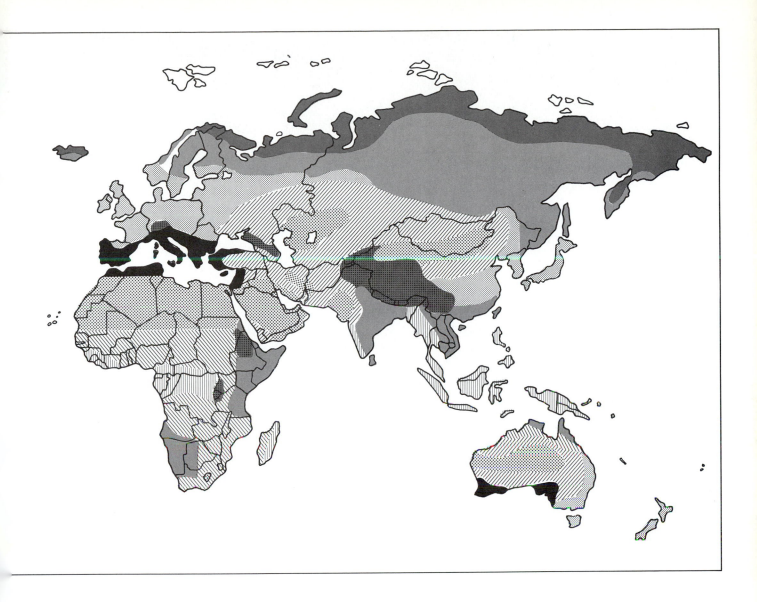

problems human beings encounter in contending with it, see Box 5.1.)

Proper route selection and seasonal timing of construction can greatly reduce many environmental problems in the tundra, such as disruptions of fish spawnings and caribou migration. But increased hunting and the destruction of the habitat resulting from the influx of people will continue to be critical problems for wildlife. For example, large amounts of gravel are needed for roadbed construction, but removal of gravel from surface deposits destroys the denning sites of foxes and wolves. The constant flow

Figure 5.3 (*Above*) Caribou trekking across the Arctic tundra. (U.S. Department of the Interior, National Park Service.)

Figure 5.4 (*Left*) A construction road parallels the Alaska Pipeline, permitting relatively easy access to Alaska's vast interior. (© J. E. Pasquier, Photo Researchers.)

of traffic on a highway poses a barrier to animal migration. Opening the interior with a public road would increase the influx of hunters, who could wipe out predator populations (wolves and foxes), thereby altering the entire food web permanently. Regardless of the precautions taken to protect the environment, the tundra shows irreversible effects that are solely the result of human activities.

Northern Coniferous Forests. The northern coniferous forest biome (the boreal forest or northern evergreen forest) lies south of the tundra in a broad band extending from Alaska to Newfoundland.

Figure 5.5 Evergreen forests on mountainsides in the western United States. (U.S. Department of the Interior, National Park Service photo by Jack Boucher.)

Evergreen forests are also found on mountainsides in the western United States. Figure 5.5 illustrates this forest environment. The climate in boreal forests is only slightly less severe than that in the tundra. The growing season is still short, lasting only from June through August), and the winters are very cold. Annual precipitation is moderate, averaging about 65 centimeters (25 inches). Although soils are better developed than in the tundra, they are still only thin layers overlying sheets of rocks exposed by glaciation. Low temperatures and decay-resistant evergreen needles contribute to a slow rate of decomposition, so a layer of needles 7–15 centimeters (3–6 inches) thick commonly covers the ground. The climate and the nature of evergreen needles make the soil acidic.

As in the tundra, species diversity is low. The tree canopy is mainly composed of two species of spruce (white spruce and black spruce) and balsam fir. Because of the dense shade and thick litter layer, shrub and wildflower flora are sparse. Successional stages containing paper birch and aspen provide habitat for mice, snowshoe hares, grouse, and moose. Major carnivores include wolves, lynx, weasels, and wolverines. Because of the low species diversity in the tree canopy, there are periodic outbreaks of bark beetles and defoliating insects such as the spruce budworm, spruce sawfly, and birch leafminer.

The northern coniferous forest has long felt the impact of human activities. In the 1700s and 1800s it was heavily exploited for its furbearing animals (wolves, beaver, and ermine). Today the softwood conifers are in great demand as wood pulp to make paper. In some areas, mining and mineral processing have destroyed the forests completely.

Temperate Deciduous Forests. South of the northern coniferous forest and east of the Great Plains grasslands is the temperate *deciduous forest,* occupying all of eastern United States except southern Florida. A typical scene from this forest biome is shown in Figure 5.6. Climate in this biome is temperate; there are distinct summer and winter seasons, and all regions are subject to freezing temperatures. Precipitation, ranging from 150 centimeters (60 inches) a year in the southern mountains to 75 centimeters (30 inches) in the north, is distributed more or less evenly throughout the year. Deciduous leaves—leaves that fall off the tree after the growing season—decompose more easily than evergreen needles. This in combination with the warm summer temperatures hastens the release of mineral nutrients into upper soil layers. Hence, these forest soils are more fertile than tundra or boreal forest soils.

The eastern deciduous forest is quite diverse. The greatest complexity is in the Cumberland Mountains in Tennessee and the cove forests of the Smoky Mountains in Tennessee and North Carolina, where about forty different tree species make up the canopy. Beech, basswood, maples, hickories, and oaks grow in dense stands. Beneath the upper canopy is a layer of understory tree species (flowering dogwood, magnolia, and ironwood are examples). The deciduous characteristic allows for the presence of a rich variety of spring wildflowers and shrubs, which complete many of their activities before the canopy closes. Various plants produce nuts and pulpy fruits, which greatly enhance the food supply for wildlife.

Box 5.1

Differential ground subsidence due to melting of permafrost caused by railroad bed construction in Alaska. (U.S. Department of the Interior, U.S. Geological Survey.)

Permafrost: Deterrent to Economic Development

Much of the tundra south to the edge of the northern coniferous forest is underlain by permafrost—permanently frozen soil. In the far north, it forms a continuous layer that in places is more than 300 meters (1000 feet) thick. During the short northern summer, the soil layer between the top of the permafrost and the ground surface thaws. This thaw zone, the active layer, is generally less than a meter (3 feet) thick. Because permafrost is impermeable, meltwater cannot drain through it. Thus, during the brief summer thaw the active layer is transformed into a quagmire.

Permafrost reduces wind and stream erosion, and so it helps protect the tundra's fragile plant communities and the otherwise erosion-prone, poorly developed soil. But because it is sensitive to changes in the insulating properties of the ground surface, the permafrost itself depends on its vegetative cover for protection against thawing. Even minor alterations in this vegetation can have a significant impact on the permafrost's thickness and, therefore, on the depth and expanse of the active layer. If the ground is disturbed by road construction, for example, the permafrost begins to melt. When ice melts, its volume decreases and the ground settles. But since the degree of melting is not everywhere uniform, ground-settling is very uneven, and smooth roadways soon develop into washboards that slow traffic and can damage vehicles. Uneven ground-settling can also shift railroad ties and tracks out of line, making passage by trains impossible.

When a pipeline is laid within or upon frozen ground, the permafrost melts irregularly, straining and sometimes rupturing the pipe. With oil pipelines the problem is especially severe because the oil must be heated to flow freely through the pipe.

To reduce the impact of construction activity upon permafrost, roads must be constructed only during the winter half-year. For the same reason, roads and buildings must be built upon gravel pads that insulate the permafrost and thereby keep it from melting. Oil pipelines and sewer lines must be insulated and installed aboveground.

Paradoxically in a climate where subzero temperatures prevail for so much of the year, refrigeration units must sometimes be installed in structures and equipment to protect them from thaws in the permafrost that holds them up. Natural-gas drillers in Canada north of the Arctic Circle have installed such units to keep the permafrost frozen around their wellheads. And in Siberia, construction engineers have had to build freezing units into the steel uprights of at least one railroad bridge.

Because tundra vegetation and the permafrost beneath it are so easily disrupted by human activity, plans for the economic development of Alaska's tundra regions have come under fire from environmentalists. These critics contend that, even if developers always take every feasible precaution to protect this fragile biome, their best-laid plans will sometimes go awry—perhaps with disastrous consequences.

One major threat cited by environmentalists is the danger of a break in the Alaska pipeline. This line, which is over a meter in diameter, began carrying heated oil south from the North Slope oilfields in 1977. If it ruptured completely, it could spew as much as 160 cubic meters (roughly 35,000 gallons) of hot oil onto the ground every minute. Even if the flow were cut off quickly, the oil spill would be massive and its contamination of virgin wilderness great.

Figure 5.6 Azaleas blooming along a country road in a deciduous forest in western North Carolina. (U.S. Forest Service.)

Common animal species include white-tailed deer, grey squirrel, fox squirrel, red fox, cottontail rabbit, bobcat, wild turkey, and many familiar songbirds.

The eastern deciduous forest biome has been extensively modified by human activity. Many major cities and industrial complexes are located within this biome. In places the landscape has been ravaged by strip mining (discussed in Chapter 12) and dissected by a vast network of highways and railroads. Only a few areas have been set aside as state and national parks and wilderness areas to preserve the remnants of the original forests. Most of the unprotected forests have been cleared for agriculture or sawed for lumber.

Once the protective tree cover is removed in such regions, proper soil conservation practices must be used to prevent severe erosion. Futhermore, agriculture and lumbering remove nutrients from the ecosystem that must be replaced if soil fertility is to be maintained. Unfortunately, as we suggested in our discussion of old-field succession, improper farming practices forced the abandonment of many farms from the Carolinas to southern Illinois. Because over two-thirds of Americans reside within this biome, the impact from human activity is not likely to diminish.

Temperate Grasslands. As we move westward in North America, water becomes a limiting factor for trees. Hence, forests give way to the grasslands that

Figure 5.7 Trees along waterways, a common configuration in grasslands. Note in the foreground the many flowering plants in addition to grasses. (U.S. Department of the Interior, National Park Service.)

stretch from Texas to Saskatchewan and from Illinois to the foothills of the Rockies (Figure 5.7). Annual precipitation in the grasslands biome is fairly light; it ranges from 25 centimeters (10 inches) in the west to 75 centimeters (30 inches) in the east. Grasslands experience a wide range of seasonal temperatures, which climb above 38°C (100°F) in summer and plunge below −32°C (−25°F) in winter. The particular combination of climate, composition of parent sediment, and prairie plants has produced soils that are exceptionally fertile. In fact, the prairie soils of Iowa and Illinois are more productive than soils in any other biome in the world. Wildfires (as well as the small quantities of precipitation) have played an important role in maintaining the natural grassland biome, particularly along its eastern boundary with the deciduous forest.

Many species of grasses occupy the natural grasslands, and the aridity of the particular habitat determines the actual species assemblage. Nonwoody plants other than grasses—prairie clovers, purple cone flowers, sunflowers, and goldenrods—are also abundant on most natural prairies. Grassland plants survive summer dry periods by becoming dormant; there is usually adequate moisture during spring and early summer to insure that they will grow enough and store a sufficient amount of food to survive periods of dormancy.

Characteristic herbivores of native grasslands include buffalo, pronghorn antelope, rabbits, ground squirrels, prairie dogs, prairie chickens, and many species of ground-nesting sparrows. Interestingly, many grassland animals live in herds or colonies that serve as protection in the open habitat. Grassland carnivores include coyotes, foxes, badgers, prairie falcons, and owls.

Cattle and sheep grazing and grain farming have destroyed all but 0.001 percent of the virgin grasslands of the United States. The eastern two-thirds of the biome have been plowed under and now serve as a major breadbasket of the world. Monocultures of corn, soybeans, and wheat have replaced native grassland species, in effect, eradicating the complex grassland ecosystems natural to this vast region. In the drier western portions, grasslands have been modified through overgrazing by cattle and sheep.

The prairie's thick vegetative cover and the large mass of intertwined roots and underground stems beneath it once greatly retarded soil erosion, but plowing and overgrazing have exposed the soil to wind and runoff thereby promoting soil erosion. The catastrophe known as the Great Plains Dust Bowl of the 1930s is a vivid reminder of the consequences of neglecting proper soil conservation measures. In the 1920s, economic pressures forced farmers to plow land that was only marginally productive in wet years and turned into deserts during droughts (Figure 5.8).

Improper land management has destroyed productive grasslands on all continents. Degradation has

Figure 5.8 A farm abandoned during the 1930s Dust Bowl. (U.S. Department of Agriculture.)

been particularly severe in semiarid regions. The United States Agency for International Development estimates that along the southern fringes of the Sahara Desert, 650,000 square kilometers (about 160,000 square miles) of land that was once suitable for agriculture or intensive grazing have become wasteland over the past fifty years. The seriousness of the problem was underlined by a special worldwide conference convened by the United Nations in 1977 in Nairobi, Kenya. This conference brought a new word to the public's attention—*desertification*. Desertification is the degradation of terrestrial ecosystems through human activity. It results in a lower crop-production potential, a reduction in the land's capacity to support livestock, increased environmental deterioration as a result of water and wind erosion, and usually a reduction in the resident peoples' standard of living. The term is primarily applied to events that take place in arid and semiarid regions. For a discussion of the causes and control of desertification, see Box 5.2.

Deserts. Deserts are biomes that occupy regions in which precipitation is less than 25 centimeters (10 inches) annually. Rainfall in deserts is not only sparse, but also infrequent and largely unpredictable, often occurring as cloudbursts. The plants and animals that live in desert biomes are adapted to undergo a burst of activity when these infrequent deluges occur. If sufficient precipitation falls, for example, the seeds of many desert plants germinate rapidly, yielding seedlings that flower and set seed within a span of only a few weeks. The appearance of this new plant growth in turn triggers reproductive behavior in such desert herbivores as grasshoppers and rabbits.

Although all deserts are arid, some are classified as hot and others as cold. In North America, hot deserts—the Mojave, Sonoran, and Chihuahuan deserts—occupy much of the American Southwest and northern and western Mexico. The North American cold desert is situated in the Great Basin, extending from Nevada to eastern Washington. Each desert has

Figure 5.9 The organ pipe cactus (in foreground), which has no leaves at all, and the ocotillo (behind it), which has quite small leaves. (U.S. Department of the Interior, National Park Service.)

its own characteristic vegetation. Although we commonly associate cacti with deserts, a rich flora of cacti is found only in the Sonoran Desert, as shown in Figure 5.9. Sagebrush and shadescale are characteristic plants of the Great Basin cold deserts. A stretch of sagebrush desert is shown in Figure 5.10.

The arid conditions determine the mix of wildlife species in deserts. Because of low plant productivity, these areas do not support large grazing animals. Herbivores consist mainly of rabbits, ground squirrels, mice, and rats. Major predators are coyotes, kit and grey foxes, and various species of snakes and owls.

Box 5.2

The world's deserts are expanding. Each year they claim some 5.6 million hectares (14 million acres) of farmland, grassland, and other fertile regions. Desertification, as this destructive process is called, is a major problem on every continent except Europe. Its immediate consequences are most tragic in lands least able to afford the losses: the developing nations, including India, Mexico, Algeria in North Africa, and the sub-Saharan lands of the Sahel region. But the developed nations are not immune to the encroachments of the growing deserts. Argentina has lost about three-quarters of the forestland it had fifty years ago. And in the United States, vast areas of New Mexico and Arizona that were rich grasslands in 1870 are now quite arid, the thick grassland having surrendered to sagebrush.

Nor is desertification in its broadest sense—the degradation of terrestrial ecosystems by human activity—limited to the fringes of arid regions. Dust storms still sweep over large areas in the Great Plains, evidence that severe soil erosion continues in America's "breadbasket" region. Wind erosion there and elsewhere is often aggravated by water erosion, which removes vegetative cover and leaves the soil vulnerable to windstorms. A U.S. government study released in 1977 found that 283 farms, chosen at random in various parts of the country, were losing more than 2 metric tons of topsoil per hectare each year. The U.S. Department of Agriculture estimates that we are losing 15 metric tons of topsoil out of the mouth of the Mississippi every second, or almost 500 million metric tons a year. Desertification is a gradual process. On cropland, yields decline year by year as the topsoil is eroded by wind and water and soil nutrients are dissipated. On rangeland, the quality of forage declines as the more palatable and productive species are overgrazed and replaced by less desirable plants. In arid and semiarid regions, desertification is accelerated by the extended droughts that are inevitable there. Where land abuse is severe and prolonged, grasslands and fields are reduced to stony, eroded wastelands or mounds of drifting sand.

Desertification in arid regions has several causes. Probably the major factor is overgrazing of rangeland. Simply put, overgrazing is a problem of too many animals feeding on a limited supply of forage. It accounts for the fact that more than 70 percent of western grazing lands in the United States are producing less than 50 percent of their forage potential. In starker terms, overgrazing also means that the losses in livestock and human life in certain developing countries are much greater than they would otherwise be during periods of extended drought. Such a drought killed tens of thousands of people in the Sahel region during the early 1970s.

Another cause of desertification is erosion on dry-farmed areas. The Great Plains Dust Bowl illustrates the acceleration of wind and water erosion when erosion-control procedures are lacking in semiarid, dry-farming areas. As population pressures force cultivation into marginal areas during good rainfall years, the stage is set for high rates of desertification when the inevitable droughts come. Improper water management on irrigated land also contributes to desertification. Waterlogging and increasing salinity are common problems. About 25 percent of the irrigated land in this country is mismanaged.

In many developing countries, desertification is accelerated by deforestation; much of the wood is burned as fuel. Dense stands of acacia trees were common around Khartoum, the capital of the Sudan, as recently as 1952. But by 1972, the nearest dense stand of acacia was 90 kilometers (55 miles) south of the city. This story can be repeated throughout many developing countries as millions of impoverished people cut firewood faster than the

The World's Growing Deserts: Can They Be Turned Back?

trees are replanted. (For more on this related problem, see Box 19.1.)

Technological and managerial means exist for halting desertification. They include improved animal-husbandry practices to reduce overgrazing and such conventional conservation measures as strip farming, contour plowing, and crop rotation to reduce soil erosion on farmland. A promising development now gaining support among U.S. farmers is minimum-tillage farming, in which the plow is done away with and seeds are planted in the stubble of a previously harvested crop. This technique not only reduces soil erosion but also makes great savings in labor, fuel energy, and water.

The planting of tree barriers and windbreaks to halt or at least slow soil erosion and windblown sand is an old technique that is generating new interest. Algeria has for some years been planting eucalyptus trees to combat desertification in a 1400-kilometer-long (900-mile-long) belt of the North Sahara; China has reclaimed tracts of the Gobi by planting wind barriers of trees that give local vegetation a chance to recover. Efforts to halt the Sahara's southward march into the Sahel region with a tree belt have not been very successful, however. So in 1977 a more ambitious plan was proposed at the United Nations Conference on Desertification. It calls for the establishment of a zone, 140–390 kilometers (90–240 miles) wide, of mixed land use—grassland upon which limited grazing will be permitted, forest plantations, farm tracts, and wildlife preserves. If and when this new "greenbelt" is completed, it will span Africa's 6000-kilometer (3700-mile) width from the Atlantic to the Red Sea.

Weapons to combat desertification, then, have been developed. But even in economically advanced countries their deployment is slow. Why? In the United States, the main reason for poor performance in this area is the emphasis on short-term rather than long-term benefits. Most people want to preserve our natural resources for future generations, but economic pressures force them to be more concerned about economic survival today. Other reasons for the persistence of poor land husbandry practices include a widespread ignorance of the fact that soil fertility is declining and the absence of public pressure to control land degradation.

The developing nations face additional constraints in combating desertification. These include inadequate training and educational facilities, limited access to new research developments, and ineffective agricultural-extension services. The main constraint is undoubtedly the need to feed a rapidly growing population. This need produces pressures to extend grazing and farming even farther into marginal lands. The poignant words of a herdsman in India's Rajasthan Desert, where the goat population has more than doubled since the late 1950s, expresses the problem in a way that mere statistics cannot:

> When I was a child, there was no competing for grazing lands. . . . Now every patch of good land is struggled over, and the earth never gets to rest. We know there are too many animals trying to live from our desert now, but how can we solve that problem?[*]

Whose responsibility is it to control desertification? Should society as a whole be concerned about protecting soil resources for the long-term welfare of the population, or is it the farmer's responsibility? And how can that responsibility, wherever it lies, be translated into effective action?

[*] William Borders, "Indian Scientists Seek to Contain Growing Desert." © 1978 by The New York Times Company. Reprinted by permission.

Figure 5.10 A sagebrush desert in the cold Great Basin of Utah. (U.S. Department of the Interior, Bureau of Land Management.)

Desert organisms have many other adaptations to survive in this particularly harsh biome besides the ability to reproduce quickly when conditions become favorable. Some animals, such as the kangaroo rat, rarely drink water, relying almost exclusively on the metabolism of food for their necessary water. To avoid the searing midday sun, many animals are only active at night or in the early morning or late afternoon, as shown in Figure 5.11. During particularly stressful times, some animals descend into cool dens where they enter a state of *aestivation*, a condition similar to hibernation in which the organism greatly reduces its level of activity and thereby diminishes its food and water requirements. To minimize water loss, most desert plants have either small leaves (for instance, coachman's whip) or no leaves at all (cacti). In these plants, photosynthesis is carried out in the green, often enlarged, stems. Those species that have

small leaves are also deciduous; that is, their leaves die and fall off during dry periods. A final example of a common adaptation to the dry conditions is the thorny or spiny surface of many desert plants. Since animals often use plants as a water source, nature has selected for many plants that can protect themselves.

In the past, human beings tended to avoid deserts because of the aridity. With the construction of dams and canals, however, water supplies to desert regions become more reliable, and people are invading them in increasing numbers. The dry, warm climate attracts people seeking relief from such respiratory ailments as asthma and from other conditions, such as arthritis. Land developers are taking advantage of that fact and are building extensive housing tracts in the desert, many of which cater to retired citizens seeking the health-restoring effects of the desert environment. Desert soils are often fertile, and where

Figure 5.11 By resting in a narrow band of shade, these jackrabbits avoid the searing heat of the desert. Their large ears help to regulate their body temperature by emitting heat energy. (© Verna R. Johnson, from the National Audubon Society.)

water is available for irrigation they can be cultivated with great success, as Figure 5.12 illustrates.

Deserts are fragile, however, sensitive to the slightest changes worked on them by humans. Because the vegetative cover is sparse, deserts are particularly vulnerable to intensive soil erosion during cloudbursts. Overgrazing and trampling by livestock accelerates erosion, and large areas that lose too much soil can be reduced to barren sand dunes. Also, improper irrigation practices can change productive desert lands into salt flats. Many people from "back East" have tried to bring elements of their native environment with them to the Southwest. By planting lawns and watering them extensively, they have increased humidity in localized areas, thereby encouraging the growth of molds and mildew. Another practice that has had a negative impact is the introduction of allergenic (allergy-producing) plants.

Ironically, these modifications are beginning to change significantly the very environmental conditions that first attracted people to the Southwest.

Temperate Rain Forests and Chaparral. Although much less extensive than the biomes we have described thus far, two other biomes in North America—temperate rain forests and chaparral—merit attention. The temperate rain forest, characterized by large evergreen trees, is found along the west coast of North America from central California to Alaska. Annual rainfall in this environment is as high as 380 centimeters (150 inches), and the condensation of coastal fogs on trees adds additional water to the total precipitation. This biome is most fully developed in the Puget Sound area. Here, dense stands of western hemlock, western arborvitae, grand fir, and Douglas fir reach upward for 60 meters (200

Figure 5.12 Highly productive citrus groves growing in an irrigated stretch of desert near Yuma, Arizona. (U.S. Department of the Interior, Bureau of Reclamation.)

feet). Figure 5.13 shows a typical scene from a rain forest in the Pacific Northwest. South of Puget Sound are the redwood forests, immortalized in song and verse, and north of it Sitka spruce becomes dominant. Epiphytic mosses abound throughout these forests.

Abundant rainfall and a narrow seasonal tempera- ture range help make these forests the richest wood producers in the world. Consequently they are ex- ploited for timber. Clear-cutting of forests (particu- larly the redwoods) is a long-standing controversy.

The chaparral biome occurs along the southwest coast of North America. Its climate is characterized by abundant winter rainfall and dry summers. Ever-

Figure 5.13 Luxuriant tree growth in a Pacific Northwest rain forest. (U.S. Department of the Interior, Bureau of Land Management.)

green shrubs such as chamise and manzanita dominate the vegetation. Fires are frequent during late summer, favoring the growth of shrubs and inhibiting that of trees. Underground parts of shrubs are fire-resistant and resprout profusely when winter rains return. Both leaves and stems of shrubs contain abundant quantities of oily sap that make fires burn hot and spread swiftly. In southern California, many people have built homes in the chaparral only to see them go up in smoke (Figure 5.14) or be buried beneath mudslides. Slopes are steep and the loosely consolidated material easily erodes, particularly during heavy winter rains.

Figure 5.14 A fire roaring through California chaparral in San Bernardino National Forest. (U.S. Forest Service.)

The Global Distribution of Biomes

As we have said, tundra, boreal forests, deciduous forests, grasslands, deserts, and chaparral—are found on all of the continents of the world. Tundra and boreal forests form nearly concentric bands around the North Pole, as you can see in Figure 5.2. There are deciduous forests in western Europe and northeastern China. Extensive grasslands are present also in the pampas of South America, the steppes of Asia, and the African veld. Major world deserts include the Sahara in North Africa, the Gobi in Mongolia, the Atacama along the west coast of South America, and the great desert that makes up the interior of Australia. Chaparral communities are found along the southern coast of Australia and along the Mediterranean Sea.

Although the particular species present in a particular biome vary from one continent to another, they nevertheless exhibit many similar adaptations for surviving and thriving. As an example, lack of shelter on grasslands favors selection for fleet-footed animals. The ostrich in Africa, the emu in Australia,

Figure 5.15 Three ecological equivalents from similar grassland ecosystems in different parts of the world. *Left*: The ostrich. (Courtesy, Field Museum of Natural History, Chicago.) *Middle*: The emu. (Smithsonian Institution.) *Right*: The rhea. (U.S. National Zoological Park.)

and the rhea in South America are all large flightless birds whose powerful legs help them out-distance their predators. You can see their similarities in Figure 5.15. Animals that exhibit similar adaptations in comparable but geographically distinct biomes are called *ecological equivalents.* The pronghorn antelope of North America, the pampas deer of South America, the wild horse of Asia, and the zebra of Africa are running herbivorous animals that are ecological equivalents. Running carnivorous animals of the grasslands include the coyote in North America, the cheetah in Africa, and Tasmanian "wolf" (actually a marsupial) in Australia.

Around the globe, biomes show the same sorts of disturbances as those on the North American continent that we have mentioned in preceding sections. Most of the world's grasslands have been modified by overgrazing and agriculture. Many of the deciduous forests in Europe and China have been replaced by major urban-industrial complexes and agricultural monocultures. Tundra and desert biomes around the globe are being invaded in the effort to increase the supply of food, fossil fuels, and industrial minerals.

Tropical Rain Forests. Rain forests and savannas, two extensive tropical biomes, are being rapidly changed by human activity. These biomes exist, for the most part, in overpopulated, developing nations. Population pressures for more living space, the need for resources, and increasing technological ability to exploit resources are severely disrupting these ecosystems.

The tropical rain forest (see Figure 5.16) once stretched over lowlands of the humid tropics of Africa, Southeast Asia, Central and South America, and the islands of Indonesia. Rain forests are the oldest remaining ecosystems on earth, perhaps over 60 million years old. Daily and seasonal temperature variations are small. Annual rainfall commonly exceeds 200 centimeters (80 inches), but one or more relatively dry periods usually occurs each year. Because of the warm temperatures and abundant moisture, decay is rapid, and nutrients are immediately taken up by a vast root network. Hence, soils are thin, contain little humus, and are nutrient poor. Vegetation, rather than the soil, is the major reservoir of nutrients.

Tropical rain forests are the earth's most diverse

Figure 5.16 A tropical rain forest in Malaya. (FAO photo.)

ecosystems. A two-hectare (five-acre) area can contain over a hundred species of trees, whereas the same area in the eastern deciduous forest typically contains less than twenty-five species of trees. Trees of all heights are found beneath the upper canopy and epiphytes are abundant. Luxuriant canopy growth permits little light to reach the forest floor, and therefore the ground flora is generally sparse except where the canopy has been broken by a fallen tree. Consequently, a large proportion of the animal life lives in trees. An abundance of fruit and insects supports a diversity of animals. Although temperatures are favorable year round, the timing of the relatively dry period influences the timing of activities. The beginning of the wet season, with its renewal of plant growth, is also the time when most animals begin their breeding activities.

Tropical rain forests have never been the site of extensive human settlement; the lushness of their growth and the presence of stinging and poisonous insects discouraged such intrusions. But overcrowded developing countries now see these forests as the key to progress. In Brazil, major highways are being hacked into the Amazon basin to open the way for economic expansion. Large areas of forests have been cut down for timber and to make way for plantations of cash crops such as oil palms, bananas, rubber, and cocoa. The world's oldest ecosystem is falling before the expansion of human numbers.

Tropical Savannas. Another significant tropical biome is the savanna, a grassland characterized by a continuous cover of grasses interrupted by scattered clumps of trees and shrubs. As in grasslands, fires are common, and, to survive, both trees and grass must be fire-resistant. The tropical savanna biome is found in warm areas that have 100–150 centimeters (40–60 inches) of rainfall annually as well as a pronounced winter dry period. The best known savannas are in Africa, where the number and variety of grazing animal species and their predators are unexcelled anywhere in the world. This ecosystem is the home of lions and cheetahs. Here, too, large herds of herbivores—zebra, wildebeest, giraffe, and numerous species of antelope—graze and browse (see Figure 5.17). Many of these herbivores are migratory, moving in response to the rainy seasons. Newborn animals must be able to walk within hours of birth or lose protection of the herd. Like the rain forests, savannas are changing under pressure of rapidly expanding human populations. People, farms, and cattle are crowding out the big game herds.

Figure 5.17 A herd of zebra and antelope on a savanna in Tanzania, (FAO photo.)

Tropical Scrub and Deciduous Forests. Two other tropical biomes are tropical scrub and tropical deciduous forests. These biomes are located where precipitation is intermediate between that needed to support savanna and that required for rain forests. The scrub forests contain small trees that are frequently thorny. Leaves are small and are lost during dry seasons. Where the rainfall is greater, well-developed deciduous forests such as the monsoon forests of tropical Asia are found. These forests and their inhabitants are adapted to alternating wet and dry seasons of approximately equal length.

Altitudinal and Latitudinal Zones. If we were to climb a mountain from its base to its summit, we would encounter ecosystems similar to those described in the preceding sections. Figure 5.18 illustrates the series of zones we would cross on our way up a mountain in the Colorado Rockies. We begin our imaginary journey in a sagebrush community,

and as we climb, we move successively through grassland, a Douglas fir forest, a spruce fir forest, and, crossing the timberline, the alpine tundra. We also note climate changes in our climb: air temperature decreases with altitude while precipitation usually increases. In effect, mountains telescope biotic and climatic zones; that is, in several thousand meters of altitude, we encounter the same biotic-climatic zones that we would in several thousand kilometers of latitude. Every 300 meters (1000 feet) we ascend roughly corresponds to a northward advance of 500 kilometers (310 miles).

Ecotones. Up to this point, for the purpose of distinguishing the various biomes we have described them as if they were distinct units with readily discernible boundaries. In reality, however, a transition zone, called an *ecotone*, commonly separates two distinct types of ecosystems, particularly terrestrial

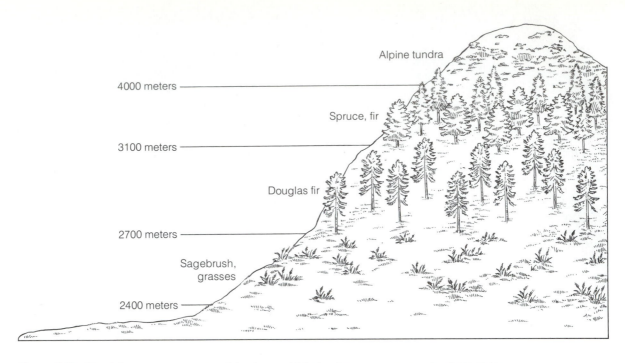

Figure 5.18 Vegetation zones on a mountain slope, which develop as a response to changes in moisture and temperature. In the central Colorado Rockies, a hiker would encounter a series of zones similar to those in the illustration.

ecosystems. An ecotone contains species of both ecosystems, and often its own unique species as well. Ecotones, therefore, often exhibit a greater diversity of plants and animals than either of the adjacent ecosystems. This tendency toward increased diversity in an ecotone is called the *edge effect*. Human settlement has greatly increased the amount of edge habitat on the planet. For example, a forest that was once continuous over many square miles may now be reduced to small isolated woodlots surrounded by pastures and fields. What were once sweeping grasslands are now fields dotted with farm houses and small towns where trees have been planted.

Ecotones may be narrow, as is the boundary between the woods and the field shown in Figure 5.19. Other ecotones are large transitional belts between biomes where one biome gradually blends into the other. Such a transition belt is the tiaga, which lies between the boreal forest and the tundra. Across this broad zone, climate changes quite gradually, and hence the vegetation also changes gradually. If we were to walk north across the ecotone from the boreal forest to the tundra, we would encounter spruce stands that become progressively less dense and more stunted. Further on, patches of tundra vegetation would become more common and spruce stands smaller and less frequent, as illustrated in Figure 5.20. Finally, we would reach a point where no trees are present, the tundra.

Ecotones, therefore, are not only boundaries between biotic communities but are also separations between major climatic zones. Hence, even a small change in climate will have its most immediate effects in ecotonal areas and cause a geographical shift in the ecotone zone. For example, a recent prolonged drought in the Sahel savanna region of Africa expanded the Saharan Desert southward into the savanna (see Chapter 17).

Figure 5.19 An ecotone between woods and an agricultural field. Such ecotones often contain bushes that provide both food and shelter for wildlife. (U.S. Department of the Interior, Bureau of Reclamation.)

Figure 5.20 Near the limits of tree-growth in Alaska, sparse open black spruce forests intermingle with bogs. (Jen and Des Bartlett, Photo Researchers.)

Aquatic Ecosystems

Aquatic habitats differ in several important respects from terrestrial habitats. Whereas the limiting factors affecting terrestrial organisms are moisture and air temperature, those influencing aquatic animals and plants are mainly related to oxygen availability and sunlight penetration. In the atmosphere oxygen is mixed with other gases, but in aquatic ecosystems it is dissolved in water. *Dissolved oxygen* concentrations in fresh water average about 0.0010 percent (often expressed as 10 *parts per million,* or ppm) by weight, which is 40 times less than the weight of oxygen in an equivalent volume of air. While the concentration of atmospheric oxygen varies little, the amounts of dissolved oxygen in aquatic ecosystems can vary widely depending upon the factors influencing inputs and outputs.

Oxygen enters an aquatic system through the air-water interface and by photosynthetic activities of aquatic plants (see Chapter 2). Thus, the amount of dissolved oxygen present depends on the rate at which these processes occur. Turbulence in waterfalls, rapids, and dam spillways and wave activity in open water increase the rate of oxygen transfer from air to water (Figure 5.21). The profile of a waterway is also a factor: a wide, shallow section of a river, for example, has a larger surface area for oxygen transfer

Figure 5.21 Turbulent rapids, which add significant quantities of oxygen to waterways. (U.S. Department of the Interior, Bureau of Reclamation.)

than a narrow, deep segment. Also, the larger the number of aquatic plants, the greater the amount of oxygen produced. Dissolved oxygen is removed from the water through respiration by decomposers, zooplankton, shellfish, and fish. Temperature also influences the amount of dissolved oxygen, because oxygen is less soluble in warm water, as shown in Figure 5.22. In addition, warm water enhances decomposer activity and hence oxygen removal.

In certain circumstances, the respiration activities of large populations of decomposers remove nearly all dissolved oxygen in surface waters. This situation is usually aggravated during late summer, when high water temperatures and low stream flow further reduce dissolved oxygen levels. Dissolved oxygen levels below 3–5 ppm cause the death of many aquatic organisms.

In contrast to oxygen levels, the water temperature of aquatic ecosystems varies less both daily and seasonally than the air temperature of terrestrial ecosystems. Furthermore, the temperature changes more slowly in water than it does in air. Water has a considerably higher heat capacity than air; that is, much larger amounts of heat energy must be added to or taken from water to raise or lower its temperature. Thus, water temperatures are less subject to change, and, because aquatic organisms are not selected for wide temperature tolerance, they have not evolved a tolerance of wide temperature variations. As a result, small changes in water temperatures have a more critical effect on the survival of aquatic organisms than comparable changes in air temperature do on terrestrial life.

Sunlight is a limiting factor for aquatic plant life because it can penetrate only to a depth of about 30 meters (100 feet) below the water surface; thus, photosynthesis is essentially confined to this area, which is called the *euphotic zone*. The depth of the euphotic zone may be significantly reduced by *suspended materials* (mostly silt and clay) that prevent sunlight penetration into greater depths.

The limiting factors identified in this section are generally applicable to all aquatic ecosystems,

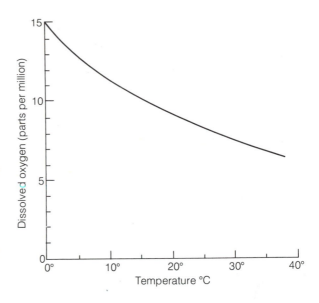

Figure 5.22 The quantity of dissolved oxygen that water can hold in relation to water temperature. The amount declines as water temperature rises.

whether ponds, lakes, marshes, rivers, or seas. Each type of aquatic ecosystem, however, does have certain unique properties. We turn now to a survey of the major types of aquatic ecosystems and their distinctive characteristics.

Freshwater Ecosystems

Lakes. In temperate latitudes, lakes deeper than about 15 meters (50 feet) become stratified into two layers in the summer. Under bright summer sunshine, surface waters warm and become lighter—or of lower density (mass per unit volume)—than the colder water below. The result is a stable layer of light, warm water overlying one of dense, cold water, with little mixing occuring between the two. The upper layer of a stratified lake is called the *epilimnion*, the lower layer is the *hypolimnion*, and the narrow transition zone between the two is referred to

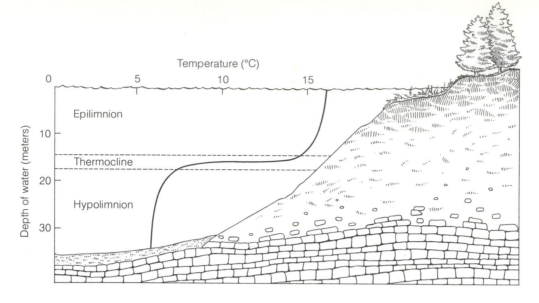

Figure 5.23 The stratification of a lake in summer as a result of temperature differences.

as the *thermocline*. Figure 5.23 shows a cross-section of a stratified lake.

When lakes are stratified, the dissolved oxygen supply of the epilimnion is replenished as usual by transfer through the air-water interface and by photosynthesis of algae. But the dissolved oxygen supply of the hypolimnion may be reduced by the decomposition of detritus that settles on the lake bottom. Some fish must stay in the cold hypolimnetic waters because they cannot tolerate the seasonally warm waters of the epilimnion (lake trout, whitefish, and cisco are examples). Ironically, these fish also have high dissolved oxygen requirements and thus could not survive unless there were a means to replenish the dissolved oxygen supply in the hypolimnion. How does oxygen replenishment occur?

The answer is that the lake does not remain stratified permanently. In autumn the surface water cools. Eventually the temperatures, and therefore the densities, of the two layers become equal. Assisted by the force of wind upon the lake surface, illustrated in Figure 5.24, water circulates, driving oxygen-rich water to the bottom and oxygen-depleted water to the surface. This mixing process, called *fall turnover*,

replenishes the dissolved oxygen supply of deeper waters to near-saturation conditions.

During winter months in midlatitudes lakes are covered with ice, and the water temperature varies from 0°C (32°F) just below the ice to 4°C (39°F) near the bottom. With the arrival of spring, the ice melts and surface waters warm. But because water becomes more dense as its temperature rises toward 4°C (39°F), the warmer but denser surface water sinks. (See Figure 5.25.) Under these conditions, the entire lake is mixed vertically, again assisted by the wind. This mixing is called the *spring turnover*.

Twice a year, then, in spring and fall, the surface and bottom waters are recirculated. Lake turnovers are essential to the replenishing of the dissolved oxygen supply of bottom waters. They thus help insure survival of fish that require cold water and high concentrations of dissolved oxygen. Furthermore, recirculation brings nutrients (primarily nitrogen and phosphorus) from bottom waters to surface waters, thereby increasing algal productivity.

All lakes share one characteristic: in the context of geologic time (thousands to millions of years) they are relatively short-lived. A common cause of aging

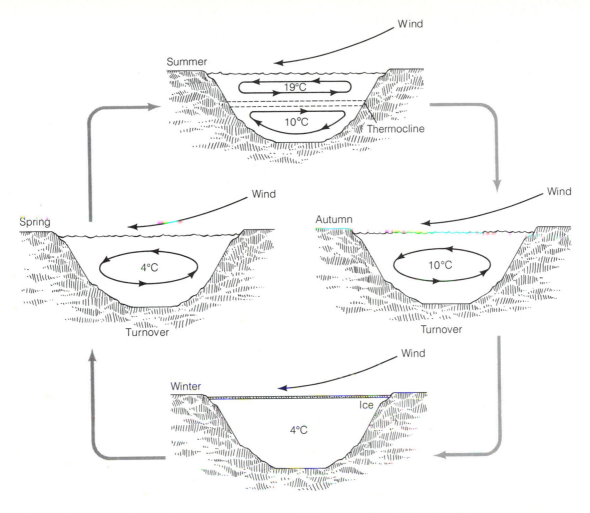

Figure 5.24 Temperature profiles of a large lake in a temperate region. The profiles vary with the season; periods of turnover in spring and autumn replenish the dissolved oxygen of deeper waters.

in many lakes is the gradual filling-in of the lake basin. Upon entering a lake, streams and rivers deposit much of their sediment load. Hence, particularly muddy rivers can fill a lake basin rapidly. Sediment layers are also built up by the accumulation of plant remains. Plant debris is particularly abundant in aged lakes where nutrient-rich waters spur luxuriant growth. This process of aging through nutrient enrichment is termed *eutrophication*. Because eu-

trophication has been greatly accelerated by human activity, this natural aging process has relevance to the overall quality of our aquatic environment, and we therefore examine the process in some detail here.

Lakes are classified according to their productivity, that is, the mass of living organisms they support. *Oligotrophic lakes* are characterized by slow rates of nutrient cycling and low productivity. An example of an oligotrophic lake is shown in Figure

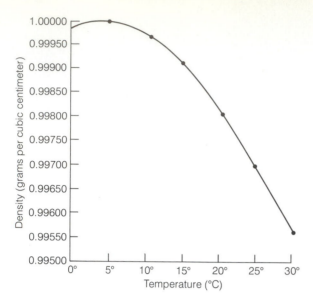

Figure 5.25 The effect of temperature on the density of water.

5.26. *Eutrophic lakes* have high rates of nutrient cycling, greater productivity, and large numbers of a relatively few species of aquatic organisms. Oligotrophic and eutrophic lakes, whose characteristics are summarized in Table 5.2, represent the extremes of a continuous range in productivity in lakes.

The quantity of living organisms that a lake can support depends upon the rate at which limiting nutrients are cycled. Sources of nutrients are the land which drains into the lake, the bottom sediments, and organisms within a lake. Some lakes are eutrophic from the time of their formation because of the influx of high concentrations of nutrients from their drainage basins. Most lakes, however, were originally oligotrophic, slowly becoming more eutrophic over many thousands of years.

Eutrophication proceeds as larger and denser growths of aquatic plants develop when greater quantities of mineral nutrients become available. In shallow near-shore waters, rooted aquatic plants grow luxuriantly. With the onset of warm weather, a sudden increase in algal populations may occur, causing the water to take on the appearance of pea soup or green paint within a few days. These blooms are usually composed of only a few species of algae—most of them members of the blue-green group of algae. These algae usually are too large to be eaten by zooplankton. Hence, blue-green algae form the base of a dead-end food chain, that is, a food chain that ends at the first trophic level.

Photosynthesis by algae adds oxygen to surface waters. But after a bloom develops the algae die, and most settle to the bottom where they are acted upon by decomposers. With such a large food supply available, bacteria (decomposers) multiply and consume large amounts of dissolved oxygen. As a result, the hypolimnion of highly eutrophic lakes becomes depleted of dissolved oxygen during the summer. When this happens, the less tolerant cold-water fish die.

As large masses of rooted aquatic plants and algae die, they contribute to the gradual buildup of the bottom sediments. The lake gradually fills in, and cold-water fish die-off while warm-water fish, such as perch, bluegills, bass, and pike, predominate. In addition, greater portions of the ever shallower lake are invaded by aquatic plants that take root in the shallow water (cattails, bulrushes, and water lilies are examples). Eventually the entire lake becomes shallow and covered with rooted aquatic plants. Thus the lake has become a marsh.

Depending on the size and depth of a lake and the nature of the soils in its drainage basin, the successional process from a lake to a marsh usually takes thousands of years. As noted earlier, however, human activities have greatly accelerated this aging process in many lakes. Accelerated soil erosion and the dumping of wastes rich in plant nutrients have speeded up the filling-in process. This accelerated aging is termed *cultural eutrophication*. It is discussed in more detail in Chapter 7.

Marshes and Swamps. At a later stage in its successional history, a lake or a bay becomes a marsh (a treeless form of wetlands dominated by grasses and sedges) or a swamp (a form of wetland domi-

Figure 5.26 A high altitude lake, characteristically oligotrophic, that is, having high water quality and low productivity. (U.S. Department of the Interior, National Park Service.)

Table 5.2 General characteristics of oligotrophic and eutrophic lakes.

Characteristic	Oligotrophic	Eutrophic
Rate of nutrient cycling	Low	High
Production by aquatic producers	Low	High
Production of animals	Low	High
Percentage of trash fish	Low	Higher (may dominate)
Oxygen in hypolimnion	Present	Often absent
Depth	Usually deeper	Tend to be shallower
Water quality for drinking and industry	Good	Poor
Number of species of aquatic organisms	Many	Fewer

nated by shrubs or trees). Examples include the Dismal Swamp of Virginia and North Carolina and the Everglades of Florida (a marsh). Freshwater marshes are especially widespread in the formerly glaciated lowlands of North America. Marshes and swamps act as sponges in holding copious volumes of water; thus, they serve to regulate streams that flow through them, absorbing waters at times of flooding and supplying water at times of low river discharge. Because nutrients and detritus tend to accumulate in these wetlands, they are very productive and support a wide variety of wildlife, including water birds and small fish.

Wetlands are disappearing in many regions due to human intervention. If artificially drained by ditching, some regions yield mucky soils that are highly productive for intensive cash-crop agriculture. Figure 5.27 shows a wetland under cultivation. In other instances, marshes have been filled in and used for industrial parks and shopping centers.

Rivers. In contrast to lakes, rivers are relatively shallow and more turbulent; thus, they expose a greater surface area per volume of water to air. Consequently, under natural conditions, the amount of dissolved oxygen varies little through the length and depth of a river. Unless large amounts of organic matter enter a river, dissolved oxygen is usually not a limiting factor. But a river's current undergoes significant variations in velocity, and is limiting for many organisms. For example, near a river's source, the channel is narrow and water flow is swift, and rapids and waterfalls are common. Aquatic plants and animals in this sector of a river are equipped with special adaptations that allow them to survive turbulent waters by anchoring or attaching themselves to the rocky channel bottom. Green algae (the principal producers in streams), mosses, and freshwater sponges grow on bottom rocks. Other organisms, such as tadpoles, hold on with suction devices or, as in the case of snails, with sticky undersurfaces.

Downstream, rapids disappear, and the current gradually slows as the river broadens. A soft sediment layer builds up on the channel floor, providing a favorable habitat for bottom feeders and burrowing animals. Such fish as bass, yellow perch, and catfish thrive in these quieter waters. Also, minute floating plants (usually algae and duckweed) that cannot survive the turbulence of river rapids may be abundant in this portion of a river.

Algae and rooted plants native to rivers generally are not present in quantities sufficient to meet the demands of river consumers. Hence, by necessity a river ecosystem is open. That is, most consumers are detritus feeders and rely upon the input of decaying organic matter from adjacent land areas or lakes that feed into streams. Runoff also contributes nutrients needed to sustain plant productivity.

Rivers have long been the traditional dumps of our society. We have always behaved as if our waterways would dilute all the wastes without affecting our well-being. But water's power of dilution is limited, and hundreds of kilometers of waterways are now essentially devoid of desirable aquatic life. For example, trout or bass are being replaced in many rivers by trash fish such as carp and gar. Although society is now much more aware of river degradation, the continual growth of urban-industrial complexes along rivers no doubt will continue to stress these aquatic ecosystems.

Aquatic Ecotones

Estuaries. An *estuary* is a unique environment created by the mixing of fresh and salt water. River mouths, tidal marshes, and embayments are examples. In an estuary water is subjected to daily tidal oscillations. Hence, organisms residing in estuaries must have wide tolerances, for they are subject to frequent changes in temperature, salinity (dissolved salts), and concentrations of suspended sediment.

Due to a unique combination of physical features, estuaries are among the most productive ecosystems on earth. Many estuaries are at the endpoint of rivers and therefore receive waters rich in nutrients that promote plant growth. River waters also deliver

Figure 5.27 Intensive agriculture on drained, wetland soils immediately south of Lake Isotokpoga, Florida. Note the fringe of cypress trees along the margin of the lake. (U.S. Department of Agriculture, Soil Conservation Service.)

large quantities of organic matter. Within the estuary, patterns of water circulation, such as the one illustrated in Figure 5.28, serve to retain and recirculate detritus and other nutrients. Because estuaries are relatively shallow, sunlight usually penetrates to the bottom; thus, the entire depth of the estuary is in the euphotic zone.

This combination of physical features supports luxuriant plant growth (both phytoplankton and sea and marsh grasses) and large populations of detritus feeders. Examples of animals that occupy lower trophic levels include oysters, mussels, crabs, horseshoe crabs, shrimp, and snails. These consumers in turn constitute a copious food source for higher trophic members, particularly fish and birds. In fact, about two-thirds of the rich commercial fishing off the east coast of the United States consists of species that spend part of their life cycle in estuaries. These include salmon, shrimp, and menhaden. The unique environmental conditions of the estuary favor devel-

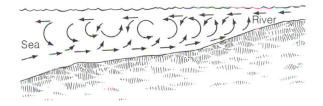

Figure 5.28 Circulation in an estuary mixing lighter fresh water with denser sea water to form a nutrient trap. (After E. Odum, *Fundamentals of Ecology,* 3rd. ed. Philadelphia: W. B. Saunders, 1971.)

opment and protection of the young: food is abundant, and low salinity and other physical characteristics serve as a barrier to many ocean predators.

The high productivity of estuaries illustrates the advantages of natural energy subsidies (see Box 2.2). The energy of water flowing downhill transports nutrients and detritus to estuaries from land surfaces. Waves and currents retain detritus and other nutrients in the estuary and keep the water stirred up, thereby cycling these materials to plants and detritus feeders. Ironically, even with our intensive uses of fossil fuel subsidies in modern agriculture, agricultural productivity has not been able to exceed (at least over an extensive area) the productivity resulting from the natural energy subsidies of estuaries.

As with freshwater wetlands, over 75 percent of estuaries have been moderately to severely modified by human activity. Leading causes of damage include cultural eutrophication, dredging to maintain navigation channels, and filling in to make way for industrial parks, shopping centers, and marinas. Figure 5.29 gives dramatic evidence of the encroachment of developments into these areas. This subject is discussed in greater detail in Chapter 15.

Deltas. Where a river enters an ocean (or a lake) its velocity decreases, and thus it deposits part of its sediment load at this point. If sediment supply is sufficient and wave activity is limited, enough sediment may be deposited to create a land surface that is slightly above the surrounding water. As long as the rate of deposition exceeds the rate of wave erosion, the land will expand seaward in a pattern that resembles the Greek letter *delta*, as you can see clearly in Figure 5.30. Nutrients are continually renewed on a delta with each increment of nutrient-bearing sediment that is spread over the lands surrounding the river mouth by flood waters. Hence, deltaic lands are extremely fertile and are often sites of intensive agriculture. But deltas are also delicate systems that are particularly vulnerable to human activity. A delta relies upon an uninterrupted supply of river sediment for its existence and if this is cut off or signifi-

Figure 5.29 The encroachment of a housing tract into marshes near Atlantic City, New Jersey. Remnants of the estuary are present in the upper right-hand corner. (U.S. Department of the Interior, Fish and Wildlife Service.)

Marine Ecosystems

Beyond the estuaries, we find saltwater, or *marine*, habitats. Although enormous, the sea is a surprisingly uniform habitat compared with the ones on land. However, productivity differs significantly from one portion of the ocean to another, as indicated in Table 5.3. These differences stem primarily from two limiting factors—light penetration of water and the availability of nutrients. Seaward from the coast, three distinctive marine habitat zones are distinguished: the intertidal, neritic, and oceanic zones, shown in Figure 5.31.

Intertidal Zone. The *intertidal zone* includes the shoreline between high and low tides; thus, land in this zone is alternately exposed to water and air daily with the tidal cycle. Although relatively rich in organic matter and dissolved oxygen, water in the intertidal zone is in a state of continual motion, and life forms are limited to burrowing organisms such as clams, crabs, snails, and worms on sandy shores and attached organisms such as green algae, brown algae, red algae, barnacles, mussels, and oysters on rocky shores. Some intertidal organisms are shown in Figure 5.32.

Neritic Zone. The *neritic zone* extends from the shoreline to the edge of the continental shelf where the water depth is about 180 meters (600 feet). While

Figure 5.30 A satellite photograph of the Nile River and delta. (NASA.)

cantly reduced—for example, by a dam on a river—the delta is cut back by wave erosion and inundated by the advancing sea. Thus, the delta is claimed by the sea and lost for agricultural activity.

Table 5.3 Fish production of the oceans.

Area	Percent of ocean	Area (in square kilometers)	Average number of trophic levels (approximate)	Annual fish production (in millions of metric tons)
Open ocean	90.0	326,000,000	5	1.6
Coastal zone	9.9	36,000,000	3	120.0
Coastal upwelling areas	0.1	360,000	1.5	120.0
Total annual fish production				241.6

Source: After J. Ryther, "Photosynthesis and Fish Production in the Sea," *Science 166:*72–76, 1969.

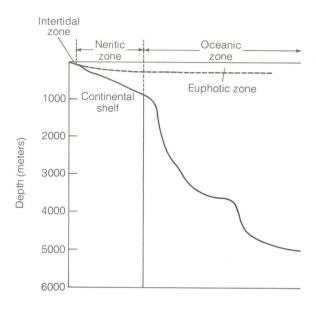

Figure 5.31 A cross-section of marine habitat zones.

this zone accounts for only about 10 percent of the ocean's surface, it is the principal site of commercial fish harvests. Sea life is abundant and very diverse throughout the depths of this zone. Together with estuaries, this coastal zone accounts for about 50 percent of the productivity of the sea. Because they are the terminus for land drainage, coastal waters are particularly rich in nutrients. Also, because the continental shelf is relatively shallow, a significant portion of the water lies in the euphotic zone. Hence, light and mineral nutrients are not significant limiting factors in the coastal zone.

Most of the remaining fish production in the sea takes place in *upwelling* areas. The major portion of upwelling fisheries are found along the west coasts of continents at the outer edge of the neritic zone. Upwellings are the result of the drag of prevailing winds on surface waters. Winds have the effect of transporting the surface layer of water offshore, away from the continents, thereby allowing nutri-

Figure 5.32 Intertidal organisms, adapted to alternate exposure to air and water. *Left*: Mud crab and barnacles. (© Jeanne White, from National Audubon Society.) *Right*: Algae-covered rocks. (© Verna R. Johnston, from National Audubon Society.)

ent-rich bottom waters to rise (well up) to the surface. The nutrient enrichment of surface waters spurs algal productivity, which in turn enhances fish productivity. Although upwelling regions make up only 1 percent of the ocean's surface, they account for almost 50 percent of the productivity in the sea, as you can see in Figure 5.33, which shows the distribution of the world's fisheries.

Some of the major food fishes found in or near the neritic zone are Peruvian anchovy, Atlantic cod, mackerel, Atlantic herring, Alaskan walleye, pollack, halibut, flounder, and salmon. These fish support huge concentrations of birds that nest near shore but feed from the sea. Common types of sea birds include comorants, petrels, shearwaters, and sea ducks.

Oceanic Zone. The *oceanic zone* covers the deep water that stretches beyond the edge of the continental shelf. Although enormous in area, open oceans are quite unproductive. Simply put, the raw materials for photosynthesis are not present together in sufficient quantities. While there is ample light near the surface, nutrients are limiting in the euphotic zone. Hence, open oceans are biological deserts: their productivity is comparable to terrestrial deserts.

Although desertlike, the open seas do have scattered oases of abundant sea life. Phytoplankton form the base of food webs in the open oceans, where trillions of tons of these microscopic plants are produced and widely dispersed. A heterogeneous group of small animals called zooplankton graze on the phytoplankton. The herbivorous zooplankton are then preyed on by carnivorous zooplankton, which are then fed upon by small fishes and squids; these in turn fall prey to a wide range of larger carnivores. The carnivores, such as tuna, dolphins, porpoises, swordfish, and sharks, are some of the largest and most superbly designed animals ever to inhabit the earth. The scattered nature of food sources accounts for the powerful swimming ability of these carnivores. Their streamlined, torpedolike shape enhances speed and efficiency of movement. The coloring of many marine organisms is quite similar and aids as a camouflage when seen from above, their dark blue-grey backs blending in with the color of the water. To predators below, their white undersides are very close to the silvery appearance of the surface. Denizens of the open oceans have developed such remarkable adaptations as sonar, extreme olfactory sensitivity, and complex senses of orientation and homing.

The sea is one of our few remaining frontiers, and until recently the impact of human activity there was relatively minor. Today, however, things are changing. Overfishing has greatly reduced the stocks of several kinds of fish, including the anchovy, the California sardine, and the Norwegian herring. Oil spills from tankers and drilling rigs and the dumping of domestic and industrial wastes threaten marine life. Many of our reserve deposits of petroleum and minerals are on the continental shelves, and our increasing exploitation of these resources is sure to have serious effects on our coastal waters.

As we have seen many times with respect to terrestrial ecosystems, marine species are interdependent, and their relationships can be dramatically disturbed by human activity. A striking example from the California coast concerns the sea otter, which preys on sea urchins that eat kelp (a giant brown algae). When sea otters were all but exterminated by fur traders in the nineteenth century, the sea urchin populations exploded and reduced the extensive kelp beds to scanty patches. Once their food source was reduced, the sea urchins began to starve. Subsequently, the reduction of sea urchins allowed the kelp to make a comeback, and the cycle began again. But because of the decimation of the sea otters, the kelp beds did not regain their former abundance.

The system was further disturbed by the increased loads of sewage being poured into the ocean. The sewage feeds sea urchins, and with the added food source, the sea urchin population exploded even though the kelp beds were greatly reduced. As a result, the sea urchins "finished off" the kelp beds in local areas where sewage was present in sufficient

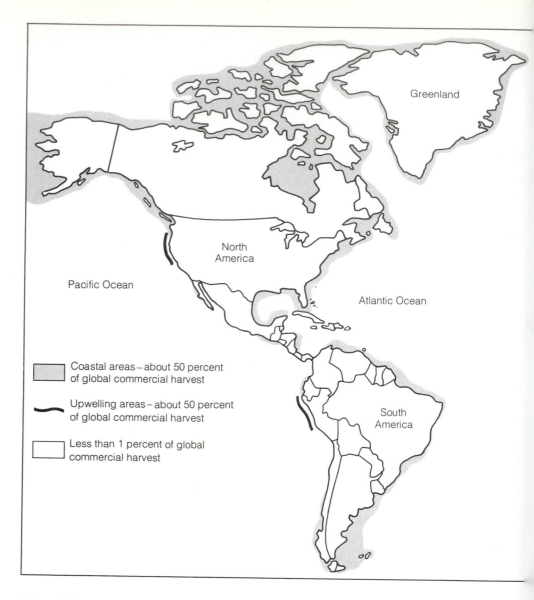

Figure 5.33 The distribution of the world's fisheries. (After *Patterns and Perspectives in Environmental Science.* Washington, D.C.: National Science Foundation, 1972.)

Coastal areas – about 50 percent of global commercial harvest

Upwelling areas – about 50 percent of global commercial harvest

Less than 1 percent of global commercial harvest

quantities to increase the sea urchin population. Because of human intervention, areas that once supported luxuriant kelp forests were reduced to biological deserts.

With legal protection, sea otter populations are slowly making a comeback. But controversy still reigns. Because sea otters also eat abalone, abalone fishermen do not want the sea otter to be protected. They forget that sea otters, kelp, sea urchins, and abalone successfully cohabited the Pacific coastline for thousands of years before humans intruded. But judging from the results of intervention, it is unlikely

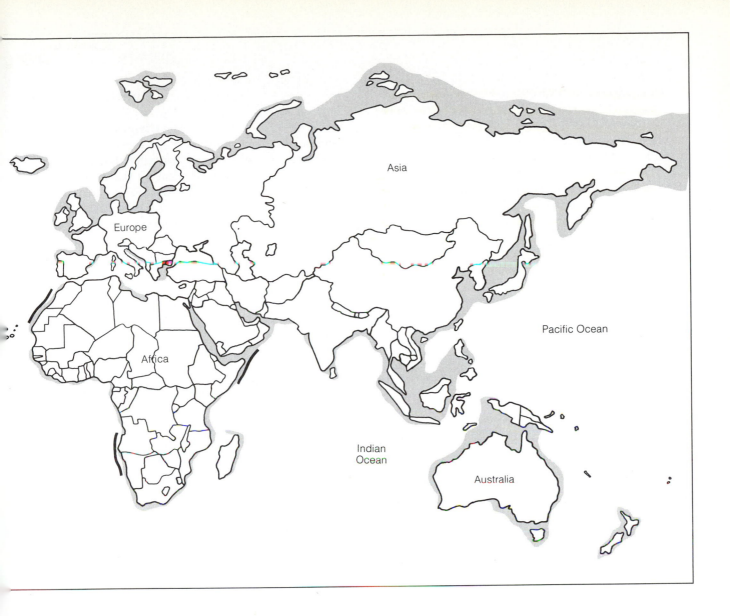

that the former balance among these four marine organisms will ever be regained.

Linkages Between Land and Water

Although terrestrial and aquatic ecosystems differ significantly, many important linkages exist between them. Aquatic ecosystems are enriched by the movement of nutrients such as phosphorus and nitrogen from the land. Decaying organic matter washed in from adjacent land areas serves as a significant food source for aquatic consumers in rivers and estuaries. Soil erosion contributes to the buildup of bottom sediments in lakes and slow-moving rivers. The rates of flow of nutrients, detritus, and soil particles are significantly influenced by human activities. The more the landscape is disturbed, the greater is the movement from land to waterways.

The flow, however, is not restricted to just one direction. Aquatic life, mainly fish and shellfish, serve as food sources for animals that primarily live on land. For example, large concentrations of sea birds feed on fish but nest on land near the shore. In addition, the wastes deposited by these birds on land help cycle nutrients such as phosphorus and nitrogen from the sea to the land.

Hence, these two diverse types of ecosystems are closely linked. Any change that occurs in one would be expected to produce changes in the other.

Conclusions

In our brief survey of ecosystems, we have caught only a glimpse of the rich diversity of life present on this small planet. But we have seen enough to know that human activities threaten to decimate this rich diversity and thereby seriously disrupt the roles played by organisms in the functioning of ecosystems. It is critical that we as a species begin to assess more accurately the degree to which our enterprises disrupt the earth's ecosystems. We must take into account that these disruptions ultimately have serious consequences for the quality of human life. How much deterioration are we willing to tolerate? How much, after all, can we tolerate and still survive?

Summary Statements

The terrestrial environment is a severe, often hostile place in which to live. Moisture and temperature are the two most common limiting factors. Soil helps to govern plant productivity and the number of consumers that an ecosystem can support.

As a consequence of such factors as climate, soil, topography, and human and nonhuman disturbance, the earth's land masses are covered by a complex mosaic of biotic communities.

The tundra has a low diversity of organisms, among the lowest plant productivity of the terrestrial ecosystems, and is readily disrupted. Long ignored, the tundra is now rapidly being exploited for its wealth of minerals and fossil fuels.

The northern coniferous forest lies south of the tundra. It has a low species diversity and is dominated by evergreens.

The eastern deciduous forest occupies the eastern half of the United States. Its diversity and productivity are high. This biome has been extensively modified by human activity.

The grasslands lie west of the eastern deciduous forest, where water becomes a limiting factor for trees. Grassland soils are highly productive. Cattle and sheep grazing and agriculture have destroyed essentially all of the virgin grasslands in the United States.

Deserts are limited by sparse, infrequent, and sporadic precipitation. Desert organisms possess many adaptations to survive in this harsh environment. This fragile ecosystem is becoming increasingly disrupted as more people are attracted to the "sun belt" in the Southwest.

Temperate rain forests lie along the northwest coast while the chaparral biome occurs along the southwest coast of North America. The temperate rain forest is extensively exploited for timber, and the chapparal has taken the brunt of urban sprawl in heavily populated southern California.

The biomes found in North America are also present on other continents. Two important tropical biomes are rain forests and savannas. Both contain a vast diversity of organisms and are falling rapidly before the pressures of growing human populations.

Steep temperature and precipitation gradients on mountainsides produce zonations of ecosystems similar to those found by moving from south to north in North America. Ecotones are transition zones between two diverse types of ecosystems, and usually display a high diversity of organisms.

The concentration of dissolved oxygen and sunlight penetration of water are often limiting factors for aquatic organisms. A rapid and marked change in water temperature can also put severe stress on aquatic organisms.

Lakes in temperate latitudes become stratified into two layers during summer. Stratification prevents the exchange of dissolved oxygen and mineral nutrients between the epilimnion and hypolimnion. Mixing occurs during spring and fall turnover.

Lakes age through nutrient enrichment. Natural aging processes are accelerated by human activities. Oligotrophic and eutrophic lakes represent extremes in lake productivity.

Marshes act as reservoirs for nutrients and water and help regulate stream flow. Many marshes have been drained for agriculture and filled in for commercial development.

Under natural conditions the amount of dissolved oxygen varies little throughout a river. Because the quantities of algae and rooted plants growing in a river are insufficient to meet the demands of river consumers, these animals feed upon the decaying organic matter that is washed in from adjacent land areas or lakes. Rivers have long been recipients of the wastes of our society.

Estuaries occur in coastal areas where fresh and salt water mix. Because of natural energy subsidies and a unique combination of physical features, estuaries are among the most productive ecosystems on earth. Estuaries have been significantly modified by human activity.

Productivity in oceans depends primarily on light penetration of water and nutrient availability. Coastal waters and areas of upwelling account for over 99 percent of the oceans' fish production. Although enormous in area, open oceans are biological deserts. The oceans constitute one of our few remaining frontiers, and they will increasingly bear the brunt of increased exploitation of fish and mineral, particularly petroleum, resources.

Questions and Projects

1. In your own words, write a definition for each of the terms italicized in this chapter. Compare your definitions with those in the text.

2. Describe the characteristics of the three horizons commonly present in a soil profile.

3. Compare and contrast cold and hot deserts.

4. What are the factors that account for the great differences in plant productivity between tundra and tropical rain forest ecosystems?

5. Compare altitudinal ecological zones with latitudinal ecological zones.

6. How might a climatic change influence your well-being and that of your community if you lived in a ecotonal area?

7. If you had to live in and obtain all of your resources from just one type of ecosystem without modifying it, which would be your choice from the following list: tundra, evergreen forest, deciduous forest, grassland, desert, temperate rain forest, chapparal, tropical rain forest, savanna? Defend your choice.

8. Ecosystems that people generally ignored for a long time because of their hostile environment (deserts and tundra, for example) are actually quite vulnerable to human activities. Discuss the reasons for this vulnerability.

9. Whether an organism resides in a terrestrial or aquatic habitat, its physical environment can be quite hostile at times. Compare and contrast the advantages and disadvantages for animals living in these two distinct environments.

10. What is the importance of spring turnover and fall turnover for aquatic life?

11. Describe how natural energy subsidies contribute to the productivity of estuaries.

12. Why are different plants and animals found in lakes, swift rivers, slow rivers, and estuaries? How are these differences related to the concept of limiting factors?

13. Why is the open ocean sometimes referred to as a biological desert?

14. Which ecosystems have the greatest species diversity? Which have the least?

15. Although all organisms require essentially the same types of substances from their environment to survive, certain substances are plentiful in some ecosystems and rare in others. Give examples of ecosystems where each of the following is plentiful and where it is rare: light, water, oxygen, soil nutrients such as nitrate and phosphate, and space.

16. Although often described as separate entities, different types of ecosystems are actually quite interdependent. In terms of the flow of energy and nutrients, describe the interdependence of the plants and animals in a deciduous forest, in a river that flows through the forest, and in an estuary at the mouth of the river. How would these relations be affected if the forest were cleared for agriculture or if a large industrial city were located on the river?

17. Discuss the relation between the productivity of an ecosystem and the degree to which it is exploited by human activity.

18. Which types of ecosystems do you believe will be most severely affected by human activities during the next twenty years? Defend your choice.

19. What types of ecosystems are present in your region? What are the important environmental factors responsible for the geographical pattern of ecosystems that you find? You may wish to arrange a field trip.

20. How have ecosystems in your region been affected by human activity? Which types of ecosystems have experienced the most severe impact? You may wish to arrange a field trip.

21. If available, visit an arboretum, botanical garden, or nature preserve. What special measures must be taken to maintain the communities being preserved there?

Selected Readings

Cloudsley-Thompson, J. L., 1975. *Terrestrial Environments*. New York: Halsted Press. An overview of the many biomes on earth. Describes the adaptations of organisms to the particular physical characteristics of the biome, concentrating on animals.

McCormick, J. 1966. *The Life of the Forest*. New York: McGraw-Hill. A well-written and beautifully illustrated account of the forest community. Describes how it changes with the seasons, and identifies the many types of forests in North America.

Niering, W. A. 1966. *The Life of the Marsh: The North American Wetlands*. New York: McGraw-Hill. A nicely written and illustrated account of activities of plants and animals in American marshes.

Richards, P. W. 1973. "The Tropical Rainforest," *Scientific American 229*: 57–68 (December). An examination of the ecological significance of one of the earth's oldest ecosystems and how it is threatened by human activities.

Scientific American. 1969. "The Ocean." *Scientific American 221*: 3. An interesting and informative look at the physical and biological resources of the oceans and human efforts to use these resources.

Smith, R. L. 1979. *Ecology and Field Biology*, 3rd edition. New York: Harper & Row. A basic, well-written ecology text containing several chapters surveying aquatic and terrestrial habitats.

Sutton, A., and M. Sutton. 1966. *The Life of the Desert*. New York: McGraw-Hill. A well-illustrated account of day- and nighttime activities of desert animals. Examines how organisms are adapted for desert survival and surveys the kinds of deserts that exist in North America.

Walter, H. 1973. *Vegetation of the Earth*. New York: Springer-Verlag. A description of the vegetation, climate, and soils found in the earth's biomes. Discusses adaptations of plants to climate and soil characteristics within each biome.

Part II

Environmental Quality and Management

In the previous section, we examined some fundamental principles governing natural processes. From our discussion, it is evident that both we as a species and our environment itself are governed by certain natural laws. For example, our surroundings have a limited capacity to absorb waste; when this limit is exceeded natural functions are impaired or even destroyed. Likewise, when more foreign matter, in the form of pollutants, enter our bodies our health may suffer.

In this section, we examine in detail the natural processes that occur in the major components of our physical environment: the waterways, the atmosphere, and the land. Then, we study how our own activities disrupt these natural processes. Finally, we discuss management techniques, including feasible alternatives to those now being practiced, and assess our progress toward restoring degraded areas and preserving environmental quality in relatively undisturbed regions.

Snow-melt feeding a mountain stream, a visible link in the hydrologic cycle.
(U.S. Department of the Interior, Bureau of Reclamation.)

Chapter 6

The Water Cycle

Of the multitude of substances on earth, water is one of the most important for life. Water accounts for 65 percent of our body weight, and serves as the medium for a vast array of life-sustaining processes. And water's role is no less crucial for society as a whole than for the well-being of each individual. Its functions in transportation, power generation, food production and processing, manufacturing, and waste treatment are absolutely basic—these activities simply could not take place without water.

However, because it is essential to us in so many ways, we sometimes naively expect water to exceed its natural capabilities. For example, we expect the water we use to cleanse itself miraculously of added wastes. Although an aquatic ecosystem can rid itself of some wastes if given sufficient time, many substances—certain pesticides, for example—do not break down once they enter an aquatic environment. The retention of these pollutants in our water supply has implications not only for us, but also for the many organisms that depend on an uncontaminated aquatic environment for their survival. Even this concern for other species is human-centered, however, for many aquatic organisms are invaluable food

sources for millions of people. In addition, particular organisms in uncontaminated aquatic detritus food webs provide a free water-cleansing service. But in a polluted environment, many of these and other aquatic organisms perish. The result is a build-up of contaminants and an increase in the numbers of less desirable organisms.

Our concerns about water focus not only on quality, but also on the possibility that our supply could diminish as our population continues to grow. We can deduce what the effects of a long-term shortage would be by recalling the hardships endured in a short-term drought—for example, the California drought in the mid-1970s, in which residents escaped water deprivation by changing their water use practices considerably. Figure 6.1 illustrates the alarming drop in water level in a California reservoir during that drought. Residents in the East, too, can recall the water shortages they suffered when reservoir levels remained abnormally low for several years in a row during the late 1960s.

In this chapter, the first of three addressing the topic of water, we examine the limiting factors affecting our supply of fresh water and survey the

Figure 6.1 Clair Engle Lake, a water reservoir for California's Central Valley, filled to approximately one-tenth its storage capacity during the drought of 1976–1977. Note the extensive exposed shoreline. (U.S. Department of the Interior, Bureau of Reclamation.)

essential links in our natural water delivery system. In the following chapters we concentrate on water quality: Chapter 7 covers water pollution, and Chapter 8, water management.

A Historical Overview

A quick perusal of a world atlas shows that most major areas of human settlement are adjacent to rivers, lakes, or oceans. This settlement pattern is not coincidental; it demonstrates the utter dependency of civilizations on adequate water supplies. A quick historical survey will reveal water's role in the development of civilization and the evolution of modern society.

As agriculture developed, people discovered the advantages of irrigation. The excess food grown on irrigated land eventually led to trade among tribes, and these first commercial transactions were the origins of the early civilizations: Mesopotamia, between the Tigris and Euphrates rivers, Egypt, in the Nile Valley, and the Indus culture (Pakistan), along the Indus River. As civilizations developed and cities came into being, the notion of a communal water supply grew more sophisticated. Public water was first drawn from nearby rivers or shallow wells. Later, governments financed the construction of elaborate aqueduct systems to transport water from remote sources to the growing cities. The most famous example known to us, of course, is the system of aqueducts constructed during the Roman Empire.

Figure 6.2 A segment of the Roman aqueduct system at Segovia, Spain, still in use today. The aqueduct, built of granite blocks, is approximately 1 kilometer (2700 feet) long. (Kenneth E. Maxwell, California State University, Long Beach.)

A few of these aqueducts are still in use today, as shown in Figure 6.2.

As civilization developed, transportation by water proved to be less expensive than overland transportation. Ships have a greater capacity for transporting goods than do camel or horse caravans. Because many European rivers were navigable by ship, European civilizations gained supremacy over the Mediterranean region. Not only did they have better routes over which to move their products, but they had easy access to water power—water wheels powered saw mills, iron forges, and flour mills—enabling them to process more raw materials.

The harnessing of water power marked the beginning of the Industrial Revolution. Industrialization received an added impetus when pressurized steam

was recognized to be a power source. Subsequently, the invention of the versatile steam engine revolutionized manufacturing and transportation. Although only a few steam engines are still in operation today, the successor to this remarkable engine, the steam turbine, is used in most electric power generating plants worldwide. With the advent of electricity, large volumes of water were needed as a coolant in power generating plants; thus, even when the source of power shifted to nuclear and fossil fuels, the role of water was not diminished. From primitive times to the present, water has been the moving force behind civilization.

With the great technological strides made during the Industrial Revolution, water came to be required for many new manufacturing processes. Paper mak-

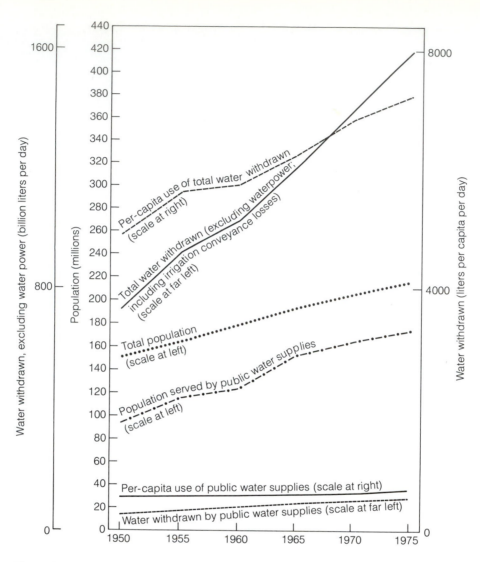

Figure 6.3 Trends in water withdrawal in the United States. (After C. R. Murray and E. B. Reeves, *Estimated Use of Water in the United States in 1975,* U.S. Geological Survey Circular 765, 1977.)

ing, steel manufacturing, petroleum exploitation and refining, food canning, and mining operations are just a few of the new processes that required—and still require—sizable quantities of water. Clearly, then, we are as dependent on a readily accessible water supply as were our primitive ancestors. In fact, our reliance on an adequate water supply has increased phenomenally, mostly during the past hundred years.

The statistics on present-day water use presented in Figure 6.3 indicate the extent to which modern society relies on water. Our bodies need 2 liters of water a day to prevent dehydration, but each of us uses 700 liters a day in and around the home. The quantity of water used by industries overshadows the amount supplied to cities and municipalities. These industrial water withdrawals, in turn, appear small when compared with the volumes of water flowing through agricultural irrigation systems. Each day the irrigation of crops in the United States requires the equivalent of 6000 liters (1600 gallons) of water per person. But, again, we should remember that this supply of water is not guaranteed. In the past, persistent drought contributed to the decline of civilizations. The Mycenaean civilization in Greece (1200 B.C.), the Harrappan civilization in India (1700 B.C.), and the American Indian Mill Creek culture (1400 A.D.), in what is now Iowa, all succumbed to drought. Even our recent experiences have taught us that it is a mistake to think of our water supply as an unlimited resource.

The Hydrologic Cycle

The total volume of water on earth has remained virtually constant over the past several million years. A very small amount of water vapor is lost yearly from the upper atmosphere, but the loss is compensated for by the addition of water vapor through volcanic emissions. Water is found on earth in solid, liquid, and vapor forms, and the total is distributed among oceanic, atmospheric, and terrestrial reservoirs. Table 6.1 shows the distribution of the earth's water reservoirs. Oceans make up the largest reservoir; and most of the remaining water is tied up as ice in polar and mountain glaciers. Comparatively speaking, the amount of water contained in lakes, rivers, groundwater, organisms, and the atmosphere is minute. Still, the water we use has its origins in this small fraction. Fortunately for us, the sources on which we rely for our water supplies are continually renewed.

Water is ceaselessly transferred among these reservoirs, and the patterns of the water's movement is called the *hydrologic cycle*. This cycle is illustrated in Figure 6.4. Basically, the pathways are simple to follow; water is taken up from the sea and land to form clouds, rain and snow fall from the clouds to earth to supply rivers, and the rivers flow back to the sea. The endlessness of the water cycle is expressed in a verse from Ecclesiastes: "All the rivers run into the sea, yet the sea is not full."

Air-Water Interaction

Solar energy sustains the functioning of the hydrologic cycle and links the atmospheric water reservoir

Table 6.1 The distribution of the earth's water reservoirs.

Mode of storage	Locations	Water volume (in cubic meters)	Percentage of total water
Surface	Fresh-water lakes	1.3×10^{14}	0.006
	Saline lakes and inland seas	1.0×10^{14}	0.008
	Average in stream channels	1.3×10^{12}	0.0001
Subsurface	Water in zone of aeration	6.7×10^{13}	0.005
	Groundwater within depth of half a mile	4.2×10^{15}	0.31
	Groundwater at great depths	4.2×10^{15}	0.31
Other	Icecaps and glaciers	2.9×10^{16}	2.15
	Atmosphere (at sea level)	1.3×10^{13}	0.001
	World ocean	1.3×10^{18}	97.2
Totals		1.4×10^{18}	100

Source: Data from U.S. Geological Survey.

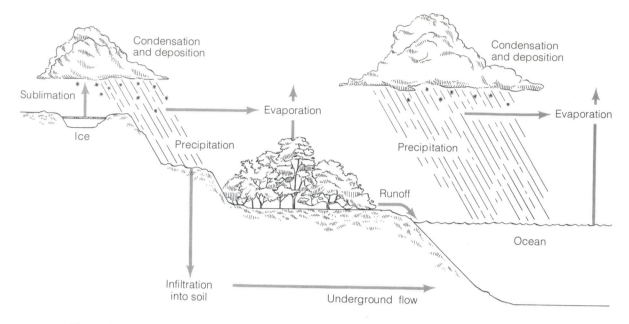

Figure 6.4 The hydrologic cycle.

with all the other water reservoirs. Energy derived from the sun moves water from the ocean and land to the atmosphere by means of two processes: *evaporation* and *sublimation*. In evaporation, water is changed from a liquid to a vapor at a temperature below water's boiling point. In sublimation, water changes from a solid into a vapor without passing through an intervening liquid state. The gradual disappearance of snow banks, even when the air temperature remains well below the freezing point, is an example of sublimation. As water evaporates or sublimes, it passes from a lower to a higher energy level, and therefore an input of energy is required. This input is received by the water via the sun.

Water is returned to the land and sea from the atmospheric reservoir by the processes of *condensation* and *deposition*. In condensation water changes from a vapor into a liquid (in the form of droplets). In deposition water changes directly from vapor into a solid (ice crystals). Figure 6.5 shows a common end-product of deposition, hoarfrost. The water droplets and ice crystals produced by condensation and deposition form clouds. In both of these processes, water passes from a higher to a lower energy state; energy is released by these processes. *Precipitation* processes—rain, snow, ice pellets, and hail—return a major portion of atmospheric water from clouds to the earth's surface.

The hydrologic cycle, then, involves a continual transfer of water from one reservoir to another. Essential to the functioning of this cycle is the capability of water to undergo change from a vapor to a liquid to a solid state. In other words, because water vaporizes (evaporates or sublimates), it moves from the land and sea to the atmosphere, and because it condenses, it moves from the atmosphere back to the land and sea as rain and snow. This evaporation-condensation sequence is termed *distillation*. It is a natural water purification process, for water that condenses in distillation is purer than in its original state, before the sequence began. As water vaporizes in the distilling process, suspended and soluble substances, including sea salts, are left behind. Through this cleansing mechanism, water from the sea eventually falls on land as freshwater precipitation, which replenishes terrestrial reservoirs. Distillation also

Figure 6.5 Hoarfrost formations, illustrating deposition. (M. L. Brisson.)

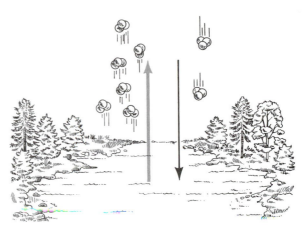

Figure 6.6 The evaporation of water, which occurs when more water molecules leave the water surface than enter.

cleanses water that has been contaminated by human activities.

The water exchange between earth and atmosphere depends both on the ability of water to change forms and the ability of air to contain water. Let us return to the concept of evaporation. A continual exchange of water molecules takes place at the interface between the atmosphere and a water body: some molecules leave the water surface and enter the air in the vapor state, while other water molecules leave the vapor state and return to the water surface as liquid. Evaporation occurs if more water enters the atmosphere than returns to the water body, as illustrated in Figure 6.6.

We measure the amount of water vapor in the air in terms of concentration. Water vapor concentrations may be expressed as the weight of water vapor per weight of dry air (grams of water per kilogram of dry air). When a given amount of air contains the maximum amount of water vapor it can hold, the air is said to be *saturated*. The amount of water vapor required to produce saturation depends on air temperature, as shown in Figure 6.7. The warmer the

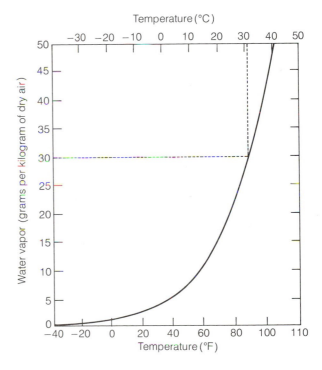

Figure 6.7 The weight of water vapor required to saturate 1 kilogram of dry air at increasing temperatures. The dashed line indicates that at 32°C, 1 kilogram of dry air requires 30 grams of water vapor for saturation.

The Water Cycle 157

air, the more moisture it can hold. Many of us are familiar with the drying effects of cold winter air and the mugginess of summer heat waves. And we have all heard the familiar saying, "It's not the heat, but the humidity that's hard to take." In actuality, heat (high temperatures) leads to a higher moisture content in the air and thus the discomfort we feel.

Citing the *relative humidity* is a more familiar way of indicating the water vapor content of air. At a particular temperature, the percent of relative humidity is determined by dividing the concentration of water vapor in an air parcel (a measurable volume of air) by the concentration of water vapor at saturation. That is,

Percent relative humidity

$$= \frac{\text{actual concentration in air}}{\text{saturation concentration}} \times 100$$

Suppose on a muggy summer afternoon the air temperature is 32°C and the water vapor concentration is 20 grams per kilogram. We can see in Figure 6.7 that the saturated concentration of air at 32°C is 30 grams per kilogram. Using our formula, we determine that the relative humidity is 67 percent. This means that the air was capable of holding 1.5 times as much water vapor before becoming saturated.

A change in air temperature causes relative humidity to change even though the quantity of water vapor present does not change. Let us assume that an air parcel above a lake is unsaturated (the usual situation), and that wind transports the air parcel upward. As the air rises, the temperature drops and the capacity of the air parcel to hold water vapor therefore diminishes. Thus, even though no water vapor is gained or lost, the relative humidity increases as the air cools. Eventually, the rising air parcel may cool sufficiently to reach saturation, that is, the relative humidity equals 100 percent. In the reverse process, descending air grows warmer, and its capacity to hold water vapor therefore increases, though the actual amount of water vapor present remains unchanged. In this situation, relative humidity is decreased.

As air parcels cool and approach saturation, water vapor tends to condense, or be deposited on *condensation nuclei*—tiny particles of matter that are always present in large numbers in the atmosphere. A typical human breath, for example, contains 50,000–500,000 such particles. These condensation nuclei are products of both natural and human activity. Forest fires, volcanic eruptions, wind erosion of soil, sea-salt spray, and effluents of domestic and industrial chimneys provide a continual supply. The most effective of these nuclei—*hygroscopic nuclei*—have a special affinity for water molecules, and condensation therefore occurs on these nuclei at relative humidities of less than 100 percent. In fact, magnesium chloride, a constituent of sea water, functions as a hygroscopic nucleus that initiates condensation at relative humidities as low as 70 percent. Cloud formation and attendant precipitation are the ultimate result of condensation, and thus occur more readily where hygroscopic nuclei are in large supply.

Precipitation Processes

Clouds in their myriad forms are the visible manifestations of condensation and deposition in the atmosphere. Cloud particles (water droplets and/or ice crystals) are very small; typically, their diameters are about half that of a human hair. The particles are so small and light they tend to remain in suspension. Hence, in order for the hydrologic cycle to continue, cloud particles must somehow grow heavy enough to fall toward the earth's surface as precipitation. Normally, the formation of a single raindrop requires about a million cloud droplets.

Both water droplets and ice crystals can be present in clouds at temperatures well below the freezing point (as low as −40°C or −40°F). Water droplets in a cloud that maintain their liquid state though the air temperature is below freezing are said to be *supercooled*. Both droplets and crystals are present in the *ice-crystal process* of precipitation formation, which accounts for most of the precipitation that falls to earth. In this process, the nuclei that form

Figure 6.8 A thunderstorm. Not all the rain leaving the cloud is reaching the ground, since the raindrops are falling through drier air. (National Center for Atmospheric Research.)

water droplets are far more abundant than those that form ice crystals, and therefore each ice crystal in a cold cloud is surrounded by hundreds of thousands of supercooled droplets. But the ice crystals grow rapidly at the expense of the water droplets, and as the ice crystals grow larger and thus heavier, they begin to descend at an accelerating rate. As the larger ice crystals fall, they intercept and capture water droplets and smaller ice crystals in their paths, thereby growing still larger. Eventually, the ice crystals are heavy enough to fall out of the cloud. If air temperatures are below freezing all the way to the ground, the crystals reach the earth's surface in the form of snow; if the air near the ground is above freezing, snowflakes melt and fall as rain.

It is important to note that the ice-crystal process occurs only in clouds whose temperatures are below the freezing point of water. Air temperatures in clouds that form in summer in the middle latitudes, and during the entire year in the tropics, are above

the freezing point. Hence, these clouds contain neither ice crystals nor supercooled water droplets. For precipitation to form in these clouds, relatively large hygroscopic sea-salt nuclei must be present. The presence of these nuclei results in the production of large droplets that fall quickly because of their weight. As they fall, the large droplets intercept and capture smaller water droplets in their paths, thereby growing even larger by a process known as the *coalescence process*.

Cloud conditions that favor the formation of raindrops and snowflakes also favor their preservation. But once a drop or a flake leaves a cloud, it enters a hostile environment in which either evaporation or sublimation begins to occur, drawing the water away from it. In general, the longer the raindrop's journey to the ground and the drier the air in its path, the greater is the amount of water returned to the atmosphere as vapor. In Figure 6.8, rain falling from a thunderstorm is entering drier air, resulting in the evaporation of some of the rain on descent. In hilly regions, this effect is partially responsible for the fact that higher land receives more rainfall than the surrounding lowlands.

Rainfall and Topography

As the result of interactions between air flow patterns and various topographic features, precipitation is enhanced in some localities and suppressed in others. For example, a mountain range lying perpendicular to the direction of the prevailing wind forms a natural barrier that results in heavier rainfall on one side than on the other, as Figure 6.9 shows. When air sweeps up the windward side of the mountain range, the air temperature falls, the relative humidity increases, and eventually precipitation—called *orographic rainfall,* when it is triggered by topographical features in this way—may develop. On the leeward side of the range, however, air descends and warms, causing the relative humidity to decrease. Hence, the mountain range establishes two contrasting climatic zones: moist on the windward side and dry on the leeward side.

This precipitation disparity is especially apparent from west to east across the states of Washington and Oregon, where the north-south oriented Cascade Mountain Range intercepts the prevailing moist air flow from off the Pacific. The result is that exceptionally rainy conditions prevail in the western portions of these states while arid conditions are characteristic in eastern portions. Figure 6.10 shows two contrasting landscapes from Oregon exhibiting this disparity. On Mount Waialeale, Hawaii, the contrast is spectacular. Annual rainfall varies from 1170 centimeters (460 inches) on one side of the mountain to only 46 centimeters (18 inches) on the other side. In every case, orographic rainfall results in markedly different plant and animal communities on each side of the mountain range. This effect also has direct effects on the domestic water supply, the type of crops that can be grown, and the type of shelter that must be built in surrounding areas. For example, the

Figure 6.9 The effect on rainfall distribution of air flow over a mountain range.

city of Denver, located on the leeward side of the front range of the Rocky Mountains, must obtain some of its water from the wet side by means of a 37-kilometer (23-mile) tunnel through the mountains.

The Hydrologic Budget

When water has been transferred from the atmosphere to the land and sea as precipitation, an essential subcycle in the hydrologic cycle has been completed. We can learn more about the cycle by comparing the movement of water into and out of the terrrestrial reservoirs with that into and out of the sea. The balance sheet for the inputs and outputs of water from the various reservoirs is termed the *hydrologic budget.*

The hydrologic budget, presented in Table 6.2, shows that each year total precipitation exceeds

Table 6.2 The hydrologic budget.

Source	Cubic meters/year (gallons/year)
Precipitation on sea	3.24×10^{14} (85.5×10^{15})
Evaporation from sea	3.60×10^{14} (95.2×10^{15})
Net loss from sea	0.36×10^{14} (-9.7×10^{15})
Precipitation on land	0.98×10^{14} (26.1×10^{15})
Evaporation from land	0.62×10^{14} (16.4×10^{15})
Net gain on land	0.36×10^{14} (9.7×10^{15})

Figure 6.10 *Left:* A rain forest in western Oregon's Silver Falls State Park. *Below:* A desert in eastern Oregon. A mountain range between the two areas accounts for the difference in precipitation. (Oregon Department of Transportation.)

evaporation on land surfaces, but that at sea, evaporation is greater than precipitation. As a consequence, the table seems to show an annual gain of water on land and a loss of water from the oceans, with the excess on land about equaling the ocean's deficit. The land, however, is not getting any soggier, nor are the world's oceans drying up. The explanation is that the excess precipitation on land drips, seeps, and flows from the land back to the sea. This fact is important in understanding the potential distribution of water pollutants, for if water in land reservoirs is contaminated, some of the pollutants ultimately reach the oceans.

The Earth's Water Reservoirs

On its journey from the land to the sea, water is often stored for varying periods of time in one of several kinds of reservoirs—in the ground, in rivers and lakes, or as glacial ice. All these reservoirs are potential sources of fresh water. So that we might clarify some of the impacts of exploitation, we now examine the characteristics and functions of these precious resources.

Groundwater

The various possible routes of precipitation that reaches the earth's surface are shown in Figure 6.11. Depending on the intensity of precipitation, the topography, and the physical properties of the soil, a portion of the moisture evaporates while the remainder either seeps into the ground, a process called *infiltration*, or runs off.

Downward-infiltrating water typically accumulates in two zones within the upper soil and rock layers of the earth's crust. In the topmost band, called the *zone of aeration*, pore spaces contain both water droplets and air. Water requirements of most land plants are supplied by moisture held in the uppermost division of this zone—the *belt of soil moisture*. Moisture in this belt is lost by evaporation

to the atmosphere and by continual withdrawal by plant rootlets. Water taken up by plants is eventually lost to air as vapor through leaf pores in a process termed *transpiration*.

Under the influence of gravity, the remaining moisture is gradually drawn down through the zone of aeration to accumulate in the *zone of saturation*. Pore spaces and fractures in rock and soil in the zone of saturation are completely filled with water. This zone constitutes the *groundwater reservoir*, and the surface separating it from the zone of aeration is called the *water table*. For the purpose of withdrawing groundwater, wells must be drilled deeply enough to penetrate the water table and enter the zone of saturation.

More than 97 percent of our nation's unfrozen reserves of fresh water are contained in the groundwater reservoir. The volume of water within 1000 meters (3000 feet) of the surface is more than nine times that of the Great Lakes. At present, 38 percent of our public water supply in the United States is met by pumping groundwater from wells. In view of the deterioration of surface water quality, maintenance of the quality of our groundwater is a high priority. In many areas, however, human activities threaten both the quantity and the quality of groundwater. One such area, the Pine Barrens of New Jersey, which lies within an easy drive of both Philadelphia and New York City, is discussed in Box 6.1.

The potential implications of excess groundwater withdrawal become evident when we consider the physical characteristics of groundwater distribution and flow. The water table generally parallels the overlying topography, as illustrated in Figure 6.12. Its level changes in response to extended wet or dry periods. In most areas, the zone of saturation is less than 30 meters (100 feet) below the surface, but in some locales, especially in arid regions, wells must be drilled several hundred meters before the water table is intercepted. When groundwater is withdrawn at rates that exceed natural or artificial recharge, the water table drops and the groundwater resource is

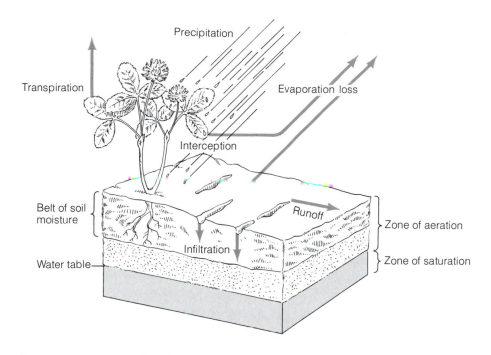

Figure 6.11 The routes of precipitation.

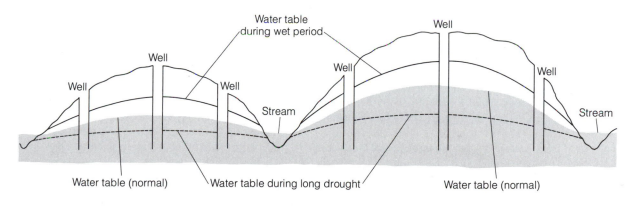

Figure 6.12 The water table. Note that it tends to follow topography and that it shows changes in level during extended wet or dry periods.

Box 6.1

The Pine Barrens of New Jersey is a 1-million-acre tract of pristine forest with exceptionally pure waters. This wilderness lies in the most heavily populated part of the state, only 50 miles from Philadelphia and 100 miles from New York City. State and local actions of the past have protected about one-fifth of the land, but the remainder has experienced heavy developmental pressure. To ease this pressure, the federal government passed a law in 1978 establishing the Pine Barrens National Reserve. This new law provides $25 million for the acquisition of 50,000 acres that are critical to the preservation of this unique paradise.

By exploring the Pine Barrens on foot and by car, the writer John McPhee came to know both the land and its inhabitants. The following excerpt from his book opens as he asks one of the "Pineys" to let him refill a canteen at his pump.

• • •

"Could I have some water?" I said to Fred. "I have a jerry can and I'd like to fill it at the pump."

"Hell, yes," he said. "That isn't my water. That's God's water. That's God's water. That right, Bill?"

"I *guess* so," Bill said, without looking up. "It's good water, I can tell you that."

"That's God's water," Fred said again. "Take all you want."

Outside, on the pump housing, was a bright-blue coffee tin full of priming water. I primed the pump and, before filling the jerry can, cupped my hands and drank. The water of the Pine Barrens is soft and pure, and there is so much of it that, like the

A selection from *The Pine Barrens* by John McPhee. Copyright © 1967, 1968 by John McPhee. This material appeared originally in *The New Yorker*. Reprinted with permission of Farrar, Straus & Giroux, Inc.

forest above it, it is an incongruity in place and time. In the sand under the pines is a natural reservoir of pure water that, in volume, is the equivalent of a lake seventy-five feet deep with a surface of a thousand square miles. If all the impounding reservoirs, storage reservoirs, and distribution reservoirs in the New York City water system were filled to capacity—from Neversink and Schoharie to the Croton basin and Central Park—the Pine Barrens aquifer would still contain thirty times as much water. So little of this water is used that it can be said to be untapped. Its constant temperature is fifty-four degrees, and, in the language of a hydrological report on the Pine Barrens prepared in 1966 for the United States Geological Survey, "it can be expected to be bacterially sterile, odorless, clear; its chemical purity approaches that of uncontaminated rain-water or melted glacier ice."

In the United States as a whole, only about thirty per cent of the rainfall gets into the ground; the rest is lost to surface runoff or to evaporation, transpiration from leaves, and similar interceptors. In the Pine Barrens, fully half of all precipitation makes its way into the great aquifer, for, as the government report put it, "the loose, sandy soil can imbibe as much as six inches of water per hour." The Pine Barrens rank as one of the greatest natural recharging areas in the world. Thus, the City of New York, say, could take all its daily water requirements out of the pines without fear of diminishing the basic supply.

All the major river systems in the United States are polluted, and so are most of the minor ones, but all the small rivers and streams in the Pine Barrens are potable. The pinelands have their own divide. The Pine Barrens rivers rise in the pines. Some flow west to the Delaware; most flow southeast directly into the sea. There are no through-flowing streams

The Untapped, Vulnerable Groundwater
of the New Jersey Pine Barrens

in the pines—no waters coming in from cities and towns on higher ground, as is the case almost everywhere else on the Atlantic coastal plain. I have spent many weekends on canoe trips in the Pine Barrens—on the Wading River, the Oswego, the Batsto, the Mullica. There is no white water in any of these rivers, but they move along fairly rapidly; they are so tortuous that every hundred yards or so brings a new scene—often one that is reminiscent of canoeing country in the northern states and in Canada. Even on bright days, the rivers can be dark and almost sunless under stands of white cedar, and then, all in a moment, they run into brilliant sunshine where the banks rise higher and the forest of oak and pine is less dense. One indication of the size of the water resource below the Pine Barrens is that the streams keep flowing without great declines in volume even in prolonged times of drought. When streams in other parts of New Jersey were reduced to near or total dryness in recent years, the rivers in the pines were virtually unaffected. The characteristic color of the water in the streams is the color of tea—a phenomenon, often called "cedar water," that is familiar in the Adirondacks, as in many other places where tannins and other organic waste from riparian cedar trees combine with iron from the ground water to give the rivers a deep color. In summer, the cedar water is ordinarily so dark that the riverbeds are obscured, and while drifting along one has a feeling of being afloat on a river of fast-moving potable ink. For a few days after a long rain, however, the water is almost colorless. At these times, one can look down into it from a canoe and see the white sand bottom, ten or twelve feet below, and it is as clear as an image in the lens of a camera, with sunken timbers now and again coming into view and receding rapidly, at the speed of the river. Every strand of subsurface grass

and every contour of the bottom sand is so sharply defined that the deep water above it seems, and is, irresistibly pure. Sea captains once took the cedar water of the Pine Barrens rivers with them on voyages, because cedar water would remain sweet and potable longer than any other water they could find.

According to the government report, "The Pine Barrens have no equal in the northeastern United States not only for magnitude of water in storage and availability of recharge, but also for the ease and economy with which a large volume of water could be withdrawn." Typically, a pipe less than two inches in diameter driven thirty feet into the ground will produce fifty-five gallons a minute, and a twelve-inch pipe could bring up a million gallons a day. But, with all this, the vulnerability of the Pine Barrens aquifer is disturbing to contemplate. The water table is shallow in the pines, and the aquifer is extremely sensitive to contamination. The sand soil, which is so superior as a catcher of rain, is not good at filtering out or immobilizing wastes. Pollutants, if they happen to get into the water, can travel long distances. Industry or even extensive residential development in the central pinelands could spread contaminants widely through the undergound reservoir.

When I had finished filling the jerry can from Fred Brown's pump, I took another drink, and I said to him, "You're lucky to live over such good water."

You're telling me," he said. "You can put this water in a jug and put it away for a year and it will still be the same. Water from outside of these woods would stink. Outside of these woods, some water stinks when you pump it out of the ground. The people that has dug deep around here claims that there are streams of water under this earth that runs all the time."

eventually depleted. This situation exists in portions of the American Southwest, where groundwater withdrawal is excessive due to intensive irrigation.

The property of rock and soil that allows it to transmit water in the zones of aeration and saturation is called *permeability*. The degree of permeability is determined by the volume of pore space (porosity) and how effectively the pores are interconnected. Materials of high permeability that contain water are called *aquifers*. In general, sandy soils and layers of sand and gravel are good aquifers, whereas finely textured clayey soils and most crystalline rocks (for example, granite), unless highly fractured, have low permeability.

In some regions, an aquifer is sandwiched between two folded layers of impermeable rock, as illustrated in Figure 6.13. If a drill taps the aquifer at a point where the water is under pressure, water flows freely from the well. Free-flowing wells—called *artesian wells*—are of more value than other wells, since they do not require an expenditure of energy for pumping.

The natural course of groundwater is through permeable layers toward points of discharge, such as rivers and seas. This movement is extremely slow; a speed of 15 meters (50 feet) per year is typical. This slow movement allows groundwater to flow in smooth continuous paths with little mixing, as shown in Figure 6.14. Part of the water that seeps in from the soil surface flows straight downward, and some of it flows toward either side, in this case supplying water to a stream and a marsh. Marshes, springs, and

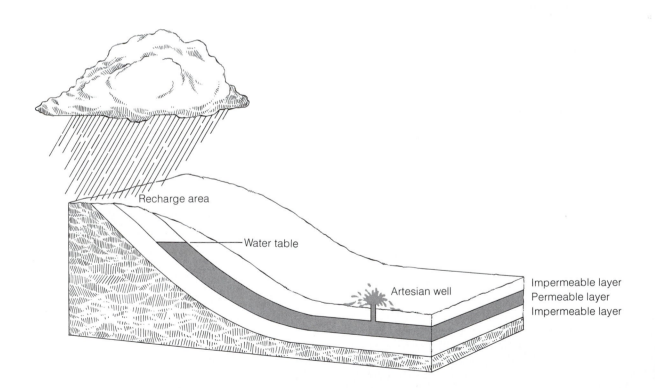

Figure 6.13 A cross-section of the types of rock strata required for artesian wells.

some streams and lakes exist where the water table intercepts or is above the land surface. In these situations the zone of aeration is absent. In arid localities where the water table is well below a stream bed, the stream may supply water directly to the groundwater reservoir.

When groundwater is pumped out, the emptied area forms a cone-shaped depression in the water table (shown in Figure 6.15). This depression does not fill quickly, because groundwater flows slowly. Where wells are closely spaced, excessive withdrawal from one well causes the size of the cone of depression to increase, sometimes resulting in the drying up of adjacent wells.

Another consequence of the slow, smooth movement of groundwater is that pollutants that do not react with soil and rock and that reach the groundwater may remain there for extended periods before being flushed out. The United States Geological Survey reports that contaminants may remain in some subsurface aquifers for hundreds of years, effectively fouling the water supply for generations to come. Also, because groundwater flow is nonturbulent, the mixing and dilution of pollutants takes place very slowly.

From our brief look at groundwater, we can see that it is possible to overexploit the groundwater reservoir causing wells to go dry. Equally serious, when groundwater becomes contaminated, it remains so for long periods of time. An example of the difficulty and expense associated with correcting a groundwater contamination problem is presented in Box 6.2.

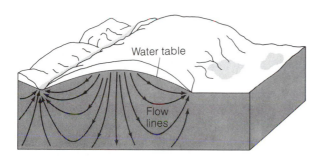

Figure 6.14 Groundwater flow.

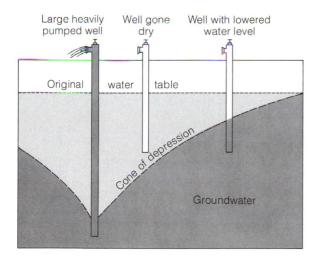

Figure 6.15 A cone-shaped depression in the water table resulting from the pumping of groundwater. Heavy pumping can cause shallow wells in the vicinity to go dry.

Streams and Rivers

Water that does not infiltrate into the ground or evaporate remains on the surface of the earth—hence the term surface water—to become runoff. Streams and rivers are the major pathways taken by the runoff component of the hydrologic cycle in the journey of water from land to sea. Water enters stream and river channels primarily by overland movement from rainfall or ice and snow-melt. Groundwater seepage, springs, and direct precipitation also contribute significantly to stream flow. Each river, plus its branches and tributaries, drains water from a fixed geographical region called a *drainage basin,* or *watershed.* The nation's major drainage basins are shown in Figure 6.16. Climate, vegetation, topography, and various activities in the drainage basin—

The Poisoned Aquifer in New Castle County

In 1972 a homeowner in New Castle County, Delaware, complained that her well had suddenly developed a foul odor. Investigators traced the contamination to a dump about 240 meters (800 feet) away that had operated between 1960 and 1968. During the dump site's development, a thin layer of protective clay had been removed unknowingly, allowing infiltrating water to seep down into the underlying aquifer. The problem was aggravated because a water company located about 1500 meters (5000 feet) away from the site pumped 15–19 million liters (4–5 million gallons) per day of groundwater to supply eighty-thousand nearby residents. The cone of depression created by this pumping helped to accelerate the flow of groundwater toward nearby wells and the water company's well. By the time this well was found to be contaminated, the polluted groundwater had already moved 460 meters (1500 feet), and a second underground plume of contaminated water was found moving toward a plastic manufacturing plant's well.

To renovate the aquifer, costly measures have been implemented. The water company cut its pumping rate by 60 percent, to 7.6 million liters per day. The deficit is now being made up by nearby water companies, and the cost is borne by the county. The plastic manufacturing company also cut back on its pumping. To remove the contaminated water, the county had to pay for the installation of twelve wells that now pump about 11 million liters per day from the contaminated aquifer. New Castle County officials estimate that it will be necessary to pump groundwater from the contaminated aquifer for about ten years to renovate the aquifer. If this remedy fails, the contents of the 60-acre dump will have to be moved to a new site, at a cost of about $20 million to county residents.

both natural and human in origin—affect both the quantity and quality of river water. These factors are discussed more fully in Chapter 7. Of particular concern to us in this section is the variety of ways in which human beings exploit river resources: while both groundwater and rivers provide freshwater for drinking water and irrigation, rivers have additional value for their role in transportation and their usable kinetic energy (the energy of motion).

Under the influence of gravity, a river's seaward travel is downhill. Where the gradient of the channel is steep, water flow is rapid and usually turbulent; where the channel slope levels off, the flow rate slackens. Along the river's course, a portion of the available energy of motion is used to erode and shape the channel. Additional energy is used to transport material loosened from the channel bed and banks and other substances entering the river, for example, waste materials.

Because the energy in a river depends not only on the rate of water flow, but also on the quantity of water involved, stream *discharge* is a better measure of the energy potential of rivers than the speed of its water. Discharge is defined as the volume of water that passes a fixed point along the river's course in a unit of time, and is usually expressed as cubic meters

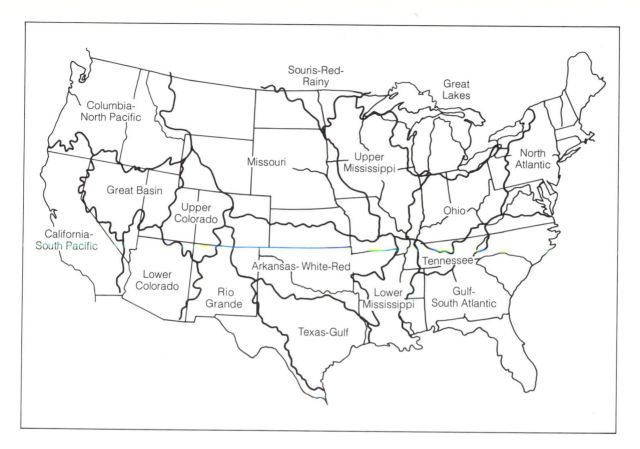

Figure 6.16 Major drainage basins of the United States. (From "Water Resource Development Map of the United States," U.S. Geological Survey, 1969.)

(or cubic feet) of water per second. Where and when discharge is great, such as during spring snow-melt in middle latitudes, channel erosion is accelerated, and the river can transport more and larger materials. More familiar to us is the flooding exhibited during great discharges, such as that shown in Figure 6.17. When river discharge drops, however, as when water is diverted for irrigation, the ability of the river to transport material decreases, and suspended sediments settle out.

Weather in the drainage basin is a primary factor in determining the discharge of a river. The discharge usually undergoes seasonal variations corresponding to rainy-dry season shifts, and in some locales it

may even undergo significant fluctuations in a single day. Rapid discharge oscillations are especially characteristic of the arroyos of the American Southwest. Arroyos are dry streambeds where discharge equals 0 until an afternoon cloudburst over the watershed results in a dramatic but short-lived increase in discharge. In contrast, in some regions, rivers flow out of heavily watered mountains only to completely dry up in the desert plains beyond—never reaching the ocean directly.

Rivers are the life-blood of many cities and of areas where irrigation is essential for agriculture. For example, in the southwestern United States, the entire economy is based on an adequate supply of water

Figure 6.17 Floodwaters invading the inhabited portions of the Snohomish Valley of western Washington. (U.S. Army Corps of Engineers, Seattle District.)

from the Colorado River. In such areas, we must be careful not to build communities and concommitant economies that overexploit the natural capacity of the river—a strong temptation, especially during extended wet periods when more than enough water appears to be available.

What basis should we use in predicting the amount of water available to us in heavily water-reliant areas? During dry periods when runoff into rivers is near zero, the discharge in rivers originates from groundwater and water that is slowly released from wetlands. This base discharge of a river is considered to be the most dependable water supply. It would therefore be a wise policy for us to match our reliance on a river with this level. Still, we must be prepared for the fact that when extended dry periods occur, even base flow levels can decline, requiring that we implement extensive water conservation measures.

Materials washed into a river from the land together with sediment eroded from the channel constitute the river's load. The river transports its load

in suspension, in solution, and along the riverbed by the pushing and rolling motions of the current. Soil erosion contributes suspended silt and clay particles and bits of detritus along with dissolved plant nutrients. The types of substances dissolved in a river vary with climate and with the rock and soil composition of the drainage basin. Also, in some localities a significant fraction of the dissolved river water components are contributed by groundwater and by rainwater that has washed soluble substances from the air. Human activities have placed an added burden on the river in the form of such waste materials as sewage and mine tailings (waste rock).

Most rivers have a tendency to meander, as shown in Figure 6.18, rather than following a straight course. They develop bends and weave back and forth within the bounds of a broad *floodplain*. Figure 6.19 shows a cross-section of a river and its floodplain. During times of abnormally high discharge, the river channel is unable to deepen and widen fast enough to accommodate all of the water; the excess spills over the riverbanks and rapidly spreads over

Figure 6.18 The meandering channel of Big Creek in east-central Louisiana. (U.S. Army Corps of Engineers.)

the flat floodplain. The extent of the area flooded depends on how high the discharge is. Usually, a floodplain is considered to be the area that would be inundated once every one-hundred years. Unfortunately, human settlement is often extensive in river floodplains: people are lured there by productive soils, the reliable water supply, the inexpensive means of transport, and the fishing and recreational potential. In many localities, a history of destructive floods has led to heavy public pressure to implement costly flood control measures such as dams and levees. We discuss these issues in the context of land use alternatives in Chapter 15.

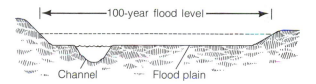

Figure 6.19 A floodplain.

Lakes

Lakes owe their existence to depressions in the landscape. The water filling these depressions comes from runoff or groundwater or both. Thus, some lakes have rivers running into and out of them, some have only an outlet, others have inlets only, and some are filled soley by groundwater.

A lake can be the consequence of any one of a variety of geologic events. Lake basins are formed by such processes as earthquakes (for instance, Lake Nyasa in East Africa), volcanic activity (Crater Lake, Oregon, shown in Figure 6.20), and glaciation (the Great Lakes). Some impoundments are the result of human activity. Abandoned rock quarries and other surface mining operations are often sites of water accumulation. Both natural lakes and artificial reser-voirs behind dams make water available for irrigation, municipal and industrial demands, recreation, and hydroelectric power.

Although lakes exist under many different climatic conditions, from Arctic, to desert, to tropical, and contain waters that range widely in both salinity and acidity, they all share one characteristic: they are relatively short-lived in the context of geologic time. Lakes that are maintained by springs or ground-water seepage may disappear if the regional water table drops. For some lakes, a climatic shift that results in increased evaporation, decreased rainfall,

Figure 6.20 Beautiful Crater Lake in southern Oregon, formed by the collapse of a volcano. (Oregon Department of Transportation.)

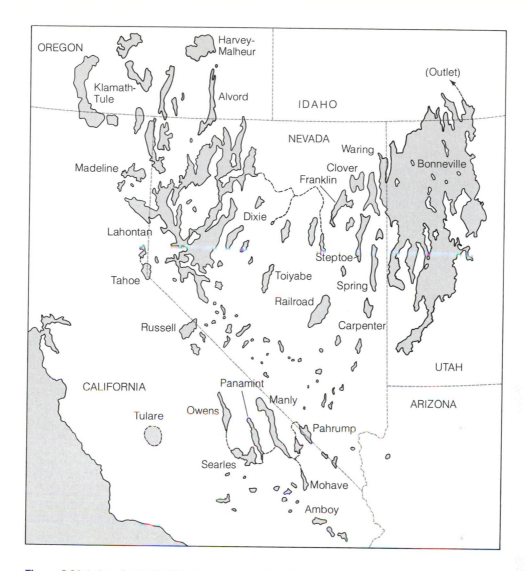

Figure 6.21 Lakes that existed in the western United States ten-thousand years ago, when the climate was cool and moist. The dotted lines indicate overflow channels. (From R. F. Flint, *Glacial and Pleistocene Geology.* New York: John Wiley & Sons, 1957.)

or both may reduce the water volume and may even cause them to completely dry up eventually. For example, about ten thousand years ago, a major climatic change from moist, cool conditions to arid, warm conditions resulted in the disappearance of numerous lakes in the western region of the United States, primarily California, Nevada, and Utah. The map in Figure 6.21 shows the lakes that existed there before the climate changed. Remnants of some of these lakes exist today—the Great Salt Lake in Utah is what remains of Lake Bonneville. Others—the Harney-Malheur in Oregon is an example—have nearly dried up, leaving wetlands. Salt flats were often left behind when these lakes dried up.

Lakes, especially large lakes, are particularly alluring to people, since they provide a source of much needed water and, in some cases, an inexpensive means of transporting manufactured goods. Today, these activities, plus the general trend toward more recreation, are resulting in more development around our lakes than ever. How tight this noose of development will grow around these delicate aquatic ecosystems will largely depend on how strictly we limit further development in their drainage basins.

Glaciers

Glaciers—which are really rivers of ice—represent another pathway taken by the runoff component of the hydrologic cycle in its journey to the sea. Today, glaciers constitute the largest freshwater reservoir, although this resource goes unused, since it is virtually inaccessible. Water in the form of glacial ice covers about 10 percent of the earth's land surface, and is confined primarily to Antarctica (85 percent), Greenland (11 percent), and some high mountain valleys (4 percent). If all of this ice were to melt, the level of the seas would rise by about 60 meters (200 feet), inundating many of the world's major cities.

In the earth's history as a whole, the presence of any glacial ice at all was a rare event. Yet during the past one or two million years glaciers alternately expanded and receded many times. As recently as eighteen-thousand years ago, more than 30 percent of the earth's land surface was covered with ice. In North America, an ice sheet perhaps 3 kilometers (2 miles) thick covered much of Canada and the northern tier states, as shown in Figure 6.22. At that time, sea level was about 100 meters (300 feet) lower than it is now, because so much water was tied up as ice.

In the northern United States the Ice Age left us a valuable legacy. Hundreds of thousands of the lakes, swamps, and bogs that dot the landscapes of the upper Midwest owe their origin to gouging action of glaciers. Under the tremendous forces of the creeping glaciers, the landscape was excavated. The excavated materials eventually were left behind as extensive deposits of sand and gravel layers; some of these now function as aquifers, while others are exploited for construction purposes. Glaciation was indirectly responsible for the rich soils of the Great Plains and central Midwest. Over geologic time, the glaciers formed or receded, depending on the climate. Although greatly diminished in size, glacial ice still represents the earth's largest reservoir of freshwater.

The Oceans

The sea, the earth's largest reservoir of water, is the goal of all terrestrial water, whether it be trickling from melting glacial ice, slowly seeping through permeable rock and soil, or rushing down the courses of stream channels. Viewed from space, the ocean is the most prominent feature of the earth's surface, covering 71 percent of its total area to an average depth of 4 kilometers (2.5 miles). And, in accord with its conspicuousness, the ocean plays an extremely important role in the environment.

The sea is an essential reservoir in most of the subcycles of material and energy movement on earth. Materials cycled into the ocean waters may take thousands or even millions of years before being cycled out again. For these reasons the sea is effectively the final resting place for much of the waterborne and airborne pollutants released by human activity. Because the sea also supports life forms that serve as essential food sources for millions of people, we are finding it necessary to determine which substances that enter the oceans pose potential hazards to marine ecosystems. In addition, given the rapid depletion of fuel and mineral resources on land, we are growing increasingly interested in the resources contained in sea water and in the rock and sediment that comprise the oceans' basins. Thus, the oceans will be receiving more and more of our attention as we seek new ways of meeting our resource needs. Up to now, the mineral resources of oceans have remained relatively unexplored sources compared with those in terrestrial environments.

Sea water contains large quantities of dissolved

Figure 6.22 The extent of glaciation in North America eighteen-thousand years ago.

salts. Sea salt is derived primarily from the weathering and erosion of soil and rock, the products of which are dissolved and carried to the sea by rivers. The saltiness of sea water renders it undrinkable; in fact, drinking salt water actually causes dehydration and an increased craving for water. The salinity of

sea water averages about 35,000 parts per million, or about seventy times the maximum concentration of salts allowed in our public supplies of drinking water. Where evaporation is excessive or sea ice is formed, salinity may range as high as 38,000 ppm, but in particularly rainy climates or where melting

Table 6.3 Concentration of the major substances dissolved in sea water.

Substance	Percentage	Milligrams per liter (parts per million)
Chloride	1.9	19,000
Sodium	1.06	10,600
Sulfate	0.27	2,700
Magnesium	0.13	1,300
Calcium	0.041	410
Potassium	0.039	390
Bicarbonate	0.01	100
Bromide	0.0067	67
Strontium	0.0008	8
Fluoride	0.00013	1.3
Iodide	0.000006	0.06

ice or stream runoff dilutes sea water, the salinity may be as low as 32,000 ppm. While sea water contains numerous dissolved materials, only a few exceed a trace percentage, as Table 6.3 shows. Sodium chloride is by far the most abundant of the dissolved salts. Although salinity varies slightly from place to place, the proportion of major salts present is remarkably constant.

The saltiness of sea water virtually excludes the oceans as a direct source of fresh water, not only for drinking, but for cleaning and irrigation as well. Through technology, human beings can separate salts from sea water to obtain the salts or fresh water or both. But economics and technological capacity are limiting factors in the exploitation of sea water. In Chapter 8, we will see that the desalination of ocean water for domestic, industrial, and agricultural purposes is technically feasible, but too costly in terms of dollars and energy consumption. And, although sea water is potentially the largest single source of numerous raw materials in addition to water, it is currently economically feasible to extract only sodium chloride (common household table salt), bromine (used primarily in the production of an antiknock agent in gasoline), and magnesium (used as a metal alloy and in pharmaceuticals).

A final note refers to the ocean's capacity to circulate energy, a function that illustrates the interaction of the earth's water supply with the sun and the wind. Ocean currents help redistribute the energy absorbed by the surface waters from the sun: in the Atlantic, the warm Florida current flows poleward and cold Labrador current flows southward. The circulation of surface ocean currents is controlled ultimately by a coupling of winds with surface waters. This link results in a clockwise movement of surface waters in the northern hemisphere and a counterclockwise movement of surface waters in the southern hemisphere.

Conclusions

Although the oceans contain the largest reservoir of water on earth, the level of salinity makes sea water unusable for domestic, industrial, and agricultural purposes. The hydrologic cycle, however, processes sea water naturally, and through distillation, evaporation, and condensation, makes it available to us as fresh water, recycled into other reservoirs. We must be sensitive to the way this recycling process works, and recognize the need to protect the relatively small amounts of fresh water it produces. Contaminants that enter the cycle at any point eventually find their way into all parts of the system. We cannot expect the natural cleansing processes to keep pace when the reservoirs are overloaded with foreign substances. All parts of the hydrologic cycle interrelate and interact, and the functioning of the system results in both our freshwater supply and a habitat for all freshwater organisms.

Excluding the water tied up in glacial icecaps, each year about 41,000 cubic kilometers (9,800 cubic miles) of fresh water become available on the continents, and we exploit these resources to meet our domestic, agricultural, and industrial demands for water. Although our freshwater supply is renewed each year through the distillation function of the hydrologic cycle, the volume of fresh water is essentially fixed. Therefore, as the world's human population soars, the amount of fresh water available for each person will decline. Hence, it is evident that we must apply conservation measures to our freshwater resources to insure that we have sufficient water of acceptable quality in the future. The types and prop-erties of pollutants that enter our water supply, their effects on the water itself and on the aquatic life it supports, and the methods of pollution abatement and freshwater conservation currently available are explored in the next two chapters.

Summary Statements

The hydrologic cycle is the continual circulation of water among the atmospheric, land, and oceanic reservoirs.

Air is said to be saturated when it will hold no more water vapor. The degree of saturation is indicated by the relative humidity. Warmer air holds considerably more water than cold air. As air parcels warm, however, they become less saturated. Cooled air parcels approach saturation and may loose water through precipitation processes.

Precipitation processes are initiated by condensation nuclei, especially hygroscopic nuclei. Precipitation forms in clouds through the ice-crystal process and the coalescence process.

Topographical barriers such as mountains can both increase and reduce rainfall.

Water flows from land to sea via infiltration (groundwater flow) and runoff (rivers and streams).

Groundwater flows through aquifers, which are composed of materials with inter-connected pore spaces. Free-flowing artesian wells are possible where groundwater is confined and under pressure.

Groundwater is our most important source of fresh water. Groundwater is also an important source of water for rivers, lakes, marshes, and swamps.

Rivers drain a defined area called a drainage basin, or watershed. The discharge (volume) of rivers varies with precipitation and activities within the drainage basin. Rivers are important agents in the sculpturing of the landscape.

Glacial ice represents the largest reservoir of fresh water, but is unused because it is virtually inaccessible.

Seawater is highly saline and therefore unusable for drinking or for irrigation. Salt, bromine, and magnesium are the only substances that are now extracted from seawater.

Evaporation of sea water by the sun and subsequent condensation are the natural processes that renew the earth's freshwater supply.

Ocean currents are an important means whereby solar energy is redistributed.

Questions and Projects

1. In your own words, write out a definition for each of the italicized terms in this chapter. Compare your definitions with those in the text.

2. How has water played a role in the location and historical development of your community?

3. From what specific reservoir of the water cycle do you obtain your water supply? Contact your local water utility for this information.

4. Give one common example of each of the following water transfer processes: sublimation, deposition, evaporation, and condensation. Which processes are most important in your region?

5. Usually, very litle difference exists between the quantity of water vapor in the air over the Southwest and that in the air over the Northeast. However, large differences do exist in the amounts of precipitation that fall on the two regions. What is the fundamental reason for this difference?

6. Cite specific attempts in your area to harness some of the energy in the water cycle.

7. In what parts of the United States would you expect high evaporation rates? Explain your answer. What happens to the dissolved salts in a lake that experiences high evaporation rates? What happens to dissolved salts in irrigation water?

8. The "residence time" for water in a water reservoir (lake, groundwater, river, glacier) is the average time that a volume of water spends cycling through a reservoir. In which reservoirs does water have long residence times?

9. What is the average annual rainfall in your region? What fraction of this amount evaporates?

10. Is groundwater being pumped faster than it is being replenished by underground aquifers in your region? If so, what is being done to minimize future water supply problems. Infomation on groundwater levels can be obtained from local water department well logs or, in some areas, from state water conservation district studies.

11. How might paved roadways, parking lots, and rooftops affect groundwater levels in an area?

12. Explain what is meant by the term floodplain. Do you live in the floodplain?

13. Does sediment transported by rivers in your region present problems? Cite specific examples of where problems occur and what is being done to minimize or alleviate these problems. Your local U.S. Soil Conservation Office will have publications or information on this problem.

14. Cite specific factors that would affect the relative position of the water table in a region.

15. Why is distillation a purification process? Where does the energy originate that fuels natural distillation?

16. Consult a map of the western United States and locate six lakes that existed 10,000 years ago (see Figure 6.21). How do these lakes compare in size now with the original lakes as shown in the figure?

Selected Readings

Ambroggi, R. P. 1977. "Underground Reservoirs to Control the Water Cycle," *Scientific American 236*:21–27 (May). A discussion of methods for increasing groundwater supplies.

Battan, L. J. 1962. *Cloud Physics and Cloud Seeding*. Garden City, New Jersey: Doubleday. A clear presentation of precipitation formation processes.

Cargo, D. N., and B. F. Mallory. 1977. *Man and His Geologic Environment*, 2nd ed. Reading, Massachusetts: Addison-Wesley. A textbook containing detailed information on the hydrologic cycle and related processes.

Davis, G. H., and L. A. Wood. 1974. *Water Demands for Expanding Energy Development*. U.S. Geological Survey, Circular 703. Projections of water needs for regions of the United States where coal and oil shale extraction will increase.

Dunne, T., and L. B. Leopold. 1978. *Water in Environmental Planning*. San Francisco: W. H. Freeman and Company. A textbook that examines methods for preventing water-related problems.

Judson, S., K. S. Deffeyes, and R. B. Hargraves. 1976. *Physical Geology*. Englewood Cliffs, New Jersey: Prentice-Hall. A textbook covering the hydrologic cycle and related processes thoroughly.

Leopold, L. B. 1974. *Water*. San Francisco: W. H. Freeman and Company. A primer on water cycling and water use.

Luas, W., and S. Beicos. 1967. *Water in Your Life*. New York: Popular Library. Covers topics from the physical properties of water to the composition of natural waters.

Murray, R. C., and E. B. Reeves. 1977. *Estimated Use of Water in the United States in 1975*. U.S. Geological Survey, Circular 765. A statistical report on total water uses in the United States.

Press, F., and R. Siever. 1978. *Earth*, 2nd ed. San Francisco: W. H. Freeman and Company. A textbook that covers the hydrologic cycle and related processes in detail.

A dead bird, one of many victims of water pollution. (M. L. Brisson.)

Chapter 7

Water Pollution

Debris bobs along a river's course and clutters the shoreline. Oil slicks slither and shimmer; refuse and dead fish foul the air and water. Signs here and there along the shore warn forbiddingly, "Danger—Water Polluted—No Fishing or Swimming." Such sights and smells are all too common along the rivers and lakes near our nation's cities. In fact, the United States Environmental Protection Agency estimates that one out of every three kilometers of the streams in this country is contaminated to some extent. Our rivers were not always in such deplorable condition, but, primarily in the interest of our industrial-technological and economic growth, we have allowed our cities and industries to impinge heavily on aquatic habitats. The consequences have been dire and new concerns continue to surface.

Unfortunately, as our technology has grown more sophisticated, the types of wastes discharged into our waterways have become more varied. Relatively new threats to our water resources include such contaminants as commercial fertilizers, pesticides, detergents, trace quantities of metals, acid mine drainage, cyanides, radioactive substances, and industrial chemicals. These substances are not only dangerous in themselves, but they can react synergistically with other materials. However, although the substances entering our waterways are more varied than ever, we have made substantial gains in the last decade in uncovering the roots of our water pollution problems. In this chapter, we survey what is known about the sources and effects of water contaminants on surface and groundwater.

A Historical Overview

In the late 1800s, when our cities were growing rapidly in response to industrialization, piles of horse manure and garbage in the streets became a deplorable problem. Occasionally, garbage collectors, who were usually highly paid political appointees, cleaned the streets and dumped the refuse into the nearest river. In New York City, platforms were specially constructed so that citizens could dump garbage directly into the river themselves. Although river dumping was discontinued in 1872, the substitute method of barging the wastes a short distance from shore before dumping them in the ocean polluted

Figure 7.1 A political cartoon from the late 1800s depicting water pollution problems of the era. (The Bettmann Archive, Inc.)

city beaches (see Figure 7.1). New Yorkers seeking relief from the summer heat flocked to the contaminated beaches in droves.

These dumping practices, plus a complete lack of the sanitary practices we now take for granted, frequently led to epidemics of waterborne diseases. City and country folk alike experienced and feared outbreaks of the so-called filth diseases, such as typhoid fever, cholera, infectious hepatitis, and dysentery. Epidemics were usually traced to water or food that had been contaminated with human feces. In fact, it was not uncommon on farms to find the barnyard, pigsty, chicken coop, privy, and cesspool in close proximity to the open well that supplied drinking water. In cities, efforts to control these diseases began with the filtering of public water supplies

through sand. Later, beginning in 1908, chlorine was added to municipal water prior to distribution.

By the early 1960s, waterborne diseases had become relatively rare in the United States, although they remain a critical problem in the less developed countries of the world, as we shall see in Chapter 8. Now the focus of concern in our country is on pollution by industrial wastes and chemical pollutants. The long-range effects of these materials in our water supply are so complicated that they have yet to be fully comprehended. The visible short-range effects, however, suggest to us the seriousness of this kind of pollution: fish kills are common in contaminated waters, beaches are made unusable, and lakes become choked with aquatic plants.

Encouragingly, the contamination of our waters is no longer continuing unchecked. At this writing, our nation is midway through a dramatic shift away from indiscriminate dumping and toward responsible pollution control. Some industries and communities have made impressive strides in minimizing their negative impact on local waterways. Others, however, seem bent on doing battle in court rather than changing their ways.

Drainage Basin Activities

Most water pollution problems originate from land-based activities carried out within drainage basins (see Figure 7.2), and not from water-based activities such as shipping, boating, and swimming. As defined in Chapter 6, a drainage basin is a geographical region drained by a river or stream. We can identify three types of drainage basins: natural basins, which are relatively undisturbed areas; rural-agricultural basins, which lie under cultivated lands; and urban-industrial basins, those occupied by cities. In each type of drainage basin, particular natural processes and human activities contribute specific contaminants to both surface water and groundwater.

Natural areas such as forests, marshes, and grasslands generally do not contribute to the pollution of waterways. Surface waters flowing through these ecosystems contain a characteristic "normal" concentration of dissolved substances, since they come into contact with different types of sediments, air, and living and decaying organisms. Streams that originate or pass through swamps or marshes are usually somewhat acidic, since they pick up organic acids generated by decomposer organisms. Naturally added material such as fallen leaves in autumn may cause temporary upsets, but these materials are rapidly cleaned out by detritus feeders. The composition of sediments varies considerably, and therefore the natural waters flowing over and through them also vary widely in dissolved mineral content. For example, the headwaters for the Wolf River in Wisconsin are low in concentrations of dissolved substances: calcium, 6.0 ppm; magnesium, 7.2 ppm; and bicarbonate, 42.0 ppm—because the rock at the head waters consists primarily of insoluble granite. Downstream, however, the granite gives way to more soluble dolomite (a rock consisting of calcium magnesium carbonate). Hence, as the river flows along its course, concentrations of the dissolved materials increase: calcium, 13.0 ppm; magnesium, 25.0 ppm; and bicarbonate, 170 ppm.

Agricultural areas contribute to the degradation of water quality in several ways. Excessive soil erosion increases a river's sediment load. Pesticides, fertilizers, and animal wastes, washed from fields and orchards, run off into streams or seep into groundwater. Regions from which contaminants are washed in this way are referred to as *nonpoint sources* of pollutants, since the contaminants they contribute come from throughout an area and not from one or two concentrated sources. In contrast to *point sources*—concentrated wastes from sewage treatment plants and industry—pollutants from nonpoint sources are especially difficult and expensive to control, because the concentration of pollutants is relatively low and the volumes enormous.

The urban-industrial drainage basins contain the greatest number of point sources of pollutants. Although the multitude of wastes from such areas are no longer dumped directly into waterways, modern-day sewer systems are still important contributors to surface water pollution. In cities, concrete, asphalt, and buildings render a large part of the urban surface impermeable to rainwater and snow-melt, thereby increasing the volume of runoff. To prevent flooding, large storm sewer pipes lying under city streets channel this runoff, usually directly to the nearest river, lake, or ocean. During a rainstorm, the air, streets, and ground surface are washed, and many pollutants—for example, hydraulic fluid, dirt, grime, oil, radiator coolant, road salt, and pet droppings—are carried into surface waterways by the storm sewer system. The components of a city sewer system, including the storm sewer pipes, are shown in Figure 7.3.

A second, smaller sewer pipe, called the *sanitary sewer,* carries waste water from homes and commercial areas to treatment plants where it is treated and discharged into the nearest surface water. We shall see later that the quality of the treated water depends on the relative sophistication of the treatment plant. When both a storm sewer pipe and a sanitary sewer pipe service an area, the system is called a *separated sewer system.*

Some sewer systems employ a single pipe to transport both urban runoff and sewage to the treatment plant. Such a system is called a *combined sewer.* During dry weather, the treatment plant receives only domestic wastes. During heavy rains, however, the volume of water flowing through the sewer often exceeds by a hundred times the amount that can be processed by the treatment plant. As a consequence, only a small fraction of the sewage water is treated, and the overflow—containing raw sewage—is bypassed and discharged directly into surface waters. Today, parts of many major cities are still being served by combined sewer systems. Every time such a sewage treatment plant is forced to bypass raw, untreated sewage, the health of the city's residents or its downstream neighbors is threatened.

The waste water of some industries is so toxic that municipal treatment plants are prohibited from

Figure 7.2 (*Left*) A schematic drawing showing a few of the many activities in drainage basins that can affect water quality. In natural drainage basins, flowing water (1) slowly dissolves and erodes rock, and wetlands (2) retain nutrients in dead vegetation. In urban-industrial basins, municipal sewage treatment plants (3) fail to remove all the wastes added during use; storm sewer water (4) contains wastes washed from city streets and lots; industrial waste water (5) contains a wide array of different water pollutants; acid water flows from mines and strip-mined land (6); heated water flows from power plants (7); industrial gases are washed out of the air by rain (8), which is thereby acidified; the transportation of oil (9) results in spills, including many minor spills during unloading; and improper landfill sites (10) contaminate groundwater. In agricultural drainage basins, crop spraying (11) adds pesticides and herbicides to rain and runoff, fertilizers improperly applied (12) are dissolved in runoff, and animal wastes (13) are washed from farmland.

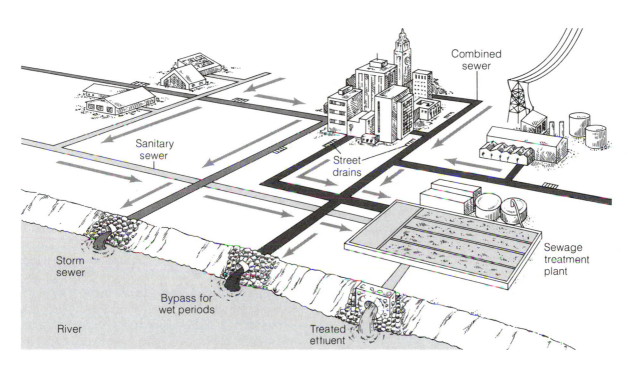

Figure 7.3 A schematic drawing of a city sewer system showing both separated and combined sewer systems.

treating it. Industries that do send their waste water to municipal treatment plants must limit the level of potentially toxic substances. The other factories must install specialized water treatment systems that are designed to remove the particular type of pollu-tants in their effluent, or liquid wastes. Not all in-dustries, however, are so equipped, and even respon-sible industries have occasional accidental dis-charges. Also, with the tightening of air and water quality standards, industries and municipalities have been forced to dispose of more of their wastes in landfill sites. These disposal systems do protect surface waters, but they pose a potential hazard to groundwater quality.

Surface Water Pollution

As we noted earlier, concerns about water pollution in the past focused primarily on the effects of contaminated water on human health. Today, in addition to our interest in human health, we are worried about the effects of polluted water on aquatic organisms and the stability of aquatic ecosystems. In this context, not only disease-producing organisms, but also plant nutrients, pesticides, heavy metals, oil, sediments, and excessive heat discharge are all water pollution threats. In the following sections, we consider the particular risks associated with specific types of pollutants.

Infectious Agents

Water is a significant vehicle in the transmission of disease when it contains *waterborne pathogens,* or disease-producing organisms. These pathogens, which can be viruses, bacteria, protozoa (single-celled animals), and parasitic worms, cause such diseases as dysentery, typhoid fever, cholera, and infectious hepatitis. Table 7.1 shows some of the more common waterborne diseases and their characteristics. Today in the United States, the *chlorination* of water supplies has greatly reduced the transmission of disease by water. But, as we noted, for a great many people in the world—35 percent, in fact—waterborne pathogens in drinking water are still a serious health hazard.

Pathogens enter the water via the feces and urine of infected people and animals. Infected body wastes can enter water in a variety of ways: as unchlorinated sewage when sewer systems become overloaded or treatment plants malfunction, as waste discharges from pleasure boats, as untreated discharges from

meat processing plants, or directly, from swimmers. If drinking water supplies drawn from contaminated surface waters are not properly treated, an epidemic may occur. In fact, the presence of a single case of a disease such as typhoid fever in a community indicates that the potential for an epidemic is present. During such periods personal hygiene and cleanliness are vital in preventing the disease from spreading. Individuals can become infected directly by drinking or swimming in contaminated water, or by eating food that has been contaminated via the food chain. For example, in the fall of 1978, Louisiana's crab industry was dealt a serious blow when some crabs were found to carry the pathogen that causes cholera.

Analyzing water for the presence of specific pathogens is time-consuming, costly, and difficult. Therefore, microbiologists usually analyze for a more readily identifiable group of bacteria called *coliforms.* Since these organisms are normally present in the intestinal tract of humans and animals, large numbers of coliforms in a water sample indicate recent contamination by untreated sewage. It is assumed that if coliforms are present in the sample, pathogens from the intestinal tracts of infected individuals could have been excreted as well. Hence, when coliform organisms are found in drinking water, municipalities either chlorinate drinking water more heavily or seek alternative sources. The Environmental Protection Agency has set an upper limit of 200 coliform bacteria per 100 milliliters of water for recreational waters. If this limit is exceeded, the waters and contingent beaches are usually closed.

As Figure 7.4 indicates, the chlorination of public water supplies and sewage treatment plant effluents has effectively eliminated outbreaks of epidemics of waterborne diseases in developed countries such as the United States. However, this technology can be interrupted or can fail completely, as during severe storms or earthquakes. Also, as we shall see further on, chlorination itself may give rise to other pollution problems, the seriousness of which is yet to be measured. Thus, though we have grown used to the

Table 7.1 Waterborne diseases transmitted through drinking water and food.

Disease	Type of organism	Symptoms and comments
Cholera	Bacteria	Severe vomiting, diarrhea and dehydration; often fatal if untreated; primary cases waterborne; secondary cases carried by contact with food and flies.
Typhoid fever	Bacteria	Severe vomiting, diarrhea, inflamed intestine, enlarged spleen; often fatal if untreated; primarily transmitted by water and food.
Bacterial dysentery	Several species of bacteria	Diarrhea; rarely fatal; transmitted through water contaminated with fecal material or by direct contact through milk, food and flies.
Paratyphoid fever	Several species of bacteria	Severe vomiting and diarrhea; rarely fatal; transmitted through water or food contaminated with fecal material.
Infectious hepatitis	Virus	Yellow jaundiced skin, enlarged liver, vomiting and abdominal pain; often permanent liver damage; transmitted through water and food including shellfish foods.
Ameobic dysentery	Protozoa	Diarrhea, possibly prolonged; transmitted through food, including shellfish.

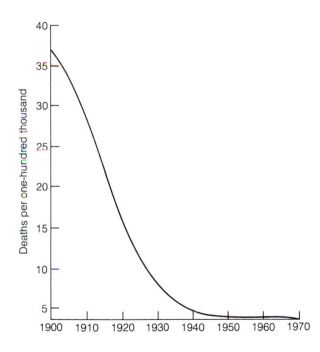

Figure 7.4 Death rates from typhoid fever, a waterborne disease. (After M. J. Pelczar, Jr., and R. D. Reid, *Microbiology*, 3rd ed. New York: McGraw-Hill, 1972.)

protection our water treatment systems have afforded us, the potential of waterborne epidemics always exists.

Oxygen-Demanding Wastes

Most aquatic organisms acquire their oxygen from the supply dissolved in water. This supply, however, quickly diminishes when organic wastes decompose. The presence of organic materials encourages the proliferation of oxygen-consuming decomposers such as bacteria and fungi. The decomposers reduce the oxygen supply, and members of aquatic communities, especially fish and shellfish, perish from oxygen deprivation.

In an undisturbed aquatic ecosystem, the quantity of organic material—detritus—is small, and therefore the amount of oxygen utilized by decomposers is also small. As a consequence, under natural conditions the concentration of dissolved oxygen remains relatively constant. This balance represents the smooth functioning of a natural purification process. When we say that a river has the ability to cleanse itself, we are referring to its natural but limited capacity to decompose and remove organic wastes.

Most industrial and municipal wastes, however, contain high—not moderate—concentrations of organic substances. Their presence encourages the growth of decomposers, and these organisms consume large quantities of dissolved oxygen. (Some inorganic industrial wastes can also react with and thereby remove dissolved oxygen.) Because the presence of the oxygen-demanding materials ultimately results in the loss of organisms through oxygen deprivation within the aquatic ecosystem, they are, by definition, pollutants. These are short-term pollutants: they are almost completely decomposed after several weeks, though in cold water the rate of decomposition is lowered considerably. By the time decomposition has been completed, the oxygen supply in the water has been severely depleted in many cases. (Most waste water, however, also contains a small fraction of organic materials that resists decomposition.) If organic wastes are eliminated at their source, oxygen is replenished in the affected area in a matter of a few days in rivers and within three or four months in the hypolimnion of lakes.

The amount of dissolved oxygen needed by decomposers to decompose organic materials in a given volume of water is called the *biochemical oxygen demand* (BOD). Thus, BOD is a measure of contamination of the waste water. Human wastes are a major source of BOD. Sewage-laden waste water entering a sanitary sewer has an average BOD level of 250 ppm. Because this water is likely to contain only about 8 ppm of oxygen, its dissolved oxygen is quickly depleted through microbial decomposition of sewage. In fact, the decomposition of the daily wastes of a single person requires all the dissolved oxygen in 9000 liters (2200 gallons) of water. Oxygen-demanding substances are also added to water used in the processing of organic materials such as vegetables and fruits, paper, meat, and dairy products. Levels of BOD in waters used in these processing techniques vary widely but are usually substantially higher than those in domestic sewage. In fact, some concentrated industrial wastes have BOD levels

greater than 30,000 ppm. Other sources of high-level BOD wastes include runoff from livestock feedlots and spoils dredged from harbors and canals.

When effluents containing high levels of BOD are released into a river, as in Figure 7.5, dissolved oxygen levels downstream follow a characteristic pattern. At the point of discharge, bacteria begin to consume organic material. The greater availability of food allows the bacteria to increase rapidly in number and consume oxygen faster than it is replenished. As the material moves downstream, its decomposition reduces the oxygen demand. Dissolved oxygen removed by decomposers is slowly replaced by transfer from air to water and by the photosynthetic activity of aquatic plants. Eventually, the oxygen replacement rate exceeds the removal rate and a stream's oxygen level is restored to its original level. This characteristic pattern, called an *oxygen sag curve,* is illustrated in Figure 7.6. Organic waste discharges have their most serious effect on aquatic life during warm summer months; during that time stream flow is low, and the warm waters, besides holding less dissolved oxygen, enhance microbial activity.

The responses of aquatic organisms in rivers receiving excessive quantities of organic wastes have been well documented. The portion of a river upstream from a discharge site can support a wide variety of fish, algae, and other organisms, but in the section of the river where oxygen levels approach zero, only a few types of sludge worms survive. As oxygen levels recover downstream, species of rough fish (carp and gar) that can tolerate low oxygen levels appear. Eventually, a normal community is restored, but a single overloading discharge of organic wastes can eliminate the natural aquatic community for some time. Also, if discharge sites are spaced too closely together, the entire downstream stretch of the river will suffer severe oxygen depletion.

The loss of oxygen is accompanied by a change in the type of decomposer bacteria present in the water, from *aerobic* (with oxygen) to *anaerobic* (without oxygen) *decomposers.* The products of *aerobic* and

Figure 7.5 Floating mats of paper-mill wastes moving downstream. (© Ron Curbow, from National Audubon Society.)

anaerobic decomposers differ as well. While the products of aerobic decay—mainly carbon dioxide, water, and nitrate—are not generally harmful, those generated by anaerobic decay include the hazardous gases methane, ammonia, and hydrogen sulfide (recognized by its rotten-egg aroma). Under anaerobic conditions, waterways become a putrid, turbid, decaying mess with bubbles of methane and hydrogen sulfide rising to the surface.

Plant Nutrients and Cultural Eutrophication

Excessive quantities of nutrients discharged into lakes and rivers from any one of many possible sources of human activity accelerate the natural aging processes of waterways (see the discussion of eutrophication in Chapter 5). Accelerated eutrophication that results from human activity is called cultural eutrophication. Cultural eutrophication of

freshwater resources is one of the most significant water quality problems facing us today. Eutrophication jeopardizes the use of water for recreation, sport and commercial fishing, agriculture, industry, and municipal supply.

Like terrestrial plants, aquatic plants require nitrogen, phosphorus, potassium, and other mineral nutrients. In aquatic systems, the two nutrients that most commonly act as limiting factors are phosphorus, in the form of phosphate, and nitrogen, in the form of either nitrate or ammonia. Thus, when levels of limiting nutrients increase, the response in some aquatic organisms is an increase in number. In the shallow near-shore waters that receive added nutrients through runoff, rooted aquatic plants grow luxuriantly, hindering swimming and boating and even reducing shoreline property values (see Figure 7.7). Masses of algae, primarily blue-green algae, release foul-smelling and unpleasant-tasting substances into the water, again, adversely affecting its recreational

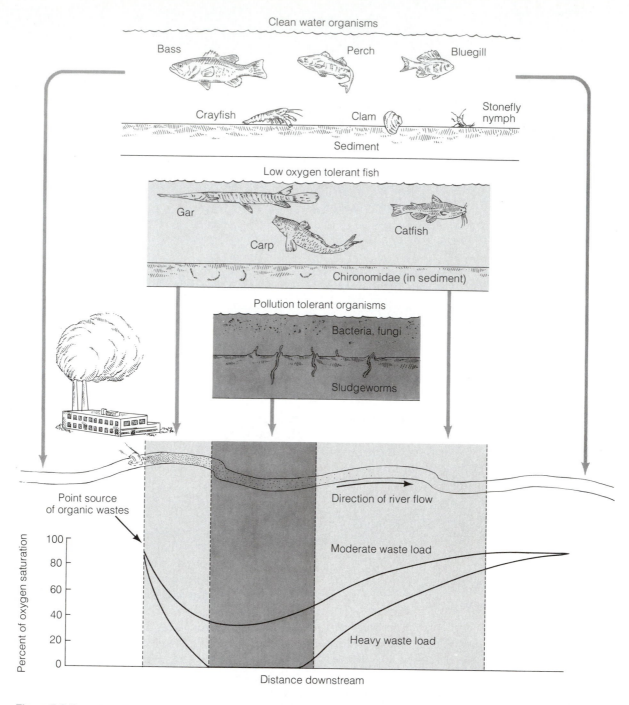

Figure 7.6 The effect of dumping organic wastes in water. Dissolved oxygen levels drop, which affects the type of organisms inhabiting various sections of a river.

Figure 7.7 Dead algae washed up on Montrose Beach, Chicago, Illinois. The extraordinary abundance of algae was the result of cultural eutrophication of Lake Michigan. (Courtesy John Hendry, Chicago.)

value. The removal of these compounds and the filtering out of the algae significantly increase the cost of supplying water to municipalities and industries. And, finally, oxygen depletions resulting from microbial decay of dead algal masses cause fish kills.

What are the sources of the nutrients that contribute to cultural eutrophication? Domestic sewage is one important point source. A modern treatment plant using secondary treatment (described in Chapter 8), removes only about 10 to 30 percent of the nitrogen and phosphorus compounds. Treated effluent from these plants may still contain about 20 ppm nitrogen and 9 ppm phosphorus. Scientists estimate that excessive growth of algae occurs when total phosphorus levels exceed 0.10 ppm (if phosphorus is limiting) or when nitrogen levels (nitrate or ammonia) exceed 0.3 ppm (if nitrogen is limiting). A major source (approximately 50 percent) of the phosphorus in sewage is household detergents.

Urban runoff is another significant source of plant nutrients. Studies in several cities across the country indicate that stormwater runoff contains several ppm of both nitrogen and phosphorus. These materials evidently enter the runoff with fertilizers, animal fecal material, dust, leached leaves, and combustion products. Tables 7.2 and 7.3 show the amounts of various pollutants in different kinds of runoff.

Industrial inputs of nutrients vary widely. Papermills, which contribute large amounts of oxygen-demanding substances, release low concentrations of plant nutrients. Industries with large surface areas to clean, such as creameries or car washes, use large amounts of phosphorus-containing cleaning agents. Phosphate mining industries are another major source.

Studies indicate that nutrient runoff from properly managed farmland is only slightly greater than that from natural areas. However, where large amounts of fertilizers are applied and drainage conditions promote runoff and erosion, the amount of nutrients washed from farmland is significant. For example, in California's Central Valley, the nitrate concentrations in irrigation return water sometimes approaches 50 ppm. In northern states, such as Minnesota, Wisconsin, and New York, the winter practice of spreading animal manure over frozen ground adds to nutrient levels in spring runoff. Feedlots that drain directly into streams, such as that shown in Figure 7.8, are another important source of plant nutrients. Groundwater and precipitation are usually minor sources of nutrients. The importance of each source varies, of course, with the activities that take place within the drainage basin.

Table 7.2 Range of concentrations of selected substances in urban stormwater.

Substance	Concentration range (milligrams/liter or ppm)
Biochemical Oxygen Demand	1.0–700
Suspended Solids	2.0–11,300
Organic-Nitrogen	0.1–16
Ammonia-Nitrogen	0.1–2.5
Total Phosphorus	0.03–42
Chloride	2.0–25,000*
Oils	0–110
Lead	0–1.9

*With highway deicing.
Source: M. P. Wanielistra, Y. A. Yousef, and W. M. McLellon, *Journal of Water Pollution Control 49*:441 (1977).

Toxic Substances

Oxygen-consuming pollutants and added nutrients affect aquatic organisms by altering their growth processes. Toxic substances are pollutants that adversely affect the organisms that ingest them, whether in a onetime dose (called acute exposure) or at low levels over a long period of time (chronic exposure).

Just as nutrients pass from one trophic level to the next, so some persistent toxic substances—for example, DDT and methyl mercury—also pass along a food chain. If these substances are not excreted or broken down biologically by an individual organism, they are retained and concentrated in the body. Thus, organisms at the top of the food chain receive the highest concentrations of these substances, since they ingest organisms that have incorporated toxic substances through their own feeding. This process is called *food chain accumulation.* Accumulation is especially pronounced in aquatic food webs, because they usually consist of four to six trophic levels, in contrast to the two to three levels normal in terrestrial ecosystems. Hence, persistent pollutants that enter the lowest trophic level become considerably more concentrated by the time they reach the topmost trophic level (the movement of a toxic substance is illustrated in Figure 7.9). Regrettably, hu-

Table 7.3 Land use and amounts of specific water pollutants washed from surface (kilograms per hectare per year).

Land use		BOD	Suspended solids	Total nitrogen	Phosphorus
Urban	Range	53–82	728–4794	3.2–18	1.0–5.0
	Average	75	1700	8.5	2.0
Pasture	Range	6–17	11.8–840	2.5–8.5	0.24–0.66
	Average	11	840*	5.3	0.30
Cultivated	Range	4–31	286–4200	15.0–37.0	0.18–1.62
	Average	18	4200*	26.0	1.05
Woodland	Range	4–7	45–132	2.4–5.1	0.01–0.86
	Average	5	98	3.1	0.10

*Data limited; therefore high value is used.
Source: M. P. Wanielistra, Y. A. Yousef, and W. M. McLellon, *Journal Water Pollution Control 49*: 441 (1977).

Figure 7.8 Domestic animals adding wastes to streams and increasing streambank erosion. (M. L. Brisson.)

mans normally eat organisms from the upper trophic levels of aquatic food chains, and thus are likely to ingest relatively large amounts of toxic substances at once. The mercury poisoning of hundreds of people in Minamata, Japan—known now as Minamata disease—dramatically illustrated the food chain accumulation of a toxic substance as it affects human beings. The Minamata tragedy is described in Box 7.1.

News of the Minamata story triggered concern over potential health hazards from mercury in the United States. Fish from some rivers receiving industrial waste, especially those downstream from chlorine manufacturing plants, were found to have levels of mercury well above the 0.5 ppm standard set by the United States Food and Drug Administration. As a result, fishing in these rivers was banned or people were warned to eat no more than one fish meal per week. Changes in the operations of plants that used mercury have by now substantially reduced the amount of mercury discharged. As a consequence, mercury levels in fish and shellfish have subsided somewhat since concern was aroused, but in many instances they still remain above the 0.5 ppm FDA limit, and commercial fishing therefore cannot be resumed. In retrospect, the Minamata incident did have its beneficial effects, since had it not occurred, people in the United States and many other countries may well have contracted this terrible disease.

The suffering and loss of life in Minamata demonstrate the serious threat inherent in mercury discharges. Mercury is only one member of a trouble-

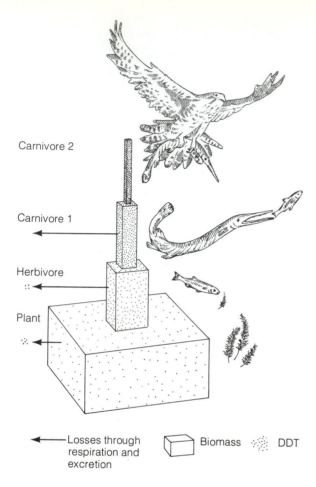

Carnivore 2

Carnivore 1

Herbivore

Plant

◄── Losses through
    respiration and
    excretion

▢ Biomass ⠿ DDT

Figure 7.9 The accumulation of DDT in a food chain.

some group of metals, however. Chemicals containing arsenic, cadmium, chromium, copper, lead, nickel, silver, and zinc are other hazardous substances. Many of these substances are known to act as carcinogens in laboratory animals and are presumed to have the same effect on people. But even the cancer-causing factor is not the worst of the story, for these chemicals, unlike most organic contaminants, are not broken down by metabolic processes in organisms. Thus, once aquatic ecosystems are contaminated by these chemicals they remain so for many years.

Pesticides are another group of substances that contain chemicals particularly toxic to aquatic life. In the United States, farmers and foresters use enormous quantities of pesticides to control weeds, insects, rodents, and disease-producing fungi. Some eighteen hundred different chemicals are used as active ingredients in pesticides, and these are combined in over forty thousand different pesticide formulations. Pesticides enter lakes and streams through the effluent of pesticide manufacturing plants, along with drifting spray mists during application, as run-off when spraying equipment is washed, and in accidental discharges. In addition, pesticides applied in the field adhere to soil particles and are washed into streams.

Of all the pesticides that enter surface waters, the class of compounds known as the chlorinated hydrocarbons pose the most consistently unacceptable risks to the environment and to human beings. The EPA has banned the widespread use of several chlorinated hydrocarbons: DDT, aldrin, dieldrin, heptachlor, and chloradane.

In spite of continuing efforts to prevent chlorinated hydrocarbons from entering aquatic environments, numerous incidents of pesticide pollution of fisheries continue to occur. The incident involving Kepone in Hopewell, Virginia, discussed in Chapter 3, is a case in point. Kepone from an industrial plant contaminated a major section of Virginia's James River, and it may be a decade before the Kepone in the environment ceases to be harmful.

Another group of industrial chemicals, also chlorinated hydrocarbons, are called PCBs (*poly*chlorinated *bi*phenyls), and these substances also contaminate aquatic ecosystems. PCBs are not pesticides, but, because they are fire-resistant and stable at high temperatures, they are used as heat-exchanging liquids, insulating materials in electric capacitors and transformers, hydraulic fluids, and in the manufacture of some plastics. During the manufacture, use, and eventual disposal of these products, PCBs enter aquatic ecosystems in a number of ways. When PCB-containing paper and plastics are burned, the PCBs, stable at high temperatures, can enter the

Box 7.1

Minamata—the Toxic Bay

People have long put up with pollution from local industries that are their economic lifeblood. Such was the case in the small coastal village of Minamata, Japan, where fishermen tolerated pollution problems from local industries for several decades. In the early 1950s, animals in the Minamata Bay region began acting strangely. Birds fell from their perches and flew into trees and buildings. Cats walked with an awkward gait; some became enraged, foamed at the mouth, and ran in circles until they died or were killed. Soon ominous symptoms appeared in the fishermen and their families: numbness, tingling sensations in the hands and feet, trembling, headaches, blurred vision, impaired speech. For the unluckiest victims, milder symptoms were followed by violent thrashing, paralysis, and even death. After some investigation it became clear that something was ravaging the villagers' brains and central nervous systems. The symptoms pointed to mercury poisoning.

Investigators found that fish and shellfish from Minamata Bay, which made up most of the villagers' diet, were contaminated by mercury compounds including mercuric chloride. They also found these mercury compounds in the effluent that flowed from the nearby chemical plant of the Chisso Corporation into Minamata Bay. And they learned that the plant had greatly increased its use of mercuric chloride not long before the residents of Minamata began falling ill.

By 1959, researchers had demonstrated that certain microorganisms in bay sediments were converting the mercuric chloride and other toxic mercury compounds into an even more toxic compound—methyl mercury. This chemical readily moves through food chains and accumulates at higher trophic levels.

How the surviving victims fared after the epidemic of Minamata disease, as the affliction had come to be called, is as tragic a story as that of the epidemic itself. The Japanese government did not officially recognize the cause of the disease until 1968, about 15 years after the first symptoms appeared. As late as 1970, government investigations recognized only 121 of the more severe cases as Minamata disease; of these, 46 proved fatal. And for years the Chisso Corporation itself managed to evade responsibility for what had happened. Not until 1973, after a four-year struggle in the Japanese courts, was the company required to pay monetary damages—$3,530,000—to just 112 victims and their families. By 1975, more detailed health surveys had recorded another 3,500 cases of the disease. But payments to so many more victims threatened Chisso's financial health. To avoid making them, the corporation established independent new companies from profitable subsidiaries. This move left the victims little hope of financial compensation if the parent company, stripped of its financial resources, filed for bankruptcy.

But monetary compensation, however generous, could not have undone the irreparable damage to the minds and bodies of many victims. On that day in 1973 when Chisso was ordered to make its first compensation payments, Shinobu Sakamoto, a 16-year-old girl whose muscular control had been gravely impaired, reacted grimly to news of the decision. "Money will not cure the disease," she said as she hobbled to school. "I want them to restore my body."

atmosphere and become attached to dust particles. The dust particles either settle to the ground or on surface waters, or are washed out of the air by precipitation. PCBs also gain entry into aquatic ecosystems via runoff, treatment plant effluent, and the leakage from improperly run municipal and industrial waste disposal sites. Problems with a related chemical, PBB, are discussed in Box 7.2.

Like the metal pollutants discussed above, chlorinated hydrocarbons such as PCBs and DDT concentrate at upper trophic levels of food chains. They are resistant to biological and chemical degradation and are particularly soluble in the fatty tissues of organisms. The highest concentrations are found in predators that occupy the top of food webs, for example, fish or fish-eating birds. Figure 7.10 illustrates how in Lake Michigan seemingly insignificant levels of DDT and PCBs become a serious problem as they move up the food chain. PCBs in sediments and water are incorporated into bottom-dwelling organisms and algae. These slightly contaminated organisms are consumed by small fish such as alewifes and young salmon and trout. Larger predatory salmon and trout feed on the PCB-laden alewifes, causing PCB levels in their bodies to increase with age to levels considered unsafe for human consumption. To protect the public from the PCB hazard, the EPA has banned the commercial sale of Lake Michigan's highly prized salmon and trout, and sport fishermen have been warned not to eat more than one meal of Lake Michigan trout or salmon per week. Levels of PCBs in Lake Michigan fish relative to the FDA safe limit are shown in Figure 7.11.

What would happen to people who did eat fish contaminated with PCBs? This question cannot be answered directly, but laboratory test animals fed diets containing PCB levels as high as those found in some Lake Michigan and Lake Ontario fish (up to 25 ppm) have suffered reproductive failure, hair loss, liver changes, and lesions in the digestive tract. And in experiments conducted by Dr. James Allen at the University of Wisconsin-Madison Medical School, monkeys fed a steady diet of only 2.5 ppm PCBs

Box 7.2

Frederic Halbert can unhappily confirm that lightning does sometimes strike the same place twice. In Mr. Halbert's case, it first hit nearly five years ago [in 1974], in the form of a mysterious ailment that sickened and in some cases killed cows on his dairy farm near [Battle Creek, Michigan]. For nearly nine months, Mr. Halbert frantically searched for the cause. When he finally identified it, the one-time chemical engineer had almost single-handedly uncovered one of the nation's worst agricultural disasters.

A toxic chemical fire-retardant, a powder known as polybrominated biphenyl, or PBB, had been mixed inadvertently into cattle feed that Mr. Halbert had bought. It was soon discovered that hundreds of other Michigan dairy farmers had received the tainted feed. And through the food products from those farms, PBB had passed into the diets of most Michigan residents.

The accident, which attracted national attention, produced a monstrous tangle of problems that still defy solution. For instance, despite continuing investigations by health researchers, nobody knows what dangers the chemical poses to people who unknowingly ate it (a study of Michigan farm families exposed to PBB indicated that it may be tied to various ills ranging from fatigue to loss of memory).

After an initial flurry of publicity about his chemical detective work, however, Mr. Halbert's name soon dropped from the headlines. Using money from insurance settlements, he quietly set about rebuilding his ravaged dairy business. He replaced his cows; about 75 had died because of

Excerpted from John R. Emshwiller, "Michigan Dairy Farmer Who in 1974 Found PBB in Cows Now Finds His New Herd Contaminated," *Wall Street Journal*, August 1, 1978.

PBB: The Environmental Poison That Won't Go Away

PBB contamination, and the remaining 725 had been quarantined and then destroyed.

But PBB has turned out to be an unexpectedly tenacious foe, and now it has struck the Halbert farm again [in 1978]. Over the past several months, the chemical has shown up in Mr. Halbert's new herd. His own farm—land and buildings—appears to be the culprit. Despite extensive clean-up efforts after the accident, tests indicate that traces of PBB persist over much of his property. Now Mr. Halbert is desperately trying to scour out the remaining contamination, but he concedes that PBB may finally be beating him. "It seems like a perpetual circle that can't be broken. I may just have to admit my land is poisoned," Mr. Halbert says. Mr. Halbert once again has company in his plight. State agriculture officials already have found 10 other farms with similar recontamination problems and are checking others.

When the accident was uncovered in 1974, little was known about PBB. However, scientists now know that it is an extremely stable and long-lived substance that will probably linger in the environment, or the human body, for years. Worries over these lingering molecules of PBB have sparked a major public-health controversy in Michigan. To clean up Michigan's food supply, state officials quarantined and destroyed tens of thousands of head of livestock. From the beginning, though, government regulators allowed food products with relatively low levels of PBB to be sold, assuming that tiny amounts of the chemical aren't dangerous.

But work by some researchers indicates that even minute quantities of PBB could harm human health. So late last year, in response to a rising public outcry, the state legislature slashed the PBB level allowed in dairy cattle to 20 parts per billion in animal fat from 300 parts per billion. It also ordered the state agricultural department to test Michigan dairy herds for remaining traces of the chemical.

And that's when Mr. Halbert's new problem surfaced. Tests show that many of his newly acquired animals exceed the new PBB limits. Mr. Halbert finds this turn of events particularly disheartening because after the accident he tried hard to protect his farm from further PBB problems. When buying his new herd, for instance, Mr. Halbert brought in most of the animals from Wisconsin, Iowa and Indiana. "I wanted to make sure they hadn't been exposed to PBB," he says. Mr. Halbert also tried to give his new herd a clean home. He purchased a high-pressure steam washer for the barns and bins where the tainted feed had been stored. He put down new concrete to cover some areas, and stopped using a 15-acre pasture where the "hot" cattle, as Mr. Halbert calls those initially contaminated, had been kept.

These efforts obviously fell short, Mr. Halbert now concedes. Today, tests show traces of PBB in his fields, where PBB-carrying manure from his original cattle was used as fertilizer. The chemical still lurks in his buildings—it has worked its way into the wood. A dust collector has found PBB even in the farm's air.

Mr. Halbert thinks that some help may finally have to come from a court of law. He is suing the companies responsible for the initial feed mix-up that led to the PBB accident. Mr. Halbert is seeking damages—amount unspecified—for his property, and says he may eventually try to get the companies to buy the farm. "I don't want sympathy," Mr. Halbert asserts. "I just want to be able to live without this knife in my back."

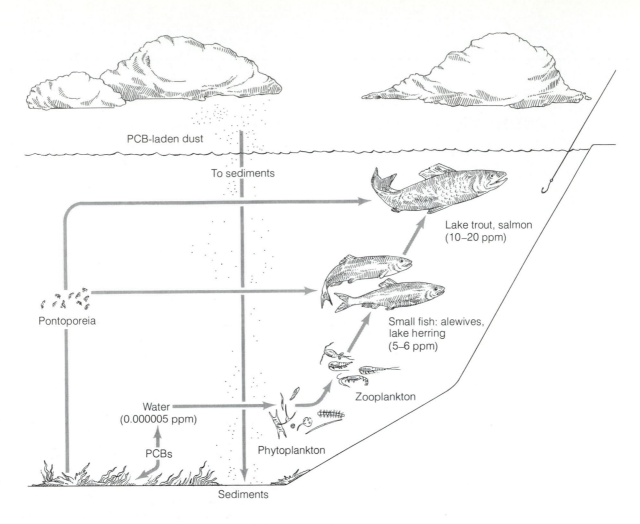

PCB-laden dust

To sediments

Lake trout, salmon
(10–20 ppm)

Small fish: alewives,
lake herring
(5–6 ppm)

Pontoporeia

Zooplankton

Water
(0.000005 ppm)

PCBs

Phytoplankton

Sediments

Figure 7.10 The accumulation of PCBs in Lake Michigan trout and salmon. The PCBs in these fish are over one million times more concentrated than those in the water.

exhibited disorders. Within one or two months, the monkeys lost hair and formed acne lesions. Reproductive failure and smaller offspring were also observed in the test monkeys.

Species from the Great Lakes are not the only ones found to have exceeded the FDA's 2 ppm standard for PCBs. Some species of fish taken from the Upper Mississippi and New York's Hudson River have shown high levels of PCBs too. The problems in the Hudson were traced to two General Electric Company plants that manufacture capacitors containing PCBs. FDA fish samples showed an average of 31 ppm PCB, and samples taken by General Electric showed 95 ppm. In the hopes of curbing the problem quickly, the company has agreed to pay at least $3 million toward the cost of cleaning up the contaminated river—a goal that no one is certain can be attained.

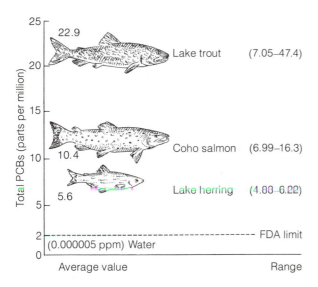

Figure 7.11 Levels of PCBs in Lake Michigan water and important fish species used for food. (Data from U.S. Environmental Protection Agency, National Conference on Polychlorinated Biphenyls, March 1976.)

Problems of PCB contamination are not likely to disappear quickly despite the fact that the Monsanto Company, this country's sole producer, ceased manufacturing PCBs in August, 1977, and halted sales two months later. These oily substances remain in use today in a large number and wide variety of devices. As these products wear out or are disposed of, some PCBs will ultimately find their way into the environment. And although United States industries have phased out the use of PCBs, products containing these substances continue to be imported from other countries. Thus, PCB contamination will undoubtedly be a problem for decades to come.

Other concentrated industrial chemicals that pose hazards both for aquatic life and human health are ammonia, cyanide, sulfide, fluoride, and strong acids and alkalis. (Acid water in the form of acid rain is discussed in Chapter 10, and acid mine drainage is covered in Chapter 12.) Industries, however, find most of these compounds indispensable. For example, compounds of cyanide such as potassium cyanide are essential for electroplating electronic components with silver and gold. Toxic chromium compounds are used to plate bumpers and other shiny ornaments. Accidental discharges of these highly toxic wastes, usually through human error, occasionally eliminate entire aquatic communities. Figure 7.12 shows the result of such an accidental discharge.

Though accidental discharges are dangerous, however, the major problem still lies in the improper treatment of industrial discharges. Spectacular industrial spills often receive the most media attention, but biologists believe that chronic exposure to toxic substances has a more catastrophic effect than acute exposure. These subtle long-term effects on aquatic organisms include lowered rates of reproduction, reduced growth rates, and abnormal behavior. Unfortunately, most sublethal effects of toxic substances on aquatic life are not well understood and further research is necessary. Similarly, we know little about the effects of long-term exposure to low levels of these chemicals in our drinking water. Nevertheless, it is not feasible for us to monitor thousands of kilometers of waterways for the vast array of industrial contaminants that could be present there. For this reason, we must institute strategies to protect public water sources from contamination rather than invest huge sums of money to develop water supplies of questionable quality. These strategies should include restrictions or bans on particularly hazardous chemicals and perhaps restrictions on the location of industries using hazardous materials.

Oil Spills

Oil pollution is an ever present threat to our surface waters, especially the oceanic and river waterways that are used to transport "black gold." Over the past decade, several dozen major spills have occurred worldwide, and one of these incidents, the grounding and breakup of the *Amoco Cadiz* on March 17, 1978, was the worst spill in maritime history. The

Figure 7.12 A massive fish kill, the result of an accidental discharge of toxic wastes. (© Robert Lamb, from National Audubon Society.)

supertanker, on charter to Shell Oil, was enroute from the Persian Gulf to English Channel ports carrying approximately 200,000 metric tons of crude oil. The vessel encountered heavy seas and experienced a steering mechanism failure. Rescue attempts by tugs failed, and the vessel went aground 2 kilometers (1 mile) off the northwest coast of France. The pounding seas broke the vessel open (Figure 7.13), and the ship's entire cargo of oil soon coated 200 kilometers (124 miles) of the French coastline and spread to a width of 60 kilometers (37.5 miles). Cleanup costs totaled $75 million, and $35 million in damage claims were expected.

Spectacular accidental spills draw a great deal of attention, but in fact they account for only 4 percent of the estimated 4 million metric tons of oil spilled

into oceans each year. Coast Guard estimates of the amount of oil spilled annually in United States waters vary widely. Reports agree, however, that oil is spilled in all the nation's waterways, from the high seas to the small inland streams. Perhaps only high mountain lakes are free from the threat of an oil spill. And the types of oil spilled range from thick, gooey, crude oil and asphalt to lighter refined products such as heating oil and diesel fuel.

The major causes of oil spills include collisions and groundings of tankers and barges, oil well blowouts, and ruptures or leakages of oil storage tanks and pipelines. Table 7.4 shows the specific causes of spills and the relative amounts of oil introduced by them into ocean waters. Spills of petroleum products from barges exceed those from tankers. Between

Figure 7.13 The supertanker *Amoco Cadiz,* breaking up in the English Channel. (Standard Oil Company, Indiana.)

1970 and 1975, barges spilled 53 million liters (14 million gallons) while tankers spilled 38 million liters (10 million gallons) in United States waters. A barge spill is illustrated in Figure 7.14. Routine tanker operations, however, such as the rinsing of ships' holds, the dumping of ballast water (water pumped into the ship's hold to stabilize it on its return voyage) and bilge water, account for much more lost oil than accidental spills from tanker breakups (though accidental spills may well have a greater impact). In addition, used oil is often dumped into sewers, only to find its way into surface water. In some waters, natural oil seeps may be important contributors.

The environmental and economic consequences of oil spills are not easy to assess. Near- or onshore oil spills, however, are known to be particularly damag-ing because of the abundance of vulnerable marine organisms they affect. Oil coats birds and aquatic furry animals (for example sea otters and seals), thereby destroying their natural insulation and bouyancy. Most affected individuals die of exposure or drown despite considerable efforts to save them. Coastlines become plastered with thick, gooey oil and cluttered with smelly dead seabed and intertidal organisms. Shellfish either die or take on an offensive odor or taste. Fishermen of the area lose a season's or more catch, vacationers cancel their plans, motel keepers lose business, and local residents cannot use their beaches.

Oil that is spilled into the ocean is dispersed via several routes. Because most of the oil in a spill is insoluble and lighter than water, it gradually spreads and thins to form an ever widening oil slick on the

Table 7.4 Sources of crude oil and petroleum products introduced into the ocean each year.

Source	Amount (metric tons)
Tanker Operations	
LOT* cleaning and ballasting	84,499
Non-LOT cleaning and ballasting	455,708
Product tankers using store reception facilities	19,492
Product tankers not using store reception facilities	63,832
Ore, bulk, and oil carriers cleaning and ballasting	119,543
Additional cleaning and disposal prior to drydocking	91,895
Tanker bilges	9,573
Tanker barges	12,787
Terminal operations	31,933
Total from these sources	889,262
Other Ship Operations	
Bunkers	9,055
Bilges, cleaning, ballasting	292,481
Total from these sources	301,536
Vessel Accidents	
Tankers and tank barges	124,071
All other vessels	48,972
Total from these sources	173,043
Offshore Production	
Total	118,126
Natural Seepage	
Total	600,000
Nonmarine Operations and Accidents	
Refinery and petrochemical plant waste oils	195,402
Industrial machinery waste oil	718,468
Automotive waste oil	1,034,588
Pipelines	25,574
Total from these sources	1,974,032
Overall Total	4,055,999

*LOT, load on top, is an oil loading technique which minimizes the disposal of oily water ballast into the ocean.

Source: W. B. Travers, and P. R. Luney, *Science* 194:791 (1976).

water's surface. As long as the slick persists—usually from a few days to a few weeks—it poses a threat to birds and animals. Some components are lost to the atmosphere via evaporation. Eventually, wave action breaks the rest into thin small patches of floating oil, and these subdivide further into small oil droplets. Wave action and currents distribute the small oil droplets throughout the upper layer of the water, and bacteria break down most of their components, although this process is slowed considerably in cold water. Oil that is not rapidly degraded by bacterial action collects to form baseball-sized tar balls, now found throughout the world's oceans and beaches. The estimated lifetime of an individual tar ball is about a year. Finally, a component of some crude oil (oil as it comes from the ground) is heavier than water, and consequently sinks, killing important plants and animals on the ocean floor. Oil that remains on the bottom is thought to have the greatest long-term impact on aquatic ecosystems.

The amount of oil that the oceans can tolerate has not been determined. Nevertheless, with more and more oil crossing the world's oceans, it is essential that we attempt to avert the economic havoc and ecological disasters associated with oil spills by adopting safer shipping methods. We examine issues associated with the transportation of oil and other hazardous substances in Chapter 8.

Thermal Pollution

Every fossil fuel or nuclear power plant discharges large quantities of heated water in the process of generating electricity. At our present level of technological development, we are unable to turn this waste heat into usable power. Waste heat originates because the steam that passes through steam turbines must be condensed before it can be repressurized. Water for cooling is withdrawn from a nearby lake or river and passed through the condensers where the steam is cooled. Heat from the steam is transferred to the cooling water, which is finally discharged as warm or hot water. Plumes of waste water being discharged from power plants are shown in Figure 7.15. Although other industries, such as the steel industry, use water for cooling, the electric power industry is by far the largest user and most important

Figure 7.14 An oil spill from a stray oil barge that was caught against a bridge on the lower Mississippi River. (U.S. Army Corps of Engineers, Vicksburg.)

contributor to the potential pollution problems associated with the discharge of heated water.

The production of waste heat, a side-effect of our insatiable appetite for energy, presents us with the problem of where and how to dispose of the vast quantities of warm water without seriously jeopardizing the environment. The problem becomes more pressing each year. By the year 2000 it is estimated that the equivalent of 25 percent of the total annual stream runoff of the United States will be used for cooling; in dry periods the equivalent of 50 percent of the country's river flow may be required.

While we do not know what the long-term impact of heated water discharges will be on aquatic organisms, we are familiar with some short-term effects. Probably the most serious problem is *thermal shock,* a condition sustained by a fish or other aquatic organism when the temperature of the water it inhabits changes suddenly. Abrupt temperature changes or the gradual occurrence of temperatures that exceed an organism's tolerance limit lead to death. Because the thermal tolerance range varies among species, the maximum temperature of discharge water permitted by government regulatory agencies depends on what specific species are to be maintained. Cold-water aquatic ecosystems (for example, trout streams) are so sensitive to an increase in temperature that the National Technical Advisory Committee to the Secretary of the Interior recommends that they not be used as a source of cooling water at all. For cooling water drawn from warm-water ecosystems this group recommends that water temperature not exceed 32° C (90° F) if largemouth bass are to be maintained, or 29° C (84° F) if perch, pike, walleye, and smallmouth bass are to be supported.

Abnormally warm water puts stress on aquatic life

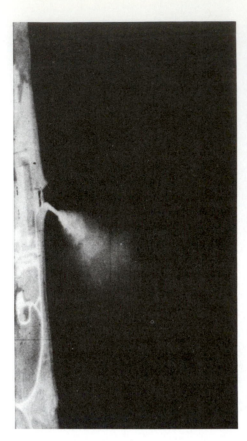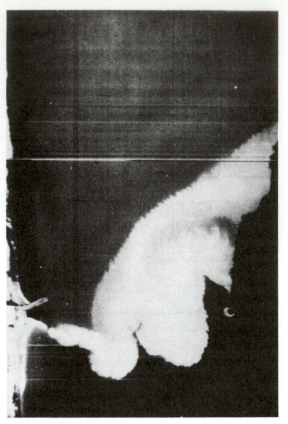

Figure 7.15 Two infrared photos of heated wastewater discharges emanating from power plants into a receiving body of water. The position of the discharge plume changes with current conditions. (University of Wisconsin-Madison, Marine Studies Center.)

in other ways. Cold-blooded animals (fish, oysters, clams), whose body temperature is the same as its surroundings, are particularly vulnerable. When water temperature rises, the body temperatures of cold-blooded animals rise too, causing increases in respiration rates that lead in turn to increases in their oxygen demand. Unfortunately, at higher temperatures water contains less dissolved oxygen, so organisms with increased demand suffer oxygen stress. In addition, toxic chemicals and diseases pose a greater threat at elevated temperatures. Unfortunately, the highest rate of thermal discharge is in summer, when the demand for power peaks. Thus, the greatest amount of hot water is discharged when organisms are the most vulnerable to a rise in water temperatures.

The withdrawal and use of cooling water causes other problems not directly related to heat discharge, as shown in Figure 7.16. Many fish die on the intake screens that prevent fish from being pumped through the power plants. Also, toxic levels of chlorine are periodically flushed through the cooling system (usually for about twenty minutes every four hours) to kill and remove bacterial slimes that build up inside pipes and reduce plant efficiency. This chlorine remains in the discharge water for a short time where it can be toxic to aquatic organisms. Also, in moderate-size lakes, the pumping of large volumes

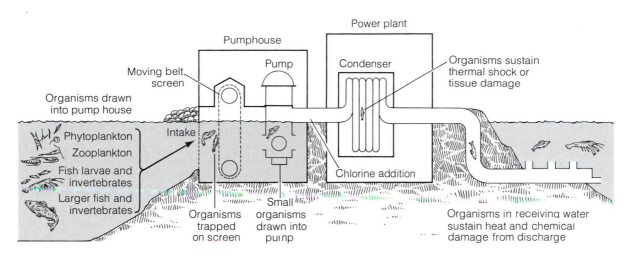

Figure 7.16 Effects of power-plant cooling systems on organisms in river, ocean, or estuarine ecosystems.

of nutrient-rich, hypolimnetic water to the surface would mechanically increase the rate of nutrient cycling and have the negative effect of speeding up the eutrophication process.

Scientists know enough about the discharges of waste heat to have reached some general conclusions regarding the siting of any industry that requires cooling water. Small and moderate-size rivers and lakes are inappropriate as cooling water sources, since large plants could divert the entire volume of such lakes or the entire flow of such rivers. Also, special precautions are necessary in tropical or subtropical settings where many organisms function just below their upper temperature tolerance limit. Studies at Turkey Point Nuclear Power Plant on Biscayne Bay, Florida, have shown that when water temperatures were increased by 4° C (7.2° F), great numbers of aquatic organisms were killed. The installation of cooling towers or cooling ponds is an alternative to the dumping of vast quantities of waste-heated water into aquatic ecosystems. An example of a cooling tower is shown in Figure 7.17.

However, cooling towers, too, can have negative effects on the environment. *Evaporative cooling towers* cool water by evaporating it. For example, at a large power plant (1000 megawatt electrical power output) about 1 cubic meter (260 gallons) of water is evaporated per second. In cold climates, the resulting water vapor may increase driving hazards by fogging highways and icing roads, and during the growing season, local vegetation such as orchard crops may be adversely affected by the increase in humidity and cloudiness. Also, disposing of the toxic chemicals used to prevent slime growth on cooling towers is another problem. The alternative to evaporative towers, *dry cooling towers,* are several times more expensive to build and require more energy to operate than wet cooling towers.

Despite the difficulties enumerated in this section, heated discharge water could conceivably be put to a number of beneficial uses. Aquaculture (fish farming) has promise in some locations where warm water can be cycled through pens to accelerate the rate of growth of such desirable food species as catfish, shrimp, and oysters. Aquacultural enterprises have been successful in Japan and are in the developmental stages in the United States. Heated discharge water also has potential for warm-water irrigation to

Figure 7.17 The cooling tower of the Trojan Nuclear Power Plant on the Columbià River near Prescott, Washington. (EPA Documerica, Gene Daniels, courtesy U.S. Environmental Protection Agency.)

extend the growing season in frost-prone regions. Heated water might also be used to heat swimming areas in cold regions, to deice canal and lock systems, and to serve as a heat source for chemical or other industrial processes.

The potential for using heated waste water in space heating is of particular interest. Heat pumps, devices that function like a refrigerator in reverse, recover heat from warm water and transfer it to buildings. Such heat pumps could help to alleviate thermal pollution problems while simultaneously saving precious fossil fuel. A few sewage treatment plants in the United States are already recovering waste heat and using it to heat buildings in the treatment complex. As the price of energy continues to skyrocket, the recovery of waste heat from major sources such as steel plants, paper mills, and sewage treatment plants, though expensive, grows more economically attractive.

Economic considerations dictate that waste heat be used in the immediate vicinity of its source. This constraint means that power plants would have to be located near urban areas if heat recovery techniques were to be used. The impacts of fossil fuel and nuclear fuel electric power plants, by far the largest contributors of waste heat, are discussed in Chapter 18. Whether the beneficial uses of recovering waste heat outweigh the increased risk of human exposure to air pollutants or radioactive wastes has yet to be decided by our society.

Pollution by Sediments

The impact of raindrops on bare ground dislodges soil particles. These particles make their way into streams, lakes, or oceans and are deposited as *sediment*. Coarse sediments settle out of water quite rapidly, but fine particles such as clays can remain in suspension for months, giving water a turbid appearance. Although rivers have always transported huge quantities of sediment to the sea, as we saw in Chapter 6, sediment loads today are greater than ever. Soils left unprotected by crop cultivation, timber cutting, strip mining, overgrazing pastures, road building and construction are subject to high rates of erosion. It has been estimated that on a per area basis, construction sites (see Figure 7.18), contribute ten times more sediment than cultivated land, two hundred times more than grassland, and two thousand times more than forested land. Drainage basins where strip mining has occurred also experience much higher erosion rates. Overall, in the United States, 85 percent of the soil erosion is estimated to originate from cropland. Although erosion rates of as

Figure 7.18 Construction sites, from which large quantities of sediment find their way to surface waters. (M. L. Brisson.)

high as 1 metric ton per hectare (3 tons per acre) are acceptable from the agricultural viewpoint, this rate may not be acceptable from a water quality standpoint.

The loss of valuable agricultural soils is one serious result of erosion, but many other problems are related to the wearing down of soil as well. For example, eroded soil particles eventually fill lakes, reservoirs, navigation channels, harbors, and river channels. Figure 7.19 shows the buildup of silt in a California reservoir. Excessive sediments greatly reduce the attractiveness and capacity of lakes and reservoirs for water-based recreational activities. They also impede navigation, cover bottom-dwelling organisms, eliminate valuable fish spawning areas, and reduce light penetration for photosynthesis. In addition, soils eroded from farmlands sweep nutrients in the form of nitrogen and phosphorus into surface waters. Thus, reservoirs that receive large sediment deposits are usually eutrophic because shallow water and high nutrient input rates result in high rates of nutrient cycling.

Although $15 billion has been spent on erosion control since the Dust Bowl days, soil erosion is still one of our most pervasive environmental problems and a major source of water pollution. Dredging and transporting sediments from reservoirs and stream channels is expensive, and the spoils of the dredging are usually dumped in valuable marshes and estuaries. In fact, each year the United States Army Corps of Engineers dredges about 300,000,000 cubic meters (400,000,000 cubic yards) of sediment from our waterways (enough to cover 360 square kilometers, or 90,000 acres of land to a depth of 1 meter) at a cost of about $250 million. From both economic and conservation standpoints, preventative measures make more sense than cleaning up after the fact. One means of controlling erosion is to encourage farmers

Figure 7.19 Silt deposits in the upper end of the San Pablo Reservoir, Contra Costa County, California. (U.S. Department of Agriculture.)

to utilize control techniques in cultivating their land (specific techniques are discussed in Chapter 17). Other strategies for reducing the amount of sediment reaching our surface waters involve minimizing the time during which construction and mining sites are left unprotected by vegetative cover, reducing the size of these areas as much as possible, building sediment retention basins, and directing water through swamps and marshes.

Groundwater Pollution

The same drainage basin activities that pollute surface waters can contaminate groundwater. Since groundwater provides 80 percent of the nation's rural water supply for domestic and livestock use and 30 percent of its municipal water supply, the impact of groundwater pollution can be considerable.

The composition of groundwater varies depending on the source of the water and its interactions with soil and rock materials along its path. For example, beneath swamps and peat bogs, groundwater is relatively acidic, since organic acids are released into it by decaying vegetation. A more familiar example is "hard" water, which contains calcium and magnesium that the water dissolved as it passed through layers of dolomite and limestone. Hard water can be a problem in residential use, since it requires large quantities of soap or detergent to achieve proper cleaning action. In response, "water-softening" detergents containing phosphorus have been widely used in hard water areas, and, as we have seen, adding phosphorus to water can precipitate the overgrowth of organisms in areas where phosphorus is a limiting factor. Also, at high temperatures the calcium in hard water comes out of solution and may clog hot water pipes or cause a buildup of scale in industrial boilers. Other naturally occurring substances in groundwater that cause problems are iron

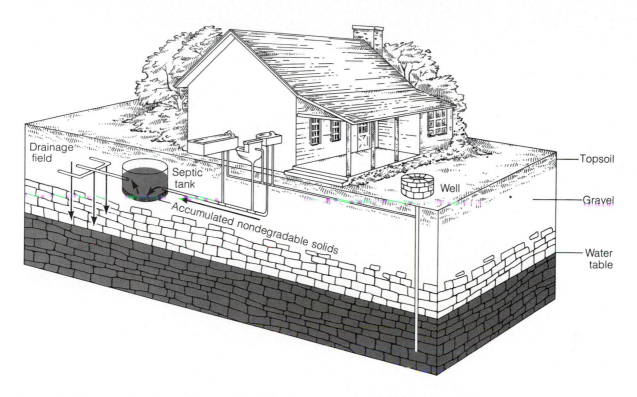

Topsoil

Gravel

Water table

Drainage field

Septic tank

Accumulated nondegradable solids

Well

Figure 7.20 Cross-section of a septic-tank drain-field system used for disposal of wastes in rural areas.

and manganese. These dissolved metals form the red-orange and black stains often seen on bathroom fixtures.

As water enters the ground it is naturally filtered through soil, and this process removes many substances such as phosphates, suspended solids, bacteria, and some viruses. In fact, infiltration through soil is so effective a purifying mechanism that septic tanks are designed to take advantage of it, as Figure 7.20 shows. In some places, industrial and municipal wastes are sprayed on the ground surface so that they may filter through the ground, thus becoming purified, and recharge the groundwater reservoir. Spray irrigation and ponding, shown in Figure 7.21, are two techniques of recharging groundwater supplies using the infiltration of waste water. In the United

States, about six hundred communities, mostly in arid and semiarid areas, reuse municipal treatment plant effluent after it has been treated in this way.

Proper filtration will occur only under certain soil drainage conditions, however. Filtration is impaired by high water tables, soils with low permeability, near-surface bedrock, and extremely sandy soils. Where these conditions exist, waste water does not spend enough time in contact with the soil before reaching the groundwater, so purification is inadequate. The bedrock of the Yucatán Peninsula of Mexico, for example, is composed of highly permeable limestone that readily allows contamination of drinking water supplies. Pollutants from cesspools and accumulated animal wastes find their way through rock cracks and crevasses into wells of the

Figure 7.21 Spray irrigation and ponding, two techniques for recharging groundwater supplies. (U.S. Department of Agriculture, Soil Conservation Service.)

region. In fact, in the mid 1970s scientists reported unacceptable levels of coliform bacteria in more than 80 percent of the samples tested from this area. Contaminated drinking water has caused infectious hepatitis to reach epidemic proportions, and dysentry now ranks as the leading cause of death in this region. Similarly, in certain areas of the United States where only a thin layer of soil overlies bedrock, contamination problems also occur. Groundwater under areas where thin soil covers limestone bedrock, as in the area illustrated in Figure 7.22, is particularly susceptible to contamination.

Furthermore, certain substances do not interact with soil and rock materials at all, but rather pass directly into the groundwater reservoir. One particularly troublesome substance is nitrate, which can seep into the ground from feedlots, septic tanks, and fertilized fields. Nitrate is potentially dangerous when it gets into well water. In concentrations of over 45 ppm, nitrate in drinking water is known to

induce abortion in cattle, and it is also linked to methemoglobinemia (commonly called blue baby disease), an often fatal condition in newborn infants that results from oxygen depletion in the blood. Nitrate concentrations can build up in groundwater where agricultural fertilizers are heavily used. In the Central Valley of California, concentrations have reached levels considered to be a public health hazard, and doctors there recommend that only pure bottled water be used in preparing infant's formulas.

Septic systems themselves, such as that shown in Figure 7.20, are a major source of groundwater contamination (and surface water contamination as well). About 22 percent of the population of the United States, primarily in rural and some suburban areas, rely on septic tank systems to dispose of human and other household wastes. In the past systems were improperly sited, and since then siting regulations have been slow to be enacted. Thus, water from systems located directly on top of frac-

Figure 7.22 An area where thin soil covers limestone bedrock, making groundwater below particularly susceptible to contamination. (M. L. Brisson.)

tured rock finds its way into wells, and systems placed on clayey soils quickly become plugged and rapidly overflow into nearby surface waters. Even when they are properly sited, however, septic systems can become overloaded, since many of the systems were designed to handle much smaller volumes of waste water than are now used in modern households. Convenience appliances such as dishwashers, automatic washing machines, and garbage disposals, as well as the greater emphasis on personal

hygiene have resulted in the generation of much more waste water per household than was the case when these septic systems were installed.

Other sources of groundwater pollution are the drainage from the waste heaps produced by mining and the runoff containing agricultural and industrial wastes. Water that seeps through municipal and industrial wastes dissolves (leaches) soluble materials and is thereby polluted. Leachate contains toxic substances, and if disposal sites are not properly engineered to collect and treat leachate, it can enter the groundwater system. (See Chapter 13.)

Conclusions

Our brief survey of the major water pollutants and their effects reveals the complex nature of our water pollution problems. The sources of pollution are many, the types are varied, and the effects are often poorly understood. Instances of catastrophic water pollution can distract our attention from the subtle effects of low-level water contamination, considered by many scientists to be the more important prob-

lem. These complexities make the control of water pollution difficult. The variety and multitude of sources challenge the ingenuity of water pollution control engineers. The lack of documentation regarding dangers of specific wastes makes industries, governments, and the public reluctant to pay for costly pollution abatement measures. Nevertheless, every effort must be made to control point sources of pollution, because once pollutants are released and become a problem, they are virtually impossible to retrieve, and it is left to the whims of nature to render them harmless. In the next chapter we survey recent efforts to clean up our waterways and evaluate their overall effectiveness.

Summary Statements

Historically, waterborne diseases were the primary water pollution concern. Chlorination of water supplies has effectively controlled this danger in the developed nations, but the concern is still relevant in the less developed countries. Coliform bacteria are used to indicate whether water is polluted with waterborne pathogens. The threat of waterborne disease is always present. To minimize this threat, wastewater effluents are chlorinated.

The decomposition of organic wastes in water by bacteria and fungi results in the depletion of dissolved oxygen in water. The concentration of organic material in water is measured by the biochemical oxygen demand (BOD). Surface waters experiencing decreased levels of dissolved oxygen are inhabited by undesirable pollution-tolerant organisms.

Municipal treatment plant effluents, some industrial effluents, and runoff from urban and agricultural areas contain significant quantities of aquatic plant nutrients. Phosphorus and nitrogen compounds are nutrients that accelerate the eutrophication process. Water quality in eutrophic lakes is impaired for industrial and municipal use as well as for desirable aquatic life.

Many chemical wastes are directly toxic to aquatic life. Some chemicals in water at extremely low concentrations become problems through food chain accumulation. PCBs and mercury are examples of substances that accumulate in fish to levels above the Federal Food and Drug Administration's limit for edible fish.

Oil spills cause direct harm to birds and aquatic organisms. Chronic effects of compounds in oil resistant to breakdown are poorly understood.

Heated water discharges may exceed the temperature tolerance limits of some aquatic organisms. Withdrawal of cooling water for power plants may kill fish on intake screens and increase rates of nutrient cycling in lakes.

Eroded sediments fill reservoirs, lakes, harbors, and navigation channels. In addition, suspended sediments impede photosynthesis, cover bottom-dwelling organisms, destroy fish spawning areas, and carry nutrients to water bodies.

The quality of groundwater is primarily controlled by purification processes that occur as water infiltrates soil. Areas with thin soils over fractured bedrock are especially susceptible to groundwater contamination. Septic tank systems in suburban areas can pose a serious threat to groundwater quality.

Questions and Projects

1. In your own words, write out a definition for each of the italicized words in this chapter. Compare your definitions with those in the text.

2. Develop a short history of water pollution problems in your area.

3. List the various types of water pollutants discussed in this chapter. Give an example and a possible source for each.

4. What are pathogenic organisms? How do scientists determine whether they may be present in water?

5. Describe the sequence of events that leads to the lowering of oxygen levels in streams receiving organic wastes.

6. What is meant when someone says that a stream or river has the capacity to cleanse itself?

7. How do the chemical products of aerobic decomposition differ from those of anaerobic decomposition of wastes?

8. Which of the following substances will decompose: chromium plating wastes? mercury? baking wastes? meat-packing wastes? PCBs?

9. How does natural eutrophication differ from cultural eutrophication? What water bodies in your area have suffered from cultural eutrophication? What are the major sources of the nutrients leading to cultural eutrophication?

10. Cultural eutrophication often leads to excessive populations (blooms) of blue-green algae. Give three reasons why blue-green algae are undesirable.

11. Are there any lakes or rivers in your region from which people are warned to restrict their consumption of fish or shellfish or where fishing is prohibited? If so, what is the economic impact of the restrictions? Which contaminants are responsible for the warnings?

12. Compile a list of toxic substances that are used by industry in your area. Historically, have any major accidental spills or flagrant dumping of these materials occurred locally? If so, are these incidents now curbed by pollution control measures?

13. Explain how a contaminant in dilute concentrations can become a serious pollution problem.

14. What properties of oil make it lethal to birds? What properties affect recreation? affect navigation?

15. What aquatic ecosystems in your region would be disrupted if a major oil spill were to occur?

16. How do fossil fuel or nuclear power plants threaten aquatic habitats?

17. Distinguish between point and nonpoint sources of pollutants.

18. What types of systems are used to dispose of household waste water in the rural areas of your state? How well do they work?

19. Does the rock and soil strata in your area have properties that are conducive to groundwater contamination? What are they?

20. What types of pollutants are most often responsible for groundwater contamination in urban-industrial areas? suburban areas? agricultural areas?

21. What single water pollution problem in your area has had the most drastic effect on the life in your community?

Selected Readings

Anonymous. 1977. "Chemical Plants Leave Unexpected Legacy for Two Virginia Rivers," *Science 198:*1015–1020 (December). An account of the impacts of mercury and Kepone pollution in two famous Virginia rivers.

Black, J. A. 1977. *Water Pollution Technology.* Reston, Virginia: Reston Publishing Company. A basic introduction to sources and consequences of water pollutants in freshwater and marine systems.

Council on Environmental Quality. *Annual Reports 1970–1978.* Washington, D.C.: U.S. Government Printing Office. A series of nine annual reports that include data on water pollution problems, sources, and clean-up efforts.

Ehrlich, P. R., A. H. Ehrlich, and J. P. Holdren. 1977. *Ecoscience: Population, Resources, Environment.* San Francisco: W. H. Freeman and Company. An in-depth survey of water supply and pollution problems is presented in Chapters 6 and 10.

Ember, L. R. 1976. "Environmental Concerns: Humans as the Experimental Model," *Environmental Science and Technology 10:*1190–1195 (December). Parts of this provocative article pertain to water pollution.

Gerba, C. P., C. Wallis, and J. L. Melnick. 1975. "Viruses in Water: The Problem, Some Solutions," *Environmental Science and Technology 9:*1122–1126 (December). Provides basic background on the fate of viruses in waste water.

Hodges, L. 1977. *Environmental Pollution,* 2nd ed. New York: Holt, Rinehart and Winston. Chapters 8 through 11 present a detailed account of the relative importance of water pollution sources.

Hutchinson, G. E. 1973. "Eutrophication," *American Scientist 61:*267–279 (May-June). Explains the scientific basis of eutrophication problems.

Miller, S. 1974. *Water Pollution.* Washington, D.C.: American Chemical Society. A collection of articles reprinted from *Environmental Science and Technology* that deal with the entire spectrum of water pollution problems.

Morton, S. D. 1976. *Water Pollution—Causes and Cures.* Madison, Wisconsin: Mimir Publishers. A factual discussion of water pollution problems without chemical or mathematical equations.

Smith, W. E., and A. M. Smith. 1975. *Minamata.* New York: Holt, Rinehart and Winston. A highly illustrated account of the tragedy at Minamata, Japan.

Travers, W. B., and P. R. Luney. 1976. "Drilling, Tankers, and Oil Spills on the Atlantic Outer Continental Shelf," *Science 194:*791–796 (November). Discusses the problems associated with the transport of oil across our oceans.

Wilson, R. D., *et al.* 1974. "Natural Marine Oil Seepage," *Science 184:*857 (May). Explains the seepage of oil from natural sources.

Warren, C. E. 1971. *Biology and Water Pollution Control.* Philadelphia: W. B. Saunders. An aquatic ecology text for undergraduate students that covers the impact of pollutants on aquatic ecosystems.

Some serious water management problems are approached by relatively simple means. Here a makeshift raft is floated in an Upper Volta stream as part of a program aimed at controlling river blindness. The can on the raft contains an insecticide that kills the larvae of the black fly, whose bite transmits the disease. (Agency for International Development.)

Chapter 8

Managing Our Aquatic Resources

In July 1977, the city of Chicago won a lawsuit against the city of Milwaukee, its neighbor 130 kilometers (80 miles) to the north. The decision requires Milwaukee to cease the occasional discharge of raw sewage into Lake Michigan—Chicago's source of drinking water. Wisconsin legislators, angered by the decision, threatened to retaliate by suing Chicago for sending polluted air into their state. Nevertheless, to meet the court order, Milwaukee has embarked on a $1.5 billion program to replace old combined sewers, mostly in the downtown area where replacement is extremely expensive and disruptive. A new sewer system that separates runoff from sewage, or a system that can treat both, is needed to prevent raw sewage from entering Lake Michigan during rainy periods. Both the city of Milwaukee and the EPA have indicated that they presently lack funds to implement the court order.

This incident illustrates the kind of conflicts that arise when a region's water resources are used for a multitude of purposes. Struggles related to quality and supply are inevitable when various interests vie for the use of any finite resource. For example, sport and commercial fishing interests are anxious to keep fish and shellfish safe for eating and free of foul tasting chemicals, while industry depends on a river's ability to degrade and dilute its waste and shies away from expensive control measures. Hunters attempt to preserve or even create wetland habitat for waterfowl, while farmers excavate ditches and drain wetlands to expand tillable lands. At the same time channels and harbors are deepened to accommodate increased barge and ship traffic and dredge spoils are spread over shoreline marshes rich in wildlife. In arid regions, cities and agriculture compete for a scant water supply, one side often diverting the entire flow of a mountain stream away from the other. In some locations, the political clout of highly populated areas is strong enough to wrangle water away from people in sparsely settled regions altogether.

Such conflicts—born of competition for a resource in limited supply—occur in virtually every community. To resolve these struggles, communities must develop plans for sound water resource management, a laudable goal that is much easier stated than achieved. Once restrictions are placed on water use in the interests of fair management, confrontations and lengthy litigation often result, for all interest groups assume that they have the right to use a public resource as they see fit. In this chapter, we examine the means whereby these conflicts may be, if not totally resolved, at least made less explosive through compromise and fair distribution.

A Historical Overview

The waterways of our nation were used by early explorers to open the American wilderness and to provide them with a route for taking stock of its vast resources. Lewis and Clark explored the Northwest on the Missouri River, and Father Marquette used the St. Lawrence, the Fox, and the Wisconsin Rivers, and the Great Lakes to gain access to the Mississippi waterway. Water transportation was viewed as the key element in the development of remote regions, since it was an inexpensive means of moving manufactured goods and natural resources. In fact, water transportation was deemed so important that between 1817 and 1840 about 5300 kilometers (3300 miles) of canals were frantically dug to connect inland cities with the world's waterways. The Erie Canal was the most famous of these; it connected the thriving port of New York with the rich agricultural, timber, and mineral resources of the Great Lakes region.

Beginning in the late 1800s, industries needed power, and growing communities actively encouraged their industries to take full advantage of available water resources for this purpose. But as industrial plants grew up along the waterways, so did the amount of waste they dumped into them. This cavalier means of waste disposal so polluted the nation's rivers that fishing and swimming became impossible in many of them. At that point, the use of the rivers was restricted to transportation and, sadly, continued waste disposal.

Today, however, we find ourselves entering a new era of water management. The uncontrolled exploitation of water resources is becoming a thing of the past; our concern now is to develop new technology for meeting our water management problems. Formerly, water resource development consisted only of hydraulic engineering efforts—the building of dams, irrigation projects, and aqueducts—but today water management is a much more complicated business. The wide variety of water resource problems we face requires the expertise of regional planners, civil engineers, biologists, chemists, economists, and recreation specialists. As we saw in the last chapter, these water problems go far beyond the fundamental one of assuring an adequate freshwater supply. Our success in meeting these complex challenges depends on our ability to educate ourselves and apply new knowledge.

Legislative Responses

We can trace the development of our water resource problems in the United States by sketching the legislative history of water quality management and planning over the years. The first action taken by the federal government toward improving water quality was the passing of the Rivers and Harbors Act of 1899, better known as the Refuse Act, because one of its provisions prohibited the direct dumping of trash into rivers and harbors. This law, however, soon faded into obscurity; it did not resurface until the 1970s.

The federal government's next response to the country's water pollution problems was the passage of the notably weak Water Pollution Control Act of 1948. This act was extended in 1956, and again in 1961. Water pollution problems grew more serious, and the Water Quality Act of 1965, more stringent than the 1948 law, was the response. This law strengthened the role of the federal government in water pollution control by establishing the Federal Water Pollution Control Administration, later reorganized under the Environmental Protection Agency. A key provision of this act required states to adopt stream quality standards by June 30, 1967, and to provide implementation and enforcement plans to meet these standards.

By 1970, the federal government was beginning to sense the technical and political complexities of water pollution control, because the states had responded to the Water Quality Act of 1965 with weak standards and had fallen far behind on implementing their abatement plans, if they had any at all. Congress reacted by passing the Water Quality Improvement Act of 1970, which strengthened federal control of discharges of hazardous substances from vessels and federal facilities. This law placed liability for oil spills from onshore or offshore facilities on the responsible party. The major strategy for dealing with water pollution continued to be the provision of federal grants for treatment plant construction. However, many industries and municipalities failed to apply for these grants or found the public coffers already emptied. Thus, pollution intensified from industries and municipalities that received no federal aid.

At this time, an attempt was made to use provisions in the Refuse Act of 1899 to combat pollution problems. This act required everyone, except municipal treatment plants, to have a permit to dump any refuse into navigable water. Administering these permits, however, proved to be a bureaucratic nightmare and litigation often resulted. To circumvent the permit system, some industries connected their disposal systems to municipal sewage treatment plants—sometimes surreptitiously. The toxic industrial effluents often poisoned or reduced the efficiency of municipal sewage treatment plants and pollution grew worse.

Vexed by the failure of prior legislation—and feeling increasing pressure from environmentally conscientious voters—Congress passed the Water Pollution Control Act of 1972. This act empowers the federal government to set minimum water quality standards for rivers and streams, although individual states are allowed to set more stringent standards if they so desire. The 1972 act makes discharging any pollutant illegal unless a permit is first obtained from the state. The act gives the EPA authority to impose effluent limitations, deadlines for compliance, and fines for industries and municipalities that do not meet the deadlines. Violations of conditions specified in a permit can result in fines of up to $10,000 per day. Repeated intentional violations can bring fines of up to $50,000 per day and a prison term of up to two years.

The 1972 law attempts to distribute more equitably the economic burdens of cleanup. Government experts set standards that are appropriate for each industry. For example, effluent limitations for steel mills are different from those for paper mills. Industries were given until July 1, 1977, to install the "best practical treatment" method. By July 1, 1983, industries and municipalities must comply with more restrictive regulations by employing the "best available treatment" technology. Where possible, the national goal is to have the nation's waters in fishable and swimmable condition by July 1983.

Strict enforcement of the stringent 1972 Water Pollution Control Act is finally reversing the long, steady trend of deteriorating water quality. Some rivers, such as the Willamette in Oregon (discussed in Box 8.1) and the Detroit in Michigan, have recovered to a point that astonishes people closely associated with the problem. Some fish-barren rivers have been repopulated and polluted beaches reopened for swimming. Figure 8.1 shows Matthews Beach on Lake Washington near Seattle, before and after recovery. And further improvement should be forthcoming, because 15 percent of industries and 67 percent of municipalities failed to meet the 1 July 1977 deadline. Many polluters who were not in compliance then are at this writing in the process of installing pollution abatement equipment. Still, the total effort is marred by some reluctant industries and municipalities that prefer to fight the legislation with courtroom battles rather than install pollution abatement equipment.

Section 208 of the 1972 act provides funds for the

Box 8.1

The Cleansing of Oregon's Willamette River

In Oregon's northwestern corner lies a 150-mile-long valley bounded by the snow-capped Cascade Range on the east and the less lofty Coastal Range on the west. This fertile green basin, drained by the Willamette River, embraces about 12 percent of the area of the state. And it is home for about 1.5 million people—approximately 70 percent of the state's population.

Despite its picturesque setting, the lower Willamette River was long plagued by massive water pollution. As early as 1910, the State Board of Health had described the river as

> . . . a conduit into which in increasing quantities in direct proportion to the increasing density of the population, along their banks is cast offal and filth until nearly all of the streams of the state have become mere sewers, the water from which is not only dangerous to drink but too filthy in many places to bathe in. Even the very fish which have no means of escape are largely becoming infected and unfit for food. This condition is rapidly growing worse and has become a peril of no mean import, and is a grave reflection upon the intelligence and degree of civilization of the entire community and should be stopped at once and forever.

How did this river, with the twelfth largest flow in the United States, come to be in such a deplorable condition? From their beginnings in the nineteenth century, the cities and industries of the Willamette Valley discharged untreated sewage and wastes into the river and its tributaries. As the state's three largest cities—Portland, Salem, and Eugene—grew, so did the flow of sewage they poured into the Willamette. And as the vegetable- and fruit-processing industries grew, so did their outpourings of wastes into the river.

In time, these discharges so badly polluted long stretches of the river that its water was unfit for domestic use, for recreation, or even for watering livestock. And by greatly reducing dissolved oxygen in the lower 50 miles of the river, particularly from July through September, the wastes also interfered with the propagation and migration of fish. Between 1920 and 1928, for example, waste-induced oxygen depletion blocked migrating salmon. Also during the summer months, large, unsightly, foul-smelling sludge mats, buoyed by bubbles of methane gas released by decomposing organic matter in the sludge, frequently rose to the surface.

Although the citizenry first became concerned about the Willamette's pollution in the late 1920s, the cleanup was slow. By 1939, less than 6 percent of the wastes flowing through the Willamette Valley's sewers was being treated before it entered the river. Moreover, most of the treatment plants served small communities on tributary streams. Urgently needed sewage treatment for the larger towns and cities was delayed for more than a decade by World War II. By 1957, all cities and towns had installed at least primary sewage-treatment facilities—that is, settling tanks or basins which removed some of the suspended solids from the effluent. The communities themselves had to pay for these facilities because state and federal grants were not then available. During the 1950s the valley's industries, too, reduced the amount of oxygen-demanding wastes they dumped into the river, especially during the critical low-flow months of August and September.

Despite these efforts, in the fall of 1957 the dissolved oxygen content of the lower river dropped to less than 1 ppm, far below the hoped-for level of 5 ppm. The problem was that pollution-abatement efforts had been offset by a rapidly expanding population and industrial base. The situation was aggravated further by an expanding fruit and vegetable canning industry that produced its greatest volume of wastes during the summer and early fall months when the river could least tolerate additional wastes.

People soon realized that the paper industry would have to upgrade its pollution-control facilities, and that cities coping with ever-growing quantities of cannery and meat-packing wastes would have to improve their sewage-treatment plants. Furthermore, the city of Portland was found partially responsible for the deteriorating conditions in the river because it had not completed its project to treat all its sewage. Legal action was filed against the city in 1959 to force it to speed up work on the project, and in 1960 Portland voters passed a five-year pay-as-you-go program to complete it. In 1961, Eugene upgraded its treatment facilities by adding secondary treatment facilities—tanks in which sewage from primary settling tanks is further processed. And in 1964, Salem followed suit.

The first signal that the valley's efforts to reduce water pollution were beginning to pay off came in 1965 when the Oregon Fish Commission counted 79 migrating Chinook salmon in the river. A thou-sand were counted in 1966, and twice that number in 1967. (In 1973, a record 22,000 salmon were counted.)

By 1968, marked improvements in water quality were apparent, and by 1969 the entire river met established water quality standards. (The improvements could not be entirely attributed to pollution-abatement efforts, however. Between 1941 and 1969, thirteen impoundments had been constructed in the Willamette River Basin. Water released from these reservoirs during the critical summer months diluted wastes, helped increase dissolved oxygen levels, and lowered stream temperatures.)

In 1968, to ensure that neither the Willamette nor any other river in the state would ever again be badly fouled, Oregon established a waste discharge permit system. Its regulations spelled out to all municipal and industrial dischargers the waste load limitations and control measures each had to meet so that the river basin's water-quality standards would be maintained. Each discharger was also told that these effluent limitations represented the maximum capacity of the river to assimilate waste, and that any future increases in waste-water discharges would have to be offset by a corresponding improvement in water treatment.

With these regulations, the flow of oxygen-demanding wastes into the Willamette River was finally brought under control. But municipalities and industries in the Willamette River Basin—like those in other heavily used river basins—will probably face other stringent pollution-control measures if the people demand further improvements in water quality.

Figure 8.1 Matthews Beach, on Lake Washington near Seattle, before and after pollution abatement by surrounding communities. (Municipality of Metropolitan Seattle.)

development and implementation of plans for water quality management of watersheds. Recognizing that water pollution problems often transcend the jurisdictional boundaries of states, counties, townships, and cities, provisions in the law allow local governments in the same drainage basin to band together to form a water quality planning agency. Such agencies have authority to set effluent standards for industries under the close watch of state regulatory agencies and to adopt other policies that will reduce pollution problems within the basin. Although this type of planning represents the first far-sighted effort at managing water pollutants on a watershed basis, it also introduces another bureaucratic step, in that both the state and the EPA must approve the agency's plans. The law also provides for public participation in the development of economically and socially acceptable plans for a specific region, given that minimum federal effluent standards are met. The success of such a watershed plan hinges to a large degree on citizen demand for cleaner water.

The 1972 Water Pollution Control Act is one of the most ecologically enlightened programs in environmental management. It recognizes that some wastes concentrate in food chains, and that emission of these materials must be halted. In addition, it recognizes that nonpoint sources are major sources of water pollutants in some watersheds. Through close monitoring of aquatic ecosystems over the next decade, we will be able to determine whether this law and related ones are sufficient to return our waterways to fishable and swimmable conditions.

A whole new set of water pollution concerns surfaced in 1974, when sixty-six industrial chemicals, some of them known carcinogens, were identified in the drinking water of New Orleans. These findings hastened the passage of the Safe Drinking Water Act of 1974, which gives the EPA responsibility for setting and enforcing minimum national drinking water standards. The act applies to public water systems having at least fifteen service connections or regularly serving at least twenty-five people.

The task given to the EPA under the Safe Drinking Water Act is formidable. Surveys of drinking water supplies across the United States turned up 253 specific organic chemicals. The large number suggests to scientists that more new chemicals will probably be identified. Furthermore, these chemicals were found to come not only from industrial sources, but also from municipal discharges, urban and rural runoff, and some natural sources. Ironically some chemicals formed as a result of chlorination.

A majority of the chemicals found in drinking supplies had never been examined for possible carcinogenic effects. Because the EPA has the legal responsibility to determine if a relation exists between the chemical composition of the nation's drinking water and chronic diseases such as cancer, the agency asked the National Research Council to address this question. The Council reported back in 1977 that although twenty-two carcinogens or suspected carcinogens had been identified in drinking water supplies, no hard evidence linked current concentrations of these compounds with a higher incidence of cancer. However, the Council also warned that these studies were incomplete and that caution should be exercised. An example of public responses to the Safe Drinking Water Act is presented in Box 8.2.

In addition to the Water Pollution Control Act of 1972 and the Safe Drinking Water Act of 1974, which together have primary authority for protecting public water supplies, seven other federal laws help to prevent water pollution. These laws are summarized in Table 8.1

Water Quality Control

As water moves through an urban-industrial system, it is pumped from a source, treated, distributed, used, collected, treated again, and then discharged into the nearest surface water. Figure 8.2 illustrates the urban-industrial water cycle. The process is made up of two treatment stages: one prior to use and one after use. Treatment of waste water is designed to protect water quality in aquatic ecosystems. Treat-

Box 8.2

Troubles on Tap:

When 45 people became ill at the crowded Safari campground [near Carlisle, Pennsylvania] during the 1977 Memorial Day weekend, they all had similar symptoms: intestinal cramps, vomiting and diarrhea. State health authorities found the cause in a place that campers in a modern American facility would never have suspected: the drinking water.

The investigators found that water from one of the campground's two wells wasn't receiving sufficient chlorine to kill harmful bacteria. State water inspectors could have alerted the campground to the problem before the outbreak, but they hadn't inspected the wells for two years. The campground quickly installed new treatment equipment. "We drink this water ourselves. We don't want it to be bad either," says Bernice Grell, the owner.

What happened at the Safari campground that weekend was by no means unusual, particularly in Pennsylvania, which for several years had led the nation in reported outbreaks of waterborne disease. Although Americans generally assume that their drinking water is guaranteed safe, in many areas of the country its quality has been declining over the past few years, federal environmental officials say.

The quality of tap water, although recently brought under the control of federal legislation, is still largely dependent on a jumble of often-conflicting regulations and on state and local officials who vary widely in their enthusiasm for maintaining water quality. Added to this is the fact that the nation's tap water is supplied by innumerable public and private sources, including thousands of very small water companies struggling

along with little money and with outdated treatment plants and old distribution systems.

Many water systems also are endangered by chemical pollutants, according to Jack Schramm, the Philadelphia-based regional administrator of the Environmental Protection Agency. "We have reason to believe that the trouble may be more extensive than is generally believed," he says.

However, the EPA does see some progress being made in improving some states' water-treatment and distribution systems and in having water samples collected and analyzed on a regular basis. This is mainly due to the passing of the Safe Water Drinking Act by Congress in 1974. Under this, the EPA was able for the first time to formulate the maximum levels of chemicals, bacteria and other contaminants that drinking water can contain.

These regulations, which went into effect in June, 1977, also spell out the extensive monitoring procedures for operators of public water systems. For example, the operator of a system serving a city of 50,000 is required every month to test at least 55 water samples.

The problem is that although all water systems in the U.S. must comply with the regulations, the EPA wants the individual states to enforce the law themselves. For a variety of reasons, several states have refused to go along with the EPA program, including Pennsylvania, Oregon, Utah, Wyoming, South Dakota and Indiana.

Some states are running into opposition from the water suppliers. In Minnesota, Gary Englund of the state's health department says that 37 communities that are "very antigovernment" are refusing to submit water samples for analysis by the state. In Virginia, regulators say they are running into roadblocks from small utilities that insist they can't

Excerpted from John L. Moore, "Troubles on Tap," *The Wall Street Journal*, October 9, 1978.

The Deteriorating Quality of Our Drinking Water

afford to make the required improvements to improve tap-water quality.

But it's heavily populated Pennsylvania that has some of the nation's worst examples of water quality. State health officials say that quality has slipped as the state's enforcement of standards has declined. They blame budget cuts, political infighting and obsolete standards.

The result can be seen in towns like Hooversville, in rural western Pennsylvania, where the tap water is sometimes "pure black" due to the high concentration of minerals such as iron, according to a state report. And the city of Philadelphia reports that algae growing in an open reservoir produced green tap water in one part of the city recently.

Similar money problems at the state level mean that inspections of utilities and of other water-supply systems are sporadic. Although some larger utilities and municipal systems are inspected an average of 2.3 times a year, smaller water supplies often are ignored.

Federal authorities say that if Pennsylvania were to police the new federal water-quality rules itself, as many as 30 new inspectors would be needed. But even though the EPA would have provided the state with $855,000 in subsidies this year [1978], little enthusiasm has been shown for enforcing the program by the state legislature or by the state's Department of Environmental Resources, which controls the two drinking-water agencies.

There are several reasons for this reluctance. Enforcing the EPA rules, it's feared, will be highly unpopular, particularly among the customers of small community systems. These systems will almost certainly have to raise their water rates to pay for the improvements required by the new standards. Added to this, the environmental movement in Pennsylvania is in decline, and there are few champions for new programs.

Thus the EPA is preparing to enforce the water-quality regulations itself, something it had been trying to avoid. "Because of its size, its industrial base and importance, it is the most important holdout state," says Mr. Schramm, the EPA official.

For many small water systems in Pennsylvania, the problem will be finding the money to conform with the federal water-quality regulations. Nearly 260 of the investor-owned companies have annual revenues under $50,000.

State engineers report continual problems resulting from antiquated treatment plants and distribution systems. Being privately owned, the small concerns aren't eligible for government grants or loans and find private loans difficult to get. Take the Hooversville Water Co.—the system that produced the black tap water. When the utility commission sent in investigators, they found the tiny concern in dilapidated condition: Breakdowns were frequent and the water mains leaked, which wasn't surprising considering they were the type of wooden lines that went out of use in the 19th Century.

That was in mid-1975. Rather than improve the service, the owner sold the system to the town of Hooversville for $26,100. But the town didn't have the money to make the necessary repairs. It was only recently that Hooversville obtained enough government grants and loans to begin an $86,000 program to install new mains, build a new treatment plant and drill a well. Until that is done, Hooversville residents will have to make do with water that, although quite safe to drink, is described by Mayor Timothy Swintosky as being "mostly brown."

Table 8.1 Federal legislation related to control of toxic substances in water.

Year	Legislation	Areas of concern
1954	Atomic Energy Act	Regulates low-level radioactive waste discharges into water.
1972	Marine Protection Research and Sanctuaries Act	Regulates ocean dumping.
1972	Ports and Waterways Safety Act	Regulates oil transport
1972	Insecticide, Fungicide, and Rodenticide Act	Classifies pesticides and regulates their use.
1974	Hazardous Materials Transportation Act	Regulates the transport of hazardous materials on land.
1976	Resource Conservation and Recovery Act	Regulates the treatment, storage, and disposal of hazardous wastes.
1976	Toxic Substances Control Act	Regulates the use of dangerous chemical substances.

ment prior to use is necessary because effluent standards are not as stringent as drinking water standards, nonpoint sources are not controlled, and not all point sources are in compliance with existing standards. We discuss wastewater treatment first, since it encompasses the techniques used to treat public water supplies.

Wastewater Treatment

The construction and operation of a sewage treatment plant represents a major investment of public funds. Therefore, citizens should be aware of what they get for their tax dollar. Sewage treatment plants are designed to remove from municipal sewage a wide variety of substances that people pump into sewers. These substances fall into three categories, each of which is treated differently within a sewage treatment plant. Insoluble materials, such as grease, oils, fats, sticks, beverage cans, and other assorted flotsam and jetsam floating on the water's surface, are mechanically screened or skimmed off the surface. Suspended substances—human wastes, paper, ground-up garbage, and so forth—are removed when water is diverted into quiescent basins where suspended substances slowly settle out. And dissolved pollutants—for example, sugar, starch, and milk—are removed by decomposers that feed on these organic wastes, the microorganisms themselves subsequently settling out in settling basins.

Activated sludge treatment is the most common technique used to treat municipal wastes and wastes from food and paper industries. This technique combines natural biological processes—aerobic decomposition—and mechanical processes to remove and degrade wastes in water. Activated sludge treatment plants take up less space than other treatment methods, an important consideration in view of high land values in cities.

A simplified diagram of an activated sewage treat-

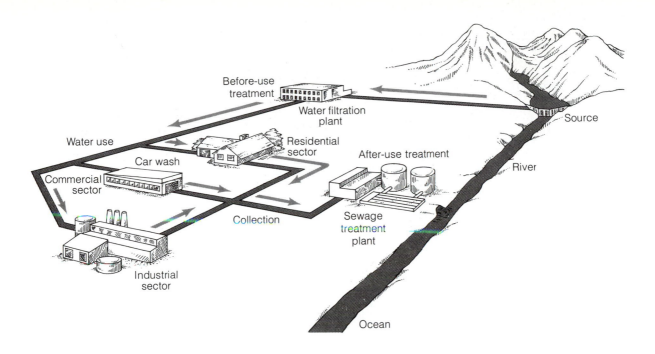

Figure 8.2 The water-use cycle in an urban area.

ment plant is shown in Figure 8.3. Sewage flows by gravity or is pumped into the treatment plant. As sewage enters a plant, it is typically 99.9 percent water and 0.1 percent impurities. In the first step, sewage is passed through a screen that removes debris such as rags, sticks, and cans. It then flows into a small grit tank where fast-settling materials such as sand and pebbles are removed. The materials removed in these first two steps are generally disposed of at a landfill. These pretreatment procedures are necessary to protect pumps and other mechanical equipment in a treatment plant. Once they are completed, the sewage enters the primary settling tank, where its velocity is reduced, and within two hours about 45 percent of the organic suspended solids settle as *sludge* to the bottom of the tank. This process of freeing waste water of its suspended solids is called *clarification*. The settled sludge is then pumped to sludge-handling equipment.

After being partially clarified in the primary tank, the sewage enters an aeration tank where it remains for six to eight hours. Here compressed air is continuously pumped through the sewage to provide oxygen for aerobic bacteria. In this tank, bacteria multiply rapidly and degrade both the remaining suspended material and most of the dissolved organic wastes. From the aeration tank, water flows into a final clarifying tank, where bacterial cells form clumps called *floc*. Floc settles out in the final clarifying tank, and some of it is recycled back into the aeration tank to provide "seed" bacteria. Because recycled bacteria are well-adapted (activated) to conditions in the aeration tank, they multiply quickly, and therefore are especially efficient at removing organic wastes. This step gives the process its name—activated sludge treatment. Unrecycled floc is pumped to sludge-handling equipment. Water leaving the final settling tank is clear and contains only about 10 to 15 percent of the organic material originally present in the in-coming sewage. But because it

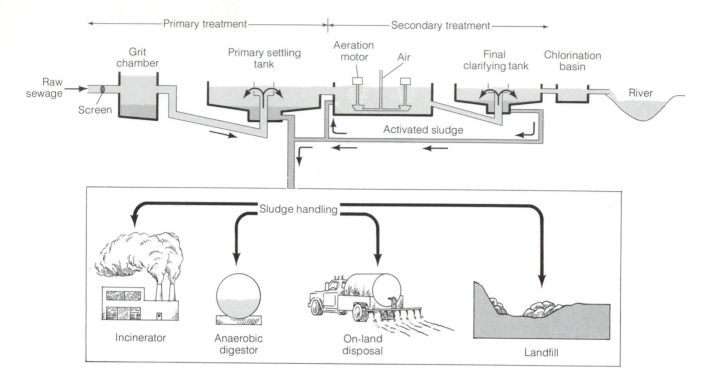

Figure 8.3 An activated sludge treatment plant for sewage.

is possible for pathogens to survive the prior treatment steps, the waste water is chlorinated before being discharged into surface waters.

The treatment process itself creates another major problem—the disposal of accumulated sludge (mostly composed of dead bacteria). Sludge-handling equipment often represents about 50 percent of the capital cost of a modern sewage treatment plant. The nature of the sludge formed is dependent to a large degree on the composition of the sewage received by the plant. In some cases, the sludge can be marketed to offset sludge handling costs. For example, in Milwaukee, Wisconsin, brewery wastes and domestic sewage produce a sludge that is sold as a commercial organic fertilizer for use on lawns and golf courses. Unfortunately, rising energy costs may force this plant to cease fertilizer production. Other plants,

though, have found uses for sludge more in keeping with the fuel supply shortage; they use anaerobic bacteria to digest sludge, thereby producing methane gas, which can be recycled back to the treatment plant buildings as an energy source.

The uses to which sludge can be put are limited by the composition of the product. For instance, not all cities can use sewage sludge as fertilizer, since their industrial waste water contains toxic metals that are concentrated in the sludge during treatment. If contaminated sludge is applied to land as fertilizer, the toxic metals can accumulate in crop plants. These metals, however, do not accumulate in seeds, so such crops as corn and wheat can be fertilized with contaminated sludge provided the entire plant is not used for cattle feed. The toxic metals in some sludge may also poison anaerobic bacteria, making the anaerobic digestion of sludge impossible. Hence, some plants dry and burn sludge containing toxic substances rather than turning it to another use. Even sewage sludge that is relatively uncontaminated is often burned, disposed of in sanitary landfills (see Figure 8.4), or dumped into the sea, and the fertilizing and soil building value of this material is lost.

Sewage treatment is classified as primary, secondary, or tertiary, according to how thoroughly the effluent is processed. *Primary treatment* takes sewage only through the primary settling stage, removing about 35 percent of the oxygen-demanding wastes—or BOD—and 60 percent of the suspended solids. In *secondary treatment,* the sewage is subjected to the additional step of biological degradation, which brings the total BOD and suspended solids removed to about 90 percent (refer back to Figure 8.3). But even after secondary treatment, about 90 percent of the plant nutrients (phorphorus and nitrogen compounds) and about 10 percent of the organic materials resistant to biological breakdown remain in the treated effluent. If these substances will aggravate water pollution problems in the receiving surface water, advanced secondary treatment procedures, sometimes referred to as tertiary treatment, are required.

Tertiary treatment is accomplished by a variety of procedures that have various treatment objectives. If cultural eutrophication is of concern, treatment plants must remove phosphorus. More efficient phosphorus removal is required by the EPA where plants empty into bodies of water in which phosphorus is considered to be the limiting nutrient for algal growth. The Great Lakes are examples. The most commonly used method for phosphorus removal is the addition of chemicals that, upon introduction into waste water, form precipitates that coagulate and settle, dragging along with them small suspended organic particles, as in Figure 8.5, step C. Alum (aluminum sulfate), ferric chloride, and lime (calcium hydroxide) function in this way. This process increases the removal of BOD and suspended

solids to about 98 percent. Up to 99 percent of the phosphorus is removed from waste water when the water is filtered after the chemicals have been added (Figure 8.5, step E).

In some instances it has been more economical to avoid cultural eutrophication by diverting treated waste water away from lakes. Lake Washington near Seattle, shown earlier in Figure 8.1, formerly received nearly 80 million liters (20 million gallons) of treated domestic wastes per day. The effluent from sewage treatment plants contained at least 50 percent of the original nutrient load, and therefore aquatic weeds and algae thrived in the lake. In response to this problem, a pipeline was installed to funnel wastes around Lake Washington to a new treatment plant on Puget Sound. Treated effluent from this plant is discharged into the offshore tidal flow, which carries remaining nutrients out to sea. Because treated effluent is no longer discharged into Lake Washington, water quality in the lake has vastly improved: algae populations are lower and waters are discernibly clearer.

Normally, nitrogen is not removed from sewage effluent, but in special circumstances it may be desirable to do so. To preserve Lake Tahoe's pristine nature, for example, local citizens agreed to add nitrogen-removal equipment (Figure 8.5, step F) to their sewage treatment facilities. In general, nitrogen removal is desirable where effluents have high levels of nitrogen and where nitrogen may be the limiting factor, the case in some lakes and estuaries. Nitrogen removal may also be required where treatment plants recharge groundwater.

Even after all these treatment procedures, some chemicals are likely to remain in the treated water. Examples of such persistent materials include color-causing chemicals (brown-stained water) and many industrial chemicals. Most of these substances are removed by the filtration of the water through powdered or granulated activated carbon (Figure 8.5, step G). When carbon absorption is used after clarification and filtration, the water is of such high qual-

Figure 8.4 Two methods of sludge disposal. *Top:* Sanitary landfill. (M. L. Brisson.) *Bottom:* On-land disposal in Fulton County, Illinois. (The Metropolitan Sanitary District of Greater Chicago.)

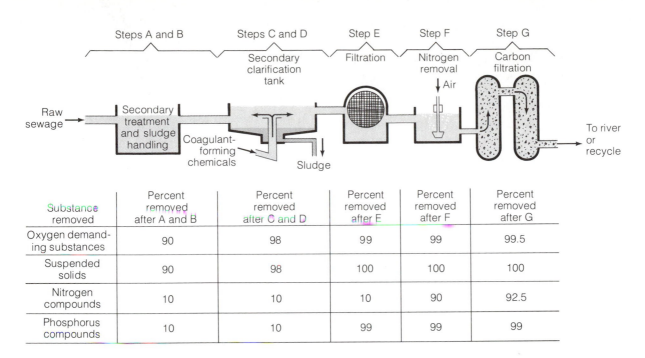

Substance removed	Percent removed after A and B	Percent removed after C and D	Percent removed after E	Percent removed after F	Percent removed after G
Oxygen demanding substances	90	98	99	99	99.5
Suspended solids	90	98	100	100	100
Nitrogen compounds	10	10	10	90	92.5
Phosphorus compounds	10	10	99	99	99

Figure 8.5 Tertiary steps available for water treatment and the resulting improvement in water quality. (After U.S. Environmental Protection Agency.)

ity that it can be reused directly after chlorination. In Windhoek, South Africa, carbon-filtered sewage effluent is directly recycled into the drinking water supply. South Lake Tahoe cycles its carbon-treated water through a recreational lake that supports trout, a fish that is especially sensitive to the presence of water pollutants.

Most large industrial plants have their own wastewater treatment facilities, which discharge into the nearest surface water. Industries that produce organic wastes rely on the same techniques employed in primary and secondary stages of conventional sewage treatment. If nondegradable wastes are present in their waste water, industries dispense with primary and secondary treatment and use only one or more of the tertiary steps we have described. For example, toxic metals are often removed through the addition of chemicals that induce coagulation and settling.

The next-to-last step in all municipal and some industrial treatment processes is chlorination. Chlorine is a chemical that kills pathogenic organisms that may have survived the treatment steps. Although this step is necessary to prevent the spread of waterborne diseases, scientists have recently discovered that chlorination also results in the formation of small amounts of potentially dangerous chemicals, called *chloroorganic* compounds. The most prevalent of these compounds are called trihalomethanes, of which chloroform is the most common. How hazardous extremely low concentrations of these chemicals are is yet to be determined. The discovery that chlorination results in the formation of chloroorganic substances poses a serious conflict for water resource policy makers. Chlorination disinfects the effluent from sewage treatment plants, but this practice results in the formation of undesirable substances. For this reason, new methods of disinfection are currently being sought. We will discuss the implications of these findings in water supply treatment in the following section.

The final step in the sewage treatment process is disposal. Most communities discharge their treated waste water into nearby waterways. As we saw in Chapter 7, however, some communities dispose of their treated water effluent on land, utilizing the natural filtering action of the soil to remove about 98 percent of the BOD, 95 percent of the phosphorus, and 85 percent of the nitrogen that remain after secondary treatment. Land disposal of waste water from secondary treatment plants cycles moisture and nutrients for crop production, thus turning a disposal problem into an asset. Generally, waste water is delivered to crops by conventional irrigation methods, such as fixed or moving sprinkling systems. In Muskegon, Michigan, over 150 million liters (40 million gallons) per day of treated waste water irrigates about 26 square kilometers (10 square miles) of corn. Other cities spread treated waste water on land to recharge groundwater supplies, again taking advantage of the soil's capacity to filter the water.

The 1972 Water Quality Act required that all communities treat their waste water at least through the secondary treatment stage. Some communities are satisfied with meeting EPA requirements, but others may choose to bear the cost of further treatment in the interest of producing cleaner water. The decision rests partly on the fact that each step beyond secondary treatment increases treatment costs by roughly 20 percent, as Figure 8.6 shows. Such costs are borne by both taxpayers and consumers. It is to be hoped, however, that in the future ecological soundness will be weighted at least as heavily as economics in decisions about treatment.

Water Supply Treatment

Approximately 240,000 public water supply systems exist in the United States. Of these, 30,000–40,000 are municipal systems that were installed decades

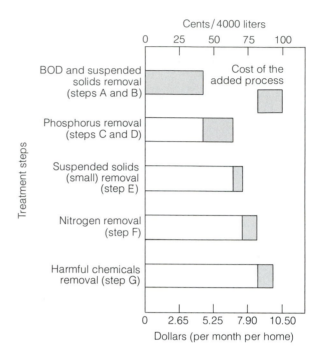

Figure 8.6 The cost of additional tertiary steps in water treatment. (After U.S. Environmental Protection Agency.)

ago, when water was relatively clean and treatment was designed only to prevent the transmission of waterborne diseases. The remainder of the water supplies regularly used by the public includes approximately 200,000 hookups in service stations, motels, and the like that are not connected to larger regulated municipal supplies.

The degradation of surface water and groundwater quality over the years has made additional precautions necessary for insuring that drinking water supplies are safe. The Safe Drinking Water Act of 1974 requires that public drinking water supplies meet standards for more than fifty separate physical, chemical, biological, and radioactive components. The law allows citizens to sue municipalities if these criteria are not met. It is probable that maximum limits for more chemicals, most likely the carcinogens found in trace amounts, will be added to the list in the near future.

Contaminants in drinking water supplies are controlled by well-established treatment procedures. But the degree of treatment necessary hinges on initial water quality. Most groundwater is sufficiently pure to require only chlorination prior to distribution. Surface water is generally stored for several days in a quiescent reservoir to allow suspended materials to settle out and the dissolved oxygen content to increase. These procedures improve water clarity and taste.

If surface waters contain algae, bacteria, or suspended clay particles, additional treatment steps are needed before chlorination. Sand filters are normally used for removing algae and bacteria, and suspended materials are removed through coagulation and settling procedures. Chemicals such as alum and lime are added to the water to form insoluble precipitates. The precipitates form a floc that gathers the suspended materials as it settles. In addition, some surface water and groundwater supplies are exceptionally hard. The calcium and magnesium, which cause hardness in water, can also be removed by the addition of alum and lime. Generally, softening water is less expensive at a central treatment facility than in individual homes by means of commercial softeners. Water purified with coagulation procedures, however, may still contain taste- and odor-causing chemicals, natural organic chemicals, and industrial chemicals.

As we noted in the preceding section, the standard practice of prior chlorination to eliminate pathogens is now under close scrutiny due to the discovery that chlorination of untreated water containing even small amounts of organic chemicals results in the formation of small amounts of chloroorganic compounds. In addition, these compounds may find their way into water supplies from industrial discharges. Some of these substances, such as chloroform, are known carcinogens. In an effort to minimize the exposure of people to these substances, the EPA is moving toward requiring public water utilities to keep the amount of chloroform and related substances in drinking water below 100 parts per billion by filtering the water through activated carbon. This plan, however, has been criticized as too costly. Costs for cleanup by carbon filtering would add $7–$26 annually to the water bill of an average family of three. Nationwide, projected costs are estimated at $616–$831 million. Opponents of the EPA requirements say it is unreasonable to spend such large sums of money when a definite link between low levels of chloroorganics in drinking water and cancer (especially bladder cancer) has not been established. The National Research Council's inability in 1978 to establish the connection between drinking water impurities and cancer added more fuel to the debate. The earlier studies reviewed by the Council did imply that a connection existed, but these studies did not take proper account of other well-known risk factors, such as cigarette smoking, alcohol use, occupation, coffee consumption, and socioeconomic level.

Thus, to date we still do not know whether intake of low levels of chloroorganic compounds causes cancer. The connection may well exist, and the EPA, in carrying out its mandate to protect public drinking water supplies, will probably continue to push for the costly treatment procedures. The public is caught in the middle: should we go along with the EPA or should we oppose these costly measures to keep local government costs down? As is often the case with environmental issues, the difficult choice is between economic consideration and an X factor, a possible but unproved risk. A wholistic approach to meeting a community's water and wastewater needs is discussed in Box 8.3.

Nonpoint Source Control

We have seen that a wide variety of techniques are available to control point sources of pollution, and that stringent enforcement of the amount of pollutants coming from these sources is technically feasible. But in some watersheds additional monies expended on point source control to meet the 1983

Box 8.3

Recycling Sewage Through Plant Communities

George M. Woodwell, author of the article excerpted here, is a major figure in the environmental sciences. His research has dealt mainly with the development, structure, and function of ecosystems, particularly forests and estuaries; the environmental cycling of nutrients; and the environmental effects of pesticides and ionizing radiation. Currently, he is Director of the Ecosystems Center at the Marine Biological Laboratory in Woods Hole, Massachusetts.

• • •

Modern cities are open systems, supplied continuously with essential resources such as water and food from the uplands and designed to release their wastes ultimately into nature for "treatment." Such a design is reasonable, of course, if the demands placed on nature are small in proportion to nature's capacity for meeting them. Problems arise when the cities increase in size and number; then the fresh water becomes more valuable and the wastes released are not "treated" by natural systems but accumulate to degrade ever-larger segments of the earth.

• • •

The problem is made worse by the admixture of industrial wastes that are released into municipal sewage systems. Some of these wastes are not rendered innocuous by natural systems; they may be accumulated to concentrations that are acutely toxic to the biota and a serious hazard to humans.

• • •

In effect the cities are using estuaries, other coastal waters, and thousands of cubic miles of ocean to dilute their wastes—and failing. The

open-system approach to the design of cities is obviously breaking down. In a crowded world the convenience of releasing wastes directly into nature is not available without unacceptable consequences. We face the challenge now and for the future of closing these open systems by designing cities that reuse their own wastes.

• • •

There is no simple solution. The construction of conventional "secondary" sewage-treatment plants such as those now being built may alleviate the problem but these are at best a temporary solution in a pattern that cannot succeed in the long run. Conventional sewage plants commonly collect the sediments, remove the oils and fats, aerate the water to hasten bacterial decay of organic matter, and chlorinate and release the partially cleaned effluent. The sludge may be digested further; the residue is dumped. This treatment is effective in removing much of the organic matter, including particles, but does not remove nutrients or certain toxins.

• • •

Is there an alternative to this pattern of water management that will assure recovery of the water, ultimately allow reuse of nutrients now wasted, and protect natural bodies of water from further corruption?

• • •

The answer seems to be a clear affirmative. . . . Recent research . . . has shown that aquatic systems, especially marshes and ponds used in conjunction with terrestrial systems that benefit from irrigation, can be used to manage sewage effectively and cheaply. A diagram of such a system is shown [at right]. . . . It now seems reasonable to assume that an area of 50 acres containing a marsh associated with a shallow pond of approximately equal

Excerpted from George Woodwell, "Recycling Sewage Through Plant Communities," *American Scientist* 65:556–562, 1977.

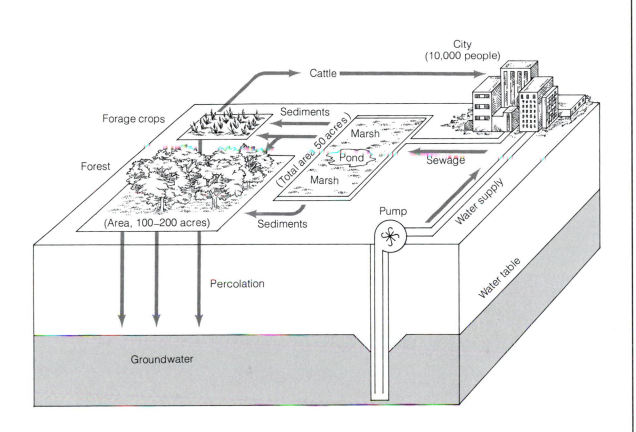

area could deal effectively with domestic sewage for 10,000 people. Sewage that contains industrial wastes requires a different system entirely; these wastes should of course be recovered by the industries themselves.

The area necessary may seem large, but the use of such an area is part of the cost of urbanization. Larger areas are being used now without controls: the Hudson estuary, the Charles River, San Fran-cisco Bay, Puget Sound, the New York Bight, and the lakes and oceans of the world. The cost is unmeasured, but it appears in higher prices for water, water shortages, the loss of fish and fisheries, and uncounted toxic hazards to man. In the light of these costs, the transition toward closed, controlled systems, even with their high requirements of land and careful design, seems not only necessary but economical.

deadline set by the 1972 legislation may not produce a noticeable change in water quality. Such is the situation in watersheds that receive a heavy pollutant load from nonpoint sources such as urban stormwater runoff, agricultural runoff, and the air (in the form of materials that settle or wash out from it). Studies carried out for the President's Council on Environmental Quality have shown that 40–80 percent of the BOD entering surface waters within cities originates from sources other than municipal and industrial treatment plants. Stormwater runoff is assumed to be the major contributor.

Some nonpoint water pollution problems can be solved by the regulation of specific chemicals. For example, since the EPA banned the widespread use of DDT, concentrations of DDT in aquatic ecosystems have declined measurably. Figure 8.7 shows the decline in DDT in Lake Michigan fish and the first indications that the PCB level might be responding to the withdrawal of PCBs from the market by Monsanto, the nation's sole manufacturer. Any decline in PCB levels, however, will be much slower than for DDT, because PCB-containing devices and materials will remain in use for several decades to come. Also, because waste PCBs were dumped illegally and before their dangers were known, they

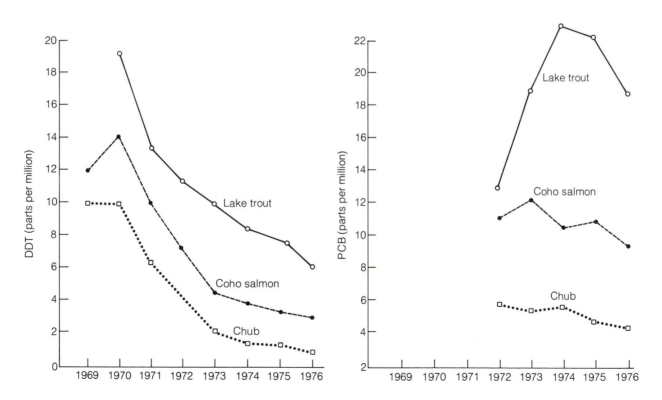

Figure 8.7 Average DDT and PCB levels in Lake Michigan. (Note the decline in DDT since it was phased out in the early 1970s.) (U.S. Department of the Interior, Fish and Wildlife Service.)

continue to enter ecosystems from many unknown sources.

Changes in land use regulations also influence water quality. In the past, the location of a septic tank was left to the whims of the landowner. Today, however, many states are beginning to implement permit systems to ensure that septic tanks are installed only where groundwater and surface water contamination is unlikely. In locales where soil conditions are not suitable for septic tanks, landowners are required to install holding tanks. Periodically, owners must pump these tanks out and haul the waste water to treatment facilities.

Some communities may be able to improve the quality of stormwater runoff by taking advantage of the "free services" provided by natural ecosystems. For example, at La Belle, Florida, stormwater runoff is diverted to marshes and swamps and the weedy edges of lakes, where vegetation filters out suspended solids and retains some of the nutrients. Stormwater is also ponded there so that it can soak into the ground and recharge groundwater supplies.

Controlling Accidental Spills

Every year, several spectacular accidents take place involving the spilling of hazardous substances into waterways. No means exist for containing water-soluble chemicals—for example, ammonia, cyanide, and phenol—and therefore, relying on dilution and degradition, however slow, is the only alternative in such circumstances. The effects of these measures are limited at best. Therefore the only real control procedure for water-soluble contaminants is accident prevention. Most accidents, however, involve petroleum or petroleum-derived products, as discussed in Chapter 7. Because oil and most related products do not readily mix with water, accidental spills of these substances can be contained.

Still, before we consider specific containment procedures, it is worth noting that prevention measures are as important in handling oil as they are in handling water-soluble chemicals. And as more oil is shipped in world trade, prevention becomes increasingly important. For example, it is essential that tankers be inspected regularly for seaworthiness. Too many old "rust buckets" are still transporting oil. Another important prevention measure, by no means utilized thoroughly at this time, is specialized training for oil tanker and barge personnel. Improved training could reduce the number of accidental groundings and collisions substantially, particularly in this era of supertankers. These gigantic vessels, such as the one in Figure 8.8, which carry ten to twenty times more oil than ordinary tankers, require about 3 kilometers (2 miles) in which to stop. Thus, with such large vessels on the sea, all captains and pilots would be well served by thorough training.

Finally, designing ships and barges with spill-prevention as a top priority would also cut the number of spills. If ships and barges were constructed with double hulls, the likelihood of ruptures of the innner containment section during a collision would be greatly reduced. However, at the International Conference on Tanker Saftey and Pollution Prevention, in February 1978, the United States received virtually no support for its proposal that this costly design alteration be made a requirement for all tankers. The conference did agree on some positive steps toward oil spill prevention, however. All new large tankers over 18,000 deadweight metric tons (20,000 tons) will be required to have separate ballast compartments (so that ballast water will not be contaminated by oil), dual radar systems, and back-up steering systems. Target dates for adopting these standards are between 1979 and 1981. Thus, it is obvious that even when a pollution problem is recognized internationally, reaching agreement about the solution takes considerable time.

When a grounding or collision does occur, the first priority is to contain the oil as quickly as possible. One procedure consists of pumping the oil out of the crippled vessel into a second ship. Often, however, empty tankers are not immediately available. For example, when the vessel *Metula* went aground in the Straits of Magellan in the mid-1970s, only 9,000

Figure 8.8 The supertanker *Arco Fairbanks* en route to the new oil terminal at Valdez, Alaska. (U.S. Coast Guard.)

metric tons of black oil were spread onto the turbulent waters initially. But ten days later, the ship broke up, releasing an additional 36,000 metric tons of oil. If an empty tanker had been available, much of the spilled oil could have been transferred and the impact of the spill greatly reduced.

Once oil is on the sea, it can be contained by a floating barrier or boom, as illustrated in Figure 8.9. A variety of devices have been developed in recent years for removing the contained oil floating on the water's surface. Some of these devices can remove up to 5700 liters (1500 gallons) of oil per minute in quiet waters. But, again, removal containment devices are not always immediately available. And containment equipment only functions properly in quiet waters. When wave heights exceed 1 meter (3 feet), containment equipment is useless, and the only hope then is that winds will keep the oil slick from coating the nearby shore.

An alternative to removing oil spilled at sea is to treat it in place by using one of several strategies: applying a material to which the oil will adsorb (hold to the surface), burning the oil, or adding a chemical preparation that will either sink or disperse the oil. Burning and adsorbing techniques are partially successful, but sinking or dispersing oil is usually undesirable, since dispersed oil can be even more harmful to fish and shellfish than a floating slick. Outside of harbors, the policy is to attempt to burn floating oil, but this is sometimes impossible, since igniting crude oil is often very difficult.

If the slick reaches the shore and beaches, mobile cleanup units can separate sand from its gooey coating, but these machines are expensive. A cheaper alternative is to soak up the oil with straw, hauling away and burning the straw once it is saturated. As you can see, prevention of an oil spill is the most important strategy we have for combatting the problem of oil pollution, since cleanup technology is primitive and costly, and all too often the necessary equipment is unavailable at the time and location of a spill.

Environmental Quality and Management

Figure 8.9 A floating containment boom used to surround oil spills. Contained oil is collected by pumping it into a barge. (U.S. Coast Guard.)

Enhancing the Freshwater Supply

In the past, when we wanted more water we either dug another well, pumped water from the nearest river or lake, or moved away and tried again. But today we have fewer options, since our growing population uses more water than ever before in every area of the world. In many regions, the rate of water withdrawal from rivers or underground reservoirs already exceeds the natural recharge rate. For this reason, water shortages are common on the mid-Atlantic seaboard, the Texas Gulf and Rio Grande regions, in lower Colorado, and in southern California. Figure 8.10 shows the surpluses and deficiencies in the nation's water supply. Worldwide, 70 percent of the human population lacks a safe and dependable water supply at this time.

The prospects for maintaining an adequate water supply as our population continues to grow are not encouraging. Some scientists have calculated that at present rates of freshwater withdrawal, the hydrologic cycle can furnish enough fresh water to support a world population of 8 billion. At the current rates of population growth and water use, this number will be reached by about 2020, well within the expected lifetime of most of us. And this estimate is based on the assumption that water will be available where and when it is needed. However, the world is plagued by droughts and floods continuously, and diverting water to places in need is often impossible.

In order that we can formulate our water management policies wisely, then, we must be familiar with techniques that can increase the availability of fresh water and that can conserve existing water resources. Presently, our primary means of increasing freshwater supplies are groundwater withdrawal, dam construction, and watershed transfer. We turn now to an examination of these methods and their far-ranging environmental consequences.

Groundwater Withdrawal

Groundwater supplies a portion of the water supply in many regions of the country. But groundwater reservoirs can be depleted if the withdrawal rate exceeds the rate of natural recharge. Such a situation now prevails in the High Plains region of West Texas. In the vicinity of Lubbock, Texas, over 14,000 square kilometers (5500 square miles) are irrigated by water pumped from the ground. Because

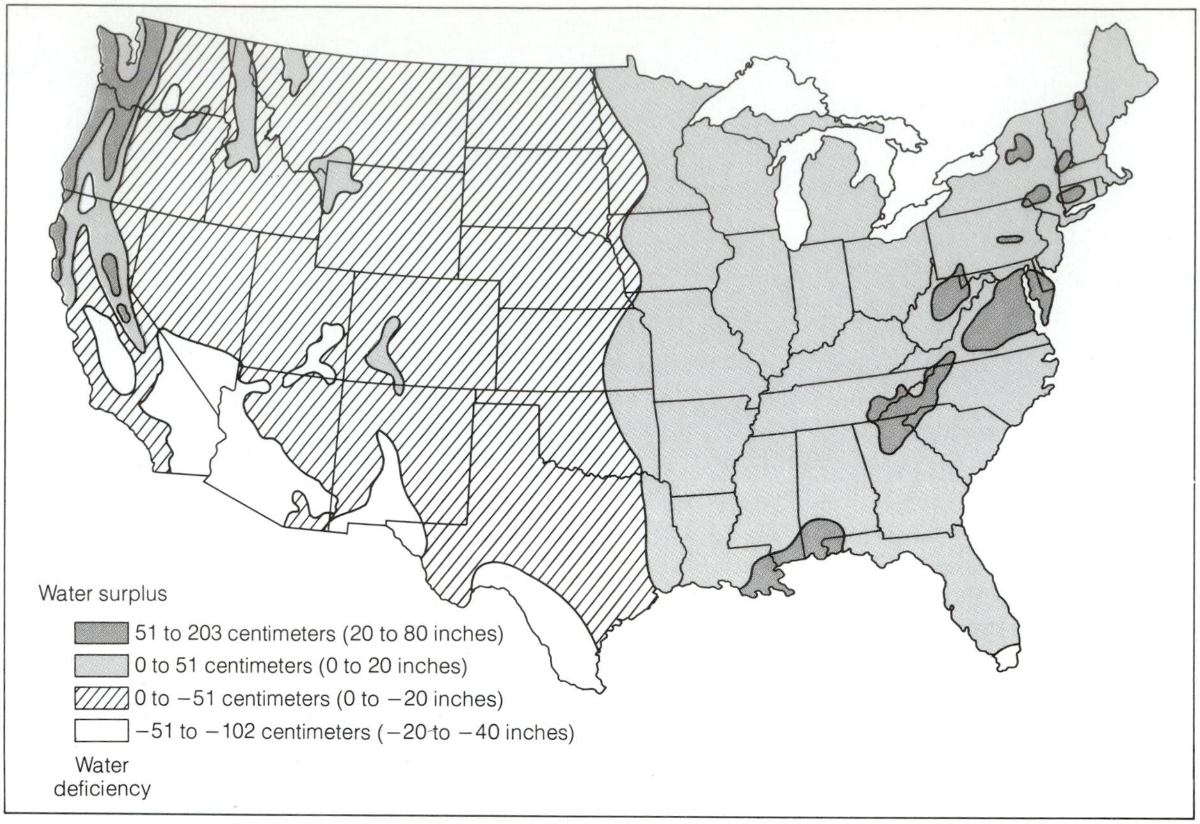
Water surplus
51 to 203 centimeters (20 to 80 inches)
0 to 51 centimeters (0 to 20 inches)
0 to −51 centimeters (0 to −20 inches)
−51 to −102 centimeters (−20 to −40 inches)
Water deficiency

Figure 8.10 Areas of water surplus and water deficiency in the continental United States, based on once-through use for the year 2000. (U.S. Department of the Interior Environmental Report. Conservation Yearbook Series, Vol. 6.)

of high evaporation and low precipitation rates in this area, little surface water is available for groundwater recharge. As a result, withdrawal is occurring fifty times faster than natural recharge. Consequently, the water table is dropping dramatically (see Figure 8.11). Around Lubbock, withdrawing water is appropriately called water "mining," since the water reservoir is being emptied. Although water conservation may ease the problem somewhat, the region must soon begin to make painful economic adjustments. A Texas Agricultural Experiment Station study of the area predicts that by the year 2015, irrigated acreage will have to be decreased by 95 percent. During this time span, agricultural production is expected to decline by 70 percent. Undoubtedly, businesses in this area will also experience economic hardships. Water is being mined in other areas, including the Mojave Desert of California; the cold-desert regions of Nevada, Utah, and Oregon; western Kansas; and along the Colorado-Nebraska border. These communities, too, will inevitably experience severe economic blows as their groundwater resources are depleted.

Another problem related to groundwater with-

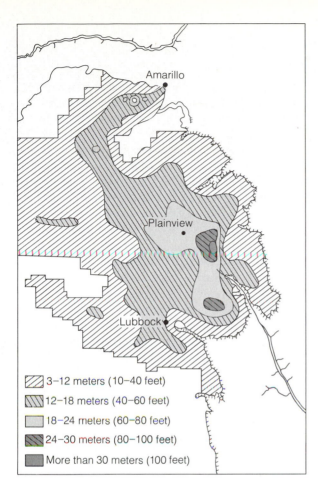

3–12 meters (10–40 feet)
12–18 meters (40–60 feet)
18–24 meters (60–80 feet)
24–30 meters (80–100 feet)
More than 30 meters (100 feet)

Figure 8.11 The High Plains area of Texas, where groundwater is pumped much faster than it is recharged naturally. The drop in the water table is caused primarily by heavy pumping of irrigation water. Numerical values represent the distance the water table has dropped from its position when pumping began. (After R. G. Kazmann, *Modern Hydrology,* 2nd ed. New York: Harper & Row, 1972.)

drawal involves the stability of the aquifer. Removal of groundwater sometimes allows the sediments that constitute the aquifer to compact, causing the overlying land to sink. This effect, similar to what happens when permafrost melts, is termed *ground subsidence,* and often accompanies heavy withdrawal of groundwater. Multimillion-dollar losses can result from ground subsidence, since repercussions often

take the form of structural damage to buildings and pipe networks, changes in the direction of flow in sewers and canals, and increased tidal flooding along seacoasts. A particularly dramatic instance occurred southeast of Phoenix, Arizona, where some 310 square kilometers (120 square miles) of land subsided more than 2 meters (7 feet). Damage was sustained by railroads, highways, and homes when cracks 3 meters deep, 3 meters wide, and 300 meters long (10 × 10 × 1000 feet) developed in the area. Further damage is expected as urbanization expands into the fissured area and as new fissures develop. In another example, in California's San Joaquin Valley, at least 30 percent of the land has subsided at least 0.3 meters (1 foot). Even this small amount of subsidence is sufficient to disrupt irrigation wells. The wells must be replaced, typically at a cost of $35,000 each.

In some areas groundwater is recharged artificially so the groundwater supply can be maintained or enhanced and the stability of the aquifer protected. As we suggested earlier, treated waste water can be spread over permeable soils and naturally filtered into the groundwater reservoir. Or water may be pumped directly into the ground through injection wells. Also, stormwater runoff can be diverted into catchment basins and allowed to infiltrate the ground. By such methods, water may be stored during periods of excess precipitation or runoff and then pumped out during periods of need. Groundwater recharge is used to enhance supply (for example, in Denver, Colorado, and California), to control saltwater intrusion (in Long Island), and to prevent land subsidence (in Mexico City and Houston).

Clearly, then, the protection of the nation's groundwater reservoirs is a major objective in our water management strategies. Overwithdrawal results not only in depletion of water supplies, but in serious ground subsidence and accompanying property damage. And the irresponsible or uninformed siting of industries or other waste-discharging concerns can result in the contamination of the water supply through aquifers. Thus, it is equally important that we keep the recharge areas of aquifers free from polluting activities.

Dams

Because precipitation is irregular, both seasonally and geographically, dams are sometimes needed to maintain a reliable water supply. Reservoirs created by dams catch water during rainy periods and store it for use during drier seasons. But dams have other equally important functions. In mountainous areas, deep canyons are dammed to create the large reservoirs necessary to generate power. Also, many dams are built as flood control structures. Clearly, we benefit from dams in many ways: they make possible electric power generation, provide many domestic and industrial water supplies, and create a multitude of recreational opportunities. For many people of the world, dams provide water for irrigation, and thus represent a means of adapting to drought. The benefits to be derived from dams, however, must be carefully weighed against the costs.

Pakistan's new Tarbela Dam on the Indus River illustrates the tradeoffs involved in carrying out a dam construction project. The world's largest earthen dam was constructed over a period of nine years at a cost of $1.3 billion. About 140 lives were lost in construction accidents. The dam creates a 260-square-kilometer (100-square-mile) reservoir to supply irrigation water for 135,000 square kilometers (52,000 square miles) of arid but fertile land. Also, the dam's hydroelectric generators are expected to double Pakistan's electrical energy output, thereby stimulating a rise in the country's standard of living.

Many problems have arisen since the dam was started. Unexpectedly, five major fractures opened in the reservoir bottom and had to be plugged at great expense. Many smaller fractures continue to form and allow water loss. This leakage is of particular concern because earthen dams can erode rapidly and collapse under the tremendous pressure of surging water, sending a wall of destructive water downstream. Reevaluation of erosion rates in the dam's drainage basin has led some experts to conclude that silt is being deposited quickly enough to fill the reservoir in twenty years. (Earlier projections of the filling time were fifty to sixty years.) If the twenty-year projection is correct, some sort of extensive soil stabilization program will have to be implemented within the watershed. But the physical construction problems are only one side of the story. The project has left some deep social scars as well. To make way for the rising waters, about 85,000 residents were displaced, some of them forcibly. Much of the government compensation for the flooded land was pocketed by corrupt local officials, and the displaced, landless farmers now idle away their lives in newly constructed government hamlets.

Many dams trap nutrients along with sediments, which accelerate eutrophication in the reservoirs, thus reducing the value of the water for recreation and drinking. And dams interfere with the spawning migration of some fish, such as salmon, unless fish ladders, illustrated in Figure 8.12, are provided. Without a way to cross a dam and continue their migration, such fish would be unable to complete their reproductive cycle. When rivers that flow into estuaries are dammed, the trapping of spring runoff can result in inadequate dilution of sea water by fresh water, and the consequent rise in salinity can prevent the reproduction of some fish and shellfish. Dams also reduce the flow of nutrients into estuaries, thereby limiting the productivity of these areas. And the land areas behind dams become flooded, with the result that vast areas of valuable agricultural land, wildlife habitat, and scenic beauty are destroyed. Finally, the safety of dams is threatened by geologic as well as structural factors, especially in earthquake-prone regions.

While dams can be greatly beneficial, their value must be determined in view of all possible risks, many of which have been overlooked in the past. However, recommending that such analyses be made is one thing; achieving agreement is quite another. People weigh risks differently according to their particular interests. The building of dams has

Figure 8.12 *Above:* A fish ladder (foreground) at the Red Bluff Diversion Dam on the Sacramento River in California. *Left:* A king salmon using a ladder to complete its migration and thus its reproductive cycle. (U.S. Department of the Interior, Bureau of Reclamation.)

always been controversial and is likely to grow more so as water supplies diminish in relation to the growing population.

Watershed Transfers

One of the most common techniques for increasing a limited water supply is transferring water from a watershed that has an excess. Several major U.S. cities, among them Los Angeles, Phoenix, and New York, rely on watersheds other than their own. In fact, Los Angeles transports water from the Colorado River through a 714-kilometer (444-mile) aqueduct system. The route of the system and a portion of the aqueduct are shown in Figure 8.13. Attempts to maximize the use of the Colorado by diversion have led to some major problems, however. As the water travels through the arid Southwest, it becomes more saline because of excessive evaporation. Hence, to meet drinking standards, it must be diluted with less salty water from Northern California sources. The waters of the Colorado are also used to irrigate crops in the Imperial and Coachella Valleys of southern California (see Figure 8.14), but only crop varieties with a high salt tolerance can be grown in these valleys.

Of the problems that arise from watershed transfer, many are political in nature. For example, irrigation water used in the Colorado watershed returns to the Colorado River with a much higher salt content. Hence, by the time the river flows into Mexico it is too saline for human consumption and the raising of most crops. To alleviate this problem, both countries signed an agreement on August 30, 1973. To meet the terms of the agreement, the United States is engaged in a $350 million program to reduce the salt content of water delivered to Mexico for irrigation in the Mexicali Valley.

Battles over water rights, especially in arid and semiarid areas, can be expected to intensify. Increased urban development, agricultural demand for irrigation water, and the mining industry's demand for water to process coal and oil shale resources cannot but create a three-way tug of war over this precious resource.

Other Enhancement Methods

Damming rivers, increasing groundwater withdrawal, or transferring watershed surpluses may be unfeasible or insufficient to meet the water needs of a particular region. In recent years, research has been directed toward other methods of freshwater enhancement. These techniques include cloud seeding, desalination, and various water conservation strategies.

Cloud seeding has been attempted often since the end of World War II. In this technique, natural precipitation processes are stimulated through the introduction of nucleating agents (artificial condensation nuclei) into clouds (see the discussion of precipitation processes in Chapter 6). In clouds of supercooled water droplets, dry ice pellets and silver iodide (a substance with crystal properties similar to those of ice) are injected to stimulate the ice-crystal process, whereas in warm clouds, relatively large water droplets and sea-salt crystals are injected to stimulate the coalescence process. These efforts are successful at initiating precipitation, and at first glance cloud seeding seems to guarantee the revitilization of farmlands parched by drought and to insure adequate water levels in urban reservoirs.

The actual amount of additional precipitation produced by cloud seeding and the advisability of large-scale seeding efforts are matters of controversy, however. Although cloud seeders claim to have increased precipitation by more than 20 percent in some instances, the question always remains as to whether the rain that falls following cloud seeding would have fallen anyway. There have been no reliable scientific studies to substantiate the cloud seeders' claims. Further, we must remember that cloud seeding merely facilitates the transfer of water from clouds to the earth's surface; it cannot produce rain without clouds. On a large scale, if successful, rain making may merely redistribute a fixed supply of rain; thus, increased precipitation in one area might be offset by a decrease in another. For example, rain making may benefit agriculture in the high plains of eastern Colorado but deprive wheat farmers of rain in the adjacent states of Kansas and Nebraska.

Even if cloud seeding is successful, it may not necessarily benefit all the inhabitants in a region. In a recent example, attempts at increasing precipitation over the Colorado Rockies triggered a considerable controversy. The Colorado River drains the melting snowpack of the Colorado Rockies. These waters in turn are used for irrigation and municipal and industrial needs in the arid lands of Arizona, southwestern California, Colorado, Nevada, Utah, and New Mexico. In 1974, a Stanford Research Institute study indicated that the water supplied by the Colorado River could be increased by 17 percent by increasing the mountain snowpack through the seeding of clouds by silver iodide generators, the process illustrated in Figure 8.15. This project would make more water available not only for irrigation and other community needs, but also for the development of hydroelectric dams and the exploitation of the region's coal and oil shale resources. But, except for those who rely upon the ski industry for a living, inhabitants of the mountains have much less to gain from cloud seeding. Making the winter snow cover thicker and longer-lasting would shorten the growing season, reduce the amount of land available for grazing, and increase the danger of avalanches. Also, tree growth would be slowed, road travel made more difficult, and logging and mining seasons shortened. And, overall, the danger that the nucleating agent, silver iodide, may enter food chains requires research. Clearly, cloud seeding poses problems that may be just as complex as the ones it attempts to solve.

Since 97 percent of all the water on earth is in the oceans, it would seem that efficient methods of *desalination*—the removal of dissolved salts from sea water—might solve some freshwater supply problems. Sea water contains 35,000 ppm dissolved salts, and because brackish groundwater can have even higher concentrations, these sources are not suitable for drinking or irrigation as they are. Dissolved salt content in drinking water should not exceed 500 ppm, and for most crops, salt content in irrigation water should not exceed 700 ppm. Thus, if sea water or brackish water is to be used, the salt content must be reduced substantially.

Many methods of desalination exist, but all require substantial quantities of energy, and are therefore expensive. In the mid–1970s, the total cost of producing desalinated water—including energy, equipment, and maintenance costs—was about four times greater than the cost of water drawn from conventional freshwater sources. Still, in regions where the freshwater supply is inadequate or surface waters are heavily polluted, desalination may emerge as an economically viable alternative for municipal supply. However, using desalinated water for irrigation purposes is, comparatively speaking, much less feasible economically. With government subsidies, fresh water from rivers or reservoirs is delivered to irrigated fields for less than 5 percent of the cost of desalinated water. Furthermore, most irrigated farmland in the United States is located far from oceans and well above sea level, so the cost of pipelines and pumping would be prohibitive for agricultural purposes. And as energy costs continue to escalate, the cost disparity between desalinated water and water from conventional sources will grow. Hence, until less expensive alternative energy sources become available, desalination will find only limited application.

As with other resources, conservation is a means of extending available water supplies that can be practiced under any circumstances. Irrigation, accounting for 80 percent of water usage in the western states, requires judicious conservation in particular, since it is what is known as a consumptive use; that is, most water (more than 60 percent) used in irrigation is lost by evaporation or infiltration and is thus unavailable for immediate reuse. In contrast, water use by municipalities and industries is considered nonconsumptive, since about 95 percent can be re-

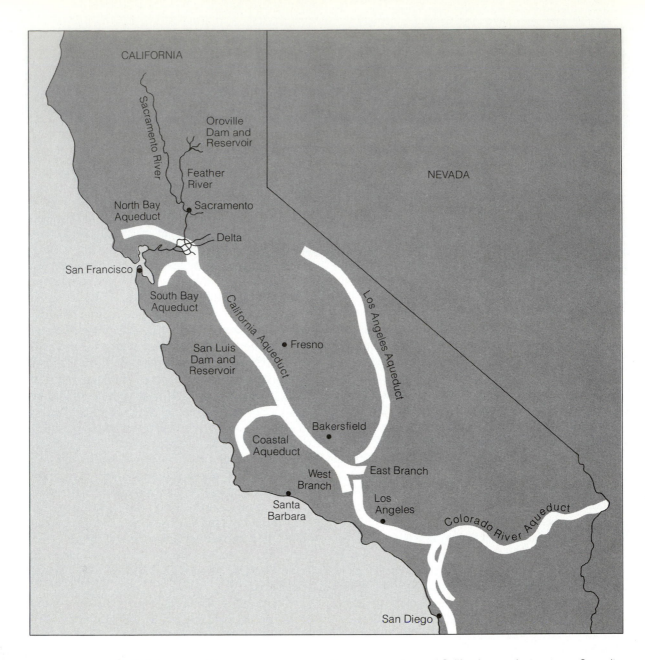

Figure 8.13 *Above:* The route of the Colorado River, Los Angeles, and California aqueduct systems. *Opposite, top:* A portion of the aqueduct. (Photo courtesy of the Metropolitan Water District of Southern California.)

Figure 8.14 (*Opposite, bottom*) Irrigating the fertile Coachella Valley of California with Colorado River water. The ridged field (center) is being flushed of salts. (U.S. Department of the Interior, Bureau of Reclamation.)

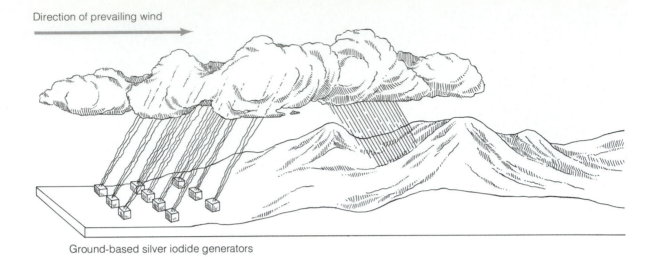

Direction of prevailing wind

Ground-based silver iodide generators

Figure 8.15 A scheme being used in some areas to increase precipitation. Silver iodide smoke is released into the air to seed clouds.

covered for possible reuse. Conservation of irrigation water is managed with this distinction in mind. Besides applying only the amount of water absolutely necessary for a crop, farmers intent on conservation, particularly during critical shortages, may find it necessary to employ costly measures, such as installing buried irrigation pipes or lining irrigation canals with plastic sheets to reduce seepage, shown in Figure 8.16. Where appropriate, farmers can anticipate droughts by growing more drought-resistant crops such as cotton or wheat.

Water conservation is also possible in industry. Such industries as paper manufacturing, for example, are beginning to identify and employ processes that utilize a treatment-use-treatment cycle. Such processes not only drastically cut water use, but make compliance with government water quality effluent standards much easier.

In the home, the prime contributor to needless water loss is leaky faucets. Householders can conserve water by mending all leaks, refraining from watering lawns, decreasing the frequency of toilet flushing and car washing, and taking short showers rather than baths. Toilets that use about 12 liters (3 gallons) of water per flush rather than the normal

20 liters (5 gallons) are now available, but people can reduce the per-flush water use of conventional toilets by installing several bricks or a water dam in the toilet's water tank. These household conservation techniques might begin to interest even the nonconservation-minded, since where voluntary water conservation is insufficient to reduce water consumption, water departments may force reduced consumption. For example, they can install restricting devices on household water meters to reduce the flow into the house to a mere trickle. Another effective method open to water companies is the imposing of penalites on injudicious users.

Global Water Problems

Our focus in this chapter has been on water supply problems in the United States, but the water supply in other countries is also of concern, and in some areas conditions are already critical. One way of determining the size of a continent's water supply is to calculate the percentage of runoff withdrawn for human use. As a general rule of thumb, where water withdrawal is less than 10 percent of runoff, a conti-

Figure 8.16 A plastic liner being installed in an irrigation canal to prevent water losses through seepage. (U.S. Department of the Interior, Bureau of Reclamation.)

nent's water resources are considered adequate, and where withdrawal is 10–20 percent of natural runoff, the supply is considered inadequate, necessitating water management strategies. When over 20 percent of the water runoff is withdrawn, water supply is considered to be the limiting factor for economic development. Water withdrawal needs for the year 2000 are given in Table 8.2. The figures show that in twenty years demand in south and east Asia, Europe, and Africa will exceed 20 percent of the total runoff. These regions will require sophisticated water management schemes to maximize the availability and efficient use of fresh water.

Continental averages, however, hide some important factors. The unequal distribution of freshwater supplies within a continent requires that water be transferred from water-rich to water-poor areas. Even continents where use is less than 10 percent of runoff—North America, for example—require

interbasin transfers. Such plans are not easily implemented, either physically or politically, when the water flows through or by neighboring countries. For example, India and Pakistan engaged in very intensive negotiations for several years before agreeing on mutual allocation rights for the water of the Indus River. And, we have seen, diversions of the Colorado River have led to conflicts between the United States and Mexico. Almost all nations share at least one major river basin with their neighbors. In fact, fifty-two major watersheds are each shared by at least three nations. When we couple this situation with the projection by the United Nations Food and Agricultural Organization that water withdrawal worldwide will increase by 240 percent by the year 2000, we can see that freshwater management calls for international cooperation. Such cooperation will not be easy to achieve. Conflicts among countries bordering the heavily industrialized Rhine River illustrate the

Table 8.2 Estimated percent of runoff required to meet water needs by the year 2000.

	No industrial recycle	90-percent industrial recycle
Europe	31	23
USSR	10	7
South and East Asia	51	36
Africa	25	18
North America	7	5
Latin America	8	6
Australia, New Zealand	2	2

Source: M. Falkenmark and G. Lindh, "How Can We Cope with the Water Resources Situation by the Year 2015?" *Ambio 4*:119–122.

problems. It is difficult to convince an upstream country such as West Germany to expend large amounts to clean up the river for the benefit of a downstream nation such as the Netherlands. Yet such cooperation is essential if the world is to have enough water to support its human population.

Even where water is available, serious problems with contamination remain. Although industrial pollution is now the most significant problem in developed countries, the serious hazards for humans in less developed nations still involve contamination of existing water supplies by infectious organisms. Particularly where the water supply used for drinking is also used for bathing and clothes washing, contamination from human excrement is common. In most developing countries, a vast backlog of planned human waste disposal facilities are yet to be completed; until they are, an estimated 1.2 billion people will live without safe drinking water. Compounding the problem is a failure of governments to educate their citizens to a modern understanding of sanitation and disease prevention. Until sanitation facilities and thorough education are available, these people will continue to suffer indignity, discomfort, and death from such waterborne diseases as dysentery and cholera.

A particularly horrible disease common to tropical African and South American countries is *onchorer-*

ciasis, or "river blindness," which blinds up to 20 percent of the adult population in some villages and leaves another 30 percent with impaired vision. The disease is feared so intensely that in some areas people have deserted fertile agricultural valleys to escape it. About 20 million people are infected worldwide. Individuals are infected by parasitic worms spread by the blackfly, which requires swiftly moving water as its habitat. In areas where dams slow river water, thereby eliminating the habitat suitable for blackflies, stagnant water connected with irrigation projects serves as a breeding site for mosquitos carrying other infectious diseases, such as malaria, yellow fever, and encephalitis.

Irrigation canals also serve as an ideal habitat for another disease, called *schistosomiasis*. This disease, virtually unknown in the United States, afflicts about 250 million people throughout the world. The disease is particularly prevalent in Africa and Asia, but it also occurs in some islands of the West Indies, where it is called belharzia. Egypt considers schistosomiasis its number one health problem.

The construction of large hydroelectric power plants and irrigation projects has promoted schistosomiasis, which is spread by a parasitic worm that requires snails as an intermediate host. Irrigation canals associated with such water projects are water-filled year-round, increasing the survival of schistosome-infected snails. Thus, any time of the year when people bathe, swim, or work in these waters they run the risk of becoming infected by the organism's free-swimming form. The worms penetrate the skin and eventually become established and reproduce in the blood vessels associated with the intestinal tract. Here the organisms sap nutrients from the bloodstream of their host. Infected individuals suffer a loss of appetite and energy, weight loss, and liver disorders. In severe cases, the disease leads to brain damage and death.

In regions of developing nations where sanitation has been improved, a marked decline in the incidence of waterborne diseases has been observed. But proper sanitation practices must be continually maintained, lest these diseases break out again. A case in point is malaria, which strikes about 100 million people each year and kills about 1 million. A few years ago, public health officials felt that the eradication of malaria was imminent, but now the best these officials hope for is an increased measure of malaria control. In Pakistan, an intense antimalarial effort beginning in 1961 reduced the number of cases from 7 million to 9500 in 1968, and health officials thought that by 1975 the number in that country would be near zero. But that year saw 10 million cases reported, and 23 million cases are expected to have occurred in Pakistan alone during 1979. Resurgences of malaria have also been reported in Thailand, India, Sri Lanka, Vietnam, Indonesia, and Bangladesh. These demoralizing turnabouts are attributable to inadequate funding and personnel for control programs, insufficient attention to urban areas, and the growing resistance of malaria-carrying mosquitos to DDT.

The World Health Organization estimates that even under the best of circumstances, establishing minimal sanitation facilities in developing nations will take at least one or two decades. Only when people are freed from the constant suffering of waterborne and foodborne diseases will they have a full sense of human dignity and the energy to better their lot. There is little room for complacency in the world community as long as these deplorable conditions exist.

Conclusions

We have seen in this chapter that adequate technology and legislation exist to control water pollution from point sources. Strategies for controlling nonpoint source pollutants, however, will require more money and more originality. Costly site-specific approaches are the only means of combatting water pollution problems from nonpoint sources. And the relative hazards posed by the chronic exposure of humans to new chemicals, especially chloroorganics, have not been determined.

Attempts to meet the future demands for water in water-poor regions of the world pose many problems. Solutions will entail tradeoffs of environmental quality if these regions are to continue to grow. Less developed nations must first address the problems posed by waterborne diseases, but many of them will also have to begin planning now to meet greater demand if they are to raise their standards of living.

Summary Statements

Water pollution control legislation was largely ineffective until passage of the Water Pollution Control Act of 1972. The 1970s also saw the enactment of other laws that directly or indirectly relate to water pollution control, among them the significant Safe Drinking Water Act of 1974.

The primary stage of sewage treatment removes insoluble and suspended materials from waste water. Secondary sewage treatment steps are designed to remove dissolved organic materials. Tertiary treatment steps can be added to treatment plants to remove phosphorus compounds, nitrogen compounds, and/or toxic and colored compounds. Sludges generated from treatment processes present a solid waste disposal problem.

Drinking water supplies are treated to remove suspended materials and are chlorinated to prevent the spread of waterborne diseases. The presence of small amounts of chloroorganic compounds in drinking water supplies presents a potential chronic health hazard.

Even if all industries and municipalities provided required treatment, nonpoint source pollutants from urban runoff and agriculture would still be significant water pollution sources. Nonpoint source pollution is especially difficult to control because it requires changing many traditional ways of carrying out our daily activities.

Accident prevention training is the best management technique for reducing the number of spills of harmful substances. Spills of most substances into surface waters are nearly impossible to clean up. Some oil spills can be cleaned up, but only with difficulty and at great expense.

Groundwater supplies are finite and should not be withdrawn at rates that exceed natural recharge rates. Heavy groundwater withdrawal can cause ground subsidence. Artificial recharge can be used to augment natural recharge.

Water supplies can be enhanced by damming rivers, transferring water from one watershed to another, desalinization, and precipitation enhancement. All methods have cost and benefits that must be weighed carefully. Because of higher energy costs, desalination is likely to provide drinking water only in regions where critical shortages exist. Water conservation practices can substantially cut the amount of water we use in homes and industries.

In less developed nations, waterborne diseases are still a constant threat. Water supplies are improperly protected and are often contaminated with human and animal wastes. A decade or more will be required for improving sanitation in these countries.

Questions and Projects

1. In your own words, define each of the italicized terms in this chapter. Compare your definitions with those in the text.

2. Determine whether a water quality management agency for your watershed has been established under section 208 of the 1972 Water Pollution Control Act. If so, what are its objectives?

3. From newspaper accounts, try to determine what several of the industries in your community are doing to meet the requirement of "best available treatment" deadline

of July 1, 1983. Which facilities are planning to meet the deadline? Which plants are fighting the requirements and why?

4. In what stage of the sewage treatment process (primary, secondary, or tertiary) would the following materials be removed: Paper fibers? Maple syrup? Table salt? Phosphate laundry detergent? Dirt in wash water? Hand soap?

5. Visit your local sewage treatment plant. Find out how efficiently suspended solids, biochemical oxygen demand (BOD), phosphorus, and nitrogen are removed from waste water and determine whether your local treatment plant meets designated effluent standards for these substances.

6. Why should sewage treatment plants prohibit industries from dumping toxic substances into sanitary sewers?

7. Should all communities or industries, regardless of size, be required to treat their waste water to the same degree? State your reasons.

8. Are sanitary and storm sewers completely separated in your city? If not, what happens during wet weather? Your local sewage treatment plant or health department will be able to provide this information.

9. What is the environmental impact of the effluent from your sewage treatment plant? Is the effluent being used for a useful purpose? Suggest some alternatives.

10. Determine how sewage sludge is disposed of in your community. Are any new alternatives, such as land disposal, being explored or implemented?

11. Use a topographic map to outline your watershed. Identify some water pollution sources (both point and nonpoint) in the watershed. Draw up a plan to improve water quality in this watershed.

12. Contact your local water supply utility to determine if your water supply has been analyzed recently for organic chemicals, especially chloroorganics. If so, which compounds are of primary concern? How or where do they originate?

13. Evaluate the dependability of your community's water supply. What circumstances might disrupt or cut off supplies?

14. Which specific nonpoint source pollutants present water quality problems in your area?

15. Develop a report on a water quality improvement or rehabilitation project for a surface water that is or has been completed. Pick a nearby example if you can. Your state's Office of Environmental Protection is a good source of information.

16. How is your state or community equipped to deal with accidental spills of toxic substances? Contact your state's Office of Environmental Protection for information.

17. Visit a dam site in your region. What are the environmental and economic benefits and costs? Be certain to consider land use patterns prior to dam construction.

18. Should large metropolitan areas be allowed to import water from more sparsely populated watersheds? Explain your answer.

19. In arid regions, who should have the right to limited water supplies—cities, industries, farmers of irrigated lands, or energy resource development interests. Set up teams in class to debate this issue.

20. What water conservation measures, if any, are used in your region?

21. Devise and implement a water conservation plan for your home. Use water meter readings to measure the effectiveness of your plan.

Selected Readings

Ambroggi, R. P. 1977 "Underground Reservoirs to Control the Water Cycle," *Scientific American 236:* 21–27 (May). A discussion of underground storage and how it can be used to increase the water supply in water-poor regions.

Barrett, B. R. 1978. "Controlling the Entrance of Toxic Pollutants into U.S. Waters," *Environmental Science and Technology 12:* 154–162 (February). A detailed review of federal legislation dealing with potential water pollutants.

Battan, L. J. 1969. *Harvesting the Clouds: Advances in Weather Modification.* Garden City, New Jersey: Doubleday. A clear discussion of attempts at weather modification.

Cargo, D. N., and T. I. Mallory. 1977. *Man and His Geologic Environment.* Reading, Massachusetts: Addison-Wesley. Presents a detailed account of ground subsidence and Colorado River salinity problems.

Council on Environmental Quality. 1970–1978. *Annual Reports.* Washington, D.C.: U.S. Government Printing Office. A series of annual reports that describe our nation's progress toward controlling water pollutants. Also provides an international perspective.

Crites, R. W., and C. E. Pound. 1976. "Land Treatment of Municipal Wastewater," *Environmental Science and Technology 10:* 548–551 (June). A discussion of methods and implications of land treatment.

Edmondson, W. T. 1973. *Environmental Quality and Water Development,* C. R. Goldman, J. McEvoy III, and P. J. Richerson, eds. San Francisco: W. H. Freeman and Company, pp. 281–298. A description of the reversal of eutrophication in Lake Washington.

Ember, L. R. 1976. "Surveying America's Lakes," *Environmental Science and Technology 10:* 862 (September). Overview of eutrophication problems in the lakes of the United States.

International Joint Commission. 1978. "Environmental Management Strategy for the Great Lakes System." Windsor, Ontario. Final report of the international reference group on Great Lakes pollution from land use activities, often called the PLUARG report.

National Wildlife Federation. "Setting the Course for Clean Water. Washington, D.C.: National Wildlife Federation. A citizen's guide to section 208 of the 1972 water quality management program. A single copy is free. Write to National Wildlife Federation, Education Division, 1412 16th Street, N.W., Washington, D.C.

Pojasek, R. B. 1977. "How to Protect Drinking Water Sources," *Environmental Science and Technology 11:*343–347 (April). A survey of problems related to drinking water contamination.

Tank, R. W. 1976. *Focus on Environmental Geology,* 2nd ed. New York: Oxford University Press. A collection of case histories that deal with water pollution problems such as erosion and sedimentation, water supply, and groundwater contamination.

United States Environmental Protection Agency. 1976. *Environmental Pollution Control Alternatives: Municipal Wastewater.* Washington, D.C.: U.S. Environmental Protection Agency. description of alternatives for the disposal of waste water and sludge generated by municipalities.

Wade, N. 1977. "Drinking Water: Health Hazards Still Not Resolved," *Science 196:*1421–1422 (June). A discussion of issues surrounding the implementation of the Safe Drinking Water Act.

Woodwell, G. M. 1977. "Recyling Sewage Through Plant Communities," *American Scientist 65:*556–562 (September–October). A summary of experiments in which effluents have been applied to fields.

A distant thunderstorm, dramatically demonstrating the awesome dynamism of weather.
(National Oceanic and Atmospheric Administration.)

Chapter 9

Weather and Climate

If you've ever stood by helplessly and watched your picnic basket blow away during an unexpected spring thunderstorm, or in winter shoveled a foot of "partly cloudy skies," off of your driveway, you are painfully aware of the vagaries of weather. Weather is a sometimes pleasant, sometimes tumultuous, but ever changeable ingredient of our everyday life. We adjust our dress and our driving habits to meet it, it influences our moods and our psychological well-being, and it can even be the deciding factor in a football game. Long-term weather conditions—which constitute the *climate* are important regulators of the geographical distribution of soils, plants, and animals. Short-term extremes in weather can, in the case of a hurricane, for instance, reduce a city to matchsticks in hours, or in the form of a hailstorm, lay to waste in minutes the fruits of a farmer's year of labor. Weather has ultimate control of our water and food supply and determines the rate at which we consume fossil fuel for space heating and air conditioning. Clearly, weather and climate deserve attention in the context of environmental science. But other reasons apply as well.

Until recently, humanity was completely subject to climate and the whims of weather. Since the Industrial Revolution, however, some of our activities have actually begun to influence weather and climate. We discussed the modification of precipitation through cloud seeding in the last chapter. Less intentional effects, those related to air pollution, are beginning to have much more profound consequences. Air pollution is changing rainfall patterns and altering temperatures locally, and effects may be found to be global as well. Before we can evaluate the impact of our activities on weather and climate, however, we must first examine the structure and composition of the atmosphere, the reasons for weather, and the role of climate in the functioning of ecosystems.

The Atmosphere: Composition and Structure

In the beginning, the earth's atmosphere was primarily a mixture of the noxious gases methane and ammonia. Through the millions of years that constitute geologic time, these gases almost entirely escaped to space. Gradually, they were replaced by other gases released during volcanic eruptions. Oxygen was added to the atmosphere initially by the breakdown of water vapor and later by photosynthetic organisms.

The atmosphere today is a mixture of many different gases and variable quantities of *aerosols*, suspended solid and liquid particles. Because the atmosphere undergoes continual mixing, atmospheric gases occur almost everywhere in roughly the same proportions up to an altitude of about 80 kilometers (50 miles). Hence, we may travel anywhere over the earth's surface and be confident that we are breathing essentially the same type of air, although in some regions pollution has altered the composition of air by adding new materials and by increasing the concentrations of some of the atmosphere's natural components.

By volume, nitrogen normally constitutes 78.08 percent of the lower atmosphere (below 80 kilometers), and oxygen, 20.95 percent. The next most abundant gases are argon (0.93 percent) and carbon dioxide (0.03 percent). The atmosphere also contains trace quantities of the gases neon, helium, krypton, hydrogen, xenon, ozone, methane, water vapor, and several others. The present volume of some of these trace gases (for example, carbon dioxide and water vapor) varies significantly from place to place and with time. Figure 9.1 shows the principal gases in the lower atmosphere.

The environmental significance of an atmospheric gas, however, is not necessarily related to its relative abundance. Collectively, water vapor, carbon dioxide, and ozone occur in minute concentrations, yet they are essential for maintenance of life. For example, no more than about 3 percent by volume of water vapor occurs in the lower 11 kilometers (6.8 miles) of the atmosphere, not even in the warm, nearly saturated air of tropical rain forests, but without water vapor there would be no water cycle to sustain terrestrial ecosystems. Also, carbon dioxide, water vapor, and ozone together act as a blanket over the earth's surface that causes the lower atmosphere

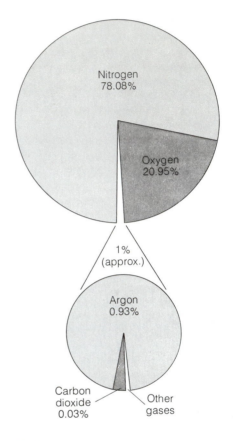

Figure 9.1 The principal gaseous components of the lower atmosphere. (After A. N. Strahler, *Principles of Earth Science.* New York: Harper & Row, 1976.)

to retain warmth, thus making it more amenable to life than it would be without these "minor" gases.

The aerosol content of the atmosphere is also quite small, yet these suspended particles participate in some important processes. Most aerosols are found in the lower atmosphere near their source, the earth's surface. They originate through wind erosion of soil, in sea salt from ocean spray, and in volcanic eruptions. Some aerosols act as nuclei for cloud and precipitation development, and some influence air temperatures by interacting with sunlight.

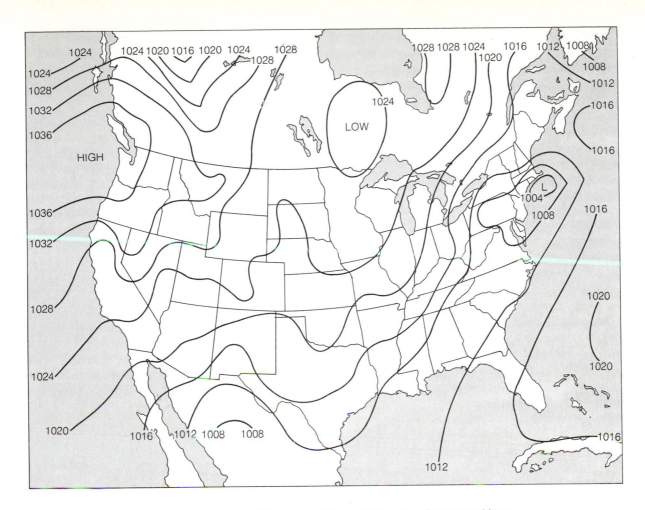

Figure 9.2 A typical weather map, showing areas of relatively high and relatively low air pressure. Lines connect places having the same air pressure. (U.S. Department of Commerce, National Oceanic and Atmospheric Administration, Environmental Data Service.)

Air Pressure

The weight of the atmosphere over a unit area of the earth's surface is called *air pressure*. At any locale, air pressure varies only slightly from day to day, but these variations can accompany some dramatic changes in weather. In the middle latitudes, our weather is dominated by a continuous procession of air masses of high and low air pressure, as illustrated in Figure 9.2. When weather is under the influence of high pressure (a "high") skies are generally fair,

but when low pressure (a "low") moves in the weather usually turns stormy.

Gravity concentrates the mass of the atmosphere close to the earth's surface, and air density and air pressure therefore drop rapidly with increasing altitude, as shown in Figure 9.3. However, it is impossible to specify an altitude at which the atmosphere definitely ends. That is, on an interplanetary journey, there would be no point enroute that could clearly be identified as the beginning of the "atmosphere" of space. Rather, we can describe the vertical extent of

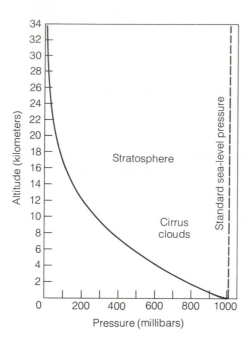

Figure 9.3 The loss in air pressure as altitude is gained. (After A. N. Strahler, *Principles of Earth Science*. New York: Harper & Row, 1976.)

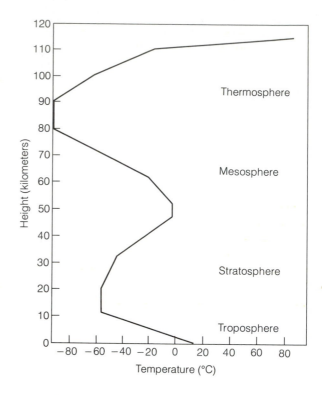

Figure 9.4 The variation of air temperature with altitude. (After M. Neiburger, J. G. Edinger, and W. D. Bonner, *Understanding Our Atmospheric Environment*. San Francisco: W. H. Freeman and Company, 1973.)

the atmosphere only in terms of the relative distribution of its mass. Half the atmosphere's mass lies between the earth's surface and an altitude of about 5.5 kilometers (3.4 miles). And about 99 percent of the atmosphere lies below 32 kilometers (20 miles). Beyond 80 kilometers (50 miles), the relative proportions of the gases change markedly, and beyond approximately 950 kilometers (590 miles), the atmosphere merges with the interplanetary gases, hydrogen and helium.

Atmospheric Spheres

For the convenience of study, the atmosphere is usually subdivided into a series of concentric shells, shown in Figure 9.4, according to the temperature at varying altitudes. Most weather events occur in the *troposphere,* and it is in the lower portion of this shell that the many terrestrial life forms, including human beings, are found. The troposphere extends from the earth's surface to an altitude that ranges between 16 kilometers (10 miles) at the equator and 6 kilometers (3.7 miles) at the poles. It is normally characterized by temperatures that decrease with height. Thus, air temperatures atop mountains are usually lower than in surrounding valleys.

The *stratosphere,* the next layer, extends to an altitude of about 50 kilometers (31 miles). It is characterized by isothermal (constant temperature) conditions in its lower portion, and by a gradual increase in temperature with height in its middle and upper reaches. Recently scientists have become concerned over the possible detrimental effects of pollutants entering the stratosphere. Since little mixing of air occurs between the troposphere and stratosphere, pollutants that do enter the stratosphere tend to stay there for extended periods. Dust thrown into the stratosphere during violent volcanic eruptions, for example, has remained there for as long as three to five years.

The *mesosphere*, the shell above the stratosphere, extends up to an altitude of 80 kilometers (50 miles) and is characterized by a decrease of temperature with height. Above this is the *thermosphere*, which is marked by temperatures that increase with height. Our activities thus far seem to be having little direct impact on these outermost atmospheric subdivisions.

The Dynamism of the Atmosphere

Our weather is a composite of a wide variety of phenomena—spectacular lightning displays, devastating tornadoes and blizzards, feathery frosts, and gentle June breezes. Although these phenomena vary in intensity, each one is an expression of the dynamic nature of the atmosphere. The atmosphere's dynamism is maintained by a ceaseless flow of energy from the sun to the earth to space. In this section, we examine this energy flow in order to establish a basis for our discussion of the weather.

More than 99 percent of the energy that results in weather phenomena comes from the sun. The sun's energy travels through space in the form of radiation, half of which is visible as sunlight. *Radiation* is a process whereby energy flows rapidly from place to place as oscillating waves. These waves are illustrated in Figure 9.5. Many types of radiation exist, and they are distinguished from one another by specific properties and by wavelength. In addition to sunlight, familiar types of radiation include x-rays, microwaves, and radio waves. The spectrum of types of radiation is shown in Figure 9.6, and key points on radiation are summarized in Box 9.1.

Of the enormous amount of energy radiated by the sun—about 4×10^{27} calories each minute—only about one-half of one billionth the amount is intercepted by the earth. (See Appendix III for a discussion of powers of ten.) The earth's movements in space distribute this energy unequally over the earth's surface. Rotation of the earth on its axis accounts for day to night variations in energy receipt while seasonal changes accompany the orbiting of the earth about the sun.

As solar energy travels through the atmosphere, it interacts with the constituents of the atmosphere. Figure 9.7 illustrates these interactions. Some sunlight is absorbed by oxygen, ozone, water vapor, ice crystals, water droplets, and dust particles, and

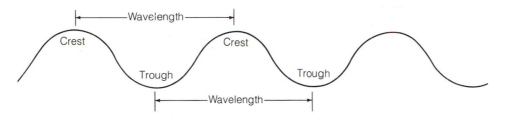

Figure 9.5 Waves, the form in which radiational energy travels. One wavelength is the distance between successive troughs or successive crests.

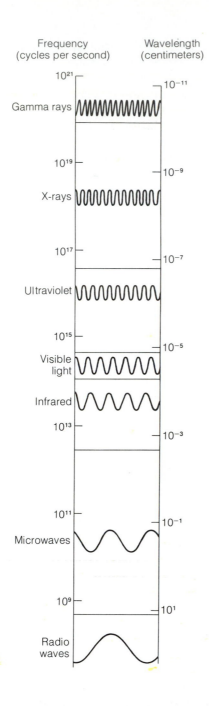

Frequency
(cycles per second)

Wavelength
(centimeters)

10^{21}

10^{-11}

Gamma rays

10^{19}

10^{-9}

X-rays

10^{17}

10^{-7}

Ultraviolet

10^{15}

10^{-5}

Visible
light

Infrared

10^{13}

10^{-3}

10^{11}

10^{-1}

Microwaves

10^{9}

10^{1}

Radio
waves

Figure 9.6 The spectrum of types of radiation.

through these interactions air is warmed to some extent. Some solar energy is reflected—primarily by clouds—back to space. Another portion is scattered, dispersed in all directions. In fact, it is the scattering of the blue portion of visible sunlight by nitrogen and oxygen molecules that gives the sky its blue color. Solar energy that is not reflected, scattered back to space, or absorbed is transmitted to the earth's surface, where further interactions take place.

A certain fraction of the solar energy that gets through the atmosphere is reflected by the earth's surface back into the atmosphere and perhaps out to space. The amount of reflection depends on the nature of the surface at the point at which the radiation meets it. Skiers are well aware that snow has a high reflectivity, so high, in fact, it can cause sunburn. The measure of a surface's reflectivity is known as the *albedo*. Fresh-fallen snow has an albedo of between 80 and 85 percent; that is, 80–85 percent of the solar energy that meets the snow is reflected. At the other extreme, dark surfaces, such as black-topped roads and green forests, exhibit albedos of as low as 5 percent. The portion of solar energy that is not reflected is absorbed, thereby warming the earth's surface.

Recent satellite measurements indicate that about 29 percent of the in-coming solar energy is reflected by the earth-atmosphere system (the earth's surface and atmosphere considered together) and lost to space. Hence, about 71 percent of the solar energy intercepted by the earth-atmosphere system is absorbed and ultimately involved in the functioning of the environment. Table 9.1 shows the proportions of radiation reflected and absorbed. Any alteration of

Table 9.1 Budget of in-coming solar radiational energy.

	Percent
Absorbed by the atmosphere	19
Absorbed at the earth's surface	52
Reflected back to space	29
	100

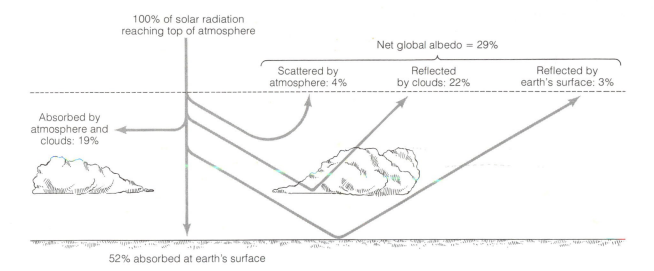

Figure 9.7 Interactions of sunlight with the atmosphere and the earth's surface. (After R. A. Anthes et al., *The Atmosphere.* Columbus, Ohio: Charles E. Merrill, 1975.)

the atmosphere's composition or the earth's surface characteristics could change the amount of energy available. Because this energy is required for maintaining the many interrelated processes occurring within ecosystems, such as photosynthesis and the water cycle, the implications of changes of this nature are far-reaching. As we shall see in Chapter 10, such changes are in fact taking place.

Of the solar energy absorbed by the earth-atmosphere system, only 19 percent is absorbed directly by the atmosphere; in other words, the atmosphere is relatively transparent to sunlight. The remaining 52 percent of solar energy (remember, 29 percent was reflected) is absorbed at the earth's surface, primarily because of the low albedo of the ocean waters covering three-quarters of the globe. Consequently, the earth's surface is the principal receptor of solar heat. The earth's surface, in turn, continuously radiates this heat back to the atmosphere, and the latter, in its turn, eventually radiates it off to space. Thus, the earth's surface

functions as the prime heat source for the atmosphere, and this factor is responsible for the normal vertical temperature profile of the troposphere. As we saw in Figure 9.4, air is warmest close to the earth's surface, generally cooling with altitude, that is, away from its main heat source.

Energy Balance

The earth's weather depends to a great degree on the flow of radiational energy described in the preceding section. But the radiation emitted by the earth's surface and the atmosphere has properties that differ from solar radiation, since a considerable temperature difference exists between the earth-atmosphere system and the sun. The warmer the temperature of a radiating object, the shorter are the wavelengths of its radiated energy. The effective radiating temperature of the sun is about 6000°C (11,000°F); thus, the waves emitted by the sun are relatively short. In solar radiation, which includes the visible portion of

Box 9.1

Electromagnetic Energy

Our world is continually bathed in electromagnetic energy. Electromagnetic energy is energy radiated by a moving electrical charge. It is, therefore, commonly called *electromagnetic radiation* or *radiation* for short. And it is said to be electromagnetic because it possesses both electrical and magnetic properties. Most of the electromagnetic energy that reaches the earth originates in the sun. Some comes from more distant stars and other celestial objects. And today a considerable amount is artificially generated here on earth. There are many types of electromagnetic energy; together they make up what is called the *electromagnetic spectrum*. Light, which is visible electromagnetic energy, makes up only a very small portion of this spectrum. Other types of electromagnetic energy, including radio waves, infrared radiation, and gamma radiation, are discussed below.

Electromagnetic radiation travels in the form of waves. Like a sound wave, an electromagnetic wave is described in terms of its wavelength and frequency. The length of a wave, or *wavelength*, is the distance from crest to crest, as shown in Figure 9.5. The wave's *frequency* is defined as the number of crests (or troughs) that pass a given point in a given period of time, usually one second. A wave's frequency is inversely proportional to its wavelength; that is, the higher the frequency the shorter the wavelength. This inverse relationship is depicted in Figure 9.6. Some radio waves have frequencies of just a few hundred cycles per second (cps) and wavelengths hundreds of kilometers long. Gamma rays, by contrast, have frequencies as high as 10^{24} (a trillion trillion) cps and wavelengths as short as 10^{-14} meters (a hundred trillionth) of a meter.

Unlike sound waves, electromagnetic waves can travel through a vacuum as well as through gases, liquids, and solids, and in a vacuum they all travel at their maximum speed—299,800 kilometers (186,000 miles) per second. All forms of electromagnetic radiation slow down when passing through materials, their rates of speed depending on their wavelength and the material.

No type of electromagnetic radiation begins or ends at precisely marked points along the spectrum. For example, red light shades off into invisible infrared radiation (infrared = below red) beneath it on the frequency scale. And on the other side of the visible radiation's portion of the electromagnetic spectrum, violet light shades gradually into invisible ultraviolet radiation (ultraviolet = beyond violet).

We find it convenient to give different names to different segments of this energy spectrum because we detect, measure, generate, and use different segments in different ways. But all types of electromagnetic radiation are alike in all respects except their wavelengths, frequencies, and energies (and these characteristics are functions of their frequencies: the higher the frequency the higher the energy level). We have already noted that all forms of this energy are produced by a moving electrical charge, and that they all travel at the same speed in a vacuum. Moreover, all types of electromagnetic radiation are reflected, refracted (bent), and/or absorbed in the same manner as visible light.

At the low-energy (low-frequency, long-wavelength) end of the electromagnetic spectrum are the waves produced by alternating-current (AC) electricity used in household wiring. The energy level of AC waves is so low that these waves are barely detectable. Their frequency is only 60 cps, and they are very long indeed—about 4800 kilometers (3000 miles).

Adjoining AC waves, and overlapping them to some extent on the spectrum, are *radio waves*. Their wavelengths range from many kilometers to a small fraction of a centimeter, and their frequencies extend from a few thousand to around a billion cps. FM (frequency modulation) radio waves, for example, range from 88 million to 108 million cps; thus the familiar 88 and 108 at opposite ends of the FM radio dial.

Next comes the *microwave* portion of the electromagnetic spectrum, ranging from around 30 centimeters to around 0.3 centimeters. Some microwave frequencies are used for radio communication. Astronomers, using radiotelescopes, study stars and other celestial objects by means of the microwaves they emit and absorb.

Between microwaves and light we find *infrared radiation,* often called infrared heat or radiant heat. We cannot see it, but we can feel the heat it generates if it is sufficiently intense—as it is, for example, when emitted by a hot stove. Actually, small amounts of infrared radiation are emitted by every known object or material, however cold. Naturally occurring infrared radiation (some of it coming directly from the sun but most of it converted by the earth and atmosphere from solar radiation at shorter wavelengths) warms the earth and powers the weather phenomena discussed in this chapter. Astronomers use infrared radiation in studying the sun, the planets, and other stars.

At its uppermost frequencies infrared radiation shades off into the lowest-frequency visible radiation, red light. Wavelengths of visible light range from about 0.35 micrometers at the violet end to about 0.80 micrometers at the red end. (A micrometer is 10^{-6} meter.) Light is of course essential to the photosynthetic activities of plants. The photoperiods of plants (that is, the relative lengths of their exposures to light and darkness) coordinate the opening of their buds and flowers and the dropping of their leaves. Animals' photoperiods

regulate their reproduction, hibernation, and migration.

Above light on the electromagnetic spectrum, in order of increasing frequencies, increasing energy levels, and shorter wavelengths, are *ultraviolet radiation, x-rays,* and *gamma radiation.* All three types of radiant energy occur naturally, and all can be produced artificially. All have medical uses: ultraviolet radiation is a potent germicide; x-rays are a powerful diagnostic tool; x-rays and gamma radiation are used in treating cancer patients. And all three are now being used by astronomers to open new windows on the universe—to obtain information about far distant objects and processes that visible light cannot provide.

But these three highly energetic types of radiation have their dark side too. Ultraviolet rays can irreparably damage the retinal cells of the eye. They can permanently blind you if you look at the sun (say, during a solar eclipse) without using a filter that blocks them while letting visible radiation through. And overexposure to ultraviolet rays, x-rays, or gamma rays can cause sterilization, cancer, genetic mutations, or fetal tissue damage.

Fortunately for us, the atmosphere blocks most of the incoming ultraviolet radiation and virtually all of the x- and gamma radiation. Without this protective shield, all life on earth would be quickly destroyed. (Astronomers using ultraviolet, x-ray, and gamma-ray radiation must employ balloons, rockets, and artificial satellites to carry their instruments above the atmospheric barrier.) But radioactive materials emit dangerous levels of x-rays and gamma rays. So do various devices including x-ray machines and "atom smashers" (particle accelerators). Overexposure to high-energy electromagnetic radiation, whatever its origin, must be zealously guarded against.

the electromagnetic spectrum (refer back to Figure 9.6 and Box 9.1), the peak energy intensity is emitted by waves of about 0.48 micrometers in length (in the green). Figure 9.8 shows the relative intensity of solar radiation. On the other hand, the much cooler earth, radiating at a temperature of about 20°C (68°F), emits longer waves, as shown in Figure 9.9. In fact, the earth-atmosphere system radiates infrared radiation with a maximum intensity at a wavelength of about 10 micrometers. Hence, the earth-atmosphere system responds to the input of solar short-wave radiation by emitting long-wave radiation.

Because solar and terrestrial radiation represent different portions of the electromagnetic spectrum, their respective interactions with atmospheric components also differ. We noted earlier that the atmosphere absorbs only about 19 percent of the sunlight intercepted by the earth. In contrast, the atmosphere absorbs a large percentage of the infrared radiation that is emitted primarily by the earth's surface. A portion of the infrared radiation absorbed by the atmosphere is then reradiated back toward the earth's surface. Reradiation by the atmosphere makes it difficult for heat to escape into space, so the temperature of the lower atmosphere is more hospitable for organisms. In fact, it is estimated that the earth's average surface temperature is more than thirty degrees Celsius warmer because of this effect!

The atmospheric gases that absorb infrared radiation and thereby impede its loss to space are primarily water vapor, carbon dioxide, and, to a much lesser extent, ozone. Because it was once thought that the glass of greenhouses retained infrared energy in a similar manner, this process is termed the *greenhouse effect*. We know today, however, that greenhouses keep plants tolerably warm primarily by preventing air movement and only to a lesser extent by inhibiting radiational cooling. Hence, strictly speaking, it is incorrect to apply the greenhouse analogue to the earth-atmosphere system. Still, reference to the "greenhouse effect" is so common that we will use it here, but always in quotations.

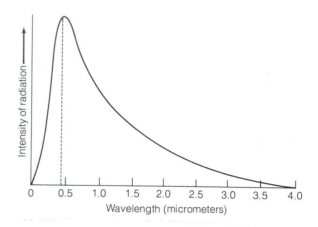

Figure 9.8 The intensity of solar radiation over a band of wavelengths. The peak energy intensity is emitted by waves 0.48 micrometers in length, in the green region of the visible spectrum. (After H. R. Byers, *General Meteorology*. New York: McGraw-Hill, 1959.)

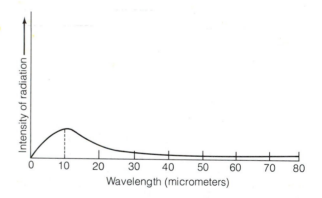

Figure 9.9 The intensity of terrestrial radiation over a band of wavelengths. The peak energy intensity is emitted by waves about 10 micrometers in length, in the infrared region of the spectrum. (After H. R. Byers, *General Meteorology*. New York: McGraw-Hill, 1959.)

To illustrate the "greenhouse effect" in the atmosphere, let us compare the typical summer weather of the American Southwest with that of the Gulf of Mexico coast. Both areas are at the same latitude and therefore receive about the same intensity of sunlight, and both areas commonly experience afternoon temperatures above 30°C (86°F). At night, however, temperatures in the two areas differ markedly. In the Southwest the air is relatively dry, allowing infrared radiation to readily escape to space; energy is lost and the earth-atmosphere system cools rapidly. Thus, the temperature there may fall below 15°C (59°F) by dawn. Along the gulf coast, however, the air is more humid, causing a greater absorption of infrared radiation. Because a portion of this energy is reradiated back toward the earth's surface, the temperature may fall only into the 20s°C (70s°F). Water in clouds also produces a "greenhouse effect"; this explains why nights are usually colder when the sky is clear than when the sky is cloud-covered.

Energy Flow

If the net radiant (solar short-wave and terrestrial long-wave) energy distribution is calculated separately for the earth's surface and for the atmosphere, we find that more energy enters the earth's surface than leaves it, and that more energy leaves the atmosphere than enters it. This calculation seems to imply a continuous accumulation of energy at the earth's surface (warming) and a continuous depletion of energy from the atmosphere (cooling). But in actuality the atmosphere does not experience long-term cooling relative to the earth's surface. Hence, a net transfer of heat energy must be occurring from the reservoir of excess energy (the earth's surface) to the reservoir of deficit energy (the atmosphere).

The movement of heat energy from earth to atmosphere is brought about primarily by *latent heat transfer* (about 80 percent) and, to a much lesser extent, by *sensible heat transfer* (about 20 percent). Sensible heat transfer is the more readily understood of the two mechanisms. The term "sensible" is used because energy redistribution brought about by this process can be monitored, or sensed, as temperature changes. In sensible heat transfer, heat is transported by *conduction* from earth to atmosphere through direct contact between air and the earth's surface, as illustrated in Figure 9.10. As the lower portion of the atmosphere is heated by the ground, it becomes lighter than the cooler surrounding air and begins to rise. The surrounding air then sweeps into its place, is heated, and also rises. The result is a transfer of energy by *convective currents* from the earth's surface to the atmosphere.

Latent heat transfer is a more subtle process. It results from energy differences in the physical states of water. (The relation between temperature and states of water is explored in Box 9.2.) With the application of solar energy, water from oceans, lakes, and rivers evaporates; conversely, when water changes from the vapor phase back to the liquid phase (condenses) within the atmosphere, this energy is released, warming the atmosphere. The energy required for evaporation is supplied at the earth's surface, the reservoir that has excess heat energy, and is released by condensation to the atmosphere, the reservoir with a heat energy deficit. This process is also illustrated in Figure 9.10.

Inequalities in radiant energy distribution occur not only vertically, between the earth's surface and atmosphere, but also horizontally, that is, by latitude. Because the earth's shape is nearly spherical, parallel beams of in-coming sunlight strike equatorial regions more directly than areas at higher latitudes, as shown in Figure 9.11. Hence, solar energy is spread out over a greater area at higher latitudes than at lower latitudes. Over the course of a year, more solar energy per unit area enters the earth-atmosphere system in the area between the equator and 38 degrees latitude than leaves by terrestrial infrared radiation. In contrast, from 38 degrees latitude to the pole, more terrestrial radiation is lost per unit area than is gained by solar radiation. Hence, polar latitudes are sites of net radiant energy loss (cooling), and tropical latitudes are sites of net radiant energy gain (warming). Figure 9.12 illustrates this concept.

Tropical latitudes are not continually warming, however, nor are higher latitudes continually cooling. Hence, transport of energy must be taking place

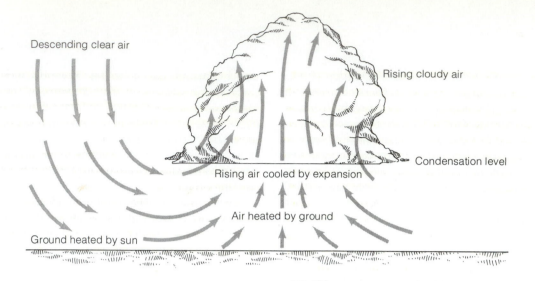

Figure 9.10 Latent and sensible heat transfer. Convective currents and evaporation-condensation transfer heat energy from the earth's surface to the atmosphere. (After M. Neiburger, J. G. Edinger, and W. D. Bonner, *Understanding Our Atmospheric Environment.* San Francisco: W. H. Freeman and Company, 1973.)

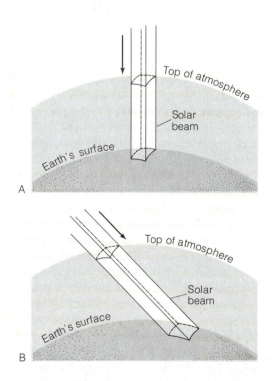

Figure 9.11 Variations in the angle of solar radiation at (A) tropical latitudes and (B) higher latitudes.

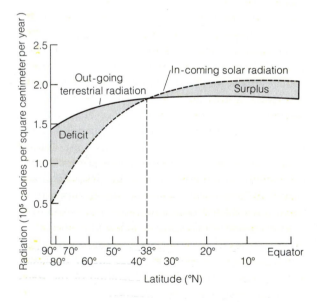

Figure 9.12 Variations in in-coming solar radiation and out-going terrestrial radiation by latitude over the course of a year. (After R. E. Newell, "The Circulation of the Upper Atmosphere," *Scientific American 210:*69, March 1964. Copyright ⓒ by Scientific American, Inc. All rights reserved.)

from south to north across the 38 degree latitude circle. This transport occurs primarily through the northward movement of the warm air that forms in lower latitudes to replace cold air flowing southward from higher latitudes, as shown in Figure 9.13. About 50 percent of the energy transported poleward is in the form of sensible heat, whereas latent heat transfer—evaporation of water in lower latitudes and condensation in storms at higher latitudes—accounts for about 30 percent. The remaining 20 percent of needed poleward heat energy transport is brought about by the flow of cold and warm ocean currents.

If all inequities in energy distribution were smoothed out, the temperature would be the same throughout the entire earth-atmosphere system, and all these energy transfer processes would cease. But neither vertical nor horizontal energy flow in the atmosphere ever achieves a uniform distribution of energy. The factors that cause the unequal distribution of heat energy operate continually, however, thereby triggering the mechanisms of transfer.

The mechanisms by which heat energy redistribution is achieved involve the movement of cold and warm air from place to place, condensation and cloud formation, storms, and convective currents—in other words, all the various weather phenomena. Thus, in summary, a cause-effect chain operates in the atmosphere, starting with the sun as the prime energy source and resulting in weather. This examination of the pathways of energy flow in the atmosphere also serves to emphasize again the importance of the water cycle in the functioning of the environment. The water cycle plays the principal role in vertical energy transport in the atmosphere and is an important contributor to poleward heat energy transport. Hence, any alteration in the operation of the water cycle can also modify the weather.

Scales of Weather

Meteorologists usually classify weather patterns according to the spatial dimensions, or scale, of the phenomena. From largest to smallest, scales of weather are designated global, synoptic, meso, and micro. The large-scale wind systems of the globe (polar easterlies, westerlies, and trade winds) and the semipermanent centers of low and high pressure are features of the *global-scale circulation*, shown in Figure 9.14. Semipermanent highs move poleward in summer and equatorward during the winter, exerting an important control on climate. These high-pressure systems determine the locations of the world's largest tropical deserts. As we shall see later, subtropical highs also contribute to the air pollution potential of some coastal regions.

Synoptic-scale weather is continental or oceanic. Migrating highs, storms, and air masses, usually highlighted on television and newspaper weather maps like the one in Figure 9.2, are important components of this scale. *Mesoscale systems* include thunderstorms, hailstorms, and sea and lake breezes—phenomena that may influence the weather in one section of a city while leaving the remainder unaffected. The flow of air within a specific environment, for example, the emission of smoke from a chimney, represents the smallest subdivision of atmospheric motion—the *microscale*.

Each small-scale circulation unit is part of and dependent on larger-scale circulation. For example, extreme radiational cooling (the rapid nighttime loss of infrared radiation) conducive to local frost formation (on the microscale) requires a synoptic weather pattern that favors clear skies and light winds (a high pressure system).

Our analysis of the linkage between solar energy input and weather allows us to appreciate the effects human beings might have on weather by tampering with the environment. Any alteration of the atmospheric composition and the nature of the earth's surface cover could alter the earth-atmosphere energy balance. Because the various scales of weather are interdependent, changes in the energy balance may ultimately influence all scales of weather. These changes, in turn, could have potentially adverse consequences for food production and fuel and water supplies.

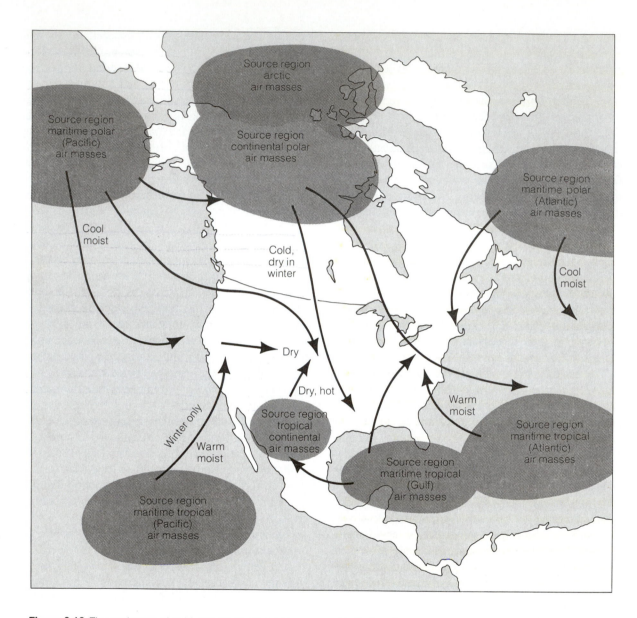

Figure 9.13 The exchange of cold and warm air masses across North America. (After A. N. Strahler, *Introduction to Physical Geography.* New York: John Wiley & Sons, 1970.)

Box 9.2

Temperature, Heat Energy, and Phase Changes of Water

Ultimately, the various weather phenomena are consequences of the flow of heat from place to place within the earth-atmosphere system. In this box, we examine how this energy is quantified and how phase changes of water result in the transfer of heat from place to place.

Many people are familiar with temperature without being aware of the physical basis for temperature differences. *Temperature* is a measure of molecular motion; the faster the motion of molecules, the higher the temperature.

Two commonly used measures of molecular activity are the Fahrenheit and Celsius temperature scales. In this country, the Fahrenheit temperature scale is commonly used in weather reports and by medical personnel, whereas the Celsius scale is used in most other scientific applications. If a thermometer equipped with both scales is immersed in a glass containing a mixture of ice and water, the Fahrenheit scale will read 32° while the Celsius scale will read 0°. In boiling water, the readings will be 212° Fahrenheit (F) and 100° Celsius (C). The Celsius scale has the numerical convenience of a 100-degree interval between the boiling and freezing points of water.

We have indicated that less molecular activity occurs in cold substances than in hot substances. There is, theoretically, a temperature at which molecular activity essentially ceases: it is called *absolute zero,* or 0 degrees Kelvin, and corresponds to $-273.15°C$ ($-459.6°F$). The Kelvin scale indicates temperature as the number of degrees above absolute zero and is a more direct measure of molecular activity, or heat energy. (Since nothing can be colder than absolute zero, there are no negative temperature values on the Kelvin scale.) A 1-degree interval on the Kelvin scale corresponds precisely to a 1-degree increment on the Celsius scale. The three scales are compared in Figure 1.

For purposes of determining or comparing amounts of heat energy in substances or systems, a unit of measurement is required. Scientists commonly measure heat energy in units called calories. The *calorie* is defined as the amount of heat needed to raise one gram of water 1 Celsius degree. This definition specifies a substance (water), the mass of water (1 gram), and the temperature change (1 Celsius degree): (Note that the "calorie" used in measuring the energy content of food is actually

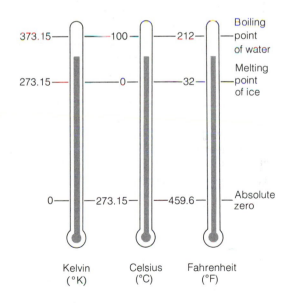

Figure 1 Comparison of three common temperature scales.

(Continued)

Box 9.2 (*Continued*)

1000 heat calories, or 1 kilocalorie.) Heat energy is often expressed in terms of another unit, *British Thermal Unit*, or BTU. A BTU of heat is defined as the amount of energy required to raise 1 pound of water 1 Fahrenheit degree. The two units differ only in magnitude; the BTU is exactly 252 times larger than the calorie.

Two different objects at the same temperature do not necessarily posess the same amount of heat energy, nor is the same quantity of heat required to raise the temperature of equal amounts of two different substances by 1 degree. The amount of heat required to raise the temperature of 1 gram of a substance 1 Celsius degree is called its *specific heat capacity*. The specific heat capacity of all substances is measured relative to the specific heat capacity of water, which by definition has a value of 1.00 calorie per gram. Water has the highest specific heat of any naturally occurring substance on earth, which is why it is the most efficient natural heat exchanger. It takes only one-fifth as much energy to raise the temperature of sand as it does to bring about an equivalent temperature rise of water. Thus, some materials (those with a low specific heat capacity) exposed to the same heat source warm up more rapidly than others (those with higher specific heat capacity). At right are the specific heats of some familiar materials, arranged in order of their specific heat capacities, from highest to lowest.

Much more energy is required to bring about a phase change than is required to change the temperature of a substance 1 Celsius degree. The energy required to bring about a change from the solid to the liquid phase is called the *heat of fusion*, whereas the energy required to cause a change in phase from the liquid to the gaseous phase is called the *heat of vaporization.*

Water	1.000
Ice	0.478
Wood	0.420
Sugar	0.274
Aluminum	0.214
Granite	0.192
Sand	0.188
Gold	0.031

As an illustration, let us follow the fate of an ice cube initially at $-40°C$ ($-40°F$) as heat is added to it (Figure 2). The first calories added to the ice cause its temperature to rise. For every Celsius degree of temperature rise, 0.5 calories must be added per gram of ice. Hence, for ice the specific heat capacity is 0.5 calories per gram per Celsius degree. Once 0°C is reached, an additional 80 calories of energy per gram must be supplied to break the forces that bind water molecules in the solid phase. The temperature of the water and ice remains at 0°C (32°F) until all the ice is melted. From that point on, only 1.0 calorie is needed to raise the temperature of 1 gram of water 1 Celsius degree (specific heat capacity of liquid water). Once the boiling point (100°C, 212°F) is reached, the temperature again remains constant until all the water is vaporized. Vaporization of water requires 540 calories per gram (heat of vaporization). If the sequence is reversed—that is, if we cool the water vapor until we have a piece of ice at $-40°C$—as equivalent amounts of energy are released, the temperature drops and the phase changes take place. Heat added or released during a phase change of water is called *latent heat.*

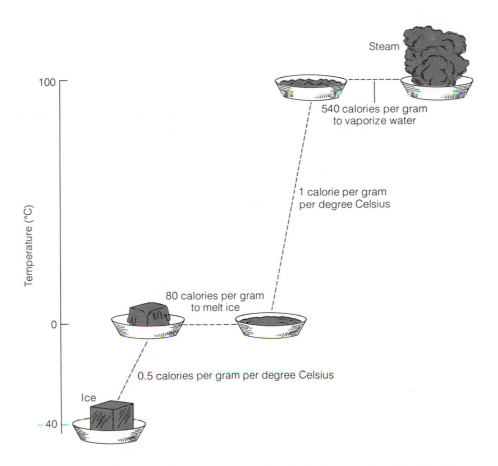

Figure 2 Energy needed to change temperature (specific heat capacity) and the physical states (heat of fusion and heat of vaporization) of water.

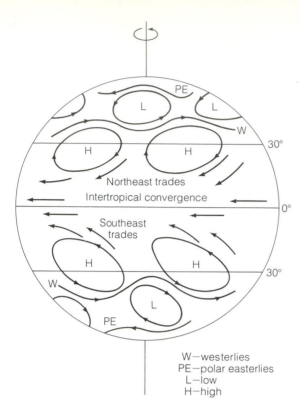

Figure 9.14 Global-scale atmospheric circulation.

W—westerlies
PE—polar easterlies
L—low
H—high

The Changing Climate

As we saw in Chapter 3, long-term average weather conditions constitute a primary regulator (limiting factor) for organisms. These weather conditions, averaged over an extended period, plus expected extremes in weather behavior constitute the climate. Climate is the principal factor controlling the geographical distribution of the various worldwide biomes and ecotones that we examined in Chapter 5. And climate is one of the critical factors in soil formation, food productivity, and energy utilization for heating and cooling. For these reasons, one of our major environmental concerns is the potential impact of climatic change on ecosystems. In fact, there is evidence suggesting that today we are in a period of

dramatic climatic shifts. In order to assess the possible ramifications of such changes, we first consider some fundamentals of climatic behavior.

Some Principles of Climate

The climate of a region ultimately depends on the characteristics of the air masses that either develop over it or move into the region from elsewhere. An *air mass* is a mass of air covering thousands of square kilometers that is relatively uniform in temperature and water vapor content. Four basic types of air masses exist: warm and dry, warm and moist, cold and dry, and cold and moist. The characteristics of an air mass are determined by the nature of the surface over which the air mass develops and travels. Air that is situated for an extended period over cold, snow-covered ground (for example, northern Canada in winter) becomes cold and dry. On the other hand, an air mass that forms over a wet, warm surface (for example, the Gulf of Mexico) becomes warm and moist. Locales of formation of the various air masses that regularly invade the United States were shown in Figure 9.13. In general, cold air masses that develop in the north push southward, while the warm air masses of the south stream northward. Recall that this air mass exchange is the principal mechanism involved in poleward heat energy transport.

As air masses move from place to place, their temperature and moisture content are modified to some extent. Thus, in winter a cold air mass loses some of its punch as it plunges southeastward from Canada into the central United States. Cold air masses tend to change much more rapidly than warm air masses. Hence, a heat wave may spread from the gulf coast into southern Canada with about the same intensity, but a cold wave usually warms considerably by the time it reaches the deep South.

The specific kind of air mass that flows into a midlatitude locality is determined by the westerly wind pattern in the midtroposphere. Partially for this reason, the upper-level westerlies are referred to as "steering winds." Aloft the westerlies girdle the

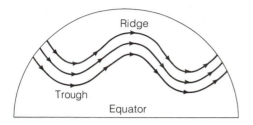

Figure 9.15 The pattern of westerly winds, which flow around the middle latitudes in a wavelike pattern of ridges and troughs.

middle latitudes in wavelike patterns of troughs and ridges, as illustrated in Figure 9.15. Thus, the westerlies actually consist of a north-south air flow superimposed on an overall west to east air flow. In regions where the westerlies blow from north to south, cold air masses are steered southward. But where the westerlies flow from south to north, warm air masses are funneled northward. Hence, the type of air mass that invades a region depends on the location of the westerly wave pattern with respect to that locality.

A particular wave pattern in the westerlies tends to persist for periods ranging from several days to weeks or longer. Then abruptly (usually in one day or less) a shift to a new wave pattern occurs. This shift is usually accompanied by a change in the regional distribution of air masses. For example, areas that have been cold and dry may suddenly turn mild and stormy, while areas that were warm and moist abruptly become cool and dry. Thus, weather behavior at a particular midlatitude locality involves a sequence of spells (episodes) of weather of varying length punctuated by abrupt periods of change.

This episodic behavior of midlatitude weather is usually much more pronounced during winter than in summer. The reason for this difference is that a greater contrast exists between air masses in winter than in summer. In addition, stormy weather is more likely to occur during times of transition from one air mass to another. As a new air mass flows into a locality, it displaces an old air mass. Depending on density differences between the two air masses, the new air mass either slides under (Figure 9.16) or moves over the old air mass (Figure 9.17). Warm, moist air is lighter than cold, dry air, so that a warm, moist air mass overrides and may displace a cold, dry one. The leading edge of the warm air is called a *warm front*. But cold, dry air displaces warm, moist air by sliding under the warm, moist air. The leading edge of cold air is called a *cold front*. The net effect of air mass displacement, then, is the lifting of air, which, in turn, leads to cooling, cloud development, and rain- or snowfall.

The climate of a particular locality is usually described in terms of its average monthly temperatures and precipitation. However, it is clear from our discussion of the episodic behavior of weather that it would be more informative to characterize the climate by citing the frequency with which the various types of air masses (or weather episodes) occur. For example, during January an upper-midwestern city might normally be influenced by cold, dry air 60 percent of the time; mild, moist air 30 percent of the time; and mild, dry air 10 percent of the time. From this description we can gain a sense of the frequency of change in weather, rather than just a notion of the precipitation and temperature range.

Describing climate in terms of air mass frequency appears to be a valid approach when we consider the apparent air mass control of certain biomes. For example, the region dominated by cold, dry Arctic air corresponds closely with the location of the boreal forest in Canada. As shown in Figure 9.18, the southern boundary of the boreal forest coincides with the normal position of the leading edge of Arctic air during winter, and the northern border of the forest closely corresponds to the average position of the leading edge of Arctic air during summer. The geographical position of the grassland biome of North America can also be explained in terms of air mass distribution. Hence, climatic change that alters air mass frequency may also change the boundaries of biomes.

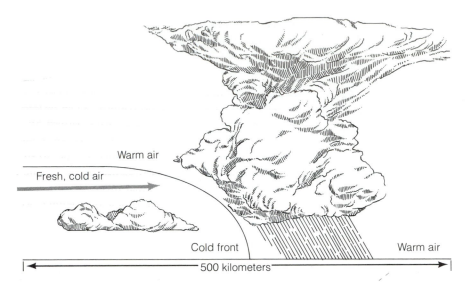

Figure 9.16 Cold air displacing lighter warm air by sliding under it. (After R. A. Anthes et al., *The Atmosphere.* Columbus, Ohio: Charles E. Merrill, 1975.)

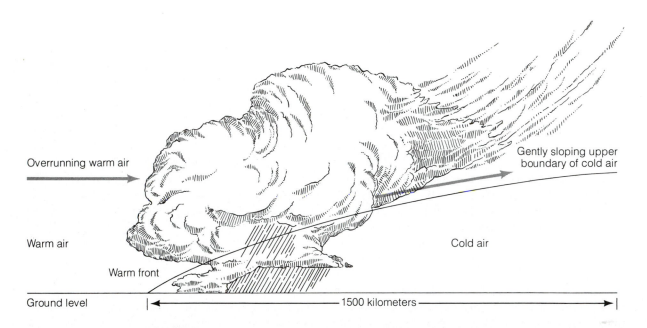

Figure 9.17 Warm air displacing heavier cold air by riding over it. (After R. A. Anthes et al., *The Atmosphere.* Columbus, Ohio: Charles E. Merrill, 1975.)

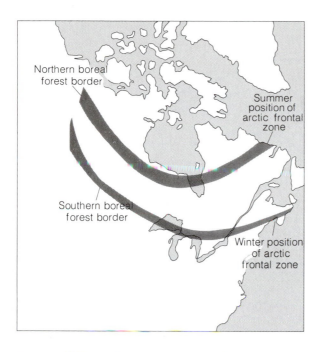

Figure 9.18 The northern and southern borders of the boreal forest of Canada, corresponding to the leading edge of Arctic air in summer and winter. (After R. A. Bryson, "Air Masses, Streamlines and the Boreal Forest," *Geographical Bulletin* 8:228–269, 1966. Reproduced by permission of the Minister of Supply and Services Canada.)

Although noting the frequency with which air masses occur is an interesting way of viewing climate, other factors must be considered if we are to have a complete picture of local climate. Water bodies such as oceans and large lakes exert a modifying influence on the temperatures of adjacent land areas. Altitude is important, since it affects temperature and the type and amount of precipitation, and latitude is also critical in that it indicates the intensity of solar radiation.

It is common practice for climatologists to compare the weather of a specific week, month, or year with climate recorded in the past. If such a comparison were carried out for the same period across the nation, we would find that anomalies (departures from normal) in temperature and precipitation are never the same everywhere. That is, the United States is so large that anomalous cold or drought, for example, never grips the entire country at the same time. The nonuniformity of climatic anomalies is linked to the westerly wind pattern. As an example, suppose a certain westerly wind wave pattern causes abnormally cold conditions in the eastern United States, exceptionally mild weather in the western United States, and normal weather in the middle of the country. Now suppose that this pattern occurred very frequently during the course of a winter, as it did during the notorious winter of 1976–1977. This pattern is represented in Figure 9.19. During such winters, average temperatures across the country mirror the characteristics of the dominating westerly wave pattern: colder than normal in the east, warmer than normal in the west, and normal temperatures in between.

The geographical nonuniformity of climatic anomalies is an important consideration in view of the fuel shortages we experience. Because the amount of fuel consumed for home heating depends to a large extent on outside temperature, it is important for us to be able to predict whether the coming winter will be colder or warmer than normal. One thing that can be said with confidence is that across the nation, some places will be warmer than normal, some localities normal, and some places colder than normal. Exactly where the winter will be colder than normal (and by how much) is the critical question for development of fuel allocation strategies. Unfortunately, given the present state of the art of long-range weather forecasting, we are unable to answer this question.

Hemispheric Climatic Trends

The geographical nonuniformity of climatic anomalies has some interesting consequences with respect to recent trends in air temperatures across the northern hemisphere. Computing the average temperature for the northern hemisphere for each year entails

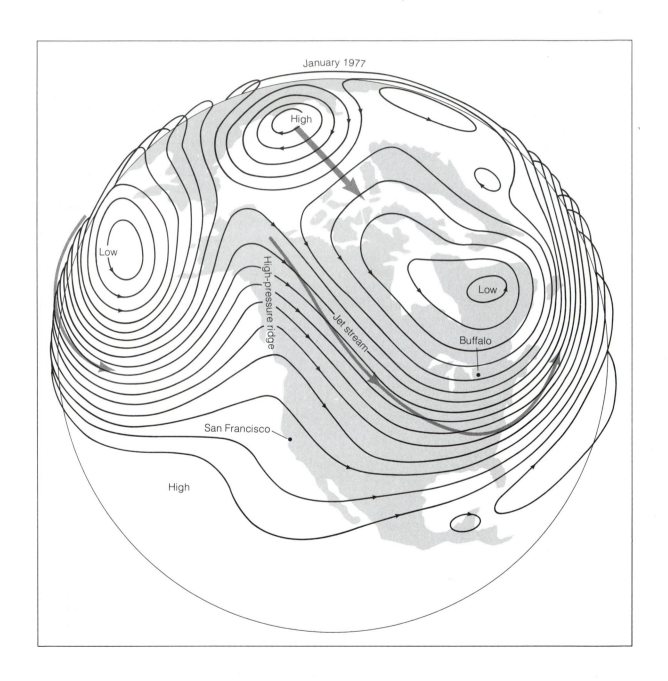

Figure 9.19 The westerly wind pattern that prevailed during the winter of 1976–1977, bringing record cold to the East and record drought and heat to the West. (After T. Y. Canby, ''The Year the Weather Went Wild,'' *National Geographic 152:*799–829, 1977.)

input of temperature data from hundreds of localities. Figure 9.20 shows the average annual hemispheric temperature from the 1870s through the early 1970s. Note that a general warming trend began in the 1890s and continued into the early 1940s. Then, beginning about 1945, an oscillatory cooling trend set in that seems to be continuing to this day. Because of the geographical nonuniformity of climatic anomalies, however, the temperature trend shown in Figure 9.20 is not representative of all localities across the hemisphere. On the contrary, while the average annual hemispheric temperature has been dropping since 1945, in some places annual temperatures have been rising, and in other locales annual temperatures have remained virtually unchanged. In the United States, for example, the cooling trend in the 1960s was quite marked in the southeastern states while the Pacific Northwest showed a warming trend.

It is misleading not only to casually apply the direction of hemispheric climatic trends to all localities, but to assume that the magnitude of climatic change is the same everywhere. In fact, a small change in average hemispheric temperature may translate into a much larger change in certain localities. It appears, for example, that the slight drop in average hemispheric air temperature since 1945 was accompanied by a considerably greater temperature drop in polar regions. It follows that in any meaningful assessment of the potential impact on ecosystems of hemispheric climatic trends, we must not dismiss as unimportant small changes in average hemispheric temperature. Locally, hemispheric trends may be amplified significantly and even reversed, thereby disrupting biotic distributions and physical processes.

"Normal" Climate

In this discussion of the nature of climatic behavior we have referred more than once to the "normal" climate. What exactly is meant by the term "normal" in this context? Through international agreement by climatologists, normal values for temperature and precipitation for each locality are averages computed from measurements compiled over a thirty-year period. Current climatic normals are based upon the period 1941–1970. The assumption is that these average temperatures and precipitation values represent a reasonable guide for future weather expectations.

Problems arise, however, when we assume that the weather of this thirty-year period actually constitutes the normal climate. First, it turns out that climate can change significantly in periods much shorter than thirty years. In addition, a thirty-year period gives us a very restricted view of the climatic record and the variability that climate can exhibit. The weather of 1941–1970 (our current climatic norm) was actually an unusually mild period of history. Hence, in the perspective of an extensive climatic record, the current hemispheric cooling trend may be nothing more than a return to a more "normal" climate. The fate-

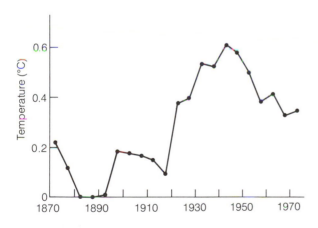

Figure 9.20 Variations in surface temperatures in the northern hemisphere from the 1870s through the early 1970s. Values are five-year averages expressed as deviations from the 1880–1884 average. (After W. A. Brinkmann, "Surface Temperature Trend for the Northern Hemisphere—Updated," *Quarternary Research* 6:355–358, 1976.)

ful impact of such a climatic reversion is described in Box 9.3.

The Responses of Ecosystems to Climatic Change

When the climate in a locality changes, plants and animals respond. If the climatic shift is severe, the tolerance limits of organisms may be exceeded and the species composition of ecosystems changes. However, while climatic change tends to be abrupt, the response of organisms is more gradual. That is, it takes time for plants and animals to disperse, become established, and come into equilibrium with a new climatic regime. This lag time in the response of organisms has a disruptive effect on the functioning of ecosystems. Consider an example. Suppose the climate in a particular region is arid. As is typical in a desert ecosystem, the soil is sparsely covered with various native plants. Climate and vegetation are in equilibrium. Now suppose that an abrupt shift to a humid, rainy climate takes place. The assemblage of plant and animal species begins to change. However, different species have different abilities to disperse and become established. The result is a series of unpredictable species assemblages that eventually lead to a different climax community. Before the climax ecosystem is reached, predatory and parasitic interactions are disrupted. Thus, as we saw in our discussion of interrupted and reversed succession in Chapter 3, it may take many years for a stable climax community to become established through succession. In the early stages of succession, large areas of the soil surface remain sparsely vegetated and subject to erosion by heavy rains. Hence, until a complete vegetative cover is established, soil erosion is severe and nutrients are cycled out and lost from that ecosystem.

The abrupt nature of climatic change also has a disruptive effect on agriculture. For example, a shift from humid to dry conditions may occur so rapidly that a farmer may not have sufficient lead time to adjust his cultivation practices to include crop varie-ties that will prove productive under the new climatic regime. Even slight climatic shifts may have a disastrous effect in localities that are already climatically marginal for farming.

The Uncertain Climatic Future

An obvious question regarding the climatic future is whether the post-1945 cooling trend will continue. Unfortunately, we cannot answer this question with certainty. Although the technological age has provided weather forecasters with many sophisticated tools, including satellites and high-speed computers, long-range forecasters as yet cannot accurately predict the weather more than about two weeks in advance. The problem stems partially from the margin of error seemingly inherent in the worldwide network of weather monitoring instruments, and the absence of observational data from vast areas of the ocean. But perhaps a more imposing difficulty involves the complex way in which weather behaves. For one thing, the episodic manner of weather behavior requires that the forecaster predict not only the type of future episodes but also their duration—a difficult if not impossible task, particularly as the forecast period lengthens.

While weather forecasters do not know enough about the workings of the atmosphere to formulate reliable long-range forecasts, we can gain at least some insight on future climatic variability by studying climatic records of the past. Unfortunately, these records reveal that the mildness of the first half of this century was an unusual event unprecedented in many hundreds of years. Long-range prospects, then, appear to favor continued cooling. This prediction has inspired some climatologists to foresee a return of the Ice Age. On the other hand, recent studies suggest that the cooling trend may be a temporary climatic fluctuation that will give way to pronounced global warming by the close of the century, if it hasn't already.

Either of these alternatives—cooling or warming—would trigger cataclysmic effects, whether they

Box 9.3

Climatic Change and the Greenland Tragedy

In the late ninth century, there began a lengthy episode of unusually mild climate in the North Atlantic region. This warming enabled Viking explorers to probe the far northerly reaches of the Atlantic Ocean. Previously, severe cold and extensive drift ice had been insurmountable obstacles to European navigators. By 930 A.D., the Vikings established the first permanent settlement in Iceland, some 970 kilometers (600 miles) west of Norway and just south of the Arctic Circle.

Among Iceland's early inhabitants was Eric the Red, a troublesome individual whose exploits eventually caused him to be banished from Iceland. He sailed west, and in 982 A.D. discovered a new land, which he named Greenland. Some historians think he called it that to entice others to follow him; then as now, much of the new land was buried under a massive ice sheet. But in the late tenth century the climate there was so mild that some sheltered areas probably were indeed quite green.

In such a coastal place, on Greenland's southwest shore, Eric founded the first of two colonies. Although far from prosperous, it developed into an agrarian society. Initially, it fared well, and its population climbed to roughly three thousand.

By the fourteenth century, however, the climate was rapidly deteriorating. Drift ice again expanded in the North Atlantic, hampering and eventually halting navigation between Iceland and Greenland. Greenland was cut off from the rest of the world. What happened to the Norse settlers in succeeding years can only be inferred from their graves and the ruins of their homes; there were no survivors.

In 1921, an expedition from Denmark examined the remains of the Greenland settlements. The expedition reported that the colony had lasted for five-hundred years and had suffered a slow, painful annihilation. There is evidence that grazing land was buried under lobes of glacial ice and that most farmland was rendered useless by permafrost. Near the end, the descendants of a once robust and hardy people were ravaged by famine: they became crippled, dwarflike, and diseased. There is also some suggestion that the malnourished colonists were attacked by pirates when they turned to the sea for food.

The Greenland tragedy may be the only historical example of the extinction of a European society. What lesson does it teach contemporary society?

The lesson is that climate changes, and that it can change rapidly—sometimes with serious, even disastrous, consequences. Nowhere are people more vulnerable to climatic shifts than in regions where the climate is just marginal for their survival. These regions include areas where barely enough rain falls to sustain crops and livestock and places where temperatures are so cold and the growing season so short that only a few hardy crops can be cultivated successfully. In such regions, even a slight deterioration of climate can make agriculture impossible. If the inhabitants have no alternative food source, and if they cannot migrate to more hospitable lands, their fate may be similar to that of the early Greenland population.

involve the relentless advance of glaciers over Chicago and New York or the inundation of coastal areas by rising seas as the Antarctic and Greenland ice sheets succumb to rising temperatures. Chances are, however, that neither of these disasters is close at hand. If these circumstances develop at all it will be hundreds or thousands of years from now. Thus, most climatologists today are less concerned with them as reasonable prospects. Climatologists are more concerned about the likelihood of increasingly unreliable weather, that is, *climatic destabilization*. The first half of this century was not only mild, but the weather was also remarkably stable and reliable. Recent climatic trends strongly suggest that there is now an increased possibility of hostile weather episodes, that is, more extremes in weather behavior, including record warmth and cold, drought, and excessive rain and snow.

Climatic destabilization could have a particularly severe impact on food production. While modern agriculture can adjust to gradual changes in climate, it will be difficult to devise an agricultural strategy sufficiently flexible to cope with successive years in which rainfall and temperatures fluctuate from one extreme to another. Food production and human population soared during the generally favorable climatic conditions of the first half of the twentieth century. But climatic destabilization threatens to reduce food productivity in some key areas of the globe. And according to some scientists, the effect of reduced productivity on human population growth may be profound. The actual and potential impact of climatic change on food supply is examined in more detail in the context of the world food shortage in Chapter 17.

Conclusions

In this chapter, we have seen that weather is far from being a capricious act of nature. Weather in its many forms is a response to unequal heating and cooling within the earth-atmosphere system. Excess heat is transported from the earth's surface to the atmosphere and from the tropics to polar latitudes by migrating storms, conduction and convection, and the exchange of cold and warm air masses.

Weather averaged over long time periods constitutes climate. Our midlatitude climate is strongly influenced by the prevailing patterns that characterize the westerly winds. When prevailing patterns change, the climate changes. And, since the well-being of ecosystems hinges on climate, a climatic change may cause serious disruptions.

We appear to have entered a period of climatic change. Average hemispheric temperatures are slowly cooling, and the occurrence of extreme weather behavior appears more likely than it has in some time. The role that air pollution may have played (if any) in recent climatic trends is among the topics we consider in our examination of atmospheric pollution in the next chapter.

Summary Statements

The atmosphere is a mixture of gases in which solid and liquid particles are suspended.

The significance of an atmospheric gas or an aerosol is not necessarily related to its relative concentration in the atmosphere.

Weather changes accompany relatively small fluctuations in air pressure.

Air pressure and density drop rapidly with altitude, but the atmosphere has no clearly defined upper boundary.

The atmosphere is subdivided into concentric spheres on the basis of vertical variations in temperature. These spheres are termed the troposphere, stratosphere, mesosphere, and thermosphere.

Solar radiation that is not absorbed by the atmosphere or returned to space by reflection or scattering reaches the earth's surface.

Depending upon albedo, a portion of solar radiation that strikes the earth's surface is reflected and a portion is absorbed.

The earth's surface radiates infrared energy and is the primary energy source for the atmosphere.

Water vapor, carbon dioxide, and ozone absorb and reradiate terrestrial infrared energy back to the earth's surface, thereby moderating the temperature of the lower atmosphere.

Energy is transported from the earth's surface to the atmosphere and from tropical latitudes to polar latitudes primarily through sensible and latent heat transfer.

Energy transport in the atmosphere results in a series of interdependent scales of weather.

The climate of a region hinges on the characteristics of air masses that regularly invade that region.

The wavelike pattern of the midlatitude westerlies is responsible for the regional distribution of air masses and the episodic behavior of weather.

Anomalies and trends in climate are geographically nonuniform in both magnitide and direction.

Locally, hemispheric trends in climate may be significantly amplified and even reversed, thereby disrupting plant and animal distributions and physical processes.

In the perspective of the long-term climatic record, the current hemispheric cooling trend may be merely a return to a more "normal" climate.

Climatic change is rapid, having a potentially disruptive impact on ecosystems.

Climate in the future may be characterized by less reliable weather.

Questions and Projects

1. In your own words, write a definition for each of the terms italicized in this chapter. Compare your definitions with those in the text.

2. Prepare a list of the varied ways in which weather influences our daily lives.

3. On a winter day, is the air temperature more likely to be higher if the ground is snow-covered or if it is bare? Explain your answer.

4. What is the purpose of the large fans used in some Florida orange groves during calm, clear winter nights?

5. The maximum poleward energy transport takes place at about 38°N latitude. Would you expect this latitude belt to be particularly stormy? Explain your answer.

6. What is the effect of spreading coal dust over the snow cover in late spring?

7. In your own words explain the following statements: (a) The sun drives the atmosphere. (b) The atmosphere is heated from below.

8. Design a house that will take advantage of the "greenhouse effect" during the winter heating season but will not overheat in summer.

9. Although in-coming solar radiation reaches its maximum intensity at noon, usually air temperature does not reach its maximum until two or three hours later. Explain this phenomenon.

10. List the major ways in which heat is added to or removed from the atmosphere.

11. Why must in-coming solar radiation balance out-going terrestrial radiation? What would be the consequences if this balance was disrupted?

12. Is there evidence from your area of the country that climate is becoming more extreme? Check with local newspapers, fuel companies, or agricultural suppliers.

13. How might climatic change (to warmer, colder, drier, or wetter conditions) influence the economy of your community?

14. The relative concentration of a gas in the atmosphere does not necessarily indicate its importance in influencing the weather. Elaborate on this statement.

15. Are weather data routinely collected in your community? If so, where and by whom? What factors are monitored (for example, temperature, precipitation, wind, solar radiation)? Where are these data stored and how are they summarized?

16. Is it possible for a slight cooling trend in hemispheric average temperature (1°C per decade, for example) to disrupt the functioning of certain ecosystems? Defend your answer.

17. In the long-term perspective, our current "normal" climate actually may be anomalous. Explain this possibility.

18. Why are convective air currents more likely to develop over a vegetated surface than over a snow-covered surface?

19. The bulk of solar radiation on earth is absorbed by the oceans, with lesser amounts being absorbed by air and land. In view of this distribution, speculate on the relative role of the oceans in any large-scale climatic shift.

20. Weather records in the United States extend back not much further than one

hundred years. This brief record provides us with a restricted view of climatic variability. Speculate on several ways in which we can lengthen the record by reconstructing the climatic past.

Selected Readings

Anthes, R. A., H. A. Panofsky, J. J. Cahir, and A. Rango. 1978. *The Atmosphere*. Columbus, Ohio: Charles E. Merrill. An excellent introduction to weather and climate.

Bryson, R. A. 1974. "A Perspective on Climatic Change," *Science 184*:753–760. A review of the nature of climatic change and possible ramifications of current climatic trends.

Hare, F. K. 1963. *The Restless Atmosphere, An Introduction to Climatology*. New York: Harper & Row. A lively introduction to principles of climatology and the factors that control the climate of a locality.

Lehr, P. E., R. W. Burnett, and H. S. Zim. 1965. *Weather*. New York: Golden Press. An exceptionally well-illustrated and lucid consideration of weather phenomena and weather forecasting.

National Research Council. 1975. *Understanding Climatic Change, A Program for Action*. Washington, D.C.: National Academy of Sciences. A report of the U.S. Committee for the Global Atmospheric Research Program on current research concerning climatic variations and the need for further research on global climate.

National Research Council. 1977. *The Atmospheric Sciences: Problems and Applications*. Washington, D.C.: National Academy of Sciences. A review of major problems under study by researchers in the atmospheric sciences.

Typical sources of air pollution, an undesirable by-product of our industrial society. (M. L. Brisson.)

Chapter 10

Air Pollution

On the morning of October 26, 1948, a fog blanket reeking of pungent sulfur dioxide fumes spread over the Monongahela Valley town of Donora, Pennsylvania. Before the fog lifted five days later, almost half the area's fourteen-thousand inhabitants had fallen ill, and twenty of them had died. This killer fog resulted from a combination of the mountainous topography and particular weather conditions that trapped and concentrated deadly effluents from the community's steel mill, zinc smelter, and sulfuric acid plant.

The tragedy at Donora first brought to the nation's attention the fact that air pollution can kill, though it was not the first such incident in the world, as Table 10.1 shows. On occasion, some extremely toxic substances have been accidentally released into the air we breathe. The discussion of the Seveso tragedy in Box 10.1 illustrates this fact well. Nonetheless, we know that such catastrophes, though noteworthy, are rare, and that our concerns are best focused on the health hazards involved in long-term exposure to low-level air pollution. And, although the impact of air pollution on human health is our most serious concern, the adverse effects on plants

and animals as well as weather and climate have serious implications for us too. In this chapter, we study the sources and effects of air pollution with an eye toward possible solutions to this far-reaching problem.

A Historical Overview

Air pollution is not new. Indeed, it is at least as old as civilization itself. The first air pollution episode probably occurred when early people, new to the art of fire making, discovered that a poorly ventilated cave is no place to cook a meal. By medieval times, air pollution had become so serious that in some localities government agencies were empowered to investigate offensive sources. In England, for example, four such commissions were established between 1285 and 1310 to study the fouling of air that accompanied a shift from wood to coal as the principal fuel in lime kilns.

It was the Industrial Revolution, however, that was the single greatest contributor to air pollution as a chronic problem in the United States. In post-Civil

Table 10.1 Some notable air pollution disasters.

Year of disaster	Location	Deaths exceeding normal mortality predictions
1880	London	1000
1930	Meuse Valley, Belgium	63
1948	Donora Valley, Pennsylvania	20
1950	Pica Rica, Mexico	22
1952	London	4000
1953	New York City	250
1956	London	1000
1957	London	700–800
1962	London	700
1963	New York City	200–400
1966	New York City	168

War days, cities in the North swelled with new industries and new immigrants to work them. Both the industries and the population flourished in an environment increasingly fouled by the fumes of foundries, oil refineries, and steel mills. In those days, a city took pride in smokestacks like those in Figure 10.1; these fixtures were considered proof that the community was partaking in the nation's new prosperity. Efforts to regulate air quality were meager, and little was known about possible health effects. In an attempt to placate the wheezing and coughing populace, some physicians even argued that polluted air had medicinal value.

Contemporary concern over polluted air does not stem from any disenchantment with the fruits of industrialism. The existence of belching smokestacks still serves to reassure many Americans that the economy is healthy and unemployment low. But the evidence that air pollution is related to illness, or at best to ill effects, is growing. Many citizens are beginning to realize that we might be trading the health of our bodies for reassurance that our economy is growing.

Air Pollution and the Quality of Life

Determining the effects of air pollution on human health is our chief concern in this section. A dramatic rise in respiratory diseases in the recent past has particular relevance. Is there in fact a direct correlation between the degree of air pollution and the incidence of respiratory disease? We have not yet succeeded in answering this question with certainty, but the fact that we are trying represents a significant advance. Of considerable concern also are the adverse effects that air pollutants have on the animals and plants we depend on for food and fiber. In this section, we examine in detail how air polluted by a variety of gases and aerosols impairs the functioning of living things.

Human Respiratory Problems

Air pollutants attack human health primarily through the respiratory system. Air that we inhale follows a long pathway, as illustrated in Figure 10.2, through the oral and nasal passages, the windpipe,

Figure 10.1 Industrial smokestacks in Pittsburgh in 1906. (Carnegie Library of Pittsburgh.)

and the bronchioles before finally ending up in the air sacs of the lungs. Within the air sacs, the oxygen in the air is taken up by red blood cells and then circulated to all parts of the body. There are many air sacs in the lungs (more than 300 million), and together they have a combined surface area of 100 square meters (about 1100 square feet). This feature facilitates oxygen uptake. If blood is deprived of its oxygen supply, the individual succumbs to asphyxiation (oxygen starvation). Air pollutants are inhaled into the respiratory tract along with air and follow the same route to the air sacs in the lungs. In sufficiently high concentrations, some air pollutants function as asphyxiating agents. Others severely irritate the tissues of the respiratory tract itself.

We are not entirely at the mercy of air pollutants,

however. Our respiratory systems are armed with defense mechanisms. Nasal hairs filter out large particles, and glands in the nose and larger bronchioles secrete a sticky mucus that captures small particles, including most dust and pollen, and dissolves some gaseous pollutants. Mucus and the pollutants it accumulates are transported upward out of the lungs to the oral cavity by the continuous beating of millions of *cilia*—hairlike structures that line the nasal passages and bronchioles. (Figure 10.3 shows the cilia in the lungs of a horse.) Once the pollutant-bearing mucus reaches the mouth, it is usually swallowed or expelled by the cough reflex system. Regarding the latter, when the lining of the respiratory tract is irritated by foreign matter, a message is sent to the cough control center of the brain and the cough reflex is activated, forcing mucus from deep inside the lungs upward to the mouth.

Very small particles (less than a micrometer in diameter) can penetrate the defenses of the nasal passage and bronchi and reach the air sacs of the lungs. Bacteria and some airborne pollen are examples. Still, although air sacs are protected by neither cilia nor a mucus covering, they are equipped with specialized, free-living cells called *macrophages*, shown in Figure 10.4, that engulf and digest foreign organic particles. In fact, each macrophage consumes up to a hundred bacteria or pollen grains before it is killed by accumulated poisons. Dead macrophages are forced up to the cilia and expelled.

The ability of air pollutants to penetrate the body's natural defenses and to disrupt respiratory processes differs from one pollutant to another. The gases that act as asphyxiating agents in the air sacs displace oxygen being transferred to hemoglobin molecules—the transporters of oxygen in red blood cells. Hence, as increasing concentrations of these gases are inhaled, the quantity of life-sustaining oxygen that the bloodstream transports from the lungs decreases. Carbon monoxide, a constituent of auto exhaust, is the asphyxiating agent most common in our environment. Most people show no symptoms when breathing carbon monoxide concentrations of

Box 10.1

At 12:27 P.M. on June 10, 1976, in a chemical factory in the North Italian town of Meda, an unattended reactor vessel overheated, rupturing its safety valve and spewing its vaporous contents into the atmosphere. As the vapor was carried downwind, it condensed, forming a white cloud. The cloud sank over the adjoining town of Seveso, enveloping houses and gardens in its path. Apparently, many people were outdoors at the time. Children ran about in the whitish mist, thinking it great fun.

Before long, however, the children and older residents who had been outside when the cloud descended began to suffer nausea, dizziness, headaches, and skin irritation that soon developed into painful sores. By nightfall, animals in the vicinity—birds, cats, livestock—had become ill. Over the next few days, many children were hospitalized for their burnlike sores, and hundreds of animals died.

Not for almost two weeks did the residents of Seveso learn from the chemical factory's owners that the white cloud contained TCDD, a chemical by-product generated in the production of 2,4,5-T, a powerful defoliant and antibacterial agent. TCDD is shorthand for 2,3,7,8-tetrachlorodi-benzo-p-dioxin; it is more commonly called simply dioxin. And TCDD, or dioxin, is perhaps the most toxic substance made by human beings. Microscopic amounts of it—measured in parts per trillion—have killed laboratory animals. Dioxin is a known teratogen, and it is also suspected of having mutagenic and carcinogenic effects.

Environmental Quality and Management

The Seveso Tragedy

A week after the danger had been publicly identified, more than seven-hundred people living in the most heavily contaminated area had been evacuated. Soon thereafter, the 114 hectares (280 acres) of the evacuated area were cordoned off. A larger zone of some 283 hectares (700 acres) was barred to nonresidents, and the eighteen-hundred children under fifteen who lived there were sent elsewhere during the day to reduce their exposure. A less contaminated area of several thousand hectares with a population of some twenty thousand was designated a "zone of caution."

Over the next several weeks, thousands of people reported various symptoms of dioxin-related illness. Most complained of minor skin disorders (blisters and discoloration), but some five hundred exhibited other symptoms of dioxin poisoning, including the above-mentioned nausea and dizziness. Many pregnant women, fearing teratogenic effects, received therapeutic abortions.

Fortunately, most of the victims were ill only briefly. Some residents did develop liver damage after the incident (dioxin is known to cause this in laboratory animals). But no deaths have been attributed to the accident. And the incidence of birth defects has been no higher than average among children born to mothers who were pregnant when the dioxin escaped.

Serious worries remain, however. The greatest of these concerns the long-term effects of dioxin poisoning on Seveso's residents. Will some of them yet develop serious, perhaps fatal, illnesses? So little is known of dioxin's effects on human health that the people of Seveso and the scientific community can only wait and see.

Another worry is the economic and social disruption of the Seveso region. Crops were destroyed, and houses contaminated by the long-lived poison, as was the top foot or so of soil. The chemical plant was permanently closed, putting 170 people out of work. And some factories outside the contaminated zones were forced to close by canceled orders.

A related worry is the formidable problem of decontamination. Dioxin can be safely broken down only by high-temperature incineration. Hence, one proposal is to construct a special incinerator that will burn all contaminated vegetation from the inner zones—and many thousands of tons of contaminated topsoil, the fertility of which will be destroyed in the process. (This solution has been vehemently opposed by many residents, some of whom have already moved back into their houses and want to simply forget about the incident and the possibility of lingering health hazards.)

The Seveso disaster points up the urgent need for fail-safe emission control devices wherever highly toxic and other hazardous substances are manufactured, transported, stored, and used. It also points up our urgent need to learn as much as possible from such mistakes once they have been made. As one scientist put it in the aftermath of the crisis, "The world missed a golden opportunity to get a handle on dioxin exposure and what it means to humans. We didn't get exposure levels. Why they didn't take fat biopsies, I don't know. The data just aren't there."

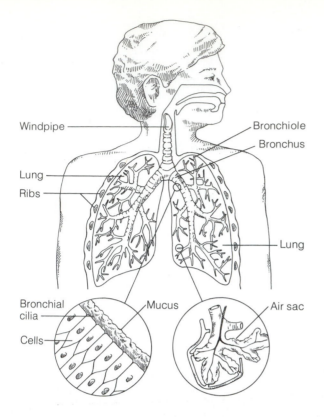

Figure 10.2. The human respiratory system.

Windpipe
Bronchiole
Bronchus
Lung
Ribs
Lung
Bronchial cilia
Mucus
Air sac
Cells

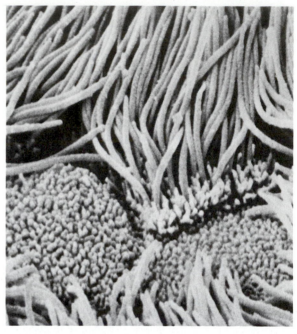

Figure 10.3 Cilia on the surface of a bronchiole in the lung of a horse. (American Lung Association.)

less than 10 ppm. But after several hours of exposure to 100 ppm, some individuals experience dizziness, headache, and impaired perception. With concentrations of 300–400 ppm, vision problems, nausea, and abdominal pain may develop, and 750 ppm can be fatal. While typical urban air contains 10–30 ppm, concentrations of carbon monoxide may exceed 200 ppm in highway tunnels and immediately behind auto exhausts, constituting a serious health hazard. Figure 10.5 shows the kind of situation that raises the carbon monoxide level in our cities.

Asphyxiating agents also put stress on the heart. Individuals afflicted with angina pectoris, a form of heart disease, appear to be particularly susceptible to this effect. Lack of oxygen in the heart muscle causes angina victims (about 4.2 million nationwide) to suffer attacks of chest pain. In late 1977, the Council on Environmental Quality reported that the fre-

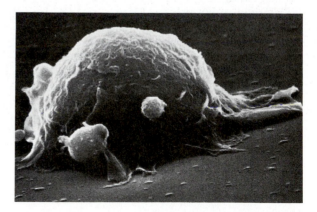

Figure 10.4 A lung macrophage, surrounded by particles of fly ash, viewed under a scanning electron microscope. (Courtesy of the Lawrence Berkeley Laboratory, University of California, Berkeley.)

quency of angina attacks increases as patients are exposed to higher levels of carbon monoxide.

Hydrogen sulfide, a decomposition product of organic material, is also an asphyxiating gas, but unlike carbon monoxide, it does not displace oxygen in the blood. Rather, in very high concentrations (achieved in confined places such as sewers or mines), hydrogen sulfide impairs that part of the brain that controls chest movements essential for normal breathing and causes almost instantaneous death. In the low concentrations that we normally encounter, hydrogen sulfide is relatively harmless, although its rotten-egg odor certainly does nothing to improve the environment.

Gases that act mainly as irritants of the respiratory tract include ozone, sulfur dioxide, and nitrogen dioxide. Although their specific reactions differ, effects of these irritants generally include impairment of macrophage activity, bronchiospasms (coughing), heavy secretion of mucus, inhibited movement of cilia, and distension of air sacs.

Particles, or aerosols, that penetrate the body's natural defenses may pose a serious health threat. Usually, the most severe effects develop after an individual has experienced long-term exposure to relatively high concentration levels. People in dusty occupations such as mining, metal grinding and the manufacturing of abrasives often suffer from long-

Figure 10.5 Rush-hour traffic, which produces high concentrations of air pollutants. (U.S. Department of Housing and Urban Development.)

term exposure to dangerous concentration levels.

Long-term occupational exposure to dust eventually can give rise to one of a family of lung diseases usually named according to the particulates involved. Serious lung diseases and their causes include silicosis (from quartz dust generated during mining), asbestosis (from asbestos fibers), and byssinosis (from cotton dust). Depending on the particulate type and the concentrations inhaled, impact on lungs may consist of irritation, allergic reactions, or scarring of tissue. Typically, victims experience coughing and shortness of breath and, in the long run, may develop pneumonia, chronic bronchitis, emphysema, or lung cancer.

Certain particles are harmful when inhaled due to their interactions with other air pollutants. Some particles interfere with the functioning of cilia, thereby slowing the flow of mucus and increasing the residence time of toxic pollutants in lungs. It is in this way that carcinogenic agents, for example, are retained in the lungs, increasing the likelihood of tumor formation. In addition, particulates may act as carriers of other pollutants. For instance, soot and fly ash, products of coal burning, can transport sulfur dioxide into the lungs. This is an example of synergism, discussed in Chapter 3.

Some toxic particles are so widely distributed in the general environment that to some extent they pose a potential health hazard to us all. Lead is probably the most notable of these pollutants. Lead is emitted into the atmosphere primarily through the exhaust of vehicles burning gasoline that contains lead as an antiknock additive. The danger of lead lies in the fact that it accumulates in the body more rapidly than it is excreted. We know little, however, of the toxic threshold of lead in people, although we are well aware that lead poisons children who eat lead-based paint chips. Lead poisoning attacks the blood-forming mechanism, the gastrointestinal tract, and, in severe cases, the central nervous system. Lead may also impair the functioning of heart and kidneys.

Only 5–10 percent of lead that is ingested is retained by the body, but 30–50 percent of lead that is inhaled is retained. Until recently, scientists thought that the primary sources of our daily intake of lead were food and water, and not the air we breathe. But in the mid-1970s, an EPA study comparing lead levels in children living in Los Angeles with those living in a rural area showed the opposite to be the case: the higher lead levels in the Los Angeles group apparently resulted from the fact that these children breathed more lead.

It can be argued that no link has been demonstrated to exist between air pollution and diseases such as emphysema and chronic bronchitis. But an examination of the symptoms of these diseases suggests otherwise. Chronic bronchitis, for example, is a long-term inflammation of the bronchi that is accompanied by enlargement of mucous membranes and heavy secretion of mucus. The walls of the bronchi contract in this condition (as a part of the cough reflex set off by the irritation), reducing the volume of air that reaches the air sacs; hence, chronic bronchitis victims experience shortness of breath. Emphysema often occurs along with or as a result of chronic bronchitis. As bronchi contract, air sacs coalesce and the surface area available for oxygen uptake by the blood is reduced. The eventual fate for emphysema victims is asphyxiation. Hence, symptoms of bronchitis and emphysema are similar to those attributed to inhalation of air pollutants. The comparison itself suggests that these diseases may be caused or aggravated by polluted air.

Another indication that air pollution is linked to respiratory diseases lies in the data showing that the incidence of respiratory diseases is greatest in cities with the highest air pollution levels. Still, direct evidence that unequivocally links elevated levels of a specific air pollutant with a particular respiratory disease is rare. Proving the existence of such a connection would be extremely complex, since the urban atmosphere is composed of numerous gases and aerosols whose relative concentrations continually fluctuate and whose interactions may be synergistic or antagonistic. The problem is further complicated, since a particular individual's reaction to polluted air is influenced by his or her age, level of physical activity, and general health, as well as the size of the dose and frequency of exposure. In addition, some individuals further aggravate the situation by smoking. This factor is discussed in detail in Box 10.2.

We have seen in earlier chapters that it is difficult to isolate simple cause-effect relationships when dealing with a complex environment. Thus, much indirect evidence can be cited to demonstrate a general cause-effect relationship between air pollution and respiratory disease, but explicit empirical data are sparse. Nonetheless, enough indirect statistical evidence exists to suggest positively that polluted air is a threat to human health. While it is prudent to be cautious in formulating policy on the basis of indirect evidence, the absence of more substantial proof is a weak excuse for inaction, particularly where human suffering is the alternative.

Effects on Other Species

Experiments involving laboratory animals are invaluable to our understanding of the effects of air pollutants on human health. However, little is known about the impact of polluted air on the well-being of other types of animals, most of which do not inhabit urban-industrial areas. In general, it appears that the two air pollutants most damaging to other species are fluoride and lead. In both cases, organisms are affected most commonly by ingesting contaminated vegetation rather than by inhaling the pollutants.

The processing of ceramics and phosphate rock releases fluorides into the atmosphere. Some plant species are damaged by hydrogen fluoride at a concentration of only 0.1 parts per billion (ppb). Others, including alfalfa and orchard grass, detoxify fluorides by combining them with organic acids, thus rendering them harmless to the plants themselves. Fluorides can reach concentrations of 40 ppm in these forage plants. When livestock consume these plants, however, the organic compounds containing fluorides break down, and the fluorides released can be lethal.

Dairy cattle are most sensitive to fluoride poisoning, called *fluorosis*. Fluorides reduce milk production and attack teeth and bones, producing lameness. Chronic fluorosis eventually leads to death. Substantial losses of cattle in Florida have been attributed to fluoride emissions from factories processing phosphate deposits for fertilizers. In fact, in the middle and late 1950s, fluoride poisoning claimed the lives of thirty-thousand cattle in two Florida counties.

Fewer documented cases of lead poisoning in animals are on record than for fluorosis, but the symptoms of this condition are no less debilitating. Afflicted animals lose their appetites, develop dry coats and muscle spasms, and frequently suffer paralysis of the hindquarters. Some zoo animals (especially members of the cat family) have been poisoned by their instinctive self-cleaning activities, their coats having been contaminated by airborne lead.

Some of the most dramatic instances of air pollution damage to vegetation have been caused by sulfur dioxide fumes from iron and copper smelters. In fact, these emissions have virtually eliminated vegetation over large areas adjacent to some smelters, as exemplified in Figure 10.6. Alfalfa, lettuce, barley, and white pine are particularly sensitive to sulfur dioxide. A white pine showing the effects of sulfur dioxide poisoning and a healthy tree of the same species are shown in Figure 10.7.

Ozone is another air pollutant that in relatively high concentrations is hazardous to plant life. The ill effects of ozone have been exhibited by grapes, sweet corn, lettuce, and pine. Windborne ozone from Los Angeles, for example, is destroying hundreds of thousands of ponderosa pine in the San Bernardino and San Jacinto Mountains 125 kilometers (80 miles) from the city.

Air pollution most commonly damages the leaves of plants. A waxy layer covers leaves to control the loss of water through evaporation. Leaf surfaces are also equipped with tiny pores, which allow carbon dioxide to enter the plant for photosynthesis. When water is limiting for the plant, as on a hot, dry, windy day, for instance, these pores close, preventing both

Figure 10.6 Remains of the vegetation surrounding a copper smelter in Sudbury, Ontario. The rest was destroyed by sulfur oxide emissions from the plant. (Courtesy of Richard Stiehl, University of Wisconsin-Green Bay.)

water loss and the entry of gases into the leaf. When sufficient water is available to the plant, the pores open to permit carbon dioxide to enter and water vapor to escape. Under these circumstances, gaseous air pollutants may enter leaf pores along with the carbon dioxide.

Gaseous pollutants, such as sulfur dioxide and ozone, enter leaves and dissolve in the water that adheres to surfaces of cell walls. Here a variety of complex, poorly understood chemical reactions takes place. The first sign that a plant has suffered air pollution damage is the appearance of *chlorosis*, a yellowing of the leaves due to chlorophyll loss. This condition is illustrated in Figure 10.8. The pattern of leaf damage can sometimes be used to identify the pollutant responsible. For example, fluorides characteristically accumulate at leaf tips and edges, and in these areas the leaf initially turns yellow. When pollution damage is extreme, plant tissues die and leaves turn brown (this condition is called necrosis). When a sufficient number of leaves die photosynthesis and plant growth is retarded. If enough leaves are killed, respiration exceeds photosynthesis; in these circumstances, the stored energy reserves are used up and the plant dies. While gases may readily enter leaf pores, particulates are generally too large to gain access and usually have little adverse effect. However, in some cases particulates may coat leaf surfaces, thereby reducing the amount of sunlight available for photosynthesis.

Figure 10.7 *Left:* A ten-year-old white pine severely damaged by relatively low concentrations of sulfur dioxide (possibly with ozone). *Right:* A healthy white pine from the same area grown in filtered air. (U.S. Department of Agriculture.)

The tolerance of vegetation to air pollution varies considerably with plant species and type of pollutant. The susceptibilities of various plants to selected pollutants are shown in Table 10.2 The reason for variations in the susceptibility of plants to gaseous pollutants remains largely unknown. In isolated cases, scientists have suggested explanations. For example, in varieties of onion resistant to ozone pollution, the leaf pores have been found to be sensitive to ozone and to close in its presence, thereby protecting the leaf interior. Generally, however, much remains to be learned about variations in the response of plants to pollutants.

The Delicate Ozone Shield

In general, air pollution does the most damage to living things in localities near pollutant sources, that

Figure 10.8 Chlorosis caused by sulfur dioxide pollution. The affected leaves have light areas between veins. The leaf on the right is healthy. (U.S. Department of Agriculture.)

Table 10.2 Susceptibility of selected plants to air pollutants.

Plant	Sulfur dioxide	Ozone	Fluoride	Peroxyacetyl nitrate (PAN)	Nitrogen dioxide
Alfalfa	S	I	R	I	
Cabbage	I	R	R		
Citrus			I		I
Corn (sweet)	R	S	S	R	
Grape	I	S	I		
Lettuce	S	S	I	S	S
Onion	R	I	I	R	
Pine	S	S	S		
Potato	R	S	I		
Ragweed	S	R	R		
Rhubarb	S	I	R		
Rose	R		I		
Soybean	S		R	I	
Sunflower	S	R	I		S

S = sensitive; I = intermediate; R = resistant.
Source: After A. C. Stern et al., *Fundamentals of Air Pollution*. New York: Academic Press, 1973.

is, in urban-industrial areas. However, some of the pollutants accumulating in the atmosphere pose a direct threat to all life forms without regard to their location. These pollutants may destroy the delicate ozone layer that shields organisms from dangerous radiation.

Ironically, while ozone is a pollutant in the lower atmosphere, its presence in minute concentrations (less than 1 part per 100,000) in the upper atmosphere is essential for the continuation of life on earth. Ozone forms primarily in the stratosphere as a product of the absorption of ultraviolet radiation (UV) by oxygen. Ozone itself also absorbs UV. These absorption processes shield organisms from exposure to this potentially lethal radiation.

However, at the long-wave end of ultraviolet radiation, called UV B, absorption is only partial and is very sensitive to changes in ozone concentration. UV B is responsible for sunburn and, according to many researchers, causes or contributes to skin cancer. The virulent form of skin cancer, melanoma, kills about a third of its victims. Disruption of absorption processes that produce ozone would increase the intensity of UV B reaching the ground, and this increase, in turn, could increase the incidence of skin cancer.

In view of this serious health threat, in the early 1970s, scientists began expressing alarm over certain activities that introduced into the stratosphere chlorine and oxides of nitrogen—chemicals that break down ozone. Oxides of nitrogen are injected into the stratosphere in the exhaust of certain aircraft, such as the supersonic transport (SST), shown in Figure 10.9, and can drift into the stratosphere following the extensive use of nitrogen fertilizers. However, recent research indicates that the net effect of aircraft exhaust and nitrogen fertilizers on ozone concentrations is negligible.

Today many scientists consider the most serious

threat to the ozone shield to be from halocarbons, commonly called Freons. Most Americans have used two halocarbons, knowingly or unknowingly, in their daily lives. One was used until recently as a propellant in common household aerosol sprays such as deodorants and hairsprays, while the second is still used as a coolant in refrigerators and air conditioners. Both of these substances are accumulating in the troposphere and gradually seeping into the stratosphere. Once in the stratosphere, they break down chemically, releasing chlorine, which reacts with and destroys ozone.

So far, halocarbons apparently have not yet reduced stratospheric ozone levels significantly. But what of the future? Continuation of halocarbon emissions at 1973 rates could eventually reduce stratospheric ozone by 14 percent. The attendant increase in UV B reaching the earth's surface would probably result in more cases of severe sunburn and skin cancer. According to a variety of estimates, the United States can expect between 2,000 and 15,000 additional cases of skin cancer per year for every 1-percent drop in stratospheric ozone. The effects of more intense UV radiation on vegetation and climate are less well known, but undoubtedly they would be harmful.

In response to the health threat posed by halocarbons, the federal government is now regulating their production and use. On October 15, 1978, the manufacture of halocarbons for nonessential propellants was banned, and as of December 15, 1978, federal law prohibits the manufacture of nonessential aerosol products that use halocarbons as propellants. However, many scientists in other countries remain un-

Figure 10.9 The Concorde, the Anglo-French supersonic transport (SST) that has been in regular operation since 1976. In the United States, SST development was halted in 1971, primarily because of high cost and projected low profits. Also, conservationists objected to high noise levels and the possible danger that SST exhaust could harm the ozone layer. While scientists recently determined that this threat to the ozone layer is not a serious one, the problem of noise remains at airports serving SSTs. (Photo from British Airways.)

convinced of the hazards of halocarbons. Hence, a worldwide ban on halocarbons appears unlikely in the near future. This means that 60 percent of the world's production of halocarbons will continue unabated.

Air Pollution and Weather

Another contemporary concern is the possibility that air pollution could have detrimental effects on weather and climate. Specifically, this concern focuses on the possibility that certain human activities may disrupt the earth-atmosphere radiation balance. Indeed, as suggested in our survey of weather phenomena in Chapter 9, human beings can influence weather and climate in at least three ways: (1) by changing the thermal properties (reflectivity, for example), of the earth's surface; (2) by releasing waste heat to the atmosphere; and (3) by altering concentrations of certain minor components of the atmosphere, for example carbon dioxide. Although many questions remain unanswered, it appears that the first two factors are important only on a local scale, that is, in big cities. Many scientists argue, however, that pollution of the atmosphere is changing concentration of critical components of the atmosphere and is thereby contributing to temperature changes on a hemispheric scale.

Urban Weather

Travelers approaching large metropolitan areas such as Chicago or New York are often greeted by an unsightly veil of dust, smoke, and haze. This dome of dirt, characteristic over many urban-industrial areas, is the consequence of a convective circulation of air that concentrates pollutants. This air motion itself is a product of temperature differences between the city and surrounding rural areas.

The average annual temperature of a city is typically several degrees warmer than that of the surrounding countryside, although on some days the

Box 10.2

There are standards for air quality outdoors and in industrial workplaces, but most people spend most of their time in places where air quality is neither monitored nor regulated—in homes, offices, classrooms, stores, cars, buses, planes, and other enclosed places.

• • •

The few studies that have been done of air pollution in homes have identified several potentially important sources of harmful contaminants. Gas stoves, for example, emit nitrogen dioxide, nitric oxide, and carbon monoxide.

• • •

The most pervasive form of air pollution indoors, however, almost certainly is tobacco smoke. The health risk that smokers impose upon themselves has been suspected for a long time and established beyond any reasonable doubt since the 1960s. The plight of some millions of American nonsmokers who are clinically allergic to tobacco smoke is well known to the victims and to the medical profession, but it has not been widely publicized. Also well established medically is the special threat that tobacco smoke represents for nonsmoking victims of asthma, bronchitis, emphysema, and heart disease.

Excerpted from P. R. Ehrlich, A. H. Ehrlich, and J. P. Holdren, *Ecoscience: Population, Resources, Environment.* San Francisco: W. H. Freeman and Company, 1977.

Environmental Quality and Management

Smoking and Indoor Air Quality

More recently, attention has turned belatedly to the possibility that tobacco smoke at concentrations commonly encountered indoors is hazardous to otherwise healthy, nonallergic nonsmokers. The risk to nonsmokers (more accurately, passive smokers, that is, people who inhale smoke *involuntarily* from their environment) should hardly come as a surprise in view of the large number of carcinogenic and otherwise toxic substances that have been identified in tobacco smoke. They include cadmium, oxides of nitrogen, benzo(a)pyrene and other hydrocarbons, carbon monoxide, and particulate matter.

Probably many passive smokers have consoled themselves with the belief that the voluntary smokers sucking the smoldering tobacco are inhaling and absorbing most of the pollutants themselves. Unfortunately, studies have shown that the sidestream smoke—that which drifts off the tip of the cigarette—contains several times as much of the main pollutants as the mainstream smoke inhaled by the voluntary smokers.

Obviously, the sidestream smoke is diluted by surrounding air before being inhaled by the passive smoker, but the dosage can still be substantial. Concentrations of particulate matter in excess of 3000 micrograms per cubic meter [of air] have been measured in smoke-filled rooms at parties, or 40 times the U.S. standard for outside air. . . . Carbon monoxide concentrations in smoke-filled rooms reach at least 40,000 micrograms per cubic meter and prolonged smoking in enclosed automobiles has produced CO concentrations of 100,000 micrograms per cubic meter [of air]. Studies of the carboxyhemoglobin content in the blood of voluntary and passive smokers indicate that the passive smoker in usually smoky conditions receives the equivalent of smoking one to two cigarettes per hour.

. . .

The likelihood that otherwise healthy nonsmokers are being damaged by inhaling other people's smoke, coupled with the certainty that passive smoking is harmful to persons unfortunate enough to be allergic to tobacco smoke or suffering from a variety of preexisting diseases, make society's tolerance for public smoking in an era of growing environmental concern an ever more striking anomaly. As of 1976, the state of California, New York City, and a few other major political entities had passed laws restricting smoking in public buildings and public transportation systems. These measures are a good beginning, but there is ample reason to go much further. [It has been argued that] in situations where nonsmokers have no escape (such as on buses and aircraft, whose ventilation systems are incapable of maintaining a real distinction between smoking and no-smoking zones), smoking should long ago have been banned outright.

There is no doubt, of course, that voluntary smoking provides a great many people with a great deal of pleasure. But it seems to us that this activity should be legally confined, as certain other pleasurable (and considerably less dangerous) activities are, to consenting adults in private.

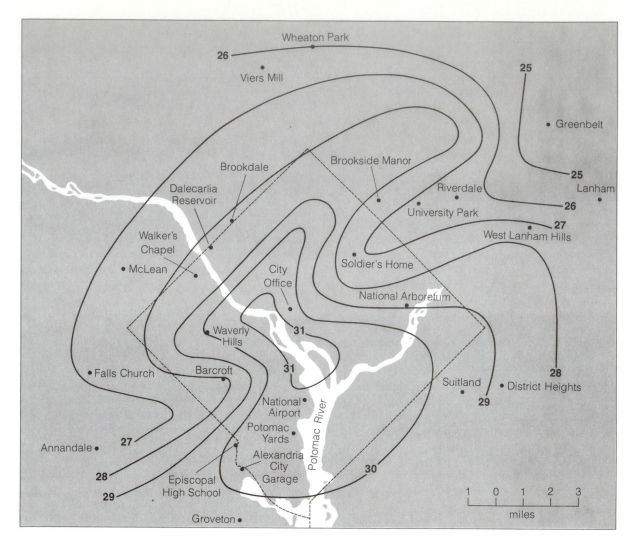

Figure 10.10 Average winter low temperatures (in °F) in Washington, D.C., illustrating the urban heat island effect. (After C. A. Woolum, ''Notes from a Study of the Microclimatology of the Washington, D.C. Area for the Winter and Spring Seasons,'' *Weatherwise 17:*6, 1964.)

thermal contrast may be 10°C (18°F) or more. As a result, snow tends to melt faster and flowers bloom earlier in a city. Among the several factors that contribute to an *urban heat island,* illustrated for Washington, D.C., in Figure 10.10, is a city's concentration of heat sources (for example, people, cars, industry, air conditioners, and furnaces). Because all

heat from every source eventually reaches the atmosphere, the air of a large city receives a considerable amount of waste heat, or thermal pollution. In fact, on a typical winter day in New York City, waste heat from urban sources amounts to more than twice the energy received by solar radiation.

The emission of heat into city air is facilitated by

Figure 10.11 The creation of the convective air circulation pattern that results in a dust dome over an urban center. (After W. P. Lowry, "The Climate of Cities," *Scientific American 217:20*, August 1967. Copyright © by Scientific American, Inc. All rights reserved.)

the thermal properties of urban building materials: concrete, asphalt, and brick conduct heat more readily than does the vegetative cover of rural areas. Thus, the heat loss at night by infrared radiation to the atmosphere and space is partially compensated for in cities by a release of heat from buildings and streets. The temperature contrast is further accentuated by a city's lack of standing bodies of water. Urban drainage systems quickly and efficiently remove most runoff, so less of the available solar energy is used for evaporation (latent heat). Thus, more solar energy is available to heat the ground and air directly (sensible heat).

Under certain circumstances, the relative warmth of a large city compared with its surroundings promotes the development of a convective circulation of air. As shown in Figure 10.11, warm air at the city's center rises and is replaced by cooler, denser air from the countryside. When light, synoptic-scale winds are blowing, the rising columns of air gather aerosols into a *dust dome* over the city. Dust may be a thousand times more concentrated over industrial areas than in the air of the open countryside. If the synoptic-scale winds increase to more than about 13 kilometers per hour (8 miles per hour), the dust dome elongates downwind in the form of a *dust plume*,

spreading the city's pollutants over the distant countryside. The Chicago dust plume, for example, is sometimes visible as far as 240 kilometers (150 miles) from its source.

A dust dome reduces the amount of solar radiation that penetrates the urban atmosphere. The reduction is particularly marked (up to 90 percent) in the ultraviolet portion of the solar spectrum. Though too much UV B is dangerous, as we saw in our discussion of the ozone shield, our bodies need adequate UV B to produce vitamin D. Depletion of UV B apparently contributed to a high incidence of rickets (a malady resulting from vitamin D deficiency) in industrialized European cities before vitamin D supplements were added to the diet. Ironically, since urban centers are so often the villains with respect to pollution-related diseases, UV-induced skin cancer is lower in cities than in rural areas.

Certain air pollutants usually found in urban air, including a variety of dust particles and acid droplets, influence the development of clouds and precipitation within and downwind from a city. These pollutants, many of which are hygroscopic, serve as nuclei for cloud droplets, thereby accelerating condensation. The tendency for cloudiness and precipitation to occur more frequently in urban-industrial areas than in the surrounding countryside is enhanced by the fact that rising urban air helps to bring air parcels to saturation.

The influence of urban air pollution on condensation and precipitation is illustrated by typical climatic contrasts between urban and rural regions. In cities, winter fogs (ground-level clouds) occur several times more frequently than in the surrounding countryside, and downwind from cities, rainfall is increased by 5–10 percent. The fact that the disparity tends to be exhibited on weekdays, when urban-industrial activity is at its peak, suggests that precipitation enhancement is at least partially attributable to pollutants of urban-industrial origin.

Because precipitation, fog, and cloudiness in urban areas often have adverse effects on both surface and air transportation, any artificially induced enhancement is potentially troublesome. Restrictions in visibility slow traffic, curtail air travel, and contribute to auto accidents. In recent years, however, improvements in local visibilities have been reported, apparently the results of stricter enforcement of air quality standards that triggered a shift (albeit temporary) to cleaner burning fuels.

Still another human activity affecting weather phenomena involves jet plane traffic, which is modifying the cloud cover, especially along heavily traveled air corridors between major cities. The visible jet contrails etching the sky—for example, those in Figure 10.12—are feathery cirrus clouds traceable to water vapor and condensation nuclei produced as combustion products in jet engines. The increased cloudiness, in turn, reduces sunshine penetration, and may locally enhance precipitation by serving as a source of ice-crystal nuclei.

Acid Rain

As we saw in Chapter 6, the distillation function of the water cycle purifies water. However, in falling from clouds to the ground, rain and snow wash pollutants from the air. Rainfall is normally slightly acidic, since it dissolves some atmospheric carbon dioxide, producing weak carbonic acid. But where air is polluted with oxides of sulfur and nitrogen, rainfall produces strong sulfuric and nitric acids. Precipitation that passes through such contaminated air may become two hundred times more acidic than normal. The range of acidity and alkalinity, called the pH scale, is shown in Figure 10.13. In the figure, the normal acidity of rainwater is compared with the pH values of some other familiar substances.

Where acid rains fall on soils or rock that cannot neutralize the acidity, regional surface waters also become more acidic. This change in turn affects the wildlife dependent on the water reservoirs. Fish

Figure 10.12 Contrails produced by the exhaust of jet aircraft. (Photo Researchers.)

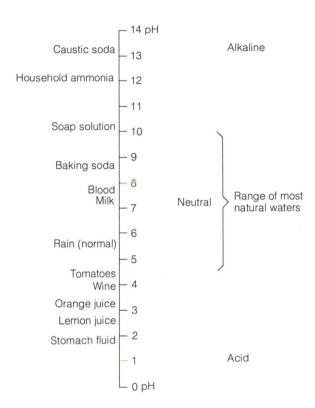

Figure 10.13 The scale of acidity and alkalinity (the pH scale) and the relative pH of some common substances.

populations in some lakes and streams of Norway, Sweden, and the northeastern United States apparently were reduced or eliminated because of increased acidity in their aquatic habitats. For example, the disappearance of brook trout from Adirondak (New York) mountain streams is attributed to acid rains. Another effect of acid rain is that it accelerates rates of weathering of building materials, especially limestone, marble, and concrete. Metals, too, corrode faster than normal when they are exposed to acidic moisture.

A sharp rise in the acidity of rain and snow in the past two decades has been recorded in the eastern United States, as shown in Figure 10.14. Although

scientists do not completely agree on the reasons for this increase, some have speculated that the reduction in the amount of particulates in the air due to pollution abatement measures may have actually added to the problem of acid rainfall. Apparently, prior to implementation of certain air quality controls, alkaline particles (fly ash) in industrial smoke plumes served to neutralize acid rain. Now, however, with the use of such devices as electrostatic precipitators and other aerosol-removing techniques (discussed in Chapter 11), rains are becoming more acidic.

Hemispheric Trends in Climate

As we discussed in Chapter 9, average temperatures in the northern hemisphere rose between the 1890s and the 1940s and have fallen since then (refer back to Figure 9.20). The magnitude of the temperature change was very small, averaging less than $0.1 \, C°$ per decade. As we noted, however, small changes in hemispheric temperature are often greatly amplified in specific localities and may signal important changes in climate. Such climatic changes may disrupt agriculture and affect water supply and energy demand for space heating and cooling. In this regard, it is noteworthy that the post-1945 cooling trend has been sufficient to return us one-sixth of the way back to the harsh climate of the period known as the Little Ice Age, which lasted from the 1600s to the 1800s. During that period, mountain glaciers advanced in the Alps, summer frosts ravaged crops in middle latitudes, and unusually severe cold weather and storms battered the land.

Recently, speculation and research have centered on the possible role of air pollution in long-term climatic changes. Investigations thus far have focused primarily on the impacts of increased atmospheric dust and carbon dioxide on temperature. Growing concentrations of carbon dioxide are generally thought to result in an increase in the intensity of the "greenhouse effect" described earlier—that is, an increase in the absorption and reradiation of in-

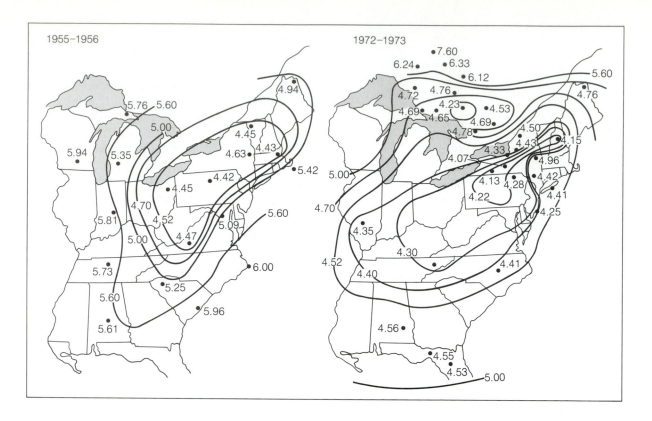

Figure 10.14 The increase in acidity of precipitation over the eastern United States from 1955–1956 to 1972–1973. The figures represent the average annual pH of precipitation. (After G. E. Likens, "Acid Precipitation," *Chemical and Engineering News 54*:29–37, November 1976.)

frared radiaton, which warms the lower atmosphere. In addition, it is possible that growing atmospheric concentrations of halocarbons and nitrous oxide may also be adding to the intensity of the "greenhouse effect."

In the view of some scientists, increased atmospheric turbidity (dustiness) has the opposite effect of increased carbon dioxide: rather than warming the earth, an increase in turbidity adds to the reflectivity of the earth-atmosphere system, reducing the amount of solar energy that reaches the earth's surface and thus resulting in a cooling of the atmosphere. Although the release of carbon dioxide associated with large-scale fossil fuel combustion may have contributed slightly to the hemispheric warming trend of the 1890s to 1940s, beginning in the

1930s, the cooling effect of the atmosphere's elevated dust load generated by a variety of human activities (overgrazing, intensive agriculture, and industrialization) began to prevail. The post-World War II cooling trend can therefore apparently be attributed to some extent to human activities. Also, dust from increased volcanic activity, especially since 1955, may be an important contributing factor to hemispheric cooling.

Other scientists hold a different view. They contend that increased turbidity may produce atmospheric warming rather than cooling. While acknowledging that the introduction of volcanic dust into the stratosphere tends to promote atmospheric cooling by increasing planetary albedo, they point out that dust resulting from human activities is generally confined to the lower troposphere and that relatively little ends up in the stratosphere above. For the most part, dust in the troposphere scatters solar radiation back to the earth's surface, where it is absorbed, and larger dust particles absorb both solar and infrared radiation. The net contribution of tropospheric dust, then, is the warming of the lower atmosphere. This opinion has prompted some scientists to consider the current hemispheric cooling as merely a temporary reversal of an overall warming trend.

It is anticipated that on-going research will eventually resolve the controversy now surrounding the impact of air pollution on hemispheric air temperatures. At the present time, climatologists have many more questions than answers. Research has been given renewed impetus recently by the growing awareness that minor changes in hemispheric temperature could potentially produce climatic changes with disastrous consequences in some areas of the globe. This possibility is illustrated by the climatic shifts that already have taken place during the current hemispheric cooling trend. Apparently, the monsoon rains that are so essential for agriculture in the half of the world in which starvation is prevalent are now less dependable than they were during the first fifty years of this century Figure 10.15 shows the pattern of these rains over a twenty-five-year period. The failure or reduction of monsoon rains could be catastrophic for the people of India and Sahelian Africa. As we shall see in Chapter 17, the latter have already suffered a terrible drought resulting from just such a failure in monsoon rains.

Air Pollution Episodes

Up to this point in the chapter, we have concentrated on the many undesirable consequences of polluted air. Now we turn to the dynamics of air pollution—the interactions that influence the concentrations and behavior of pollutants once they are expelled into the air. In general, the higher the concentrations of pol-

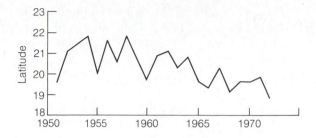

Figure 10.15 The penetration of monsoon rains northward in North Africa from the 1950s to the early 1970s. The trend suggests that drought is more likely in the Sahelian countries located just south of the Saharan Desert. (After R. A. Bryson and T. J. Murray, *Climates of Hunger.* Madison: University of Wisconsin Press, 1977; ⓒ by the Regents of the University of Wisconsin, p. 103.)

lutants in air, the greater is the hazard. Weather conditions exert the most significant control on air pollutant concentration, and therefore they are the focus of this section.

Once pollutants are emitted into the atmosphere, their concentrations begin to decrease. The rate of concentration decrease, or dilution, is partially determined by the extent to which pollutants mix with cleaner air. The more effective the mixing, the more rapid is the dilution. At times, conditions in the atmosphere favor rapid dilution, and in these instances the impact of pollutants is minor, but on other occasions—termed *air pollution episodes*—conditions in the atmosphere minimize dilution and the impact is severe, particularly on human health. The weather factors influencing the severity of air pollution are wind speed and air stability.

Wind Speed

Intuitively, we know that air is likely to mix more vigorously on a windy day than on a calm one. As a general rule of thumb, the doubling of wind speed cuts the concentration of air pollutants in half. This estimate is illustrated in Figure 10.16. Certain weather patterns favor light winds, and thus inhibit the dispersal of contaminants. When the weather of a locality comes under the influence of a high pressure system, for example, winds near the center of the high are very light or calm and pollutants fail to disperse.

We have seen how temperatures can differ between a city and the surrounding countryside. Wind speed, too, can differ in adjacent rural and urban areas under the same regional weather regime. In a city, winds are slowed by the rough surface created by the canyonlike topography of tall buildings and narrow streets. In fact, wind speeds are normally 10 percent slower in a city than in the surrounding countryside. When light regional winds (less than 16 kilometers, or 10 miles, per hour) prevail, the discrepancy is even more pronounced, amounting to a wind-speed reduction of up to 30 percent. Hence, the dilution of air pollutants by wind is particularly impeded in the urban localities where, ironically, most of these substances are generated.

Air Stability

When we consider the influence of *air stability* on pollutant concentrations, it is convenient for us to distinguish between polluted air and the relatively clean air into which polluted air is introduced. We refer to polluted air as air parcels, discrete volume units, whereas we speak of clean air simply as surrounding air.

If a parcel of polluted air is warmer than surrounding denser air, it rises. Such a parcel is said to be buoyant with respect to the surrounding air. Air pressure on a rising air parcel steadily decreases, so the parcel expands the way a helium-filled balloon expands as it ascends. Figure 10.17 shows the decrease in air pressure as an air parcel gains altitude. As long as no heat is lost through the imaginary walls of the parcel, the parcel's heat content is unchanged. But in expanding while rising, the parcel does work. This work is accomplished at the expense of the parcel's heat energy; hence, the parcel's tem-

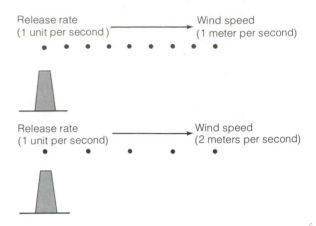

Figure 10.16 The doubling of wind speed increases the spacing between particles by a factor of two, thereby reducing concentrations.

perature drops. A parcel of air normally cools at a rate of about 9.8°C per 1000 meters (about 5.5°F per 1000 feet) of lifting (as long as condensation does not take place).

If the vertical temperature profile of a layer of air is such that the temperature at every elevation is lower than the temperature of an air parcel rising through it (see Figure 10.18), the air parcel will continue to rise. In such a situation, the surrounding air layer is said to be *unstable*. If, on the other hand, a parcel enters a layer of air whose temperature at every elevation is warmer than the temperature of the air parcel (see Figure 10.19), the parcel is no longer buoyant, and it begins to sink. In this case, the air layer is said to be *stable*. A parcel experiences more mixing when emitted into unstable air than when emitted into stable air. Stable air inhibits the upward transport of air pollutants and acts as a lid over the lower atmosphere to trap air pollutants. Continual emission of contaminants into stable air results in the continual accumulation and concentration of pollutants.

We can make a rough estimate of the stability of air layers by observing the behavior of a plume of

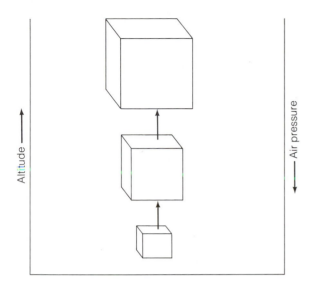

Figure 10.17 The decrease of pressure and expansion of an air parcel as it rises through the atmosphere.

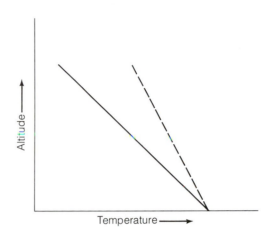

Figure 10.18 An unstable layer of air. At every elevation the temperature of an air parcel (broken line) is warmer than the temperature of the surrounding air (solid line).

smoke belching from a stack. If the smoke is entering an unstable air layer, looping occurs in the plume. This motion causes the polluted air parcels to mix readily with the surrounding air, facilitating dispersal. This situation is illustrated in Figure 10.20. If, however, a smoke plume flattens and spreads out, as in Figure 10.21, stable conditions are indicated and relatively high pollutant concentrations are maintained downwind from the stack.

In summary, air stability influences the rate at which polluted air and clean air mix. If air layers are stable, dilution is inhibited; if air layers are unstable, dilution is enhanced.

Temperature Inversions

On some occasions, layers of air within the troposphere are not merely stable; they are extremely stable, and can trigger severe air pollution episodes. This situation arises when the temperature of air layers increases with altitude, shown in Figure 10.22.

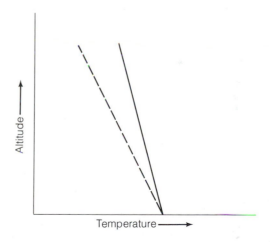

Figure 10.19 A stable layer of air. At every elevation the temperature of an air parcel (broken line) is colder than the temperature of the surrounding air (solid line).

Figure 10.20 A looping smoke plume, indicating emission into an unstable layer of air.

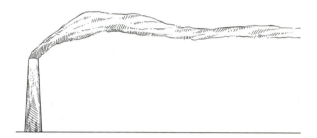

Figure 10.21 A smoke plume forming a narrow ribbon, indicating emission into a stable layer of air.

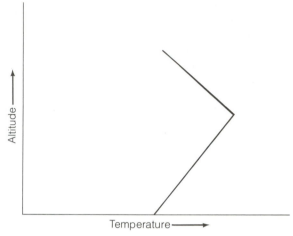

Figure 10.22 The temperature profile of an extremely stable air layer, a temperature inversion.

Such a condition is known as a *temperature, or thermal, inversion.* In a temperature inversion, warm, light air overlies cooler, denser air in a persistent stratification that strongly inhibits mixing and dilution. There are two important ways whereby a thermal inversion may develop: by the sinking (or subsidence) of air, or by radiational cooling.

A subsidence temperature inversion forms a lid over a wide area, often encompassing several states. It develops during a period of fair weather when the hemispheric weather pattern causes a high pressure system to stall. A high is characterized by descending (hence, warming) air currents that spread out near, but not at, the earth's surface. The warm air is prevented from reaching the surface by a shallow air layer at ground level known as the *shielding layer.* In the shielding layer, air is mixed by convection and a fixed temperature profile is maintained. Air just above the shielding layer, however, is warmed because of its descent, and may be significantly warmer than the air immediately beneath the top of the shielding layer. A thermal inversion forms, separating the air of the shielding layer from the modified air above. The development of this type of temperature inversion is shown in Figure 10.23.

Radiation temperature inversions are perhaps more common, and are often more localized, than subsidence temperature inversions. At night, under clear skies, the earth's warmth is rapidly lost to space through infrared radiation as the ground cools. Surface air layers are then chilled by contact (conduction) with the cooler ground. Because the air at the surface is coldest, a thermal inversion develops. Usually, after sunrise, as solar energy is absorbed by the ground and reradiated to the lower troposphere, the inversion gradually disappears and a normal temperature profile is restored. However, in winter in places where snow covers the ground, and when

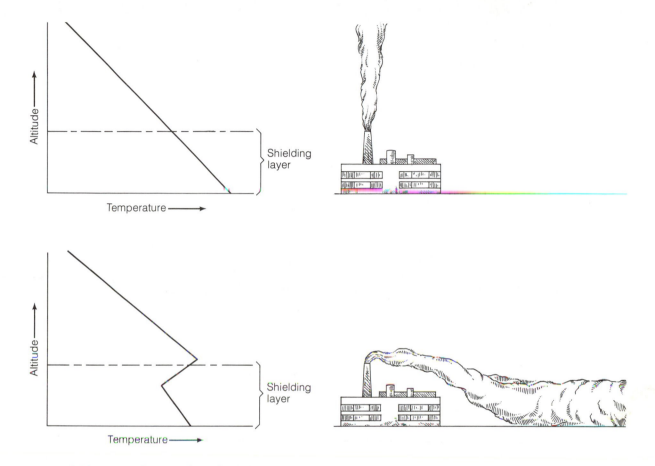

Figure 10.23 The development of an elevated temperature inversion by subsidence (sinking) of air.

the sun's rays are low, a low-level radiation inversion may persist for several days or even weeks at a time, severely inhibiting the dispersal of air pollutants.

The Location and Frequency of Air Pollution Episodes

The frequency with which atmospheric conditions occur that favor air stagnation, and hence reduced air quality, varies from place to place and with the seasons. Regions of the United States with particularly high air pollution potential include southern and coastal California, portions of the Rocky Mountain states, and the mountainous portions of the mid-Atlantic states. In general, winter conditions in the western United States are the least favorable for pollutant dispersal, while in the East, autumn appears to be the season in which air quality is lowest.

Many regions with high air pollution potential are also locales of rugged topography. Hills and mountain ranges can act as barriers, blocking horizontal

Figure 10.24 An air pollution episode in Los Angeles. (Fred Lyon, from Rapho-Guillumette, Photo Researchers.)

winds that disperse polluted air. In addition, thermal inversions that form in such lowlands as river valleys are often strengthened by an accumulation of cold, dense air that drains downward from nearby highlands at night. The result is a persistent stratification of light air over dense air. Such a situation contributed to the Donora tragedy described at the beginning of this chapter.

The combination of topographic features, a high concentration of pollutant sources (more than 6 million cars!), and frequent periods of air stability make Los Angeles particularly susceptible to severe air pollution episodes (see Figure 10.24). Figure 10.25 shows the climatic and topographic features influencing air quality in that city. The weather of Los Angeles, like that through much of California, is strongly influenced by the eastern edge of the semipermanent Pacific high pressure system. This high is responsible not only for California's famous fair weather, but also for descending air currents that generate a subsidence temperature inversion at 700 meters (2300 feet) over Los Angeles. This lid to vertical air mixing forms on about two-thirds of the days of the year. The exceptionally high incidence of extremely stable atmospheric conditions is aggravated in Los Angeles by topographic barriers. The city is situated in a bowl that opens to the Pacific and is rimmed on three sides by mountains. Consequently, cool breezes that sweep inland from the ocean are unable to flush the pollutants out of the city; mountains and an elevated temperature inversion effectively encase the city in its own fumes. And it is within this crucible that a complex chemistry takes place to produce smog, discussed in detail in a succeeding section.

Natural Cleansing Processes

Conditions that favor the accumulation and concentration of pollutants in air are countered to some

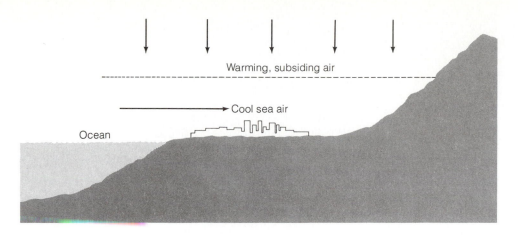

Warming, subsiding air

Cool sea air

Ocean

Figure 10.25 A diagram of the climatological and topographical features that give Los Angeles an unusually high air pollution potential.

extent by natural removal mechanisms. Some particles are removed from the air by becoming attached to buildings and other structures. Others are subject to gravitational settling, or sedimentation, which is most effective for aerosols with radii greater than 10^{-5} centimeter. The relative sizes of various kinds of particles and their settling velocities are given in Figure 10.26. Obviously, heavier (and hence larger) aerosols settle out more rapidly than do smaller aerosols. Hence, larger aerosols settle out near their sources while small aerosols may be carried for many kilometers and to great altitudes before finally settling to the ground.

The most effective natural removal mechanism is the scavenging of pollutants by rain and snow, called washout. In fact, in localities of moderate precipitation, as much as 90 percent of aerosol removal is accounted for by precipitation scavenging. Although gaseous pollutants are somewhat less susceptible to scavenging than are aerosols, they dissolve to some extent in raindrops and cloud droplets.

As we have already noted, though air pollutant scavenging is advantageous from an air quality perspective, it has an adverse effect on the quality of precipitation, creating acidic rains, which have a potentially deleterious impact on surface water qual-

ity and aquatic life. It is quite evident, therefore, that in localities of heavily polluted air, natural cleansing mechanisms may merely transfer pollutants from one component of ecosystems (air) to another (water).

Air Pollutants: Types and Sources

Where do the pollutants come from that are the ingredients of air pollution episodes? Sources of natural air pollutants include forest fires, areas of high pollen dispersal, wind erosion, organic decay, and volcanic eruptions. Table 10.3 shows the relative significance of different types of human-induced pollution. The single most important source of atmospheric pollutants fostered by human activity is the motor vehicle. According to the EPA, transportation vehicles yearly emit more than 100 million tons of the major air pollutants. Many industrial sources contribute as well; pulp and paper mills, iron and steel mills, oil refineries, smelters, and chemical plants are steady producers of pollutants. Additional pollutants come from fuel combustion by industrial and domestic furnaces, refuse burning, and various agricultural activities (for example, crop dusting and plowing). Overall, more than 200 million tons of

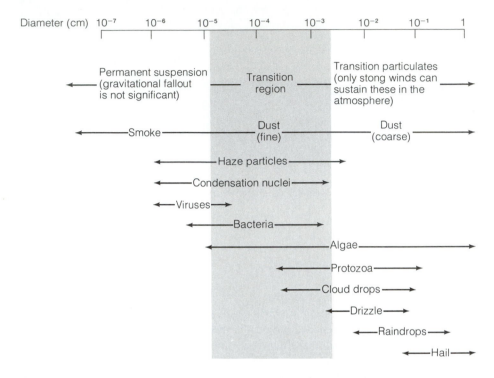

Figure 10.26 The relative diameters and settling velocities of particles. As diameters of suspended particles increase, the rate of fall increases and gravitational settling becomes important. (After *Patterns and Perspectives in Environmental Science.* Washington, D.C.: National Science Foundation, 1973.)

contaminants are emitted into the atmosphere each year as the result of human activity in the United States—almost a ton for each person.

The substances listed in Table 10.3 are called *primary air pollutants.* Some of these materials undergo chemical reactions in the atmosphere to become *secondary air pollutants,* which include acid mists and smog. In some instances, the environmental impact of individual primary pollutants is less severe than that of the secondary pollutants they form. In this section, we consider some of the characteristics of environmentally important primary and secondary pollutants, turning first to the gaseous pollutants and then to the aerosols.

Oxides of Carbon

The burning of fossil fuels releases carbon dioxide into the atmosphere. Some of this carbon dioxide cycles into oceans or is taken up by vegetation through photosynthesis. Currently, the amount of carbon dioxide in the earth-atmosphere system is rising at the rate of 10 ppm per decade. The increasing combustion of coal, oil, and natural gas coupled with the worldwide destruction of forests is responsible for the rise. As noted earlier, this trend may have important ramifications for global climate.

By far the most important natural source of atmospheric carbon monoxide is the combination of

Table 10.3 Emission of air pollutants by source in the United States in 1974 (in millions of tons per year).

Source	Sulfur oxides	Nitrogen oxides	Hydrocarbons	Carbon monoxide	Particulates
Transportation	0.8	10.7	12.8	73.5	1.3
Stationary fuel combustion	24.3	11.0	11.7	0.9	5.9
Industrial	6.2	0.6	3.1	12.7	11.0
Solid waste disposal	0.0	0.1	0.6	2.4	0.5
Miscellaneous	0.1	0.1	12.2	5.1	0.8
Total	31.4	22.5	40.4	94.6	19.5

Source: After Council on Environmental Quality, *Sixth Annual Report,* 1975, p. 440.

oxygen with methane, a product of the anaerobic decay of vegetation. Carbon monoxide is removed from the atmosphere by the activity of certain soil microorganisms. The net result is a harmless hemispheric average concentration of less than 0.14 ppm. The principal human-made source of carbon monoxide is motor vehicles. Because it is odorless and tasteless, carbon monoxide defies direct detection by humans and constitutes a serious health threat where sources are concentrated (for example, in highway tunnels and underground parking garages).

Hydrocarbons

Hydrocarbons include numerous compounds of hydrogen and carbon. Methane is the hydrocarbon that occurs naturally in the atmosphere at the highest concentration (about 1.5 ppm), but it is relatively unreactive at this concentration. Occurring in somewhat lower concentrations but much more reactive are the terpenes, a variety of volatile hydrocarbons that are emitted by vegetation. (Volatility is the quality of vaporizing readily.) These compounds are responsible for the aromas of such trees as pine, sandalwood, and eucalyptus. Terpenes form particles that scatter sunlight, producing the blue haze that is often seen above forests. In urban areas, the principal source of reactive hydrocarbons is the incomplete

combustion of gasoline in motor vehicles. In addition, because gasoline is very volatile, a significant fraction of urban hydrocarbons (as much as 15 percent in some cities) enters the air during gasoline delivery and refueling operations.

Although little is understood about the natural cycling of hydrocarbons in the atmosphere, in the low concentrations typically encountered in city air these substances pose no serious environmental problems. Some hydrocarbons, such as benzene (a component of many consumer goods, including rubber cement and furniture stripper) and benzo(a)-pyrene (a product of fossil fuel combustion), however, are carcinogenic. And in combination with other pollutants (especially oxides of nitrogen) and sunlight, hydrocarbons produce smog.

Oxides of Nitrogen

Nitric oxide and nitrogen dioxide are the principal compounds of nitrogen that reduce air quality. Less than 15 percent of the total amount of these substances in the atmosphere is contributed by human activity. The remainder is apparently produced by biological processes in the soil. A major human-made source of nitric oxide is the internal combustion engine, within which develop the very high temperatures needed for oxygen to combine with nitrogen. In addition, all industrial and domestic combustion processes contribute some nitric oxide to the air. Once nitric oxide is exhausted into the atmosphere, it combines with oxygen to produce the more toxic nitrogen dioxide. In rural areas, concentrations of nitrogen dioxide are usually less than 0.004 ppm, but in heavily polluted urban air they may exceed 0.1 ppm. At this concentration, nitrogen dioxide is a serious irritant of the respiratory tract. And because nitrogen dioxide is visible as a brownish haze, it restricts visibility. When nitrogen dioxide combines with water vapor, nitric acid, a highly corrosive substance, is formed. Oxides of nitrogen are also important ingredients in the generation of smog.

Compounds of Sulfur

Under natural conditions, sulfur enters the atmosphere as sulfur dioxide from volcanoes, as sulfate particles from sea spray, and as the hydrogen sulfide produced in the anaerobic decay of organic matter. These sulphur compounds are washed from the air by rain and taken up directly by soil, vegetation, and surface waters. Today, many human activities result in the emission of sulfur into the atmosphere—they add nearly one-third the amount of sulfur cycled into air by natural processes. Most of the sulfur dioxide we contribute comes from the burning of sulfur-containing fuels, coal and oil. In addition, the smelting of sulfur-bearing ores—lead, zinc, and copper sulfides—add significant amounts of sulfur dioxide to the atmosphere.

In air, sulfur dioxide combines with oxygen to produce sulfur trioxide. In concentrations greater than 5 ppm, both of these compounds irritate respiratory passages and can aggravate asthma, emphysema, and bronchitis. Sulfur dioxide concentrations in urban air may reach 0.5 ppm, but average annual concentrations as low as 0.03 ppm have long-term adverse effects on some vegetation. In addition, sulfur trioxide dissolves in water droplets to produce a

sulfuric acid mist, a corrosive mixture that restricts visibility. Industrial activities such as paper and pulp processing emit hydrogen sulfide as well as a family of organic sulfur-containing gases called mercaptans. Even in extremely small concentrations, these compounds are foul-smelling. In addition, hydrogen sulfide damages statuary, as illustrated in Figure 10.27, and tarnishes silverware, lead-based paints, and copper facing.

Smog

When vehicular traffic is congested, for example during morning and evening rush hours, *photochemical smog* is likely to form. Oxides of nitrogen and hydrocarbons in auto exhaust react in the presence of sunlight to produce a noxious, hazy mixture of aerosols and gases. While this reaction is common in urban areas, winds occasionally transport auto exhaust into suburban and rural areas where the sun's rays trigger smog development.

Constituents of photochemical smog include PAN (peroxyacetyl nitrate) and ozone. PAN is deleterious to vegetation and stings the eyes. While normal levels of ozone at the earth's surface average only about 0.02 ppm, during very smoggy periods ozone concentrations may exceed 0.5 ppm. At these relatively high concentrations, ozone irritates the eyes and causes coughing and headaches. It also degrades rubber and fabrics and, as we saw in Table 10.2, damages some cash crops. Topography and climate combine to make Los Angeles particularly prone to smog days. But photochemical smog is a serious problem within and downwind from many other metropolitan areas as well.

Aerosols

Up to this point in our survey of the major air pollutants, we have focused primarily on gases. Now we turn our attention to the millions of tons of tiny solid particles, also called particulates, and liquid droplets (acid mists, for example), collectively termed aerosols. Sea salt spray, soil erosion, volcanic

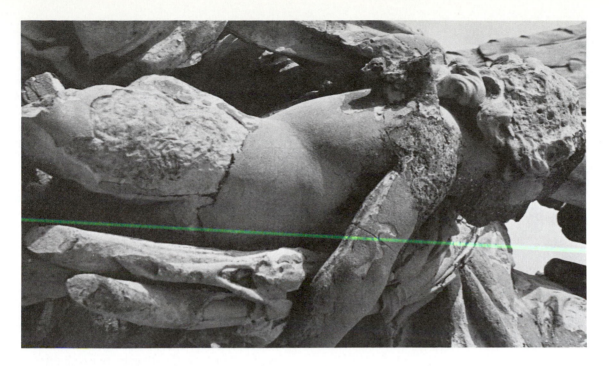

Figure 10.27 Statuary corroded by years of exposure to air pollutants. (© De Sazo, Photo Researchers.)

activity, and various industrial emissions account for about one-half of the atmosphere's total aerosol load. The other half is primarily the consequence of atmospheric reactions among various gases.

Perhaps the most common aerosols are dust and soot. Dust is chiefly the product of wind erosion of soil accelerated by agricultural activity. Soot (tiny solid particles of carbon) is emitted during the incomplete combustion of fossil fuels and refuse. The composition of a wide variety of other particles is governed by the type of activities carried on in a given area, that is, specific types of mining, milling, and manufacturing. In urban-industrial air, particulates usually include a diverse array of trace metals such as lead, nickel, iron, zinc, copper, magnesium, and cadmium. These particulates pose a significant health hazard, since their typically small size allows them to be readily inhaled. In addition, air may contain asbestos fibers and pesticide and fertilizer dust.

Lead is an unusually widespread particulate, and poses a serious health threat, as we have seen. The burning of gasoline containing lead anitknock compounds is the prime contributor to the atmosphere's lead content, which has increased dramatically over the last century, as Figure 10.28 shows. In the northern hemisphere, airborne lead concentration is now about a thousand times greater than the natural level. Recent studies show that concentrations of lead in the air tend to be highest next to highways, but that they extend beyond the bounds of well-traveled roads and pose a significant hazard to roadside livestock and vegetation. Since the 1975 model year, however, automobile manufacturers have been required to produce engines that burn unleaded gas, and there is no doubt that this shift will eventually reduce the atmospheric lead problem significantly.

Air normally also contains various viable particles, including fungal spores and pollen. The disturbance of the land by agricultural and construction activities

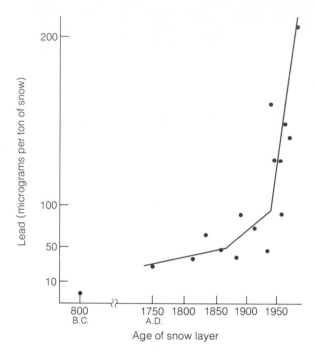

Figure 10.28 Lead content of snow cores in the Greenland ice sheet. Most of the dramatic increase during the twentieth century reflects the burning of leaded gasoline. (After C. C. Patterson and J. P. Salvia, "Lead in the Modern Environment," *Environment 10:*72, April 1968.)

causes abundant growth of weeds (for instance, goldenrod and ragweed) whose pollen evokes allergenic reactions such as hayfever in roughly one out of every twenty people. Given their adverse affects, these pollens too fit our definition of a pollutant.

Conclusions

Although the case against low-level air pollution is incomplete, more than enough evidence is available to suggest that polluted air has detrimental effects on living organisms. Many questions are unanswered and many new lines of inquiry are yet to be fully explored, but continuing research promises to provide us with a more complete assessment of the total environmental impact of air pollution. In the meantime, it is prudent—if not imperative—that we monitor closely the sources of air pollution and the weather conditions that contribute to the making of air pollution episodes with an eye toward identifying the means of reducing air pollution's toll. In the following chapter, we look to the management techniques currently available to us for rendering our air safer for all living things.

Summary Statements

In sufficiently high concentrations, some air pollutants are asphyxiating agents and others severely irritate the respiratory tract.

The human respiratory system is armed with mechanisms that defend against air pollutants. These include nasal hairs that filter air, glands that secrete mucus to trap pollutants, cilia in the bronchi that beat pollutant-laden mucus up to the mouth for expulsion, a cough reflex, and macrophages in the lungs to engulf and digest foreign matter.

Much evidence indicates that air pollution threatens human health, although specific cause-effect relationships are difficult to isolate.

Little is known about the impact of polluted air on the well-being of nonlaboratory animals, although cases of lead and fluoride poisoning have been reported.

Pollutant damage to vegetation primarily takes place when certain gases enter leaf pores. The tolerance of vegetation to air pollution varies considerably with the plant species and type of pollutant.

Formation of ozone in the stratosphere protects life on earth by filtering out harmful intensities of ultraviolet radiation. This ozone shield is threatened by the halocarbons accumulating in the atmosphere.

Human activities influence the climate of large cities by altering the local radiation balance. In some cases, this effect gives rise to an urban dust dome.

Human activities modify weather and climate on a hemispheric scale primarily through the release of carbon dioxide and dust into the atmosphere. The precise effects of these activities are subjects of controversy.

Air pollution can enhance cloudiness and rainfall.

In localities where air is polluted by oxides of sulfur or nitrogen, rain becomes strongly acidic. Acid rainfall in turn threatens aquatic life and corrodes structures.

High wind speeds and unstable air enhance the dilution of air pollutants, while weak winds and stable air reduce dilution.

A temperature inversion consists of an extremely stable stratification of light air over heavy air that develops through the subsidence of air or through radiational cooling.

The frequency of atmospheric conditions that favor air stagnation and hence reduced air quality varies from place to place with the seasons.

Conditions that favor the accumulation and concentration of pollutants in the air are countered to some extent by gravitational settling and washout by rain and snow.

Air pollutants are classified as primary and secondary pollutants. Primary pollutants are materials entering the atmosphere directly. Secondary pollutants are the results of chemical reactions undergone by the primary pollutants in the atmosphere.

Important air pollutants include carbon dioxide, carbon monoxide, hydrocarbons, nitric oxide, nitrogen dioxide, sulfur dioxide, sulfur trioxide, hydrogen sulfide, mercaptans, smog, and a wide variety of solid and liquid particles.

Questions and Projects

1. In your own words, write a definition for each of the italicized terms in this chapter. Compare your definitions with those in the text.

2. The kinds of pollutants that foul the air of a particular locality depend on the specific types of industrial, domestic, and agricultural activities taking place there. Prepare a list of sources and types of air pollutants in your community. Is information

available on the quantities of pollutants emitted by each source? You may wish to consult your local health department.

3. Evaluate the air pollution potential of your community in terms of the frequency of stable atmospheric conditions, location of major pollution sources, and topography.

4. Has your community ever experienced a severe air pollution episode? If so, did an increased incidence of respiratory illnesses accompany the incident? You may wish to refer to past issues of local newspapers or consult with your local public health agency.

5. Speculate on steps that city planners might take to reduce the intensity of urban heat islands, thereby decreasing the frequency of dust domes.

6. Collect a sample of snow, melt it, and filter the meltwater. Examine the residue on the filter paper under a microscope. Describe what you see and speculate on its origins. Compare the appearance of samples taken from different locations in your community.

7. With reference to your understanding of atmospheric stability, explain the following observation: In spring, in the Great Lakes region, convective clouds tend to develop over land but not over lake waters.

8. Maintain a daily log describing the appearance of local industrial smoke plumes. Determine whether the shape and behavior of the plumes indicate unstable or stable conditions. If a temperature inversion is evident, does it usually form at the surface or aloft? Taking daily photographs of plumes may aid your analysis.

9. It is possible for temperature inversions to form simultaneously at the surface and aloft in the atmosphere? Speculate on the sequence of weather events that would give rise to this situation.

10. In the course of a day's activities, each of us is responsible, directly or indirectly, for the emission of air pollutants. Enumerate your contributions as to types and sources.

11. Comment on the notion that at least some air pollution is an inevitable consequence of our way of life.

12. Speculate on the major sources and types of air pollutants in nonindustrialized nations.

13. How does the size of particulates relate to their potential as a health threat?

14. Describe the mechanisms that protect the human respiratory system from the intrusion of particulates.

15. Why is it difficult for scientists to isolate specific cause-effect relationships between air pollution levels and human illnesses?

16. Summarize the ways in which air pollution may be contributing to climatic changes (a) in large cities and (b) on a hemispheric scale.

17. Describe how wind speed and air stability influence the dispersal of air pollutants. Why do frequent temperature inversions in a region contribute to a high air pollution potential?

18. Distinguish between primary and secondary air pollutants and provide examples of each.

19. Why are air pollution levels so high in Los Angeles?

20. Prepare a list of natural air pollution sources in your vicinity.

Selected Readings

Battan, L. J. 1966. *The Unclean Sky, A Meteorologist Looks at Air Pollution.* New York: Doubleday. A well-written, popular account of the meteorological aspects of the air pollution problem.

Council on Environmental Quality. *Annual Reports, 1970–1978.* Washington, D.C.: U.S. Government Printing Office. Includes a yearly summary of progress and surveys the major issues involving air quality.

Panofsky, H. A. 1978. "A Progress Report on Stratospheric Ozone," *Weatherwise 31*:60–65 (April). A summary of research on air pollutant threats to the ozone layer.

Stern, A. C., H. D. Wohlers, R. W. Boubel, and W. P. Lowry. 1973. *Fundamentals of Air Pollution.* New York: Academic Press. An in-depth study of air pollution, its causes and effects.

Waldbott, G. L. 1978. *Health Effects of Environmental Pollutants.* St. Louis: C. V. Mosby Company. Contains a lucid discussion of the impact of air pollutants on the human body.

Williamsen, S. J. 1973. *Fundamentals of Air Pollution.* Reading, Massachusetts: Addison-Wesley. A basic text on air pollution, intended for science majors.

Woodwell, G. M. 1978. "The Carbon Dioxide Question," *Scientific American 238:*34–43 (January). A discussion of the impact of worldwide forest destruction on atmospheric carbon dioxide concentration.

Sulfur removal installation at a petroleum refinery near Houston, Texas. This unit is part of a $200-million expansion of pollution control equipment. It represents one strategy in industry's efforts to reduce the air pollution that attends the combustion of petroleum fuels. (Exxon Company, U.S.A.)

Chapter 11

Air Quality Management

The nation's largest fossil fuel power plant, the Four Corners plant—located in northwestern New Mexico near the point where the Arizona, Utah, and Colorado borders meet—became operational in 1963. At that time, the plant had distinct advantages over most other coal-fired power plants then in service. Not only was it located near abundant coal reserves, but, because the area is sparsely populated, the plant was not at first subject to the strict air pollution regulations that apply to its urban counterparts. Consequently, the plant's stacks liberally spewed dark plumes of particulates over a formerly pristine countryside.

Due to the relaxed pollution control policy, the dust output at Four Corners each day exceeded the combined daily particulate emissions of the cities of Los Angeles and New York! In supplying power to southern California, the plant was seriously degrading the ancestral home of the Hopi and the Navajo peoples. Eventually, New Mexico could no longer tolerate the dirty air. Empowered by new federal laws, the state forced the Four Corners power plant to install more effective pollution control devices.

The improvement in air quality at Four Corners is not unique. Many of our recent efforts to improve air quality have been successful. Air quality in general has begun to improve, and much of this improvement stems from new control strategies and stronger control legislation. But advances in air quality have entailed significant costs and have stirred up conflicts and resistance in some quarters. In this chapter, we discuss air pollution control and the problems associated with implementation.

Air Quality Legislation

After the Industrial Revolution, many efforts were made by state and local governments to control air quality, but generally these ordinances regulated only smoky nuisances arising from the burning of coal. By the late 1940s, however, smog-plagued Los Angeles began to make progress in dealing with other kinds of air pollution. Since then, California has led the way in enacting strict legislation to control automobile emissions. Federal law has more or less been modeled on the pioneering example of California.

The first piece of federal legislation directed at air pollution control was enacted in 1955. This law provided funds for research on polluted air by the United States Public Health Service, and required that states be supplied with the technical information compiled in that effort. In 1963, the Clean Air Act was passed, which expanded federally sponsored research on air pollution somewhat. The 1963 law also provided grants to the states for shoring up their meager control programs.

The first nationwide standards for automobile exhaust emissions were established in 1965, through amendments to the 1963 Clean Air Act. Again, modeled after California regulations, this legislation called for the reduction of carbon monoxide and hydrocarbon emissions by the 1968 automobile model year. It also directed the Surgeon General to study the effects of auto emissions on health. Air pollution regulations were revised again in 1967, with passage of the Air Quality Act. This law set goals for air quality, and called for state and federal cooperation in establishing standards, although enforcement rested with the individual states. This law also divided the nation into air quality control regions (AQCRs) for the first time.

Although the modifications of air quality laws that were enacted during the 1960s were well intentioned, they actually accomplished very little. But in the late 1960s, bolstered by the new public concern over environmental quality, Congress developed a new set of strict and comprehensive amendments to the Clean Air Act. These amendments, signed into law by President Nixon on December 31, 1970, called on the Environmental Protection Agency to establish uniform air quality standards for five *"criteria" air pollutants*. These standards, listed in Table 11.1, apply to ambient air (open air) in 247 nationwide air quality control regions. The amendments also charged the states with the responsibility of devising implementation plans and enforcing air quality standards for stationary sources (incinerators, factories, and power plants, for example). Congress also specified standards for motor vehicle emissions as-

signing responsibility for enforcement in this area to the federal government. The vehicle emissions standards required that carbon monoxide and hydrocarbon emissions be reduced by 90 percent of their 1970 values by 1975, and that oxides of nitrogen be reduced by 90 percent of their 1971 levels by 1976.

Two sets of air quality standards for ambient air were established by the 1970 amendments: primary and secondary. *Primary air quality standards* are maximum exposure levels that can be tolerated by human beings without ill effects, and *secondary air quality standards* are the levels of air pollutants allowable with reference to impact on materials, crops, visibility, personal comfort, and climate. Both types of standards are enforced, but overall compliance with air quality standards means meeting primary standards in cases where primary and secondary standards differ. Emission standards for stationary or mobile sources were so set to ensure, theoretically at least, that pollutants entering the open air would not exceed ambient air quality standards.

Control standards for the five criteria pollutants were based on the assumption that ill effects are likely to be sustained only after concentrations reach a certain threshold value. Some pollutants, however, pose a danger to health in any concentration. The 1970 amendments addressed these "noncriteria" pollutants by enabling the EPA to designate asbestos, beryllium, mercury and vinyl chloride as hazardous pollutants and imposed strict controls on their sources. It is likely that other substances will soon be added to the list of pollutants under special control.

In addition to the regulations for criteria and noncriteria pollutants, the ambient air quality standard for lead has been set by the EPA. Lead levels must not exceed 1.5 micrograms per cubic meter. This regulation became effective in late 1978, but it is likely to be revised, since cost of compliance with the present standard (estimated by EPA to be $530 million by 1982) could force the closing of lead smelters and thus seriously disrupt the industry.

Besides controlling particular pollutants, the 1970

Table 11.1 National ambient air quality standards for criteria pollutants.

Pollutant	Averaging time*	Primary standard	Secondary standard
Particulates (micrograms per cubic meter)**	1 year	75	60
	24 hours	260	150
Sulfur oxides (ppm)	1 year	0.03	—
	24 hours	0.14	—
	3 hours	—	0.5
Carbon monoxide (ppm)	8 hours	9	9
	1 hour	35	35
Nitrogen dioxide (ppm)	1 year	0.05	0.05
Ozone (ppm)†	1 hour	0.12	0.12

* Averaging time is the time period over which concentrations are measured and averaged.
** A microgram is one-millionth of a gram.
† Revised January 26, 1979.
Source: After Council on Environmental Quality, *Seventh Annual Report*, 1976, p. 215.

legislation also called for limits on further deterioration of air quality in some specific regions of the nation. In effect, Congress sought to regulate the siting of new industrial sources and the expansion of existing sources. In accordance with this provision of the law, and in response to prodding by conservationists, in 1974, the EPA drew up a classification scheme imposing constraints on future changes in air quality by area. But this program was not fully implemented until August 7, 1977, when President Carter signed into law the Clean Air Act Amendments of 1977.

Among the provisions of the 1977 law is one that classifies each national air quality control region and certain public lands according to the extent to which air quality within its boundaries is allowed to deteriorate. In air quality control regions where ambient standards have not been attained, the 1977 law rigidly limits the expansion of industry. The EPA will not allow new sources of pollution to be created unless it is shown that additional air pollution will be offset by reductions in emissions from other sources. States may obtain a waiver from this requirement, however, if they develop a source emissions inventory, an enforceable permit system, and an effective plan for overall emissions reduction. The latter must be designed to achieve primary ambient air quality standards by December 31, 1982, although extensions to 1985 can be granted.

In addition to regulating ambient air pollution, federal legislation also attempts to control air quality in the workplace (see Figure 11.1). The importance of controls in the workplace was stressed when a recent survey of more than forty-five-hundred industrial plants showed that 25 percent of the nearly one-million workers involved were exposed to substances capable of causing illness or death. Congress took steps to reduce occupational exposure to dangerous substances by enacting the Occupational Safety and Health Act in 1970, which called on the the Labor Department to set acceptable exposure levels, to monitor causes of worker illness and injury, and to establish training programs related to on-the-job safety. These responsibilities are now carried out by the Occupational Safety and Health Administration (OSHA), which receives technical guidance from the National Institute of Occupational Safety and Health (NIOSH). Among the dangerous substances under

Figure 11.1 Examples of the kinds of safety precautions in the workplace mandated by the Occupational Safety and Health Administration (OSHA). (Kennecott Copper Corporation photograph by Don Green.)

strict OSHA regulation are asbestos, vinyl chloride, benzene, and more than a dozen other carcinogens. In this regard, OSHA's efforts are complemented by the Toxic Substances Control Act of 1976, which requires that chemicals be studied for potential negative effects on human health.

Although OSHA's efforts have been a step in the right direction, progress in controlling dangerous exposure levels has been slow. OSHA standards are generally less rigorous than EPA air quality standards. This relative laxity may be the result of political pressures, since some labor unions claim the standards are not strict enough while most industries hold the opposite view. Such conflicts are typical of all quality control efforts. Because all methods of improving air quality are expensive, the responsible party usually resists implementing them until forced.

Depending on the political clout of the offenders, standards often represent compromises in quality rather than the highest quality possible. For industry's response to one set of OSHA regulations, see Box 11.1.

Air Quality Control

Though air and water quality control programs are analogous in effect, efforts to clean the air often pose more of a challenge than water quality programs, since air is more mobile than water. Hence, while it is convenient and appropriate for us to consider water quality control in the context of a watershed, it is limiting and inappropriate to attempt to define

Industry has often been critical of the regulations promulgated by the Occupational Safety and Health Administration (OSHA) to control hazardous materials in the workplace. In many instances, specific industries affected by OSHA's regulations have counterattacked by lobbying in Congress, which oversees the work of federal agencies including OSHA. Frequently, too, affected companies and industry representatives have gone to court to gain relief from OSHA rules they object to. Such legal gambits have often been effective. In 1978, for example, OSHA's air quality standards for seven pollutants were either challenged or overruled by the courts.

That same year, industry representatives tried a new tack. More then ninety companies and thirty trade associations joined forces under the banner of the American Industrial Health Council (AIHC) to fight the strict regulation of carcinogenic substances that OSHA had proposed. The AIHC argued that OSHA's rules were unreasonable on scientific, legal, and economic grounds.

In an effort to make OSHA's recommended regulations more acceptable to its members, the AIHC drafted a set of counterproposals regarding carcinogenic substances. AIHC recommended that:

1. A panel of experts be appointed by the National Academy of Science to identify substances that are proved carcinogens;

2. Greater care be exercised in applying to humans research findings based on the responses of laboratory animals to carcinogens;

3. Acceptable risk levels be established for individual carcinogens (the AIHC charged that OSHA had greatly exaggerated the risks associated with exposure to certain carcinogens);

4. Industry be allowed the option of supplying its workers with protective devices—clothing, respirators, and the like—instead of having to control exposure levels by making costly changes in plant structures and operating procedures;

5. Regulations be tailored to the type of workplace (the AIHC argued that rules appropriate for factories may be inappropriate for research laboratories).

What is your opinion of each of these recommendations? What are the advantages and the drawbacks of each?

---

an "airshed." Global winds and the pollutants they transport are not restricted to rigid geographical boundaries as are the tributaries of a river. The air pollution plume of Chicago, for example, may extend into Wisconsin on one day and into Indiana or Michigan on the next.

But if we are unable to define an airshed on which to focus our efforts, how shall we approach the problem of air quality control? The only feasible method is to focus on individual point sources. One proposal for alleviating air pollution is to build taller smokestacks, while another is to site new industry in rural rather than urban areas. These alternatives would no doubt reduce local ground-level air pollutant concentrations, but they would have little effect on overall air quality. Even the world's tallest chimney

(381 meters, or 1150 feet) would not penetrate the elevated temperature inversion over the Los Angeles basin. And plans to site industry beyond the city could meet with severe economic and aesthetic restraints. Displacing industry to rural areas would erode the tax base of cities and add to industrial transportation costs, not to mention eliminating the production potential of valuable agricultural land.

These limitations aside, the alternatives of building taller smokestacks and reducing source density both fail to address the real problem of pollutant emissions—neither would reduce the quantity of pollutants that enter the air. Several air quality control techniques, however, do significantly reduce emissions from point sources. In this section, we consider these techniques in relation to the major air pollutants emitted by industry and motor vehicles.

Industrial Control Technology

A major objective of industrial air quality control is the reduction of aerosol and gaseous contaminants in stack emissions. Three methods of removing particulates from stack emissions are currently available to us: electrostatic precipitation, filtering, and gravitational settling. In *electrostatic precipitators*, particles are electrically charged and then collected on plates of opposite charge, a process that can remove up to 99 percent of the fly ash emissions from power plant stacks. In the filtering approach, a particle-laden airstream passes through a series of filters—composed of such materials as spun glass, cotton, or cellulose-asbestos—that are sufficiently porous to permit air to flow through while particulates are trapped. Filters may be shaped in the form of cylindrical bags. A portion of a filtering system is illustrated in Figure 11.2. Larger particulates may be separated from the effluent stream by a *cyclone collector,* a device that induces the gravitational settling of heavier particles. A cross-section of a cyclone collector is shown in Figure 11.3. By using these techniques, some industries have lowered their particulate emissions by as much as 90 percent.

Industries can remove noxious gases that are water soluble from their stack effluents either by spraying water directly at the effluent stream, as shown in Figure 11.4, or by bubbling the gases through water. Such processes are called scrubbing. In another scrubbing technique, the effluent is passed through a slurry, or water mixture, of ground limestone to remove sulfur oxides. Calcium in the limestone chemically combines with sulfur to produce calcium sulfite, and it is this material that is subsequently collected. Up to 90 percent of the sulfur oxides in an effluent can be removed in this way. Some *scrubbers,* such as the one in Figure 11.5, are equipped with devices that remove particulates as well as gases. In spite of the effectiveness of scrubbing techniques, however, the National Wildlife Federation in early 1978 reported that only 12 percent of the nation's 970 fossil fuel power plants (major sources of sulfur oxides) used scrubbers.

Filters composed of substances to which certain gases adsorb (adhere)—activated carbon, for example—are also effective at removing gaseous pollutants. Adsorbed substances may be subsequently recovered and recycled. In other processes, chemical reactions are induced to convert gaseous or liquid contaminants into less hazardous products. Acidic pollutants are neutralized by this means, for example.

Although these air pollution abatement techniques can clean up stack effluents to some extent, they do not constitute a panacea for the problem of industrial emissions. On the contrary, many difficulties remain. For example, the particulates that cannot be readily removed from an effluent are often the very small ones (smaller than 5 micrometers in diameter), and these are the aerosols that pose the greatest health hazard, since they can penetrate the defenses of the respiratory tract. Also, control devices operate at maximum efficiency only if they are properly maintained, but many of these devices come into contact with corrosive substances that can damage or destroy them, and therefore they deteriorate and lose efficiency rapidly.

Figure 11.2 The exterior of a bag house that filters out and collects industrial dust emissions. (The W. W. Sly Manufacturing Company.)

Some polluting emissions can only be controlled through the modification of industrial processes. Industries can lower carbon monoxide concentrations, for example, by supplying more air during combustion. And they can reduce oxides of nitrogen levels by decreasing combustion temperatures. In some oil-fired power plants, the latter modification has cut emissions of nitrogen oxides by 30–50 percent. In another example, the petroleum industry can control the vaporization of hydrocarbons by installing condensers that liquify and collect them.

Before we began experiencing energy shortages, many industries switched to clean-burning fuels to reduce air pollution. Less carbon monoxide and sulfur oxide is emitted, for example, when natural gas or oil is burned rather than coal. Dwindling supplies of these cleaner burning fuels, however, now preclude this option. The current trend is toward the use of coal with a high sulfur content. In fact, as we shall see in Chapter 18, the federal government is actively encouraging fossil fuel power plants to shift from oil or natural gas to the more abundant coal.

Ironically, if adequate air quality control programs were implemented universally, and if they were successful at improving air quality, other environmental problems would arise. After all, air contaminants removed from effluents do not disappear; they must be put to use or disposed of. In some cases these products have economic value and can be recycled (a subject we deal with in Chapters 12 and 13). Too often, however, extracted air pollutants are improperly disposed of—they are dumped into wetlands, for instance, or onto sites where groundwater is affected. Thus, although the air pollution problems

Figure 11.3 A cut-away view of a cyclone particulate collector. (Research-Cottrell, Inc.)

Figure 11.5 A device that removes both sulfur dioxide and particulates. (FMC Corporation.)

Figure 11.4 A cut-away view of a scrubber. (The W. W. Sly Manufacturing Company.)

may be abated, solid waste and water pollution problems are aggravated. An effective air quality control program must not only require adequate control strategies, but also provide for proper disposal or reuse of collected air pollutants. The road in Figure 11.6, for instance, is composed of processed scrubber sludge. At this point, the road is a rare model of good recycling policy. It is to be hoped that such applications of recycling technology will lose their novelty and become a matter of course very soon.

Automotive Emissions Control

In terms of the total quantity of pollutants emitted into air in the United States, the primary offender is the motor vehicle. Hence, considerable attention has been, and continues to be, directed at reducing harmful vehicle emissions. As indicated in Figure 11.7, autos powered by internal combustion engines

Figure 11.6 A road surface composed of processed scrubber sludge. (IU Conversion Systems, Inc.)

emit several types of air pollutants. Carbon monoxide, oxides of nitrogen, and hydrocarbons are significant constituents of exhaust gases. Hydrocarbons also escape from the crankcase and carburetor and evaporate from the fuel tank. In addition, lead is emitted in exhaust when leaded gasoline is burned.

Various control measures are used to control the specific pollutants produced by different parts of the automobile. The *positive crankcase ventilation* system (PCV), for example, reduces hydrocarbon emissions by channeling crankcase blow-by gases back to the engine. PCV systems were installed by automobile manufacturers voluntarily on all new cars beginning in 1963. The escape of hydrocarbons from the carburetor and fuel system was reduced by modifications introduced on 1971 models (1970 in California). Exhaust emissions, however, have been more difficult to control, since concentrations of exhaust gases hinge on many variables that relate to engine performance. In the late 1960s and early 1970s, automobile manufacturers attempted to control exhaust emissions primarily by making engine adjust-

ments. As an example, they increased the air to fuel ratio by making carburetor settings leaner. These alterations did reduce hydrocarbon and carbon monoxide exhaust, but they also produced higher combustion temperatures and thereby encouraged the formation of nitrogen oxides. The refinement of engine adjustments eventually did reduce exhaust emissions significantly, but at the expense of fuel economy. A new approach was needed.

Exhaust controls were improved in the 1975 model year when *catalytic converters* were introduced on about 70 percent of the new autos manufactured. (This figure rose to 85 percent of new cars in the 1976 model year.) A catalytic converter, shown in Figure 11.8, chemically changes the hydrocarbons and carbon monoxide in exhaust into carbon dioxide and water vapor. This device works only when unleaded gasoline is used; in fact, lead actually destroys the catalytic converter. Overall, the catalytic converter significantly reduces a number of harmful emissions, and even contributes to fuel economy. However, the catalytic converter in its present form

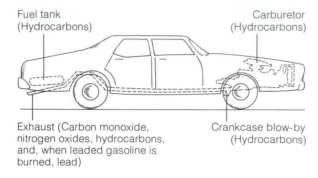

Fuel tank
(Hydrocarbons)

Carburetor
(Hydrocarbons)

Exhaust (Carbon monoxide,
nitrogen oxides, hydrocarbons,
and, when leaded gasoline is
burned, lead)

Crankcase blow-by
(Hydrocarbons)

Figure 11.7 Sources of air pollutants in the automobile.

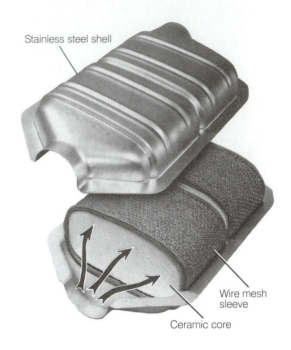

Stainless steel shell

Wire mesh
sleeve

Ceramic core

Figure 11.8 A catalytic converter. Arrows indicate the flow of the exhaust. (Chrysler News Relations Photos, Chrysler Corporation.)

may not meet exhaust standards of the early 1980s. Thus, new, more effective converters must be developed, especially to control emissions of oxides of nitrogen.

Until now, the strategy for controlling automobile emissions has primarily involved minor modifications of the internal combustion engine. Now, however, other strategies are under study, for example the development of new, cleaner fuels for the internal combustion engine as it now exists. One suggestion is that alcohol—either mixed with or substituted for gasoline—be burned as fuel, a common practice during World Wars I and II. Both ethanol (grain alcohol) and methanol (wood alcohol) are more clean-burning than gasoline; both result in markedly lower emissions of carbon monoxide and hydrocarbons. In addition to reducing emissions, powering cars with alcohol would reduce our dependence on imported petroleum and would create a new market for farmers. Only minor modifications are necessary to make internal combustion engines capable of burning alcohol, but this seemingly perfect solution does have serious drawbacks. Alcohol lacks the lubricating properties of gasoline, and it produces less energy per unit volume. The latter restriction means that larger fuel tanks would be required on alcohol-fueled autos.

The major limitation to using alcohol as a substitute for gasoline, however, is supply. At current U. S. alcohol production rates, we would be able to meet less than 1 percent of our motor vehicle fuel demands. Still, we could boost domestic alcohol production by converting coal, timber, or organic agricultural and municipal wastes into methanol, techniques discussed in Chapter 13. Some studies show that more than 200 liters of methanol could be generated from every metric ton of urban waste.

Another proposal for reducing motor vehicle emissions is the phasing out of the internal combustion engine entirely in favor of steam-, battery-, or turbine-powered engines. Air pollution problems would not be eliminated in such engines, but they would be significantly reduced. However, nontraditional motor vehicle engines require more study before they are likely to gain widespread acceptance.

At present, they fail to match the internal combustion engine in terms of durability, performance, and cost.

Whatever steps are taken by the automobile industry to reduce emissions, individuals have the power to improve air quality at present by voluntarily restricting their motor vehicle use. And municipalities with extreme air pollution problems might see their way clear to enforcing restrictions. Encouraging car pools and establishing bus lanes (see Figure 11.9), restricting downtown parking, expanding mass transit and, if necessary, rationing gasoline are all strategies that could help to alleviate air pollution problems in metropolitan areas.

Compliance:
Problems and Progress

One distinctive feature of the 1970 Clean Air Act Amendments is an overall strictness with polluters: this legislation was the first to set a schedule for compliance and to provide stiff fines for violators. Nevertheless, a combination of factors, including technological problems and the deterioration of the nation's energy supply, has kept industrial polluters from complying fully with the air quality standards set by the law.

Some people object to government-imposed air

Figure 11.9 A carpool and bus-only lane along Kalanianacle Highway, Honolulu, Hawaii. (U.S. Department of Transportation.)

quality standards, arguing that such standards are inappropriate, since no relationship between adverse health effects and concentration levels of specific pollutants has been established beyond doubt. On the other hand, some conservationists feel that existing standards are not strict enough and do not allow an adequate margin of safety. In response to both objections, the 1977 Clean Air Act Amendments call for a scientific review of the nation's ambient air quality standards. This review will no doubt lead to the further refinement of standards. In the case of ozone, standards have already been revised.

Partially because of economic considerations, the EPA in early 1979 relaxed its national ambient standard for ozone, considered an index of smog level. The maximum allowable one-hour exposure concentration was raised from 0.08 ppm to 0.12 ppm. Industry representatives had argued that the cost of compliance with the original standard was excessive. Their case was supported by recent medical studies indicating that ground-level ozone is somewhat less hazardous to health than was previously believed.

Some industries claim that the costs of control in general are so exorbitant that compliance would force them out of business. And, indeed, some industrial plants have had to close their doors because they were unable to afford control costs. For example, in the fall of 1977, Youngstown Sheet and Tube Company, a steel manufacturer, closed down—laying off five-thousand employees in the process—because the cost of pollution controls plus declining market conditions made its continued operation unfeasible.

Fearing the economic consequences of plant closings, states on occasion have supported their home industries' objections to air quality regulations. Consider an example involving the copper industry of Arizona, the nation's largest copper-producing state. The 1970 Clean Air Act Amendments required each state to prepare implementation plans for stationary sources. One EPA goal was to cut sulfur oxide emissions from copper smelters by 90 percent. But operators of Arizona's copper smelters argued that the expense of meeting this requirement would force them to close. In response to industry pressure, Arizona did not include sulfur oxide emission regulations in its implementation plan. This and similar strategies by other copper-mining states prompted the EPA to impose uniform nationwide regulations for sulfur emissions from copper smelters.

In addition to economic considerations, another objection to enforcement is that appropriate technology is not yet available for meeting standards. Industries taking this position have gone to court to do battle with conservationists in an attempt to weaken or avoid regulations. Nevertheless, in spite of resistance in some quarters to full compliance with air quality standards, ambient air quality has been improved and emissions reduced considerably. In the next two sections, we survey this progress.

Trends in Ambient Air Quality

The 1970 Clean Air Act Amendments required that the nation's air quality control regions achieve primary ambient air standards prior to June 1, 1975. Only a third of the nation's 247 AQCRs were able to comply by this deadline for all criteria pollutants (refer back to Table 11.1). Nonetheless, national and local efforts to regulate the criteria air pollutants did score some gains during the 1970s. In late 1977, the Council on Environmental Quality reported that in many urban areas, levels of total suspended particulates and concentrations of sulfur dioxide were below maximum allowable levels and still dropping. Between 1970 and 1976, particulate levels dropped 12 percent and sulfur dioxide fell 27 percent nationwide.

During the 1970s, carbon monoxide and hydrocarbon levels also dropped, primarily because of emission controls on automobiles. But carbon monoxide pollution is still a serious problem in congested urban areas such as New York City, where exhaust

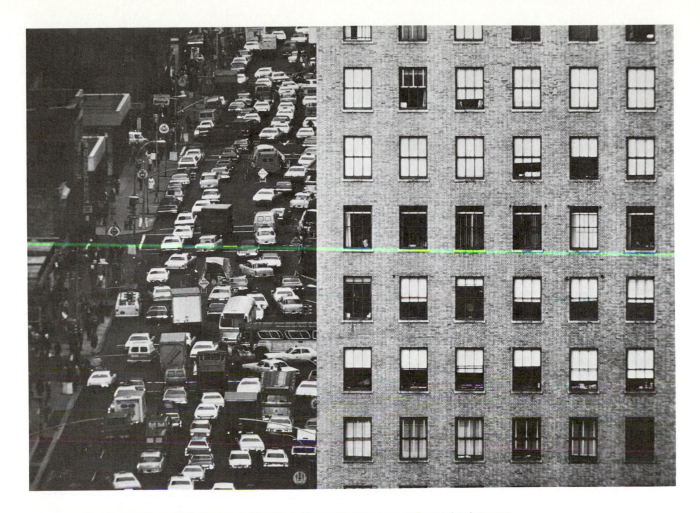

Figure 11.10 Idled traffic on Sixth Avenue in New York City generating exhaust fumes that become trapped by surrounding buildings. (U.S. Environmental Protection Agency.)

from extremely heavy traffic is trapped by the canyonlike topography of tall buildings and narrow streets (see Figure 11.10).

The data on ground-level ozone are insufficient for determining national trends. Where data are available, the outlook is generally encouraging. Exceptions are Los Angeles, Houston-Galveston, and Philadelphia, cities in which meeting ambient ozone standards will be difficult. As we noted in Chapter 10, the problem in Los Angeles, and in the rest of coastal California as well, is an unfortunate combination of climatic and topographic factors that inhibit the dilution of pollutants. In the Houston-Galveston area, hydrocarbon emissions from a growing petrochemical industry are likely to continue to stimulate ozone production. And in Philadelphia, industrial and motor vehicle emissions are not likely to tail off enough to achieve ozone standards in the near future. However, recent relaxation of the ozone standard will make compliance easier even for these cities.

Levels for oxides of nitrogen are difficult to monitor, and available data may therefore be unreliable. Nevertheless, it appears likely that nearly all AQCRs will comply with primary standards for nitrogen oxides in the next several years if the trend toward more stringent control of automobile emissions continues. The Council on Environmental Quality expects that only Los Angeles will remain significantly above the standard by 1990.

Although ambient air quality trends are moving in the right direction, the air that most urbanites breathe is still so polluted that it constitutes a health hazard. In 1978, the EPA reported that of the 105 urban areas with populations greater than two hundred thousand, twenty still violated standards for four of the five criteria pollutants, and all but two were plagued by smog.

Trends in Automotive Emissions

Originally, the 1970 Clean Air Act Amendments set 1975 as the target year for attaining new carbon monoxide and hydrocarbon auto emission standards, and 1976 for lowered levels of nitrogen oxides. In 1973, at the request of the automobile manufacturers, the EPA delayed compliance by one year. In 1974, the auto industry requested and was granted an additional year of grace. This time auto manufacturers argued that compliance could only be attained at the expense of fuel economy. Again, in 1975, compliance was postponed to 1978 for carbon monoxide and hydrocarbons and 1979 for oxides of nitrogen with the stipulation that manufacturers would have to improve on fuel economy. In the summer of 1977, Congress granted a fourth delay, until 1981 and 1982. And in areas where carbon monoxide is a particular problem, compliance deadlines may be extended to 1987.

Encouragingly, even though the original schedule for reducing automotive emissions was not met, progress has been made in emissions control. Hydrocarbon and carbon monoxide levels dropped steadily from 1970 into the mid-1970s, as Figure

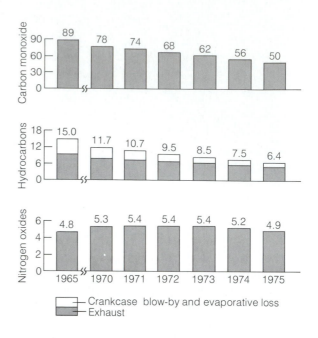

Figure 11.11 Recent trends in average emissions from highway vehicles (in grams per mile). (After Council on Environmental Quality, *Sixth Annual Report,* 1975.)

11.11 shows. But, as noted earlier, engine adjustments that controlled these pollutants proved to increase emissions of nitrogen oxides. Refinements in adjustment procedures in the 1973 and 1974 models and the subsequent introduction of the catalytic converter has begun to lower oxides of nitrogen emissions.

Assessing the Dollar Cost

A wide range of estimates are available as to the total dollar cost of air pollution. The corrosion of structures and the killing of crops are obvious effects whose cost in dollars can be estimated with relative ease. But much of the cost of air pollution is tied to elusive, less obvious, and less direct effects. And estimating the value of, say, aesthetic losses resulting from dirty air is virtually impossible.

The indirect effects of air pollution are often inseparably linked with other contributing factors, particularly where health effects are involved. For example, while high levels of air pollution may contribute to emphysema and other lung disorders, factors such as age, diet, and general health are also important. It is not possible to assign a dollar value to each contributing factor in accounting for the costs of hospitalization and treatment. Such problems of cost assessment arise wherever the link between polluted air and environmental damage is relevant but not measurable—a common circumstance. Nevertheless, in the late 1970s, while acknowledging the difficulties inherent in making such estimates, the American Lung Association indicated that the health costs of air pollution may exceed $10 billion each year in the United States.

Monitoring and Abatement

Damage due to air pollution accounts for only a portion of the total monetary costs. Another significant amount of money goes for monitoring air quality so that progress in cleanup can be documented. In fact, in the mid-1970s, the EPA spent $31 million annually for the monitoring of ambient air and $18 million for the monitoring of individual sources. However, monitoring can also save money, indirectly at least, by functioning to warn the public of potential air pollution episodes. For example, city health departments issue air pollution alerts when ozone levels approach hazardous concentrations and warn people prone to respiratory difficulties to stay indoors and avoid strenuous activity. Thus, in such cases expensive monitoring procedures result in savings associated with human health.

The costs involved in meeting federal air quality standards will continue to be considerable. A great deal of money will be required for researching and developing new control techniques, and industries will have to spend huge amounts on the purchase and maintenance of the devices themselves. In the late 1970s, projections for 1978 convey a sense of the amounts of money involved: industry typically spent almost 5 percent of its total capital expenditures for antipollution equipment. About half of this amount went for air quality control equipment, 40 percent for water pollution abatement, and 10 percent for solid waste management. Auto manufacturers in particular may complain about the money they are forced to spend in the interests of reducing vehicle emission (see the cost escalations in Figure 11.12), but a 1974 National Academy of Science study notes that the expenditures have paid off: annual benefits from automobile emission control ($2.5–10 billion) are commensurate with the anticipated costs of implementation ($5–8 billion yearly).

According to the EPA, the cost of pollution abatement is particularly burdensome for electric utilities. Between 1975 and 1985, these utilities are expected to spend $25 billion for pollution control, most of which will be directed at air pollutants. In late 1976, the Council on Environmental Quality reported that a new coal-fired power plant costs $400 per kilowatt, and that air quality control equipment adds $60 to $90 to this figure. In addition, these control devices use up to 7 percent of the plant's electrical production, thereby adding more to the total costs of producing electricity. Obviously, these added costs are passed on directly to the consumer.

Air Quality and the Economy

Some environmentalists assign full responsibility to industry and big business for the less than rapid pace of our progress toward improved air quality. But the situation is not that simple. Because ours is an industrial society, our nation's economic well-being is intimately tied to the well-being of its key industries. We depend on business and industry for employment and for the goods and services that enhance the quality of life. Our increasing demands encourage more production, and, indirectly, the result is an increase in attendant pollution.

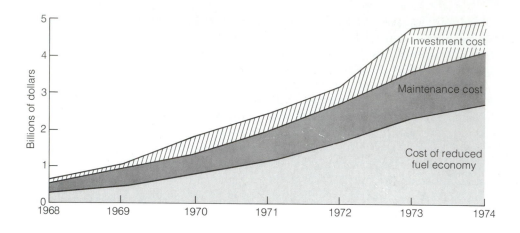

Figure 11.12 The trend in the costs of automobile pollution abatement from 1968 to 1974. Costs stem from the investment in control devices, automobile maintenance, and losses in fuel economy. (After Council on Environmental Quality, *Sixth Annual Report*, 1975.)

Still, each of us is entitled to breathe air that will not impair our health or shorten our life expectancy, no matter who bears the economic burden. Can we reconcile our cleanup efforts with the need for a healthy and growing economy? Can we continue to improve air quality without creating serious economic hardships? While we cannot answer these questions at present, we do know that improving communications is an obvious first step toward resolving conflicts between economic growth and air quality. We must hope that the on-going dialogue between conservation groups and industrial interests will generate carefully reasoned strategies for managing the atmospheric resource that are acceptable to both groups. For one industry's plea for compromise, see Box 11.2.

If we were able to maintain our present success rate in cleaning up the air, it is possible that we would meet primary ambient air standards by 1990 in all but a few urban areas. But unimpeded progress may be impossible. Only three months after passage of the 1977 Clean Air Act Amendments, the EPA granted utilities a delay in their compliance with emissions restrictions on new fossil fuel power plants. The original intent of the 1977 law was to prevent significant deterioration in ambient air quality by requiring that utilities install the best available technology (scrubbers, for example). Many conservation groups fear that the decision to make an exception for the electric utilities, and similar ones for the auto industry, will pave the way for more and longer delays in enforcement of air quality regulations in general. Ironically, given inflation, with further delays the costs of air pollution control can only escalate.

Other major obstacles to the continued improvement of national air quality involve the crunch in energy supply and our efforts to cope with it. For example, the emphasis laid on fuel economy by automobile manufacturers may work at cross-purposes with emissions control efforts, since autos equipped with adequate pollution control devices may burn more gasoline per kilometer. And the trend on the part of industries toward using more abundant but dirtier fuels, particularly coal, poses a direct threat to air quality. In this regard, the Council on Economic

Box 11.2

Pollution Control and the Steel Industry

Less rigid environmental mandates can save steelworker jobs... without forsaking environmental goals.

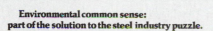

Environmental common sense: part of the solution to the steel industry puzzle.

The job of controlling air and water pollution in the steel industry is a tough one. And the costs are inordinately high.

So far, Bethlehem has spent about $550 million for hardware to clean up pollutants from the air and water we use. In addition, it costs us about $75 million a year to operate and maintain these control facilities, including the cost of valuable energy.

We're proud of what we have accomplished. Today we have in place or in progress facilities and plans designed to effectively control the major sources of pollution at our operations.

But federal and state governments want us to do more.

The road gets rockier—and costlier

In order to comply with existing regulations, it is estimated that Bethlehem must spend about $500 million *more* for environmental control. Grand total: more than *$1 billion.* Our operating and maintenance costs will also increase as more control facilities are installed and as energy costs rise. *We have no clear estimate of what the ultimate tab will be. Nor does anyone else.*

Based on the scientific data available, we question the stringency of many of the mandates we currently face. In some cases, proven technology does not exist to do the job. It takes time and money to develop control mechanisms that will be effective. In other cases, it is simply not feasible to do what needs to be done to meet the required deadlines.

Jobs are at stake

Bethlehem is now spending 25% of its capital funds for environmental controls. During the next five years, we expect this will increase to about 30%. Such capital investments do *not* produce income, but *do* increase the cost of making steel.

Expenditures like these erode the dollars we need to improve production facilities and provide job opportunities.

We are not crying "wolf"

Last year Bethlehem shut down certain facilities at our Johnstown and Lackawanna plants and laid off thousands of employees. That action was painful but necessary. Continued efforts to restore the profitability of these operations could not be justified—not when we included the huge expenditures for pollution controls that would have been required to continue operation of those facilities.

Action needed now

We support our nation's goals for clean air and water. And we endorse the recommendation of President Carter's Inter-Agency Task Force on Steel that calls for a review of EPA standards and regulations to provide more flexibility and to reduce barriers to steel industry modernization.

We also support the following: (1) rational enforcement of environmental laws and regulations; (2) greater flexibility in compliance timetables; (3) accurate determination of significant sources of pollution, their effect on public health, and the most cost-effective control techniques; (4) amortization of expenditures for pollution control facilities, including buildings, over any period selected by the taxpayer, including immediate write-off in the year the funds are expended.

Make your views known where they count

We believe a more reasonable balance between jobs and environmental cleanup is urgently needed. If you agree, tell that to your representatives in Washington and your state capital.

Bethlehem Steel Corporation, Bethlehem, PA 18016.

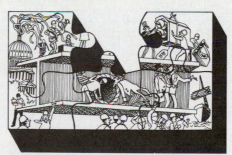

Bethlehem

In search of solutions.

Priorities, a public interest group, reported in late 1977 that the effects of increased coal-burning in power plants were beginning to offset gains made through the installation of emissions control devices.

Serious though they are, however, the conflicts between air quality control and energy supply problems are not necessarily unresolvable. The development of cleaner and more abundant fuels (solar energy, for example), and new, more effective air quality control devices would make our energy technology compatible with clean air goals. But until these alternatives become available, we are faced with adapting to the conflicts at hand. To do so we may soon have to alter our individual habits regarding energy use. For example, personal automobile use may come under strict regulation, and we may find that we are not allowed to drive when and where we please, under penalty of law. Also, prices for goods and services will undoubtedly rise along with the costs of air pollution control. These escalating pollution control costs could slow industrial expansion and in some cases force the closing of plants, thereby adding to the number of unemployed workers in our society. Clearly, then, the costs of pollution control will affect us greatly, but before we refuse to pay them outright, we must weigh them against the costs that would arise from uncontrolled polluted air: lost work hours, reduced productivity, higher medical expenses, agricultural losses, and changes in climate.

Conclusions

We are making considerable progress in our efforts to improve the quality of the air we breathe. Strict federal regulations are reducing pollution levels in the ambient air and the workplace. However, our progress has cost money and inspired great controversy, and the many problems that remain will continue to do both in the future. The dollar cost for abatement is a heavy burden for industries and consumers alike. Serious questions have been raised as to the scientific validity of certain existing standards, already leading to revisions (of ozone, for example). And we must overcome many major technical obstacles before we can adequately control certain pollutants at all. In view of these difficulties, we may very well decide that, in the interest of economic stability, we must learn to live with tolerable levels of air pollutants—that is, pollutant concentrations that pose an acceptable risk to public health rather than no risk at all.

Summary Statements

The 1970 Clean Air Act Amendments set rigid standards for ambient air quality and for emissions from automobiles and stationary sources.

Primary air quality standards are based upon potential human health effects, while secondary standards are designed to minimize air pollution impact on crops, visibility, climate, materials, and personal comfort.

The Clean Air Act Amendments of 1977 limit the amount of deterioration of air quality allowable in national air quality control regions.

The Occupational Safety and Health Act of 1970 sets standards to reduce hazards that attend exposure to dangerous substances in the workplace.

Two possible alternatives, building taller smokestacks and dispersing industry to rural sites, are ineffectual air quality control strategies. The only effective alternative is the reduction of point source emissions.

Scrubbers effectively remove water-soluble gases, including sulfur oxides, from industrial effluents. Electrostatic precipitatiors, filters, and cyclone collectors can reduce particulate emissions by up to 99 percent. Particulates that cannot be removed, however, are often the very small ones that pose the greatest health hazard.

An effective air quality control program must not only require adequate control techniques, but also provide for proper disposal or reuse of collected air pollutants.

Control of automobile emissions so far primarily has involved modification of the internal combustion engine. But auto makers have yet to comply with emissions standards set by the 1970 Clean Air Act Amendments. Because of technical constraints and the need to improve fuel economy, the government has moved the deadline for compliance forward four times.

National and local efforts to regulate levels of criteria pollutants in ambient air have scored some gains since 1970, especially for sulfur dioxide and hydrocarbons. Also, auto emissions of hydrocarbons and carbon monoxide dropped between 1970 and the mid–1970s.

The total monetary cost of air pollution is difficult to assess. Costs stem from pollution damage, monitoring, and cleanup.

Future attainment of ambient air quality standards may be hampered by increased industrial combustion of dirty fuels and by the sacrifice of automobile emissions control in favor of fuel economy.

Questions and Projects

1. In your own words, write definitions for each of the terms italicized in this chapter. Compare your definitions with those in the text.

2. What are industries in your area doing to control emissions? Do special obstacles to adequate abatement exist in your area?

3. How and where is ambient air quality monitored in your community? Are summaries of pollution levels available? If they are, how does your community's air compare with national ambient standards for criteria pollutants, shown in Table 11.1? Have there been discernible trends in local air quality?

4. How do you rate your community's air quality control efforts? What are the successes and the failures? Is the overall feeling one of complacency or is there

evidence of a well-directed effort? What are you personally doing to improve local air quality?

5. How might air pollution control regulations influence land use and community population growth rates?

6. Large corporations may be more able to absorb the costs of emissions control than small companies. Can you think of ways in which this apparent inequity could be avoided?

7. Enumerate some of the "hidden" indirect costs of air pollution. How do these costs compare with the obvious expenses resulting from polluted air (damage, cleanup, and monitoring)?

8. Does current air quality control legislation address the problem of interstate transport of air pollutants? Support your response.

9. One proposal for air pollution control is the siting of new industry in suburban and rural areas. Comment on the economic and social implications of this strategy.

10. Identify the factors that should be taken into account in designing a community air pollution index.

11. Some communities have ordinances that ban cigarette smoking in such public places as restaurants and buses. Are such ordinances a form of air pollution control? Do they violate individual rights?

12. Distinguish between primary air quality standards and secondary air quality standards. Pollutant emission standards for stationary and mobile sources are based on what fundamental assumption?

13. Why is it inappropriate to design air quality control measures in the context of an "airshed"?

14. What are the advantages and disadvantages of tall smokestacks as an air quality control strategy.

15. Identify and briefly describe the various air quality control techniques and devices used by industry.

16. How might effective air pollution abatement technology actually aggravate water pollution problems?

17. Describe some automobile emissions control strategies other than modification of the internal combustion engine.

18. Why has full compliance with the strict air quality standards specified by the 1970 Clean Air Act been delayed in numerous instances?

19. Why does progress in achieving ambient air quality standards in urban areas hinge on successful control of automobile exhaust emissions?

20. Explain how efforts to cope with shortages in energy supply may conflict with efforts to improve national air quality.

Selected Readings

American Chemical Society. 1978. *Cleaning Our Environment. A Chemical Perspective.* Washington, D.C.: American Chemical Society. Reviews the air pollution problem and recommends courses of action for improved understanding.

Conservation Foundation. 1972. *A Citizen's Guide to Clean Air.* Washington, D.C.: Conservation Foundation. Surveys the legislative background for citizen involvement in air quality policy making.

de Nevers, N. 1973. "Enforcing the Clean Air Act of 1970," *Scientific American 228:*14–21 (June). Reviews the provisions of the 1970 law and the controversies surrounding implementation and enforcement.

Johnson, W. H. 1976. "Social Impact of Pollution Control Legislation," *Science 192:*629–631 (May 14). A brief summary of some of the economic and social implications of federal air quality control laws.

Miller, S. 1977. "Building Air Monitoring Networks," *Environmental Science and Technology 11:*544–549 (June). Discusses air quality monitoring in the United States and abroad.

United States Environmental Protection Agency. Washington, D.C. The following guides are available: *Air Quality Criteria for Particulate Matter; Air Quality Criteria for Sulfur Oxides; Air Quality Criteria for Carbon Monoxide; Air Quality Criteria for Photochemical Oxidants; Air Quality Criteria for Hydrocarbons; Air Quality Criteria for Nitrogen Oxides; Control Techniques for Particulate Air Pollutants;* and *Control Techniques for Sulfur Oxide Air Pollutants.*

The Twin Buttes open-pit copper mine near Tucson, Arizona. Extracting resources by surface mining can seriously disturb the land, but in some cases, with effort, the land can be reclaimed. (EPA Documerica photo by C. Keyes.)

Chapter 12

The Earth's Crustal Resources

In June 1859, one of the world's richest deposits of gold and silver, the Comstock Lode, was discovered near Virginia City, Nevada. Overnight, the town was transformed into a booming mining camp as prospectors, merchants, and other entrepreneurs in search of a quick fortune arrived from far and near. By the early 1870s, Virginia City's population had swelled to thirty thousand and silver and gold production totaled hundreds of millions of dollars. But suddenly the Comstock Lode met the fate of all mineral deposits: the gold and silver ran out. Miners, merchants, and others fled the town, the local economy collapsed, and by the 1880s Virginia City had become a virtual ghost town with fewer than a thousand residents. Today the population is even smaller, and the town relies on tourists who come to see the artifacts of a more prosperous era. Figure 12.1 shows the remains of the Comstock mines.

The boom and bust pattern illustrated by the Virginia City story was repeated again and again during our early years. In those days, another Virginia City always seemed to be on the horizon to supply the mineral and fuel needs of the growing population. Today new discoveries are rarer, and we are aware of the finite nature of the earth's mineral and fuel resources. Nevertheless, our demands for mineral supplies continue to climb. In this chapter, we survey the earth's crustal resources and the methods by which they are extracted. We also examine the pressing problem of managing these resources in the face of growing demands and dwindling reserves.

A Historical Overview

When the Americas were discovered, they seemed to hold resources without limit. From colonial days into the early part of this century, mineral resources were extracted in this country with little or no regard for conservation or environmental impact. Reflecting the spirit of the times in which it was written, the Mining Law of 1872 (still in force today for metals) essentially gave away the government's mineral rights to anyone who located a minable deposit on federal lands.

The rapid growth of the manufacturing industries during the first half of this century created a strong dependency in the nation on a continuous supply of

Figure 12.1 The Comstock mines, Virginia City, Nevada.
(U.S. Geological Survey photo by T. H. O'Sullivan.)

raw materials. During World War I, however, when equipment production rose steeply, people in some parts of the country began to feel materials shortages. In response, people began to question the notion of limitless resources. After the war, the mining of largely nonmetallic resources on federal lands (coal and phosphate, for example) was regulated to some extent by the Mineral Leasing Act of 1920. Although some mineral rights were now leased instead of virtually given away under this law, no provisions were made for environmental protection.

Throughout the 1920s and 1930s, the pace of resource exploitation was governed by prevailing economic conditions—growing consumer demand in the 1920s gave way to the Depression in the 1930s. With the beginning of World War II, not only did the need for raw materials rise sharply, but also the

increasingly sophisticated weaponry being developed demanded a greater variety of materials. By this time, the United States had begun to rely on imports from other countries for some of its mineral needs. Later, in 1952, when the growing dependency on foreign sources seemed dangerous to national security, a presidential commission was formed to study the status of our domestic reserves. But the commission's recommendations failed to stem the growth of our dependency on a wide variety of materials. In fact, following the birth of space age technology in the 1960s, more products (computers and spacecraft, for example) were produced from a wider variety of materials having special properties—some quite exotic—than ever before.

By the late 1960s, Congress was reflecting the new environmental awareness sweeping the nation in its concern over the environmental impact of mining and the processing of crustal resources. In 1970, after investigating the situation, Congress passed the Resource Recovery Act. One provision of this law established a National Commission on Materials Policy (NCMP). The NCMP recommended that mining procedures be modified to minimize waste and environmental disruption. The commission also encouraged the government to escalate conservation efforts, and reduce national dependency on foreign imports. It was not until the summer of 1977, however, that federal legislation was passed to reduce the environmental impact of surface mining—a concern of lasting significance to be discussed in forthcoming sections.

Generation of Crustal Resources

Industrial nations are intimately dependent on the properties of a multitude of rock, mineral, and fuel resources extracted from the earth's crust—the relatively thin outermost layer of the solid earth (see Figure 12.2). Crustal resources include metals, nonmetallic minerals, rock and crushed stone, and fuels.

For the most part, commercially valuable resources are concentrated in certain places within the upper crust of the earth. This distribution pattern is important, since if these resources were uniformly disseminated throughout the crust, their concentrations would be so low that extraction would be neither technically nor economically feasible. For example, only three metals account for more than 1 percent of the total weight of the earth's crust: aluminum (8 percent), iron (5 percent), and magnesium (2 percent). Fortunately, however, metals and other resources occur here and there in high concentrations, and, hence, they can be mined.

A variety of processes have acted over the long span of geologic time to selectively concentrate crustal resources into minable deposits. These same processes are also responsible for forming the diverse features of the landscape. In surveying the evolution

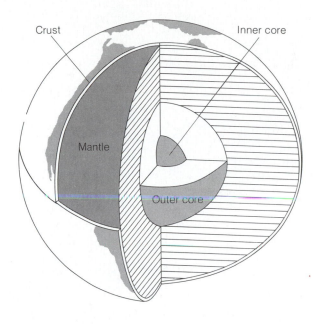

Figure 12.2 The earth's interior.

of the landscape, we not only learn how these processes work, but also we come to appreciate the finite nature of the resources they produce.

Landscape Evolution

Dynamism is a fundamental characteristic of the earth's crust; the earth's crust undergoes change continually. The landscape's appearance is a response to stresses acting on the crust both from the earth's interior and from outside. Internal stresses include volcanic eruptions and the folding and fracturing of rock under pressure—processes sustained by energy generated in the earth's interior. External stresses include interactions between the crust and the atmosphere, moving water, glaciers, and organisms, all of which are ultimately maintained by solar energy. In general, stresses acting from within the earth tend to lift up continents, while stresses acting from the outside wear down (erode) continents.

Figure 12.3 An erupting volcano, serving as a vent for the lava created in the earth's interior. (Photo Researchers.)

Igneous activity includes volcanism and the flow of masses of molten rock, called *magma.* Magmas form below the crust and migrate into and within the crust along zones of weakness, such as fractures. Some magmas eventually cool and solidify within the crust, forming masses of new rock that can be hundreds of kilometers across and form the cores of mountain ranges. In some areas, magmas feed volcanoes (see Figure 12.3) or flow through fractures as *lava* over the earth's surface. The Hawaiian Islands, for example, consist of five huge overlapping volcanoes that tower up to 10,700 meters (35,000 feet) over the

Pacific Ocean floor. And in the Pacific Northwest, the Columbia Plateau is covered by many layers of solidified lava, in places more than 1,400 meters (4,500 feet) thick. Figure 12.4 shows the distribution of these solidified flows.

Throughout geologic time, stresses within the crust have alternately increased and waned. Some stresses bent, or folded, rock layers; others exceeded the strength of rock layers, shattering, or fracturing them. Resulting *folds* range from microscopic crinkles to broad warps hundreds of kilometers across. Fracturing caused rock displacements (*faults*) rang-

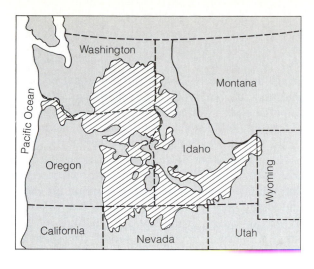

Figure 12.4 Exposures of solidified lava flows (hatched area) in the Pacific Northwest. (After *Geologic Map of the United States,* U.S. Department of the Interior, U.S. Geological Survey.)

ing in length from less than a centimeter to many kilometers. The great mountain systems of the globe are products of large-scale folding and faulting.

The stresses responsible for igneous activity and rock deformation are apparently linked to movements of the gigantic crustal plates that compose the earth's crust, as illustrated in Figure 12.5. These plates are rigid and are in motion, drifting at an extremely slow rate (several centimeters per year). Continental land masses are carried along on top of the moving plates. Geologic evidence suggests that 200 million years ago just one continent existed—referred to now as Pangaea—which subsequently broke up. Its constituent land masses, the continents we know, slowly drifted apart, eventually reaching

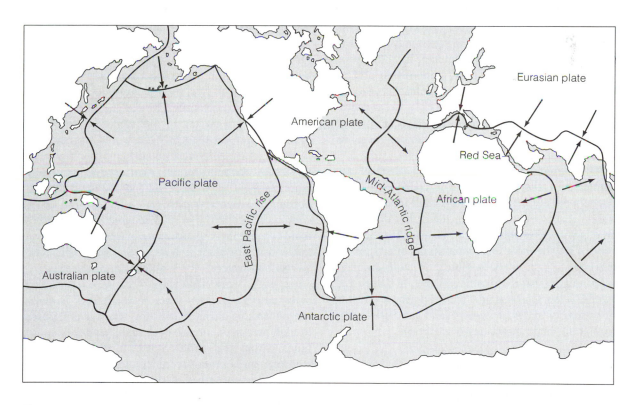

Figure 12.5 The earth's drifting plates.

their present positions. Figure 12.6 shows the pattern of continental drift that resulted in the opening up of the Atlantic Ocean.

In some locations, the crustal plates are still drifting apart—note on Figure 12.5, for example, the mid-Atlantic ridge being created by the drifting apart of the American and African plates. In these areas, lava flows to the surface to form new oceanic crust. In other localities, one plate is sliding under another to form deep oceanic trenches. Plates may also slide past one another horizontally, in the process creating large-scale fractures (the San Andreas fault in California, shown in Figure 12.7, is such a zone) and folds in the rock.

In addition to these large-scale shaping influences, more localized effects of weather and water alter the face of the landscape. For example, where *bedrock* (solid rock) is exposed to the atmosphere, it is subjected to *weathering*. A number of mechanical processes function to break rocks down: the freezing and thawing of water trapped in crevices; the alternation of extremes in air temperature, primarily in deserts; and the wedging of rocks by plant roots. Weathering by such mechanical processes is referred to as physical weathering. The decomposition of rocks through their interaction with moisture, oxygen, and carbon dioxide is called chemical weathering.

The products of weathering, both physical and chemical, are the rock fragments called sediments, and, collectively, sediment. Very small sediments are eroded, that is, removed from their places of origin and deposited elsewhere. As we discussed in Chapter 6, rivers are the most important agents of erosion by far: rivers transport the sediment in their drainage basins and deposit it along floodplains, in deltas, and in the sea. Wind is also an effective agent of erosion, but primarily in deserts, where sand and dust storms redistribute huge quantities of sediment.

Rocks

The diversity in the processes that mold the landscape—igneous activity, deforming stresses, weathering, and erosion—favors the development of a multitude of rock types, many of which are put to important uses by human beings. Rocks are classified as igneous, sedimentary, or metamorphic, according to the way in which they were formed.

Igneous rocks are created when magma, the hot, molten rock that forms below the earth's crust, cools and solidifies. Magma may remain within the earth and solidify slowly to form coarse-grained rocks, or, as we saw earlier, it may be spewed forth as lava through volcanoes or fractures, solidifying rapidly to form fine-grained rock. Granite is a particularly durable, coarse-grained igneous rock found within many mountain ranges. It is used extensively as monument stone and as a building material.

Sedimentary rocks are composed of compacted and cemented sediments. In this case, the term sediments refers both to the products of weathering and to the skeletal remains or excretions of aquatic organisms. In an example of a common sedimentary rock-forming process, waves breaking against a rock-bound coast grind shoreline rocks into fragments that are sorted by size, carried away, and eventually deposited elsewhere as beach sand. As sand accumulates, grains are packed by the weight of overlying sediment and subsequently cemented together by migrating fluids. The result is the common sedimentary rock called sandstone, often used as a building material. Another widespread and economically important sedimentary rock is limestone, which is composed primarily of the calcium carbonate remains of marine organisms—for example, shellfish and coral. Limestone is widely used in concrete and as roadbed material for highways and railroads.

As with many sedimentary rocks, *metamorphic rocks* are derived from other rocks. A rock is metamorphosed, or changed in form, when it is exposed to high pressures, intense heat, and chemically active fluids within the earth's crust. The metamorphic environment is never severe enough to melt the rock back to magma; rather, it simply causes the rock to change form. During metamorphism, the mineral components of rock sometimes become aligned, giv-

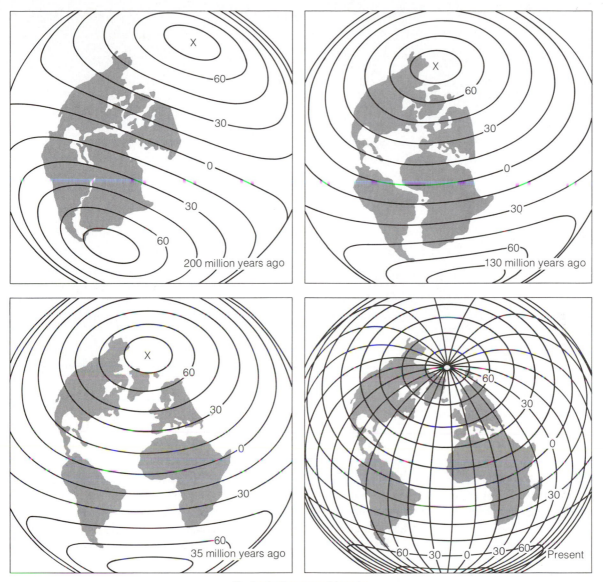

X = Ancient geographic pole

Figure 12.6 The opening up of the Atlantic Ocean over the last 200 million years. (From J. D. Phillips and D. Forsyth, "Plate Tectonics, Paleomagnetism, and the Opening of the Atlantic," *Bulletin of the Geological Society of America 83:*1584–1585.)

Figure 12.7 An aerial view of the San Andreas fault in California. Note the sudden change in the course of the streams along the fault. (U.S. Geological Survey photo by R. E. Wallace.)

ing the rock a banded appearance. Such alignment may facilitate the cleaving of these rocks into plate-like slabs. Slate, sometimes used as a roofing or flooring material, is a good example. When some rocks are metamorphosed, their constituent particles recrystallize, becoming a different size, and the overall quality of the rock is thereby changed. For example, metamorphism of limestone produces the coarser rock marble, an attractive material valued by sculpters and builders.

Although the earth's crust is composed of rocks belonging to all three rock families, igneous rocks are by far the most abundant. Sedimentary rocks, however, are the most conspicuous; they form a relatively thin veneer over nearly 75 percent of the earth's surface. In some areas, weathering and erosion have exposed igneous and metamorphic rocks that were formerly situated deep in the earth's crust. In other regions, volcanic activity has covered or is in the process of covering the surface with new igneous rocks. In most land areas—for example, that shown in Figure 12.8—bedrock is hidden under unconsolidated sediments, soil, and vegetation. Although bedrock is usually exposed in mountainous regions, in nonmountainous areas, geologists often must search for scattered outcrops or drill through the surface to determine the composition of local bedrock.

Figure 12.8 Bedrock overlain by unconsolidated sediment, soil, and vegetation. (U.S. Department of the Interior, National Park Service.)

Minerals

Rocks are sometimes valued for themselves, but more often they are sought for the *minerals* they contain. A rock is an aggregate of minerals, usually consisting of a few dominant and several accessory minerals. Minerals are solids characterized by an orderly internal arrangement of atoms, a fundamental unit of all materials. This is another way of saying that minerals are crystalline. The atoms in a given mineral are arranged in a particular order, and the internal structure and composition of a mineral determine its chemical and physical properties. Table 12.1 shows the mineral components of some common rocks.

Oxygen atoms account for about half the total mass of atoms contained in common rock-forming minerals of the earth's crust. Silicon is the next most abundant atom, accounting for about one-quarter of the total. These two elements, along with aluminum, iron, calcium, sodium, potassium, and magnesium, amount to 99 percent by weight of the earth's crust. Table 12.2 shows the relative amounts of the components of the earth's crust.

Frequently, a highly valued mineral is merely a minor component of a large rock mass. Hence, to obtain this mineral, miners must extract huge quantities of rock and then separate the desired material from its host rock. For example, the average grade of copper deposits mined today is about 0.6 percent copper. This means that every 1000 metric tons of copper-bearing rock mined will yield an average of 6 metric tons of copper and 994 metric tons of waste rock. The mechanical processes such as crushing and sorting that separate the copper out require a considerable input of energy and water, which contributes substantially to the overall costs of copper mining.

While nonmetallic minerals (calcite, gypsum, and salt are examples) typically are useful in their natural state, most metals are not. Rather, raw metals are usually chemically united with other elements such as oxygen or sulfur. For example, iron is mined as the minerals hematite and magnetite, chemical combinations of iron and oxygen, and zinc as the mineral sphalerite, zinc united with sulfur. In order for the desired metallic element to be liberated, the mineral must be broken down chemically through smelting or refining processes. (A copper-smelting plant is shown in Figure 12.9.) These processes typically require a huge energy input and usually produce

Table 12.1 Some common rocks and their mineral components.

Rock type	Important component minerals
Granite	Quartz (silicon dioxide) Muscovite (potassium aluminum silicate) Feldspars (potassium, calcium, and sodium aluminum silicates)
Dunite	Olivine (iron magnesium silicate)
Basalt	Feldspars Pyroxenes (iron magnesium silicate)
Sandstone	Quartz
Dolomite	Dolomite (calcium magnesium carbonate)
Quartzite	Quartz
Marble	Calcite (calcium carbonate)

Table 12.2 The most abundant elements of the earth's crust.

Element	Percentage by weight
Oxygen	46.60
Silicon	27.72
Aluminum	8.13
Iron	5.00
Calcium	3.63
Sodium	2.83
Potassium	2.59
Magnesium	2.09
Titanium	0.44
Hydrogen	0.14
Total	99.17

solid, liquid, and gaseous waste products that must be disposed of. In addition, however, mining operations usually recover substances of value other than the target mineral, especially in the case of metals. Often, the extraction of small amounts of valuable accessory minerals, such as silver, means the difference between a marginal and a profitable enterprise.

Not all valuable minerals are found as components of rock; some occur mixed with unconsolidated sediment. Gold, diamonds, and platinum, for example, are sometimes found in sand and gravel deposits in streambeds. Such sedimentary deposits of heavy minerals are called *placers*.

As we noted earlier, minerals can be extracted economically only from concentrated deposits. Therefore, mineral prospectors searching for new deposits focus on those areas where geologic forces have favored concentration. Some of the most important deposits of beryllium, lithium, chromium, and nickel occur in areas where magma and lava once flowed. In some instances, important minerals precipitate from groundwater (iron and copper are examples) or sea water (manganese and table salt), thereby forming concentrated deposits. Weathering and erosion also concentrate resources, either by adding desired mineral matter or subtracting unwanted host rock or sediment. Placer deposits and some copper and aluminum ore develop in this way.

Fossil Fuels

Perhaps the most valuable crustal resources are the fossil fuels—coal, oil, and natural gas—which represent the major sources of concentrated energy ex-

Figure 12.9 A copper smelter in Arizona. (Kennecott Copper Corporation, photo by Don Green.)

ploited by human beings. Coal is found in the earth's crust as distinct layers, and oil and natural gas are trapped in permeable sedimentary rock and held there between impermeable layers. As with all other crustal resources, a special set of circumstances was necessary to generate these invaluable energy sources.

When we burn coal, we are actually releasing solar energy that was stored in vegetation through photosynthesis millions of years ago. At that time, localized stands of luxuriant vegetation existed in swampy terrain that probably resembled the mod-ern Dismal Swamp of North Carolina and Virginia. As giant ferns and other plants died in such ecosystems, a thick vegetative mat accumulated in the swampy waters. Partial decomposition of the matted plant matter by anaerobic bacteria gradually resulted in high concentrations of carbon. Eventually, the mass of highly carbonized, partially decomposed plant materials—called peat—was further compacted under the weight of more sediment and plant remains. Increasing heat and pressure changed the peat into *lignite* (brown coal), and then into *bituminous coal* (soft coal). In regions of particularly intense heat and pressure, bituminous coal was transformed into *anthracite* (hard coal). These three stages in the sequence from peat to anthracite are called *ranks of coal.*

The higher the rank (from lignite to bituminous to anthracite), the greater is the carbon content and the amount of heat energy that can be generated per unit mass. Also, the higher the rank the smaller is the quantity of ash that is released upon combustion. Thus, anthracite yields the most heat per gram burned and has the lowest potential for air pollution; it is thus a "clean" fuel. Anthracite, however, is difficult to ignite and is in short supply. Also, what little anthracite exists is in demand by the metallurgical industry. For these reasons, bituminous coal, constituting about half the world's fossil fuel reserves, is the rank of coal primarily used in electric power plants and in some industrial processes. Unfortunately, bituminous coal can contain by weight up to 7 percent sulfur, which upon combustion combines with oxygen to produce sulfur dioxide, a very harmful air pollutant.

Like coal, oil is organic in origin. However, the process by which the original organic material, which included both plant and animal remains, was converted into oil is extremely complex and poorly understood. Oil is a mixture of thousands of hydrocarbon molecules (molecules containing hydrogen and carbon) and other organic molecules. Most oil occupies pore spaces in marine sedimentary rocks and is confined to its reservoir rock by impermeable strata, as illustrated in Figure 12.10. Normally, oil is under pressure, and if the overlying rock is fractured the oil may rise to the surface, forming a natural oil seep. While the pressurized nature of the oil reservoir is normally advantageous for oil extraction, care must be taken during drilling to prevent oil from bursting to the surface as a gusher, or blowout.

About 90 percent of the world's extracted oil is used for fuel. The remainder is used in the production of lubricants and a variety of products called petrochemicals. The air pollution potential of burning or processing oil is much less significant than that

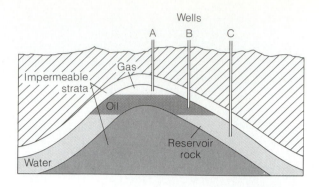

Figure 12.10 Layers of rock forming a trap for oil and natural gas. Well *A* yields gas, well *B* yields oil, and well *C* yields water. (After A. N. Strahler, *Principles of Earth Science.* New York: Harper & Row, 1976.)

of coal. The sulfur content of oil varies with the source locality but is usually less than 2 percent. In Indonesian oil, the sulfur content may be as low as 0.5 percent.

Natural gas is actually a mixture of gases consisting primarily of methane (up to 99 percent by volume), the commercially marketed fuel, and small quantities of ethane, propane, and butane. Natural gas is an even more desirable fuel than oil from an air quality perspective, since it contains only trace amounts of sulfur, which are readily removed prior to combustion. Natural gas is derived from oil, and occurs either mixed with or on top of oil, as shown in Figure 12.10, or in separate gas reservoirs.

Nuclear Fuels

Nuclear fuels, primarily uranium, power nuclear reactors. In the United States, the primary source of uranium is certain sandstone strata located in the Colorado Plateau, Wyoming basins, and along the Texas coastal plain. Apparently, uranium minerals precipitated from groundwater that circulated through these ancient sandstones, which are 70 to 350 million years old. Hence, uranium minerals mainly occupy pore spaces within the sandstones. In other parts of the world, uranium minerals occur as fillings within rock fractures and as placer deposits.

The Rock Cycle

A rock in any of the three rock families may be altered by geologic processes into another rock type, as illustrated in Figure 12.11. This *rock cycle* implies a continual generation of rock, mineral, and fuel resources. For example, weathering reduces an igneous rock mass to sediments, and the sediments are subsequently eroded and redeposited elsewhere. Accumulating sediments are compacted to form sedimentary rocks, and as the sedimentary rocks are buried beneath the increasing weight of more sediment, rising temperatures and pressure metamorphose the rock. Eventually, pressures and temperatures become so extreme that rocks melt to form magma. Magma may then cool and solidify to produce igneous rock, thereby completing the rock cycle.

The transformations involved in the rock cycle, however, are extremely slow processes. Typically, the regeneration of crustal resources takes thousands to millions of years. These materials cannot be renewed within a human generation, nor even within the lifespan of our civilization so far. Thus, from our point of view the supply of crustal resources is essentially fixed.

Geologic processes responsible for the distribution

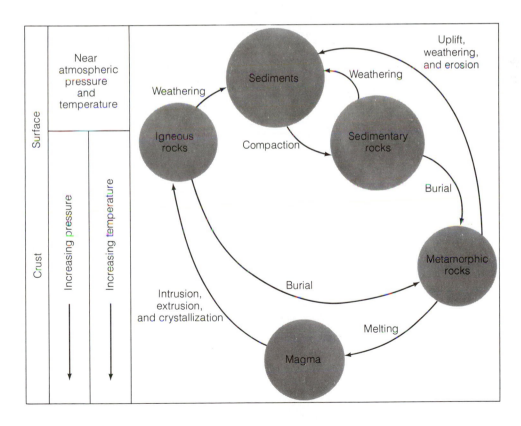

Figure 12.11 The rock cycle.

of crustal resources in minable concentrations occurred irregularly around the globe. Consequently, no nation houses within its borders all the desirable crustal resources. Some nations are favored geologically more than others, but none is entirely self-sufficient in metals, nonmetallic minerals, and fuels. Thus, countries engage in international trade to acquire needed resources. Clearly, then, as the supply of these invaluable resources diminishes, the need for international cooperation will become ever more pressing.

The Extraction of Crustal Resources

We need an adequate supply of rock, mineral, and energy resources if we are to maintain our high standard of living, but the impact we have on the environment when we extract these materials deserves serious attention. Air and water are often polluted by refining and smelting processes. Lands being mined are stripped bare of vegetation and subsequently eroded by wind and water, as shown in Figure 12.12. The severity of damage varies with the specific methods of mining used, the physical and chemical properties of the resource, and the site of the deposit. In this section, we examine the major techniques of surface and subsurface resource extraction and consider their environmental impact.

Surface Mining

About 90 percent of the rock and mineral resources consumed in the United States and more than 50 percent of the nation's coal output are extracted by surface mining methods. Surface mining has a particularly disruptive effect on the landscape, primarily because it requires the removal of the waste rock (called overburden), soil, and vegetation that lie over the deposit. But even in situations where underground mining techniques are feasible, surface mining is often favored by mine operators because it permits more complete extraction and poses far fewer safety hazards. The danger of collapse or explosion, considerable in underground efforts, is virtually nonexistent in surface mining. Furthermore, surface mining is less expensive than underground mining. For example, the surface mining of a coal deposit typically costs 40 percent less than subsurface mining would.

Several techniques of surface mining are practiced, depending on the deposit. Sand and gravel are removed from small pits, while limestone, granite, and marble are taken from *quarries,* such as that shown in Figure 12.13. In parts of the West and in the Mesabi Range near Lake Superior, metallic ores—copper and iron—are removed from *open pit mines,* huge, gaping excavations that are dug to a considerable depth. The Liberty open pit copper mine in Ruth, Nevada, for example, is 300 meters (900 feet) deep and more than a kilometer across. In another surface technique, *hydraulic mining,* powerful jets of water remove the overburden, soil, and vegetation and wash out the deposit. While hydraulic mining, illustrated in Figure 12.14, has limited application today because it washes a considerable amount of sediment into nearby drainageways, in the past it was employed for the extraction of gold and other precious metals. Another surface technique, used in streambed sand and gravel and placer deposits, is *dredging,* performed with chain buckets and draglines, as in Figure 12.15.

Strip mining is the surface mining method that most extensively disrupts the landscape, primarily because it affects a much larger surface area for the resource recovered than the other surface techniques. In this method, huge power shovels or stripping wheels literally chew up the land in gulps of 10–100 cubic meters, as shown in Figure 12.16. Strip mining currently consumes hundreds of hectares per week. It is used primarily to extract coal from seams averaging 2 meters thick that occur within 30 meters of the surface. Strip mining is also used to recover phosphate rock and gypsum deposits.

There are two basic types of strip mining: area and contour. *Area strip mining* is carried out in flat ter-

Figure 12.12 Severe soil erosion in an unreclaimed strip mine area.
(U.S. Department of Agriculture, Soil Conservation Service.)

Figure 12.13 A small dolomite rock quarry. (Photo by J. M. Moran.)

Figure 12.14 Hydraulic mining of Marvel Creek gold placers in the Akiak district, Kuskokwim region of Alaska. (U.S. Geological Survey photo by R. E. Wallace.)

Figure 12.15 A dredge used for extracting placer deposits of gold in the Yukon region of Alaska. (U.S. Geological Survey photo by H. L. Foster.)

Figure 12.16 A huge stripping wheel at the Glen Herald Coal Mine south of Stanton, North Dakota. (U.S. Department of Agriculture, Soil Conservation Service.)

rain, for example, in Florida, for phosphate, and in the western and midwestern coal fields. This method involves the stripping away of overburden to form a series of parallel trenches. When extraction in one trench is completed, that trench is filled in with the overburden removed from the adjacent trench. The technique results in the rugged topography resembling a washboard shown in Figure 12.17. In hilly or mountainous terrain, such as the Appalachian coal fields, near-surface deposits are mined by the *contour strip mining* technique. Contour strip mining requires the cutting of a series of shelves, or benches, into the steep flanks of a mountain. The overburden is then dumped downhill from each successively higher terrace onto the one below. Figure 12.18 shows a series of terraces created by contour strip mining. In some places strip mining is supplemented by huge drills, called *augers*.

Most strip mined land is in the Appalachian coal fields, where more than 16,000 square kilometers (6,000 square miles) have been disturbed. In the future, however, western coal fields are likely to be the principal site of strip mining. Almost 400,000 square kilometers (150,000 square miles) of land in the Rocky Mountain states are underlain by thick deposits of strippable low-sulfur coal, which represent two-thirds of the nation's estimated coal resources.

The severity of the impact of surface mining on the environment depends on many factors, including the topographic relief of the mined lands, the amount of rainfall in the area, the specific mining technique employed, and the chemical characteristics of the deposit and waste products. Thus far, less than two-tenths of one percent of the nation's total land area has been disturbed by surface mining, but even

Figure 12.17 Area strip mining for coal near Nucla, Colorado. (EPA Documerica photo by W. Gillette.)

Figure 12.18 Severely eroded strip-mine terrances on Bolt Mountain, West Virginia. (U.S. Department of Agriculture.)

this amount of activity has had severe local effects, especially in coal mining regions.

Another environmental problem associated with mining is how to dispose of mine spoils, or undesirable residue. Coal mining produces more spoils than other types of mining. Typically, 8 metric tons of waste are generated for every metric ton of coal mined. Waste produced by the surface mining of metallic minerals amounts to only 5 percent of coal mine spoils. Overall, in the United States, about 7.4 metric tons of mine waste are generated yearly per person. And most of the spoils produced are left at mine sites. The lack of nutrients and excessive stoniness in waste piles inhibit the establishment of potentially stabilizing vegetation. Thus, rapid erosion and dangerous sliding occur on these piles. For example, in 1966, in Aberfan, Wales, a 125-meter (400-foot) pile of coal mine wastes suddenly slid down into the town, smashing buildings and taking 144 lives, many of them children.

The seepage of rainwater through mine wastes rich in sulfur compounds produces sulfuric acid. More than 60 percent of the acid drainage in the United States comes from runoff in sulfur-rich coal mine spoils. Acid runoff that drains into rivers eliminates aquatic life and contaminates water supplies. It has been estimated that in Appalachia 16,000 kilometers (10,000 miles) of waterways have been seriously polluted by acid runoff from coal wastes. In some localities, mine drainage also contains zinc, arsenic, lead, and other metals that even in minute amounts are toxic to aquatic organisms. In addition, dust lifted from mine dumps by winds adds to atmospheric pollution, sediments washed from mines make drainageways turbid, and the piles of mine wastes themselves foul wildlife and human habitats.

Subsurface Mining

When a mineral, rock, or fuel deposit lies at a depth so great that the cost of removing the overburden is prohibitive, the resource is extracted by subsurface mining. Obviously, in the case of fluid resources (oil or natural gas), subsurface methods are the only ones that are technically feasible. To extract deep ore and coal deposits, mine operators generally create a system of subsurface shafts, tunnels, and rooms (see Figure 12.19) by drilling and blasting the rock. In most cases, a portion of the deposit must be left behind to support the mine roofs. In the room-and-pillar system of subsurface coal mining, as much as 50 percent of the coal is left in place to serve as supporting pillars, although the pillars are commonly "robbed" in the end.

Certain soluble minerals (potash and salt, for example) are removed from the subsurface by *solution mining*. In this technique, water is pumped down an injection well to the deposit where it dissolves the minerals. The solution is then brought back to the surface via extraction wells. One of the major problems with solution mining is its potential for contaminating groundwater reservoirs.

Oil and natural gas are extracted by wells drilled to depths as great as 9 kilometers (5.6 miles). Typically, less than 50 percent of the oil present is recovered from a well initially; the rest remains trapped in pore spaces of the reservoir rock. A second step, called secondary recovery, is necessary for extracting the trapped oil. In secondary recovery, water and other fluids are pumped into the well and reservoir rock to drive more oil toward production wells. Tertiary recovery, utilizing the injection of light hydrocarbons or subsurface explosives, can boost oil well production even more. But 100 percent recovery is never feasible, simply because the cost of removing the last fraction of oil would exceed the value of the oil itself.

Although oil rigs usually have a minimal impact on the landscape, the danger of well blowouts and oil spills during extraction always exists. The potential for such accidents is particularly high in offshore drilling operations, where oil spills that threaten marine life and beaches are difficult to contain. Blowouts and oil spills are also dangerous in the severely cold climate of the Arctic tundra, where the breakdown of oil by microorganisms takes place very

Figure 12.19 Workers installing a hydraulic jack to support the roof of a subsurface uranium mine. Netting below the roof and mats on the wall protect the miners from falling rock. (U.S. Department of Energy, Grand Junction Office.)

slowly. Still, although the danger is real, such incidents are rare: only eleven major oil spills have occurred in the United States since 1953, although more than eighteen-thousand wells were drilled during that period.

Although subsurface mining generally produces less waste than surface mining, the spoils that are produced are heaped on the ground at mine sites where they cause the same problems that attend surface mining spoils—acid runoff, landslides, and air pollution. Since in both, fluid and ore mining materials are extracted without being replaced, mine collapse and ground subsidence over mines are constant dangers. These developments, in turn, can have serious and costly effects on the landscape: damage to buildings as illustrated in Figure 12.20, disruption of

surface and subsurface drainage, and disturbance of wildlife habitats. Most cases of mine collapse occur in abandoned coal mines, of which there are more than ninety thousand in the United States. Pillars of coal left to support ceilings of subsurface caverns are weak and subject to failure. To prevent collapse, mine operators can inject a mud slurry into abandoned tunnels and rooms and then drain water away, leaving behind a material that reliably shores up mine tunnels, preventing collapse. This preventive measure, however, is not taken nearly often enough.

Ground subsidence resulting from the withdrawal of oil and natural gas from subsurface reservoirs can be especially costly when it occurs beneath urban areas. One particularly dramatic example occurred in

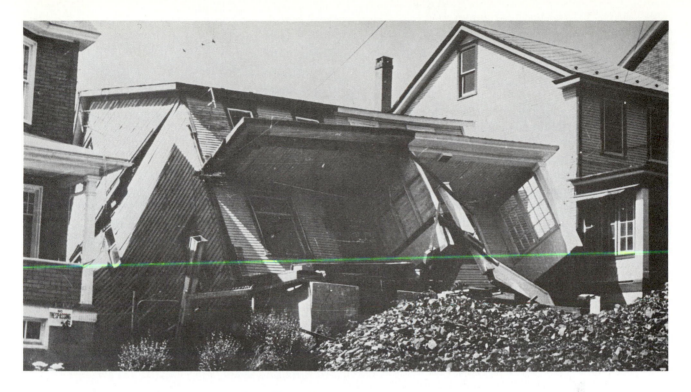

Figure 12.20 Structural damage resulting from the sinking of land overlying an abandoned underground mine. (U.S. Department of the Interior, Bureau of Mines.)

the Wilmington and Signal Hill oil fields, in the Los Angeles–Long Beach harbor area. There, beginning in 1928, large-scale oil withdrawal triggered severe land subsidence. In places the land over well sites had dropped as much as 9 meters (29 feet). This development necessitated elaborate flood control measures, including the construction of levees and sea walls. Subsidence was not halted until 1968, after reservoirs had been recharged through the injection of sea water; by then damage had totaled $100 million. Power plants, railroad terminals, docks, and much of the water and sewer systems of the city of Wilmington had to be rebuilt.

As you can see, the extraction of crustal resources can seriously impair the quality of air, water, land, and life. Can we reduce these negative effects without disrupting our way of life? This question is considered in the next section in the context of resource management strategies.

Resource Management

We have seen that processes involved in the rock cycle provide us with a fixed quantity of rock, mineral, and fuel resources, and that the extraction of these crustal resources involves considerable sacrifices in environmental quality. If we are to insure to future generations an adequate supply of those critical resources, already in short supply, we must develop sound management practices. Ideally, the principal objectives of our management effort would be to optimize conservation of resources, maximize exploration for new reserves, and minimize attendant environmental disruption. In this section, we discuss these aspects of crustal resource management, focusing primarily on nonfuel resources. The management of fossil and nuclear fuels is covered in Chapters 18 and 19, where it is considered in the context of the energy shortage.

The Mineral Shortage

Modern society relies on the properties of more than ninety metallic and nonmetallic minerals. In the past, our ability to assess the quantities of the materials available to us was often clouded by imprecise terminology. Recently, an attempt was made to standardize the terms used to classify mineral resources in accordance with their availability now and in the future. Figure 12.21 shows the standard classification system. Rock and mineral deposits are labeled *resources* when they occur in concentrations sufficient to make extraction economically feasible now or in the future. Some resources have been *identified* as to quality (grade) and quantity by actual field measurements. *Undiscovered resources* are deposits whose existence is possible (according to geologic theory) but whose specific location, grade, and size are unknown. The portion of a resource that can be extracted immediately, both legally and economically, is termed a *reserve.*

Our reserves and resources (both identified and undiscovered) of many rocks and minerals are so abundant that we are assured an adequate supply far into the future. Crushed stone, sand, and gravel are examples. But the reserves of some minerals essential for the maintenance of our highly technological society are small and dwindling under the pressures of accelerating demand. Herein lies the mineral shortage.

While production of eighteen critical minerals has soared worldwide, as Figure 12.22 indicates, production in the United States since the late 1930s has not kept pace with consumption, and the gap between the two factors continues to grow. As a consequence, the United States has turned to importing more and more minerals. In fact, we now rely on foreign nations for most of our chromium, tin, manganese, cobalt, aluminum, and titanium, and more than half of our needed zinc, mercury, nickel, and tungsten. Figure 12.23 shows the percentage imported and countries of origin for minerals imported by the United States in 1975. Our dependency on for-

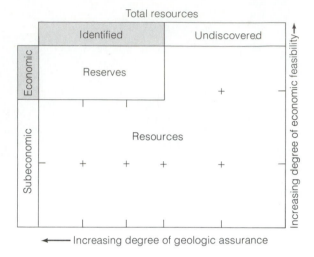

Figure 12.21 Classification of mineral resources. (After Council on Environmental Quality, *Seventh Annual Report,* 1976.)

eign imports is expected to grow by the close of the century, when we will be importing more than half of our nonfuel mineral requirements.

With respect to domestic reserves, some reports project that, at current United States use and mine production rates, we have an 87-year supply of lead, a 61-year supply of zinc, a 57-year supply of copper, and a 24-year supply of iron. These estimates are conservative, however, since they do not account for possible new discoveries or changes in the economic climate and technological innovations that could make feasible the extraction of lower grade deposits (deposits containing less target material and more waste rock). The market value of each crustal resource ultimately determines the grade that can be mined economically. As demand for a mineral increases, its price rises, and therefore expending more energy and other resources, if available, to mine and process lower grade deposits becomes worthwhile. Inevitably, however, the mining of low-grade minerals will result in greater environmental degrada-

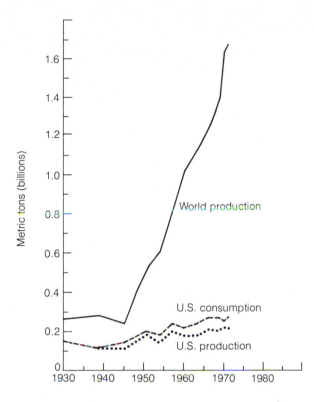

Figure 12.22 World and U.S. production and U.S. consumption of eighteen critical minerals: iron ore, bauxite, copper, lead, zinc, tungsten, chromium, nickel, molybdenum, manganese, tin, vanadium, fluorspar, phosphate, cement, gypsum, potash, and sulfur. (After E. N. Cameron, ed., *The Mineral Position of the United States, 1975–2000.* Madison: University of Wisconsin Press, 1973. © by the Regents of the University of Wisconsin.)

tion: more air and water pollution and more waste heaps.

The principal obstacle to mining lower grade deposits is energy cost. In the past, fossil fuels were abundant and inexpensive. Thus, as advances in mining and processing techniques were made, miners could extract leaner and deeper deposits than ever before, since they had the fuel resources to do it. For example, in 1900, it was not feasible to mine copper

deposits that contained less than 3 percent copper. Today the cutoff grade for copper extraction is about 0.35 percent. Thus, through much of this century, reserves of nonfuel minerals have steadily risen (although not fast enough to keep pace with demand).

Inspired by past trends of increasing reserves, some scientists dismiss the likelihood of a serious crisis in mineral supply. They point out that the quantity of critical minerals contained in the crust in dilute concentrations is enormous, and believe that, given appropriate technological innovations, the supply will carry us thousands of years into the future. This optimism may be misplaced, however. We have just cause to feel confident that human ingenuity will come up with the means for mining of exceptionally lowgrade deposits. But time—and fuel—may run out before we can apply these innovations. Unless we develop new extensive and inexpensive energy sources, the mining of very low grade mineral deposits will be impossible due to the high fuel demand of such efforts. Indeed, we may discover new technologies but be stymied by a lack of sufficient energy to employ them.

Exploration

Until new energy sources are developed, the prudent course appears to be to maximize our search for minable deposits of crustal resources. Exploration, however, has always been a financially risky venture requiring a heavy capital investment and, traditionally, a free-wheeling spirit. Usually, of, say, ten-thousand sites where a deposit might exist according to geologic theory, only a thousand merit detailed scientific study, and of these, only a hundred warrant costly drilling, trenching, or tunneling. And, typically, only one of these excavations will prove to be a producing mine.

Exploration efforts have been aided in recent years by the development of new instrumentation and the evolution of new geologic concepts. Orbiting satellites (Landsat, for example) have proven invaluable in the selection of likely exploration sites for possible

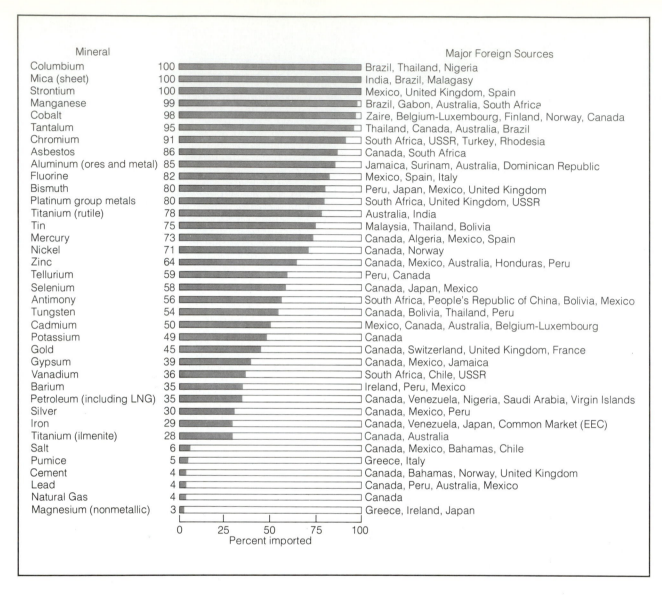

Mineral		Percent imported	Major Foreign Sources
Columbium	100		Brazil, Thailand, Nigeria
Mica (sheet)	100		India, Brazil, Malagasy
Strontium	100		Mexico, United Kingdom, Spain
Manganese	99		Brazil, Gabon, Australia, South Africa
Cobalt	98		Zaire, Belgium-Luxembourg, Finland, Norway, Canada
Tantalum	95		Thailand, Canada, Australia, Brazil
Chromium	91		South Africa, USSR, Turkey, Rhodesia
Asbestos	86		Canada, South Africa
Aluminum (ores and metal)	85		Jamaica, Surinam, Australia, Dominican Republic
Fluorine	82		Mexico, Spain, Italy
Bismuth	80		Peru, Japan, Mexico, United Kingdom
Platinum group metals	80		South Africa, United Kingdom, USSR
Titanium (rutile)	78		Australia, India
Tin	75		Malaysia, Thailand, Bolivia
Mercury	73		Canada, Algeria, Mexico, Spain
Nickel	71		Canada, Norway
Zinc	64		Canada, Mexico, Australia, Honduras, Peru
Tellurium	59		Peru, Canada
Selenium	58		Canada, Japan, Mexico
Antimony	56		South Africa, People's Republic of China, Bolivia, Mexico
Tungsten	54		Canada, Bolivia, Thailand, Peru
Cadmium	50		Mexico, Canada, Australia, Belgium-Luxembourg
Potassium	49		Canada
Gold	45		Canada, Switzerland, United Kingdom, France
Gypsum	39		Canada, Mexico, Jamaica
Vanadium	36		South Africa, Chile, USSR
Barium	35		Ireland, Peru, Mexico
Petroleum (including LNG)	35		Canada, Venezuela, Nigeria, Saudi Arabia, Virgin Islands
Silver	30		Canada, Mexico, Peru
Iron	29		Canada, Venezuela, Japan, Common Market (EEC)
Titanium (ilmenite)	28		Canada, Australia
Salt	6		Canada, Mexico, Bahamas, Chile
Pumice	5		Greece, Italy
Cement	4		Canada, Bahamas, Norway, United Kingdom
Lead	4		Canada, Peru, Australia, Mexico
Natural Gas	4		Canada
Magnesium (nonmetallic)	3		Greece, Ireland, Japan

0 25 50 75 100
Percent imported

Figure 12.23 Strategic minerals imported by the United States in 1975. (After U.S. Department of the Interior, Bureau of Mines, 1976. *U.S. Imports of Strategic Minerals,* 1975.)

deposits. Figure 12.24 is an example of the sort of satellite view that aids geologists in analyzing surface structures for exploration sites. Also, the concept of drifting continental plates has revolutionized geologic thought regarding the formation of metallic ores. The notion that ores may be concentrated in those regions where plates converge has led to recent discoveries of important copper and molybdenum deposits. Another approach to exploration is represented in current proposals to embark on a detailed inventory of United States mineral resources through a systematic program of drilling.

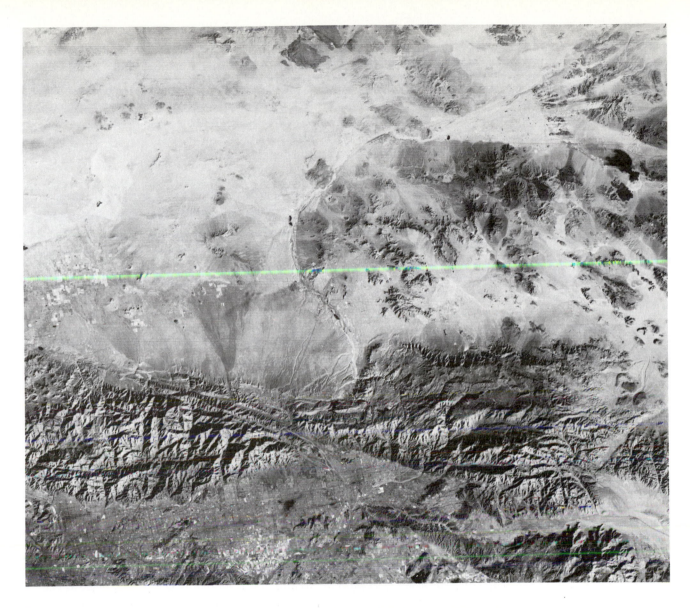

Figure 12.24 A Landsat-2 photograph of the Los Angeles area, taken from 920 kilometers (570 miles) above the earth's surface. (U.S. Department of the Interior, Geological Survey, EROS Data Center.)

Efforts to expand exploration have been hindered to some extent by conservationists and other individuals concerned with preventing the enviromental degradation associated with resource exploitation. In response to mounting pressure from these groups, the government has prohibited exploration and mining activities on federal lands with increasing frequency. By the late 1970s, miners had been excluded from 40 percent of public lands—national parks, wildlife refuges, and native lands. This figure could increase in the 1980s to encompass nearly one-third of the nation's total land area.

Seabed Resources

Some scientists are looking to the seabed as a possible source of minerals to offset shortages. Because continental shelves are actually submerged extensions of the continents, it makes sense to suspect that they contain many of the same mineral deposits that are found on land. The economic climate, however, has thus far discouraged extensive mining of the shelf. Today, some dredging of phosphorite rock (the source of phosphorous for fertilizers) is taking place, but, although deposits on the continental shelf are enormous, they are of a lower grade than those on land, and therefore terrestrial mining of phosphorite rock is still favored. In some shelf regions, wave action and ocean currents have concentrated heavy minerals in sediments, and these marine placer deposits may be profitably mined on a limited scale. Besides the known and exploited deposits of petroleum on the continental shelf—primarily off North America, the Middle East, and in the North Sea—it is also possible that petroleum is trapped within the thick sedimentary accumulations found on the continental slope and rise.

Seabed manganese nodules have received considerable attention in recent years, primarily because of the metallic elements that they contain (copper, nickel, and cobalt, along with manganese). These nodules, shown in Figure 12.25, cover about a quarter of the deep ocean bottom. Although the technology is at hand for harvesting the nodules, considerable controversy has arisen over mineral rights in the international waters of the open ocean. This obstacle, along with the high costs of mining, is likely to delay commercial exploitation of nodules at least until 1990.

Conservation Strategies

Prospects for the immediate recovery of substantial quantities of high-grade minerals from the seabed and from new discoveries on land are not very promising. Obviously, resource conservation is of prime importance. The most fundamental approach to crustal resource conservation is a reduction of our personal consumption. The extravagance of our national consumption is suggested by the fact that, though our population represents only 6 percent of the globe's population and inhabits only 6 percent of the globe's land area, we consume 23 percent of the globe's nonfuel rock and mineral resources. Only by reducing consumption, either voluntarily or under government-imposed regulations, will we gain the chance to buy the time necessary to develop the technology and energy sources required for exploration and extraction of new reserves.

Efforts are under way to make more efficient use of existing crustal resources and to reduce present rates of depletion. Formerly, mine operators often by-passed low-grade deposits in the quest for more valuable grades. Mines were frequently abandoned because extraction of low-grade deposits was not economically feasible. Now, however, miners are being encouraged to remove both high-and low-grade deposits together, stockpiling the low-grade materials until industrial demand or technology makes processing feasible. In fact, today some metals in high demand but in limited supply (copper, for example) are being recovered from old mine dumps at a significant energy savings (for copper, about 20 percent).

Another conservation strategy is the substitution of renewable for essentially nonrenewable crustal resources: for example, using wood in place of metal in light construction, an alternative discussed in Box 12.1. Also, one depletable resource may be replaced by another that is in greater supply—as when aluminum is substituted for copper, for example, or glass for tin-plated cans, or stone for copper as a building facade. In addition, research is underway to increase the lifespan of metals by developing more corrosion-resistant alloys.

The *recycling* of metals is essential to a resource conservation program. Although metals are not destroyed in manufacturing processes, they are dispersed over the earth's surface as the components of a

Figure 12.25 Manganese nodules on the floor of the Pacific Ocean. (Kennecott Copper Corporation.)

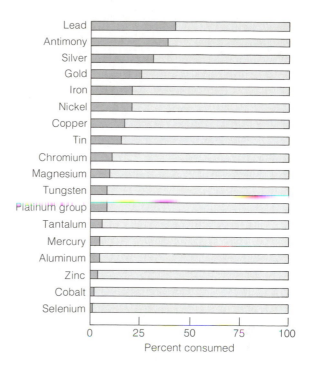

Percent consumed

Figure 12.26 Recycled scrap metal in the United States expressed as percent of total consumption of each metal. (After U.S. Department of the Interior, Bureau of Mines, *Mineral Facts and Problems:* 1975 Edition. Washington, D.C.: U.S. Government Printing Office, 1975, Bulletin 667.)

multitude of items, ranging from airplanes to zippers. When these items lose their value, their metal components are lost to us unless efforts are made to recover them. Hence, recycling initially requires the recovery and collection of metallic components. Many metals including aluminum, iron, zinc, and lead, are already recycled in significant quantities, as Figure 12.26 indicates. But some metals, such as tungsten, are so widely dispersed that the energy required virtually prohibits their recycling.

Some scrap metal is now favored over raw materials because the cost of the latter has sharply increased, partially because of the high cost of energy. The energy saving may be considerable—5–30 percent—when recycled metals are used in place of virgin ores. As an illustration, the Institute of Scrap Iron and Steel reports that energy consumption is reduced by 74 percent when scrap iron is utilized in place of virgin iron ore in the production of a thousand metric tons of steel. Thus, steel manufacturers

are installing furnaces that use 100 percent scrap iron. By the mid-1970s, such furnaces were responsible for 23 percent of the nation's steel output, up 5 percent from a decade earlier. The utilization of scrap metal in steel making is illustrated in Figure 12.27.

One major problem with recycling is its sensitivity to fluctuations in the economy. After a metal recycling boom in the early 1970s, recycling suffered serious setbacks with the onset of economic recession. Many small recycling centers went out of business and, in 1975, the use of recycled metals declined precipitously: the use of recycled copper dropped 30 percent; zinc, 27 percent; stainless steel, 43 percent;

Box 12.1

The Latest Thing in Building Materials Is . . . Wood

Money doesn't grow on trees. Iron, aluminum, and other nonrenewable resources used in the construction industry don't either. But wood does—and because it does, this oldest of construction materials is still relatively economical. It also represents one of our best hedges against future shortages of both materials and energy.

Today wood is again competing with steel, aluminum, and other nonrenewable resources, which had earlier outcompeted lumber in many construction applications. More and more, massive laminated wood beams are being used in such long-span structures as sports stadiums, footbridges, and even highway bridges. Laminated beams can be made much larger than single pieces of lumber can. They are stronger, too—just as the familiar plywood sheet, which is also made of laminated layers, is stronger than a board of the same dimensions. Laminated beams can be made in almost any shape, including graceful curves, to accommodate different design requirements. And they can be manufactured from small wood pieces of poor quality.

"Glue lam" beams, made by gluing thick wooden pieces together, have been in use since the 1930s. Glue lam has recently been joined by "press lam," a material built up of half-inch wood layers that are peeled from logs, flattened in a press dryer, sandwiched with glue, and then pressed together while still hot. Press lam is as strong as glue lam, and it can be made from low-cost, small-diameter logs.

Other recent developments in wood building materials include structural flakeboard, made from branches and other forest residue as well as from low-grade logs. Flakeboard, which is much stronger and more durable than particle board, is being used for jobs that formerly required lumber or plywood. Like glue lam and press lam, flakeboard conserves lumber because it substitutes low-grade wood for high-grade wood.

Laminated and pressed wood building materials, as well as conventional lumber, have another great advantage: their "energy effectiveness." A study prepared under the auspices of the National Academy of Sciences/National Research Council compared the total energy inputs used in the production of various wood and nonwood structural materials. The study found that the wood materials were far more energy-efficient. As proof it cited the following figures among others: (1) it takes five times as much energy to make a piece of aluminum siding as it does to produce the metal siding's wood counterpart; (2) eight times more energy goes into a steel interior-wall stud than into a wood stud; (3) producing brick siding requires twenty-five times the energy of producing plywood siding; and (4) the making of a steel floor joist takes fifty times more energy than the creation of its wood equivalent!

Wooden building materials offer two other significant energy advantages. One is that they can be—and often are—produced at a cost of zero net energy, because the energy used in their manufacture comes from residual fuel (branches, bark, sawdust, and the like) obtained as the wood is harvested. The other big advantage is that the insulating qualities of wood are far superior to those of metals, brick, or concrete; therefore wooden structures cost less to heat in winter and cool in summer.

In Box 2.2 we introduced the concept of net energy analysis, and we noted that it is a valuable tool in deciding how to maximize the returns on our precious energy investments. Net energy analysis seems to have a message for the construction industry: think wood.

Figure 12.27 *Left:* A scrap weigher at U.S. Steel's Duquesne Works near Pittsburgh directing the lowering of bundles of tightly compacted steel cans into a charging box at the plant. *Right:* The box containing thousands of cans being dumped into a steel making furnace. (Courtesy of United States Steel Corporation.)

and aluminum 7 percent. But by early 1976, with an improved economic climate, the downward trend had begun to reverse and scrap metals were in greater demand.

In many cases, recycling is a marginal operation that is not profitable unless operated at maximum efficiency, utilizing an uninterrupted flow of large quantities of recyclable materials. Hence, recycling has the best chance of success in a large urban-industrial area that has both a reliable source of appropriate waste and a market for recycled products. In such areas, a steady stream of recyclable waste of a specific composition can be assured on a contractual basis. But for recycled materials to compete favorably a large-scale recycling operation equipped with centralized facilities for separation, cleaning, and shredding is needed. This economic fact of life seriously limits the value of small-scale recycling efforts.

Land Reclamation

An integral component of crustal resource management is the rehabilitation, or *reclamation*, of surface mined lands. Examples are discussed in Box 12.2. The term reclamation describes activities that foster natural succession on land disturbed by mining. Generally, these activities include contouring, or shaping, mine spoils and overburden to minimize erosion, applying topsoil and fertilizer, and planting and maintaining vegetation to match the species assemblage natural to the area. Specific procedures depend on the physical and chemical properties of the spoils. For example, the type of vegetation planted on acidic waste heaps may not be appropriate on spoils that have high salt content.

Reclamation activities are most effective and economical when they are made an integral component of the mining operation rather than put off until mining is completed. For example, contour strip mining is not nearly so destructive of the landscape when miners employ what is known as the haulback method. With this technique, trucks haul spoils back along the contour terrace to fill in and restore the original slope of the land rather than dumping spoils downhill. This procedure virtually eliminates the danger of landslides and disturbs considerably less land than does conventional contour strip mining. It also lowers the cost of other reclamation techniques.

In the coal fields of the eastern and central United States, reclamation research has been underway for four decades. But progress in actually reclaiming strip mined land has not been as rapid as conservationists had hoped. In fact, perhaps 3 million hectares (1.2 million acres) of strip mined land in these regions remain unreclaimed. In the western coal fields, reclamation promises to be even more difficult than in the East, primarily because the region gets less rainfall and because very little revegetation research has been carried out there. Rainfall is particularly marginal in rangeland, making the reestablishment of seedlings quite difficult. Rehabilitation of open pit mines in the mountainous West is a special problem because of the enormous quantities of spoils and because revegetation is a very slow process in mountainous ecosystems. Temperatures are low, growing seasons are short, soils are shallow, and few plant species can be used.

Reclamation was given a boost nationwide in August 1977, with the long awaited passage of the Surface Mining Control and Reclamation Act. First formulated in 1971, the law was passed in both houses of Congress but was vetoed twice by President Ford, first in 1974 and again in 1975. In explaining his actions, the president cited the prospect of reduced coal production and increased unemployment if strip mining were regulated.

The 1977 federal law is the first to regulate strip mining nationwide, and sets standards for leasing, mining, and reclamation. It requires that land be restored to "a condition capable of supporting the uses it was capable of supporting prior to any mining." Before they are granted a permit to mine, mine operators must demonstrate that they can reclaim the land. To minimize erosion and contamination, they must store A and B soil horizons, and for prime farmland, they must store the C soil horizon as well. After mining, the land must be regraded to approximately its original profile and the topsoil must be replaced and reseeded. Where mined land was originally agricultural, restored land must be as productive as it was before mining. The law also requires that landowners give written consent for mining in those cases where mineral rights are held by someone other than the landowner. Finally, the law encourages miners to use the haulback method of contour strip mining and to avoid ground and surface water contamination.

Mine Reclamation:
Turning a Liability into an Asset

What is to be done with an abandoned surface mine? The excavations made for open-pit mines are so huge that some of them can never be put to beneficial use. But, for small quarries and pits, it is sometimes possible to transform a liability into an asset. Where ground water quality is not threatened, abandoned quarries and pits may be used for sanitary landfilling (described in Chapter 13). In other instances, they may be filled with water and converted into recreational ponds and lakes. For example, a series of fishing and swimming ponds line Interstate 80 in Nebraska. The ponds are actually pits that once supplied the sand and gravel used to construct the highway. In other cases, these pits turned ponds have become the focus of housing developments: they actually add to real estate values.

Abandoned subsurface mines can also become a community asset, as was demonstrated in Pennsylvania by the U.S. Department of the Interior's Bureau of Mines. McDade Park, situated on the outskirts of Scranton, Pennsylvania, is the former site of numerous strip-mining trenches, an underground anthracite mine, and ugly mine spoils.

When the coal ran out in 1966, the mines were abandoned. Acidic runoff from the spoils polluted a nearby creek and spring-fed pond. The 50-hectare (125-acre) site became a safety hazard to children playing in the area.

Then representatives of the Bureau of Mines stepped in. They began by carefully studying the site. Next, using local contractors, they embarked on an extensive rehabilitation project that in three years transformed the area into an attractive recreational, historical, and educational complex. First the land was regraded and roads and parking lots were constructed. Topsoil was then hauled in, fertilized, and planted with grasses, shrubs, and trees. Recreational facilities, including hiking trails, picnic grounds, and playgrounds, were built. Funding for these efforts was provided by county, state, and federal agencies.

The park features an anthracite museum that preserves a section of the abandoned subsurface mine and traces the history of anthracite mining. The museum, funded by the Pennsylvania Historical and Museum Commission, has an auditorium and a library as well as an extensive exhibit area.

Conclusions

According to the National Research Council, efforts to explore and exploit the earth's crustal resources are likely to conflict more and more with those to protect environmental quality. Demand for these resources will more than double over the next two decades, and we will therefore be forced to mine lower grades of minerals, consequently disrupting the environment ever more severely. The mining of lower grades of minerals will also stress our dwindling supply of fossil fuels.

Our alternative is to develop new techniques of exploration, recovery, and processing of crustal resources that will reduce environmental degradation. Of equal importance is our need to develop new inexpensive and abundant energy sources. But because implementing any new technology takes time, resource conservation now has a higher priority than ever before.

Summary Statements

Industrial nations are heavily dependent on the properties of a multitude of rock, mineral, and fuel resources extracted from the earth's crust.

A variety of processes have acted over geologic time to selectively concentrate crustal resources. These processes include igneous activity, rock formation and deformation, erosion, and weathering.

Rocks are classified according to their mode of origin as igneous, sedimentary, or metamorphic. Igneous rocks are the most abundant, but sedimentary rocks are the most conspicuous.

A rock is an aggregate of minerals usually consisting of a few dominant and several accessory minerals.

Metallic or nonmetallic minerals may occur as minor components of large rock masses. Hence, energy and water are needed for separation, smelting, and refining. Solid, liquid, and gaseous waste products are generated by these processes.

Coal deposits were developed by the anaerobic decomposition of plant matter in swampy terrain millions of years ago. With increasing heat and pressure, peat layers changed to lignite, then to bituminous coal and, in some regions, to anthracite.

Oil is organic in origin, and occupies pore spaces in certain marine sedimentary rocks.

Natural gas is derived from oil, and is found either mixed with or on top of oil as well as in separate reservoirs.

In the United States, uranium minerals are found primarily in certain sandstone strata in the West.

Rocks in any of the three rock families may be altered by geologic processes into other rock types. The typically slow pace of these processes, however, essentially fixes our supply of crustal resources.

Surface mining removes vegetation, soil, sediment, and rock that overlie mineral deposits. Surface mining techniques include quarrying, open pit, hydraulic, dredge, and strip mining.

Rain seeping into mine waste heaps triggers erosion, dangerous sliding, and the pollution of drainageways.

Deep ore and coal deposits are extracted from subsurface shafts and tunnels blasted and drilled into the rock. Mine collapse and costly ground subsidence are potential risks.

Although the danger of oil well blowouts is real in offshore drilling operations, blowouts are actually rare.

For many rocks and minerals, reserves and resources are so abundant that we are assured an adequate supply far into the future. But for some materials necessary to our highly technological society, the situation is less favorable.

The principal obstacle to mining lower grade deposits is energy costs. Until new energy sources are developed, exploration and conservation efforts must be maximized.

Conservation strategies include reducing depletion rates, substituting renewable for essentially nonrenewable resources, and recycling.

Recycling saves energy but is very sensitive to fluctuations in the economy.

Reclamation encompasses those activities that foster natural succession on land disturbed by mining.

Questions and Projects

1. In your own words, write definitions for each of the terms italicized in this chapter. Compare your definitions with those in the text.

2. What resources are mined in your community? Describe the mining methods employed. Do local or state laws require restoration of the landscape after mining is completed? If so, are these laws enforced?

3. Suggest some uses for an abandoned quarry that would not threaten groundwater quality.

4. Identify the geologic processes that are primarily responsible for the appearance of the landscapes in your locality.

5. How does the unequal distribution of crustal resources influence world trade and our national security?

6. Dollars and energy are used to redistribute mineral resources across the face of the earth, and dollars and energy must be expended to recover these materials for recycling. Suggest measures that might be taken to reduce this double expenditure.

7. What are the basic reasons for the marked contrasts between lunar and earth landscapes?

8. How might more strip mining in the Midwest influence the price of food?

9. Modern industrial societies rely on the properties of numerous minerals, yet some minerals exhibit the same or similar properties. Does this observation suggest a resource conservation strategy?

10. Compare and contrast the advantages and disadvantages of surface mining and subsurface mining.

11. How does an understanding of geologic processes aid our search for mineral and fuel resources?

12. The classification of rocks as igneous, sedimentary, or metamorphic is actually an environmental classification. Explain this statement.

13. Explain why deposits of rock, mineral, and fuel resources do not occur uniformly over the earth's surface.

14. Describe the implications of the rock cycle for our supply of rock, mineral, and fuel resources.

15. Explain how subsurface mining might threaten water quality.

16. Identify some of the environmental problems caused by wastes from surface and subsurface mining. Also, describe some of the measures that reduce impact.

17. Explain the differences between reserves and resources.

18. What is the primary obstacle to mining very low grade mineral deposits?

19. What are the prospects for recovering high-grade minerals from the seabed and new discoveries on land?

20. List conservation strategies that make more efficient use of resources and reduce present rates of depletion.

Selected Readings

Bonatti, E. 1978. "The Origin of Metal Deposits in the Oceanic Lithosphere," *Scientific American* *238*(2): 54–61 (February). A well-illustrated discussion of how metal deposits are generated in the midocean ridge.

Cameron, E. N. (ed.). 1973. *The Mineral Position of the United States, 1975–2000.* Madison: The University of Wisconsin Press. A collection of papers on future trends in U.S. mineral demands.

Cargo, D. N., and B. F. Mallory. 1977. *Man and His Geologic Environment.* Reading, Massachusetts: Addison-Wesley. A thorough treatment of geologic aspects of environmental problems.

Press, F., and R. Siever. 1978. *Earth,* 2nd ed. San Francisco: W. H. Freeman and Company. A basic up-to-date text on the principles of physical geology intended for the beginning student.

Tank, R. W. (ed.). 1976. *Focus on Environmental Geology.* New York: Oxford University Press. Includes selected readings on U.S. mineral resources and the environmental impact of mining.

Uyeda, S. 1978. *The New View of the Earth.* San Francisco: W. H. Freeman and Company. Reviews evidence for continental drift.

Wyllie, P. J. 1976. *The Way the Earth Works: An Introduction to the New Global Geology and Its Revolutionary Development.* New York: John Wiley & Sons. An unusually lucid account of geologic processes developed in the context of continental drift.

Thoughtless disposal of waste, hazardous to health and threatening to environmental quality.
(U.S. Department of Housing and Urban Development.)

Chapter 13

Waste Management

At about the time that the City of Saugus, Massachusetts, was ordered by the state to close its overflowing garbage dump, the General Electric plant in nearby Lynn needed new steam boilers. Saugus and General Electric solved their problems by cooperating with the operators of a private resource recovery facility. Saugus, along with ten other municipalities, contracted to deliver more than a thousand metric tons of refuse per day to the recovery facility. At present, the facility receives the waste, separates out the combustibles for burning to supply steam for the General Electric plant, recovers and recycles the metals, and hauls the unusable residue to a landfill site.

The Saugus resource recovery center has many attractive features. It exploits the long-neglected energy and materials potential of solid waste at a time when supplies of natural resources are declining. It reduces the amount of land needed for waste disposal when suitable dumping sites are becoming rare and land prices are soaring. And apparently the operation is economically sound: most construction and operating expenses are covered by the income gained from the marketing of steam and recovered metals.

Resource recovery facilities such as that at Saugus represent a relatively new approach to the growing problem of solid waste disposal. As we extract, process, and utilize more resources, we generate ever increasing quantities of solid waste, as Table 13.1 indicates. Formerly, we arbitrarily disposed of wastes in rivers, at sea, in the air, or in open dumps with little thought of consequences. But today, in accordance with our dawning realization that many of these wastes are hazardous to human health and deleterious to environmental quality, we are seeking to manage solid waste so that damage is minimized and usuable material is recovered. The objectives and methods of these waste management efforts are the subjects of this chapter.

Sources of Waste

Virtually all of our activities generate waste. We create waste when we extract and process raw materials and fuels, and when we use goods and services. When products wear out, become outmoded, or otherwise outlive their utility, we merely throw them

Table 13.1 Projected trend in solid waste generation in the United States (in kilograms per person per day).

	Estimated		Projected		
	1971	1973	1980	1985	1990
Total gross discards	1.60	1.70	1.94	2.12	2.27
Resource recovery	0.10	0.11	0.21	0.37	0.59
Net waste disposal	1.50	1.59	1.73	1.75	1.68

Source: After Council on Environmental Quality, *Sixth Annual Report,* 1975.

out. Fundamentally, air and water pollution are problems of waste disposal. Thus, a wide variety of materials are referred to by the term waste. For convenience these wastes are classified according to source: municipal-industrial, mining-mineral, and agricultural. The special problems, associated with nuclear wastes are discussed in Chapter 18.

The constituents of municipal-industrial waste are as diverse as those of our surroundings: they range from bricks to tree branches, tin cans to newspaper, explosives to sludge, cinders to food. Figure 13.1 shows the variety of municipal-industrial solid waste by percentage found in a typical city dump. If we exclude junked motor vehicles from our calculations, we find that each of us is responsible, directly or indirectly, for generating about 3 kilograms (6.6 pounds) of solid waste per day. Of this amount, about 1.7 kilograms (3.8 pounds) are collected by municipalities, while the remainder is disposed of by other means—through littering, for instance, or in private incinerators. Typically, the cost of collecting and disposing of municipal waste is a major expenditure for communities—in some localities ranking only behind the cost of public education and highways.

Mining and mineral wastes include mill tailings, slag and fly ash, and various mine wastes exclusive of overburden (see Chapter 12). These wastes are generated at a rate more than four and a half times greater than that at which municipal-industrial waste

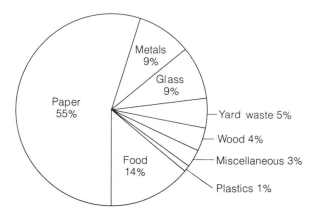

Figure 13.1 The components of a typical city dump. (After J. G. Abert et al., "The Economics of Resource Recovery from Municipal Solid Waste," *Science 183:*1052, 1974. Copyright 1974 by the American Association for the Advancement of Science.)

is produced. But because mining and mineral processing usually take place in remote areas, this enormous quantity of waste is not generally as evident to us as municipal-industrial waste; hence, we may not be particularly concerned about it, having an "out of sight, out of mind" attitude. We can gain perspective on the problem of mining-mineral waste disposal by considering the example of open-pit mining of copper ore in the southwestern United States, illustrated

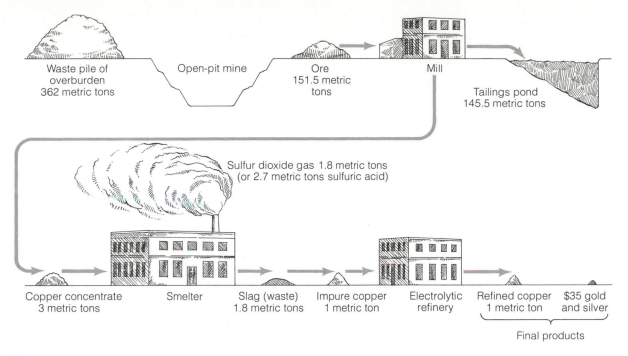

Figure 13.2 A chart illustrating the sources of waste involved in the extraction and processing of copper ore. (After National Research Council, National Academy of Sciences, *The Earth and Human Affairs.* San Francisco: Canfield Press, 1972.)

in Figure 13.2. The production of just 1 metric ton of copper (including extracting, grinding, separating, smelting, and refining) typically generates more than 500 metric tons of solid waste. The tons of air pollutants also generated by these processes are not included in this figure.

Agricultural waste is produced at a rate about six and a half times that of municipal-industrial waste production. Again, city dwellers unfamiliar with farm activities may find this figure surprisingly large. It is estimated that the total waste generated in agricultural operations amounts to more than 2 billion metric tons annually. Most of this total (75 percent) consists primarily of animal manure, crop residue, and various by-products of food production.

Wastes are also classified as to whether or not they are biodegradable. Biodegradable substances are broken down in time by the action of aerobic or anaerobic bacteria, whereas nonbiodegradable mate-

rials are not broken down by biological processes. Agricultural wastes consist primarily of biodegradable materials, and mining and mineral wastes are nonbiodegradable materials. Municipal-industrial wastes are a complex mixture of biodegradable and nonbiodegradable components.

Toxic and Hazardous Wastes

Wastes that pose a direct threat to human health have received considerable attention in recent years. Our dumping of hazardous and toxic wastes in convenient but not necessarily safe sites over the decades is now beginning to haunt us. Tragedies similar to the Love Canal incident described in Box 13.1 are likely to occur elsewhere. A November 1978 survey by the EPA revealed that perhaps 32,000 old disposal sites containing potentially dangerous materials exist

Box 13.1

The Legacy of the Love Canal

More than a hundred years ago, work began on the Love Canal in the town of Niagara Falls, New York. The canal was to link Lake Ontario with the Niagara River, thereby providing water and hydroelectric power to the growing community. But in 1888, the alternating current (AC) motor and long-range AC transmission were developed, making the canal unnecessary and uneconomical to industry. So construction was halted and the partially completed canal was abandoned.

From 1930 until the early 1950s, the 6-hectare (15-acre) canal site was used by the Hooker Chemical Company as a dump for tons of industrial chemicals, including pesticides and cleaning solutions—and perhaps chemical-warfare materials as well. In 1953, the company covered the dump with clay and sold it to the Niagara Falls Board of Education. Later, a school was built on part of the site and the rest was sold as house lots.

In 1976, after several years of unusually heavy rains, foul-smelling leachate began seeping through the topsoil onto the school playground and into the yards and basements of houses in the area. Soon there were reports of dying pets and serious health problems among human residents. In 1978, New York State Health Department investigators discovered a variety of alarming afflictions among residents, including headaches, sores, rectal bleeding, liver ailments, epilepsy, miscarriages, and birth defects. In July of that year, the state health commissioner called the former dump site a "great and imminent peril" and urged that pregnant women

and children under the age of two leave the area. That same month, President Jimmy Carter declared the Love Canal neighborhood a federal disaster area. In August, 239 families were evacuated. Meanwhile, state and federal investigators had discovered 82 different chemical compounds, 11 of them known or suspected carcinogens. One of the carcinogens was dioxin, perhaps the deadliest of all manufactured chemicals. (For more on dioxin, see Box 10.1.)

That fall, the state began a massive clean-up of the dumpsite. First, it bought the homes of the 239 evacuated families and built a chain-link fence around them to keep out trespassers. Then it sent in gas-masked workers to dig drainage ditches within the contaminated area and around its periphery that would collect the leachate for removal and treatment.

By early 1979, the state had spent more than $20 million, part of it in federal funds, on the clean-up. How much more would be necessary to complete the job, nobody could say.

The Love Canal episode forced the Environmental Protection Agency to consider regulating former dump sites more strictly, making a federal inventory of hazardous waste dumps, and establishing "buffer zones" around such sites. Shortly after the Love Canal danger came to light in 1978, the regional director of the EPA was quoted as saying of the thousands of potentially dangerous landfills in the country, "We've been burying these things like time bombs."

across the country. Of these, 638 dumps were identified as potential sources of substances that pose an immediate threat to human health.

An estimated 30–40 million metric tons of hazardous wastes, exclusive of radioactive wastes, are generated each year. This amount represents almost 15 percent of all industrial wastes and includes a very diverse array of substances. One in four of the metals in common use, such as lead and zinc, is considered hazardous. And, as discussed in some detail in Chapter 7, the manufacture of organic chemicals, such as pesticides, produces thousands of metric tons of waste products. Flammable and nonflammable solvents, explosives, and chemicals that do not meet manufacturers' specifications add to the mountain of hazardous wastes.

The EPA estimates that up to 90 percent of all hazardous wastes are improperly disposed of and that existing hazardous waste disposal facilities can handle only 40 percent of the hazardous waste stream. Refusal by municipal landfill sites to accept hazardous wastes has helped to bring the issue to national attention.

The initial disposal problem is compounded by our need to clean up old disposal sites. The total cleanup cost for all sites, according to EPA estimates, could exceed $3 billion. But no cleanup effort will solve the perplexing problem of how to deal with "midnight dumpers." For example, in North Carolina in the summer of 1978, a tank truck spread PCB-contaminated wastes over 400 kilometers (250 miles) of back-country roads in an effort to avoid high disposal costs. The cost of collecting and properly disposing of the contaminated soil along the affected roadbed is expected to total $12 million.

A Historical Overview

Through much of our nation's history, we liberally and indiscriminately tossed garbage and other refuse into our backyards, onto the street, and into marshes or rivers. Beginning in colonial days, hogs were allowed to wander city streets freely, rooting through garbage and leaving their excrement and stench behind. In post-Civil War days, as cities swelled with new industry and workers, the waste disposal problem became acute. For example, Chicago's population soared from only five thousand in 1840 to almost a million by 1890; the meager sanitation efforts then made could not keep pace, and city streets became clogged with trash. In New York City in the late nineteenth century, refuse accumulated in such large heaps along some streets that it impeded pedestrian and vehicular traffic (see Figure 13.3).

As we saw in Chapter 8, public concern over the growing health hazard eventually forced the improved collection of trash in cities. Hogs were banned from city streets, and garbage and other refuse was either hauled out to hog farms, open dumps adjacent to cities, or nearby rivers, or it was barged out to sea. But the waste problem was not solved by these actions; it was merely displaced. Not until recently have health and environmental quality regulations been implemented to control waste disposal.

We have seen that water quality regulations became increasingly strict over the years, and that dumping in surface waters was finally outlawed, especially for hazardous wastes. Industry and municipalities are now required to use land disposal sites as an alternative. Nevertheless, enforcing the law governing waste disposal often involves dispute and compromise. Let us consider a recent example.

In June 1973, asbestoslike fibers (asbestos fibers are a known carcinogen) were found in the drinking water supply at Duluth, Minnesota. The fibers were traced to taconite (low-grade iron ore) tailings that were being discharged in large quantities into Lake Superior—also the source of the community's water supply. In April 1974, citing a potential health hazard, a federal district court directed the Reserve Mining Company's processing plant, which was producing the tailings, to halt the discharges. After several stays of the lower court order, a court of appeals ruled that the health threat was potential but

Figure 13.3 Heaps of refuse clogging New York City streets in the 1880s. (The Bettmann Archive, Inc.)

not imminent. The plant was allowed to continue discharging, but the company was ordered to locate a suitable land disposal site by July 7, 1977. Meanwhile, Duluth installed a special filtration system to protect its drinking water.

In ensuing years, state regulatory agencies vied with the Reserve Mining Company over the selection of a suitable dumping site. Because of the potential health hazard, the state favored a site away from populated areas and chose the Superior National Forest. But the mining company argued on economic grounds for a site 21 kilometers (13 miles) closer to its processing plant. The Reserve Mining Company stated that it would be forced to shut down if required to use a site farther away, an eventuality that would eliminate three-thousand jobs. State courts resolved the issue by ruling in favor of the dumping site selected by the company. The federal court then lifted its 1977 closure order on the condition that preparation of the new disposal site begin by June 1, 1977. Work is currently underway (at a cost of about $370 million), and discharge of the tailings into Lake Superior should finally cease when the new disposal site is in full operation by the spring of 1980.

In cases such as the Reserve Mining controversy, it is essential that the land disposal site be chosen carefully so that the health hazard is not merely transferred from water to land (or air). Unfortunately, most of our nation's land disposal sites are still open dumps that pose the dangers of surface and groundwater contamination and air pollution. As of the late 1970s, only 29 percent of all the solid waste collected in the United States was deposited in sanitary landfills; of the rest, only 8 percent was incinerated and 6.5 percent recycled. Most of the remainder (56.5 percent) ended up on ugly open dumps, such as the one shown in Figure 13.4. Open dumps are breeding places for disease-bearing vermin, sources of noxious odors from rotting garbage, and often sites of smoldering fires. Recent legislation, however, promises to change this situation dramatically.

A few states, notably Massachusetts and California, have had standards for the collection and disposal of solid waste for many years. Typically however, most states have exercised meager control over land disposal sites. The first federal legislation on waste disposal was not enacted until 1965. The Solid Waste Disposal Act of that year called for research on the solid waste problem and provided funds to states and municipalities for planning and developing waste disposal programs. The 1965 law was amended with the passage of the 1970 Resource Recovery Act, which provided funds to states for constructing waste disposal facilities. Also, the 1970 law was the first federal legislation to encourage the recycling of solid waste.

The strict regulation of solid waste disposal is a major objective of the Resource Conservation and Recovery Act of 1976. This law requires the states to close all open dumps by 1983, and assists them in developing waste reduction programs. The 1976 act also calls on the EPA to draw up guidelines for the siting of sanitary landfills in order to minimize the environmental impact of solid waste disposal. Provisions also seek to strictly regulate hazardous wastes, from generation to ultimate disposal. The law requires the states to identify hazardous wastes and sets national standards for the generation, transport, treatment, storage, and disposal of hazardous wastes.

Hazardous materials became subject to further regulation in 1976 with passage of the Toxic Substances Control Act. The law requires premarket screening of new chemicals (excluding pesticides) for potential toxicity. Chemicals that pose an unreasonable risk in their manufacture, distribution, use, or disposal can be regulated through the EPA. While the intent of this law is well placed, its provisions have generated considerable controversy that is likely to delay full compliance until the early 1980s. Disagreements have developed over the degree of risk of different chemicals. Also, chemical companies fear that the law will force public disclosure of trade secrets and aid their competitors. These problems are compounded by the great number of chemicals currently in commercial production (about 70,000), which insures that the screening process will be both expensive and time consuming.

Waste Disposal Alternatives

Recent federal legislation is forcing states and municipalities to identify and implement alternatives to open dumping. But unless properly regulated, these alternatives too may prove hazardous to health and destructive to environmental quality. In this section we survey the alternatives open to us and the special requirements of each.

Sanitary Landfill

When successive layers of compacted refuse are sealed between layers of clean earth each day, most of the deleterious effects of open dumps are eliminated. In such facilities, called *sanitary landfills*, no open burning occurs to foul the air, disease-bearing insects and rodents do not proliferate, and the odor problem is minimal. Furthermore, sanitary landfills can restore landscapes scarred by mining or put an idle area to use. An example of the latter is Mount Trashmore in Virginia Beach, Virginia, a hill composed of solid waste that is the focal point of a recreation area.

There are two modes of sanitary landfill: trench and area. *Trench landfills,* such as that in Figure 13.5, are appropriate only in regions where the water table is low and the soil deep. In this method,

Figure 13.4 An open city dump. (U.S. Department of Housing and Urban Development.)

solid waste is spread into a trench and compacted and sealed each day with the dirt that was excavated from the trench. This process is repeated until the trench is filled. *Area landfills* take advantage of natural depressions, such as canyons or valleys, as well as abandoned pits and quarries. In this method, waste is repeatedly spread, compacted, and sealed in until the depression is filled. Area landfills are less desirable than trench landfills, since usually the cover dirt must be hauled in from elsewhere.

Sanitary landfills, while far more desirable than open dumps, do have some potential drawbacks. Rainwater seeping through wastes can contaminate the groundwater reservoir. Several precautions can be taken to avoid this problem, however. Above all the site selected for a sanitary landfill should be located well away from streams, lakes, or wells. Also, rapid runoff or drainage ditches may be installed to carry off surface waters. Impermeable clayey soils may be used to line the sides and base of the landfill,

and a sandy-clay soil used as the top of the landfill to inhibit the development of cracks that would allow entry of moisture. If water infiltrates a landfill site, the leachate must be collected and subsequently treated. In areas where the water table is high and these measures are ineffectual, the layers of waste and dirt must be mounded on the surface to form a hill.

Another adverse feature of sanitary landfills is the possible accumulation of the gases formed during the anaerobic decomposition of waste. In addition to carbon dioxide, poisonous hydrogen sulfide and methane are two important decomposition products. The latter is explosive if allowed to accumulate in enclosed spaces. Hence, where impermeable cover materials are used, vent pipes or trenches must be installed to allow these gases to escape. In large-scale sanitary landfill operations, however, it may be feasible to collect methane (the chief component of natural gas) and use it for fuel. In a landfill near Los

Environmental Quality and Management

Figure 13.5 *Left:* The spreading and compacting of refuse in a trench-type sanitary landfill. *Right:* At day's end, a layer of clean fill being spread to fill in the trench. (Deere & Company.)

Angeles, an accumulation of 13 million tons of refuse has been producing enough methane to meet the needs of two-thousand homes since 1975. In mid-1978, a landfill site run by the city of Mountain View, California, with EPA support, started a large experimental methane-producing operation with a production capacity of 28,000 cubic meters (1,000,000 cubic feet) per day.

Once a sanitary landfill operation has been completed, the site must be inspected periodically for differential settling of the ground. Settling occurs primarily as a consequence of anaerobic waste decomposition, although the weight of the refuse is also important in compacting the ground. Normally, about 90 percent of the total settlement takes place

within five years; therefore, additional filling and grading may be necessary during that period.

Finally, litter and dust blowing around on windy days during dumping and compacting, though a minor problem, can be a nuisance nonetheless. Maintenance personnel can minimize such scattering by installing litter fences around the site and regularly applying water from sprinkler trucks. Still, the staff will have to do some hand collection of litter to maintain a neat site.

Although a completed sanitary landfill does not make a suitable building site, it may be used for recreation, as shown in Figure 13.6, or as a botanical garden, a small airport, a pasture, or cropland. Although sanitary landfills offer many advantages over

Figure 13.6 A little league baseball park built on a completed sanitary landfill. (EPA Documerica photo by W. Shrout.)

open dumps for solid waste disposal, a major obstacle to wider use is public opposition. Even though a sanitary landfill may eventually be turned into a beautiful park benefiting the entire community, many people equate landfill sites with dumps, and simply refuse to accept the idea of a landfill of any kind in or near their neighborhood.

Municipal Incineration

Municipal incinerators burn combustible solid waste and melt certain noncombustible materials. Incineration is more advantageous than open dumping, since the high temperatures involved destroy disease vectors (flies and rats) and pathogenic organisms. However, because installation, maintenance, and operational expenses are higher, the average cost of incinerating solid waste is generally somewhat higher than that of maintaining a sanitary landfill. This difference does not apply in localities where appropriate sites for sanitary landfills are unavailable and

where land prices are soaring. Ironically, these very circumstances usually prevail in or around urban areas, where most household, commercial, and industrial waste is generated. The solution in such situations may be a combination of incineration and sanitary landfilling. Incineration can reduce the volume of solid waste by up to 80–90 percent. Hence, disposing of the material that remains after incineration requires considerably less land than would the nonincinerated waste.

Incineration has additional advantages over landfills of any kind. Incinerators do not directly endanger groundwater quality. And, if they are equipped with adequate air pollution control devices and attractively housed and landscaped, incinerators are more likely to gain public acceptance than are landfills. Unfortunately, most of the more than three-hundred municipal incinerators operating in the United States in 1968 were constructed before air quality control devices were required. The added cost of installing this equipment has forced the clos-

Figure 13.7 A municipal incinerator in Green Bay, Wisconsin, now closed down because of its failure to meet air quality control regulations. (M. L. Brisson.)

ing of most conventional municipal incinerators, such as the one shown in Figure 13.7. Another drawback of conventional incinerators is their inability to handle certain types of waste—for example, explosives and potentially smoky wastes such as tires.

Considerable progress was made in the 1970s in incineration technology and in appropriate air pollution control devices for incinerators. Hence, in some large metropolitan areas new incineration techniques are a central component in resource recovery facilities. These facilities are described later in this chapter.

Deep-Well Injection

Deep-well fluid injection is the pumping of fluids into pore spaces and fractures in subsurface rock layers. Deep-well fluid injection is used for a variety of purposes, including the storage of natural gas, the recharging of groundwater, and the recovery of oil from depleted wells. In the context of waste disposal, the technique offers an alternative to the use of land and surface waters for the disposal of industrial fluid waste.

Liquid wastes that are disposed of by deep-well injection include acidic and caustic chemicals, pulping liquors from paper mills, and some radioactive fluids from uranium processing plants. In general, if these materials were disposed of by other methods, water quality effluent standards would be exceeded. In 1975, in the United States, more than 175 deep wells, most of them located on sites adjacent to industrial plants, were used by industry to dispose of about a million barrels of waste daily. In addition, thousands of injection wells are utilized to dispose of oil field brines.

Probably the most significant hazard posed by deep-well injection is that of groundwater contamination. Hence, to protect groundwater quality, injection wells should be drilled into a rock layer that is sufficiently large and permeable to act as an adequate reservoir for the waste. Also, the rock into which waste is injected should be at least 300 meters (1000 feet) below the land surface and separated from fresh groundwater above by impermeable rock, as shown in Figure 13.8. Ideally, a barrier of impermeable rock should also be present to prevent the horizontal seepage of fluids.

While oil field injection wells have caused few problems, industrial injection wells pose serious risks, especially to groundwater quality. Certainly, deep-well injection is not advisable in those regions where a full understanding of the local subsurface geology is lacking.

Ocean Dumping

More than 80 percent of the waste material barged out to sea and dumped consists of spoils generated by the dredging of ship canals in harbors and rivers, as indicated in Figure 13.9. Typically, one-third or more of these spoils are contaminated by industrial and municipal effluents and runoff from farmlands. Most of the remaining waste dumped at sea is industrial waste and sewage sludge.

Sludge from sewage treatment facilities poses a special hazard to ocean life, because it usually contains heavy metals that can concentrate in marine food chains (see Chapters 7 and 8). Also, sewage

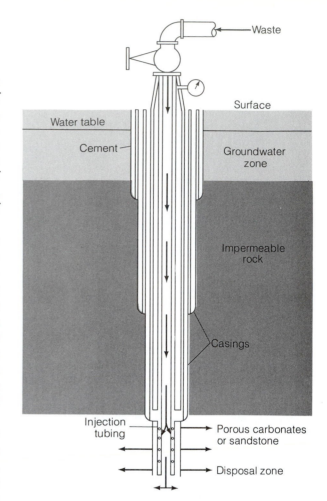

Figure 13.8 A cross-section of a deep injection well. (After "Disposing of Liquid Wastes Underground," *Environmental Science and Technology* 9:24, 1975. Copyright by the American Chemical Society.)

sludge usually contains pathogenic microorganisms. The industrial wastes dumped in the oceans encompass a wide variety of substances, many of which may be toxic to sea life, for instance, arsenic compounds, strong acid and alkaline solutions, and PCBs. Often, these hazardous materials are dumped at sea because water quality regulations prohibit them from being dumped in surface waters, and because no appropriate land disposal site is available

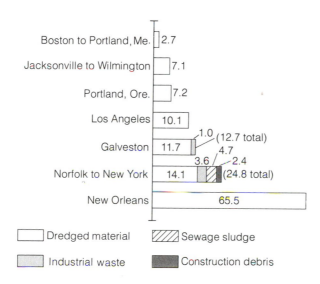

Figure 13.9 Waste disposal in the oceans by source. Numbers indicate percent of the total waste dumped at sea. (After Council on Environmental Quality, *Seventh Annual Report,* 1976.)

to receive them. Concern over the dumping of such wastes into the ocean or onto the seabed focuses not only on toxicity to organisms, but also on the possibility of food chain accumulation. Also, because mixing processes in the oceans are poorly understood, there is a possibility that ocean dumping that seems safe enough at the present time will cause unsuspected problems in the future.

With the enactment of the Marine Protection, Research, and Sanctuaries Act of 1972, industries were required to obtain an EPA permit to transport wastes on and dump them in the ocean. This law also empowered the EPA (and the U.S. Army Corps of Engineers, in the case of dredge spoils) to locate proper ocean disposal sites (beyond the continental shelf, where possible) and to regulate the types and amounts of substances disposed of. Thus, toxic industrial wastes must be dumped several hundred kilometers offshore. In the past, most ocean dumping sites were situated within 40 kilometers (25 miles) of shore. At these locations, waste could be transported back to shore by ocean currents where it could adversely affect commercial shellfish beds and force the closing of beaches. The long-term objective of the 1972 law is the eventual phasing out of all ocean dumping.

In January 1977, the EPA called for an end to all sludge and industrial waste disposal in the Atlantic Ocean (site of 90 percent of the nation's ocean dumping) by December 31, 1981. This regulation is forcing municipalities along the Atlantic seaboard to seek on-land alternatives. In some cities, however, compliance has been less than enthusiastic. In the spring of 1978, the EPA levied a $225,000 fine against the city of Philadelphia for violating provisions of its ocean dumping permit. This fine is the largest ever assessed for such an offense. The EPA charged Philadelphia with dumping more than its monthly quota of sludge and failing to develop a land-based dumping alternative.

Though restrictions on traditional methods of ocean disposal—that is, direct dumping from barges—are growing stricter other techniques of ocean dumping are being considered that will minimize impact on marine life while still allowing the sea to be used as an alternative to on-land disposal. One scheme is the emplacement of submarine pipes that will transport treated municipal waste directly from a metropolitan area to the edge of the continental shelf. Another strategy, already realized, consists of outfitting ships with incinerators, loading them with combustible materials, and sending them offshore to burn their cargo. In the late summer of 1977, this method was utilized to dispose of Agent Orange, a mixture of herbicides used as a defoliant during the Vietnam War. More than 4 million liters of this substance were taken from storage and put aboard the Dutch-owned incinerator ship *Vulcanus*. With EPA approval, Agent Orange was incinerated in the Pacific Ocean about 1600 kilometers (1000 miles) west of Hawaii.

We must recognize that while ocean dumping may be economically attractive over the short term, unquantifiable future risks remain. Government control of the allowable levels of toxic wastes currently being dumped are a step in the right direction. And limiting dump sites where feasible to deeper waters off the continental shelf is important for reducing potential future problems. But in order to determine whether or not ocean dumping is a viable long-term waste disposal alternative, we must encourage more research on transport, cycling, and the ultimate fate of pollutants in the oceans.

Summary

We have seen in this section that drawbacks and risks are associated with each of the conventional waste disposal methods. Risks are especially great when the wastes are toxic or hazardous. But certain measures can reduce these risks to some extent. The pooling of certain industrial wastes can minimize danger, for example, as when acidic wastes from an ore industry are neutralized by the alkaline wastes of another industry. Also, certain chemical treatment procedures can be used to detoxify certain waste products prior to disposal. These alternatives are expensive but they may prove to be economically feasible in view of the potential costs of unsafe toxic and hazardous waste disposal.

Solid Waste as a Resource

Provisions of the 1976 Resource Conservation and Recovery Act recognize two ways of alleviating the problems associated with solid waste: reducing the waste potential of products before those products reach the consumer, and recovering resources from solid waste, that is, recycling. Both strategies ultimately reduce the amount of land needed for disposal. Waste reduction lowers collection and disposal costs and lessens demands on raw materials and energy. Recycling, in effect, converts a community liability into a community asset by adding to the supply of available materials and energy.

Waste Reduction

One *waste reduction* strategy is to encourage the manufacture of reusable rather than disposable products. This approach is exemplified by the returnable beverage container. Following Oregon's lead in 1972, five other states—Vermont, Maine, Michigan, Connecticut, and Iowa—have passed "bottle laws" that impose a deposit on beverage containers. Deposits are refunded when containers are returned for subsequent reuse. Such legislation is aimed at stemming the proliferation of one-time-use containers, and encouraging the circulation of recyclable ones. Oregon's law, in an attempt to address related issues of unnecessary gimmicks in packaging and unsightly litter, also prohibits the distribution of flip-top or pull-tab cans and bottles.

One advantage of returnable beverage container laws is the consequent reduction in litter. For example, the Vermont State Highway Department reports that in the year following enactment of the state's bottle law, the beverage container fraction of highway litter dropped by two-thirds. But bottle laws do have some minor drawbacks: today, Oregon beer drinkers have a choice of only nine nonlocal beers as compared with the twenty-nine available prior to passage of the law.

Another waste reduction strategy is to redesign products so that they require less material per unit. With respect to automobiles, this approach has a double advantage, since smaller and lighter cars not only require less metal in production, but also use less fuel in their operation. Spurred by foreign competition, United States auto manufacturers have trimmed the length of new full-size models and installed smaller engines and fuel tanks. The newspaper industry is moving in the same direction by reducing its use of newsprint. Nationwide, newspapers are changing their traditional 8-column news format to a 6-column one. Consequently, an 88-page newspaper in the old format is equivalent to an 84-page newspaper in the new format. The net annual savings in newsprint is about 5 percent.

Current trends in the packaging of products in the United States results in the consumption of prodigious amounts of raw materials and the creation of tremendous amounts of waste. Packaging of consumer goods uses 75 percent of our glass, 40 percent of our paper, 29 percent of our plastic, 14 percent of our aluminum, and 8 percent of our steel. Typically, 30 to 40 percent of a city's solid waste consists of discarded packaging materials. One strategy for reducing this type of waste is to develop packaging methods that conserve raw materials. For example, the Campbell Soup Company has developed a can that uses 30 percent less materials than the traditional can and results in a 36 percent savings in manufacturing costs.

Directly related to the problem of packaging wastes is that of planned obsolescence. This concept has governed the operations of many important industries, particularly since World War II. It is typified by the automobile and fashion industries, which encourage consumers to purchase new products (and dispose of old products) by changing styles yearly. We could counter the effects of this practice to some extent by developing more durable products that are easily and economically repaired. But probably the most fundamental waste reduction strategy is for each of us to cut wasteful consumption and strive to maximize use efficiency.

Recycling

Reclamation as a recycling method is the separating out of materials such as rubber, glass, paper, and scrap metal from refuse and reprocessing them for reuse. In the United States, waste reclamation is a multibillion-dollar-a-year business, based primarily on the recovery of scrap metal from the more than 8 million motor vehicles junked each year. Also, as we saw in Chapter 12, a significant fraction of the lead, copper, and zinc used by industry is derived from the recycling of these metals. With the new awareness among the general population of the limited nature of our planet's resources, other metallic wastes are beginning to be reclaimed. The recycling of aluminum, for example, is currently proceeding briskly. In 1976, about 25 percent of all aluminum cans were recovered by the manufacturers. And in 1977, more than thirteen-hundred aluminum can recovery cen-

Figure 13.10 An industry-sponsored recycling center. (Reynolds Aluminum, Reynolds Metals Company.)

ters, one of which is shown in Figure 13.10, were operating nationwide. Reynolds Aluminum even provides a toll-free telephone number that callers can use to locate their nearest aluminum collection center. Recycling aluminum requires less than 5 percent of the energy needed to produce aluminum from ore.

Reclaiming metals not only reduces the energy needed for metal production, but also reduces the environmental impact of processing techniques. For example, the use of scrap iron by steel mills and foundries cuts air pollution by 86 percent, water pollution by 76 percent, and water use by 40 percent as compared with the manufacture of new steel.

The reclamation of nonmetallic wastes, examples of which are listed in Table 13.2 along with their recovery percentages, is generally hampered by apathy, scarcity of recycling centers, and a lack of economic incentives. In fact, in 1977, the EPA reported that less than 7 percent of the total nonmetallic waste generated is recovered and that most of this consists of waste paper. Of the waste paper that is recycled, most ends up in paperboard; much smaller amounts go into the production of newsprint and other writing and printing paper. Thus, for example, recycled glass accounts for less than 3 percent of the total annual glass production, and only a negligible quantity of plastics is reclaimed. In order for glass to be recycled, it must be cleaned, separated by color, crushed, and transported back to glass manufacturing plants. Recycling is generally perceived as too much bother by consumers. And for the processor, low market values do not warrant the time and money necessary for sorting and cleaning. Finally, the marketing of materials recovered from solid waste is very sensitive to fluctuations in the national economy. Thus, during the recession that began in 1974, waste paper reuse plummeted to its lowest level in a decade. In 1976, as the economic climate improved, the demand for recycled paper began to recover.

Reclamation is not limited to the recycling of materials for reuse in production. Some refuse is reclaimed directly as a fuel, termed *refuse-derived fuel* (*RDF*). When municipal waste is shredded or milled, combustibles are separated from noncombustibles, and the combustible fraction is then incinerated or burned with coal in mixtures of 5–25 percent to generate steam for space heating. Currently, only a dozen small-scale facilities employ RDF in the United States. The major obstacles to using RDF are that communities are reluctant to invest the capital necessary for recovery facilities and reliable markets for RDF are difficult to identify.

Another method of recycling is *conversion*, that is, the chemical or biochemical transformation of waste material into useful products, such as fertilizers, fuels, and a variety of industrial chemicals. In one

Table 13.2 Residential and commercial solid waste generation and resource recovery in 1973 (in millions of metric tons).

Material	Gross discards	Recycled materials		Net waste disposal	
		Quantity	Percent	Quantity	Percent
Paper	48.1	7.9	16.5	40.2	32.8
Glass	12.2	0.3	2.1	11.9	9.9
Metals	11.5	0.2	1.6	11.3	9.3
Plastics	4.5	0.0	0.0	4.5	3.7
Rubber	2.5	0.2	7.1	2.3	1.9
Leather	0.9	0.0	0.0	0.9	0.7
Textiles	1.7	0.0	0.0	1.7	1.4
Wood	4.4	0.0	0.0	4.4	3.6
Food waste	20.3	0.0	0.0	20.3	16.6
Yard waste	22.7	0.0	0.0	22.7	18.5
Miscellaneous	1.7	0.0	0.0	1.7	1.4
Total waste	130.5	8.6	6.5	121.9	100.0

Source: After Council on Environmental Quality, *Sixth Annual Report*, 1975.

such process, *pyrolysis,* organic wastes are broken down by exposure to intense heat in the absence of oxygen. The yield of this conversion technique depends on the composition of the refuse, but normally includes char (similar to charcoal), alcohol, light oils, and combustible gases. All these substances are potential fuels. This process can convert nonbiodegradable refuse (plastics and rubber) and emits no air pollutants. Commercial pyrolysis facilities are currently in the developmental stage.

Composting is an example of a conversion process known as bioconversion. It is a technique whereby aerobic bacteria break down organic waste—for example, leaves, grass clippings, and newspaper—under controlled conditions into a humuslike substance that may be used as a soil conditioner or fertilizer when supplemented with nutrients. The latter may be supplied by sewage sludge, for example. Unfortunately, commercial inorganic fertilizers are more economical than compost; hence, all but a few of the commercial composting plants in operation during the past twenty-five years have closed. In the United States, the principal compost heaps are those maintained by home gardeners.

In addition to composting, a variety of other bioconversion techniques involve the aerobic or anaerobic treatment of organic waste. When these techniques are applied to agricultural waste or the organic fraction of municipal refuse, they yield such fuels as alcohol and methane. But these methods are at present still in the small-scale testing phase. The most common methods of disposing of agricultural wastes are the plowing under of crop residues and animal waste and the open burning of plant waste. An increase in the extraction of fuel and other raw materials from agricultural and other organic wastes, such as sewage sludge, awaits a more favorable economic climate. Ironically, however, research on these methods is being spurred on by the continued deterioration of the energy supply.

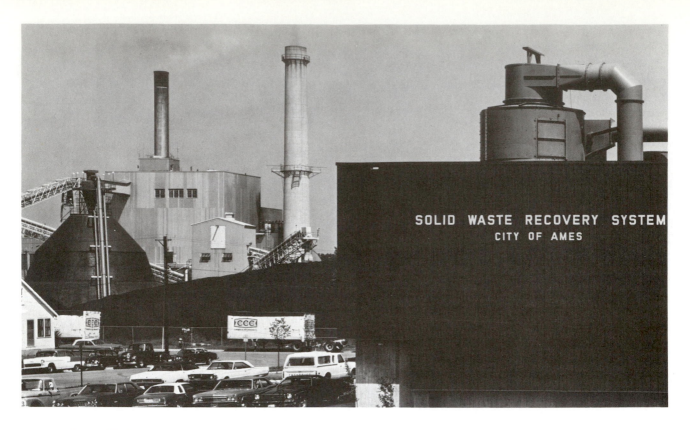

Figure 13.11 A resource recovery facility in Ames, Iowa. (Henningson, Durham & Richardson, Omaha, Nebraska.)

Resource Recovery Facilities

The ideal approach to solid waste recycling is the development of centralized *resource recovery facilities* that combine reclamation and conversion processes. We cited an example of such a system at the beginning of this chapter, the successful venture at Saugus, Massachusetts. Another is the nation's first municipally owned resource recovery facility, shown in Figure 13.11, which has been operating in Ames, Iowa, since late 1975. This facility receives solid waste from the city of Ames, Story County, Iowa State University, and several other nearby communities. At the plant, solid waste is shredded and sorted, metals are recovered, and combustibles are separated from noncombustibles. The combustible fraction is then burned with coal to fuel the city's nearby power plant. Though the Ames facility is located in the city's center several blocks from the business district, litter, odor, and air pollution are successfully controlled. For a detailed discussion of the operation of a larger resource recovery facility, see Box 13.2.

In the spring of 1978, the National Center for Resource Recovery reported that fewer than thirty resource recovery facilities were in the start-up phase or fully operational nationwide, and that most of these were intended primarily to recover fuel. Many of these operations are still pilot or demonstration plants. But the number of resource recovery facilities is expected to grow substantially in the future. Although resource recovery plants will probably never contribute more than 1 percent of the nation's total energy supply, they may constitute important local sources of heat and power in large cities.

The principal obstacles to the immediate deploy-

Box 13.2

Resource Recovery in Milwaukee

Milwaukee, Wisconsin, is the largest city in the nation to participate in a comprehensive resource recovery program. Refuse collected by the city is trucked to a downtown resource recovery facility operated by the Americology division of the American Can Company. Built at a cost of $18 million and opened in the spring of 1977, the Americology plant is reducing the city's need for landfill disposal by 80 percent and adding substantially to local supplies of energy and materials.

Refuse dumped by sanitation trucks is pushed by front-end loaders onto conveyor belts. Trash is then inspected, and hazardous materials and bundled paper are removed from the conveyor. Refuse is delivered to a massive shredder that hammers it into pieces about the size of a half dollar. By various separation techniques (see figure), iron, steel, aluminum, glass, and the remaining combustibles are recovered from the refuse stream. Residual organic matter is hauled to a sanitary landfill.

Bundled newspaper and cardboard is baled and sold to paper mills for reprocessing into pulp. The rest of the combustible refuse, representing about 60 percent of the total weight, is compacted and shipped to a nearby coal-fired power plant, where it is burned as a supplemental fuel. Recovered iron, steel, and aluminum are sold to mills and foundries for reuse. Pulverized glass and ceramics are cleaned and mixed with asphalt, forming an aggregate that is strong and durable enough to be used by the city of Milwaukee as a road paving material.

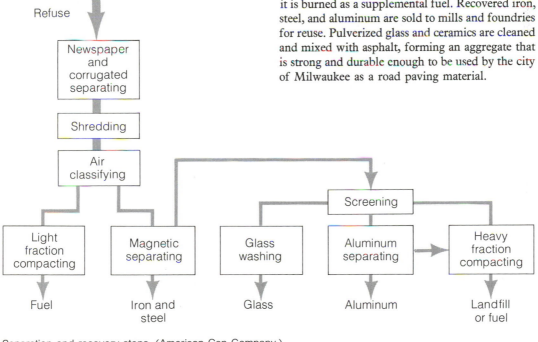

Separation and recovery steps. (American Can Company.)

ment of large numbers of resource recovery centers are technical, economic, and social. In most cases, recovery facilities employ new technologies that require more research and development before they can be applied. In some instances, major mechanical problems have delayed start-up. Technological problems are compounded by typically high capital requirements and operating expenses. For example, inflation, design changes, and the need for additional pollution abatement devices nearly tripled projected costs for a resource recovery facility in San Diego County, California.

In only a few localities in the country is resource recovery more favorable economically than landfill disposal. This situation is expected to reverse, however, as the cost of both energy and potential landfill sites continues to climb. If improvements are made in technology in the next five years, by the mid-1980s, the economic climate for resource recovery facilities should be much brighter than it is today.

Still, even if technological and economic progress is made in the near future, the establishment of resource recovery centers may be stymied by public opposition. For these facilities to be successful, the public and private sectors of the community must cooperate. But many people vehemently oppose the siting of resource recovery facilities in or near their neighborhoods, even though such plants can be designed to blend unobtrusively into their surroundings, as evidenced by the Ames center.

Conclusions

Concerns over human health risks and environmental quality are triggering improved methods of waste disposal. New strict federal legislation is phasing out open dumps and regulating the disposal of toxic and hazardous waste. But appropriate sites for landfill disposal are becoming rare, land prices are rising, and antipollution measures are growing more expensive. Hence, it appears that we must cut back our per capita waste generation. Beverage deposit laws and the redesign of products to reduce resource use are steps in this direction, but other strategies are needed. Also, because our material and energy supplies are dwindling, increased recycling of reusable components of solid waste is beginning to appear necessary.

Summary Statements

Our activities generate a wide variety of municipal-industrial, mining, and agricultural wastes.

Stiffer air and water quality regulations are forcing increased reliance upon land sites for waste disposal.

Alternatives to open dumping on land include sanitary landfilling, incineration, and deep-well injection. But, unless properly regulated, these measures may also prove hazardous to health and environmental quality.

Trench landfills are appropriate in regions where soil is deep and the water table is low. Area landfills take advantage of abandoned quarries and natural depressions such as canyons for waste disposal.

Incineration offers many advantages over sanitary landfilling, especially where land prices are soaring. But the cost of air quality control devices has forced the closing of most of the nation's municipal incinerators.

Deep-well injection is the pumping of fluid wastes into pore spaces and fractures within deep rock layers. Unless properly designed and constructed, deep-well injection may contaminate groundwater. Also, there is a danger of a well blow-out if the storage capacity of subsurface rock is exceeded.

More than 80 percent of the waste dumped at sea consists of dredge spoils and, typically, a third or more of these spoils is contaminated.

Waste reduction strategies include the use of reusable products in place of disposable items, the redesign of products so that they require less material per unit, the regulation of product packaging, and the development of more durable products.

Recycling reduces the volume of waste that must be disposed of and adds to our supply of energy and other resources. Recycling methods include reclamation and conversion processes (for example, pyrolysis and composting). The recycling of nonmetallic waste is proceeding slowly, however, because of public apathy, a scarcity of recycling centers, and a lack of economic incentives.

Resource recovery facilities, combining reclamation and conversion, are stymied to date by technical problems and high capital and operating expenses.

In view of strict air and water pollution control regulations, rising costs of landfilling, and shortages in resources and energy, it appears prudent that we cut back our per capita waste generation and optimize recycling efforts.

Questions and Projects

1. In your own words, write a definition of the terms italicized in this chapter. Compare your definitions with those in the text.

2. What is the ultimate fate of your garbage and other refuse? If it is collected by public or private haulers, where is it taken? If your trash is incinerated, what precautions are taken to prevent air pollution? If refuse is disposed of on land, you may wish to visit the site and observe the operation. If your community operates a sanitary landfill, find out what measures are taken to protect groundwater quality. Your city's sanitation department should be able to furnish such information.

3. Is any reclamation activity going on in your community? You may wish to consult with local industries to assess their waste recycling efforts. Did your community

participate in the popular recycling campaigns of the late 1960s and early 1970s? If these efforts were terminated, find out why.

4. Have conflicts occurred in your community regarding the siting of landfills or resource recovery centers? If so, what factors contributed to these conflicts?

5. How would you assess your community's solid waste management strategies? Is progress evident or are conflicts interfering? What can be done to help resolve difficulties? How are you contributing?

6. Should metals be given priority over glass and paper in recycling efforts? Support your opinion.

7. What is the role of planned obsolescence in our nation's economy? Is planned obsolescence compatible with efforts to reduce our output of waste?

8. Which one of the following recycling practices is most attractive economically: recovery of resources from (a) industrial waste heaps, (b) municipal waste, or (c) sanitary landfills? Support your choice.

9. Suggest several reuses in the home for each of the following common household items: (a) newspapers, (b) metal cans, and (c) nonreturnable glass bottles.

10. How does the proper management of solid waste reduce our air pollution problem?

11. Identify some incentives that would encourage public acceptance of sanitary landfills and resource recovery centers in or near residential neighborhoods. Identify some incentives that would reduce per capita waste generation.

12. Enumerate the advantages and drawbacks of sanitary landfilling.

13. Distinguish between area and trench landfills. Under what conditions are trench landfills inappropriate? What is the alternative, if any, in such circumstances?

14. Why have most conventional municipal incinerators been closed in recent years? What advantages do incinerators have over sanitary landfills?

15. What precautions must be taken during deep-well injection to protect groundwater quality?

16. How is ocean dumping regulated? Identify some proposed alternatives to direct dumping from barges.

17. Enumerate the advantages of measures that reduce the waste potential of products before those products reach consumers.

18. Suggest some possible major drawbacks to "bottle laws."

19. Describe several waste conversion techniques. What is the principal obstacle to their widespread use?

20. Why have most commercial composting facilities closed down in the past decade? What factors could alter this situation in the future?

Selected Readings

American Chemical Society. 1978. *Cleaning Our Environment—A Chemical Perspective,* 2nd ed. Washington, D.C.: American Chemical Society. Discusses in detail the problems of and alternatives to solid waste disposal. Ocean dumping is also discussed. An excellent overview of problems.

Cargo, D. N., and B. F. Mallory. 1977. *Man and His Geologic Environment.* Reading, Massachusetts: Addison-Wesley. Includes a chapter on the geologic aspects of waste disposal.

Council on Environmental Quality. *Annual Reports, 1970–1978.* Washington, D.C.: U.S. Government Printing Office. Includes yearly summary of progress and major issues involving solid waste management.

Hayes, D. 1978. *Repairs, Reuse, Recycling—First Steps Toward a Sustainable Society.* Washington, D.C.: Worldwatch Institute. Identifies the major issues involved in resource recovery and conservation.

Hughes, G. M., et al. 1971. "Summary of Findings on Solid Waste Disposal Sites in Northeastern Illinois," *Environmental Geology Notes,* No. 45. Urbana: Illinois State Geological Survey. Presents guidelines for evaluating the pollution potential of existing and proposed sanitary landfill sites.

Keller, E. A. 1976. *Environmental Geology.* Columbus, Ohio: Charles E. Merrill. Includes a chapter on waste disposal alternatives from a geological perspective.

Maugh III, T. H. 1979. "Toxic Waste Disposal: A Growing Problem," *Science 204:*819–823 (May 25). Discusses the many factors involved in disposal of hazardous waste.

U.S. Environmental Protection Agency. 1977. *Resource Recovery and Waste Reduction, Fourth Report to Congress.* Washington, D.C.: U.S. Environmental Protection Agency. Reviews the status of waste management strategies in the United States.

The bald eagle, our national symbol and a major symbol in the battle to save endangered wildlife. (Gordon S. Smith, from National Audubon Society.)

Chapter 14

Managing Endangered Species

The enormous flocks of passenger pigeons—sometimes more than 2 billion strong—that once darkened the skies are gone forever. The last passenger pigeon, an aging female named Martha, died in the Cincinnati zoo in 1914. It was the savory flesh and gregarious nature of these birds that led to their demise; they were good to eat and easy to kill. With the advent of railroads, which could transport dead birds before they spoiled to major markets, thoughtless commercial harvests of the pigeons began. In the 1840s, several thousand people made their livings by hunting these birds—shooting them as they flew, as illustrated in Figure 14.1, or clubbing them to death while on their nests. In Michigan alone, more than 1 billion birds were killed in a single year. In only a few decades, the highly prized species was finished. It had become *extinct*.

Endangered species are those species considered to be in immediate danger of extinction. Today, many species of plants and animals throughout the world fall into this category. In 1979, the United States Fish and Wildlife Service listed 440 foreign species of animals as endangered, including the Siberian tiger, the mountain gorilla, the Asian elephant, the white rhinocerous, the orangutan, and the blue whale. The list of endangered species at home for that year consisted of 177 species of mammals, birds, reptiles, amphibians, fish, shellfish, and insects. The domestic list included the gray wolf, the black-footed ferret, the peregrine falcon (Figure 14.2), the bald eagle, the whooping crane, and the American crocodile. Population levels for these animals are precariously low. For example, only 60–70 Siberian tigers remain in the wild, while only 10–20 breeding female crocodiles linger on in Florida's Everglades National Park.

The Fish and Wildlife Service issues figures for what are called *threatened species* as well as for endangered species. Although still abundant in parts of their territorial range, these species are classified as threatened because their numbers have declined significantly. The service classifies 39 domestic wildlife species and 18 foreign species as threatened. These listings include the grizzly bear, the chimpanzee, and the African elephant.

The public has shown a good deal of concern over the plight of many endangered animal species, but the fact that many plant species are under a similar

Figure 14.1 Passenger pigeons being hunted for market. (American Museum of Natural History.)

threat is less well known. In 1977, the U.S. Fish and Wildlife Service proposed to list approximately 1850 plant species as either endangered or threatened in the United States. These species include the beavertail cactus, century plant, desert poppy, Tennessee cornflower, and fifteen species of flowering hibiscus. By early 1979, twenty plant species from this list had been officially deemed endangered or threatened. Four endangered plant species are shown in Figure 14.3. No worldwide listing is currently available, but 10 percent of the world's estimated 250,000 species of flowering plants are thought to be endangered.

Although considerable effort has been made in recent years to protect endangered species, many people remain concerned about the future prospects of these plants and animals. Growing human needs for food and space continually threaten the survival of our fellow passengers on planet earth. In this chapter, we explore the many reasons why species are endangered, and identify and evaluate our efforts to save them.

Why Preserve Them?

Why should we be concerned about the loss of a few species? After all, we seem to be getting along fine without the passenger pigeon, the heath hen, and the eastern elk—all now extinct. Citing persuasive reasons for preservation is more difficult for living

resources than for the other resources we have discussed. Nevertheless, many reasons for species preservation exist, not the least of which are economic. For example, billions of dollars are generated by tourism and recreation, and generally the most popular tourist attractions are not human-made marvels, but national and state parks and forests. Sport hunters and fishermen, not to mention their commercial counterparts, spend millions of dollars in pursuit of their quarry, and have a deeply vested interest in the preservation of their target species. These economic factors may not represent the most significant reasons for species preservation, but they often hold the most sway among people who need to be convinced that preservation is worthwhile.

Figure 14.2 Examples of animals on the domestic endangered-species list. *Above left:* A gray wolf. (© Leonard Lee Rue III, Photo Researchers.) *Top:* A black-footed ferret. (U.S. Department of the Interior, Fish and Wildlife Service.) *Bottom:* A peregrine falcon. (U.S. Department of the Interior, Fish and Wildlife Service.)

Figure 14.3 Examples of plants on the domestic endangered-species list. *Above left:* Texas wildrice. (W. A. Silveus.) *Above:* Persistent trillium. (J. D. Freeman.) *Far left:* Santa Barbara live-forever. (Reid Moran.) *Left:* Furbish lousewort. (Photo by Douglas Gruenau from *Garden* magazine, published by The New York Botanical Garden.)

Another argument has more relevance to the human race as a whole. Plants and animals can possess undiscovered or undeveloped values that are irreplacable and that have survival value for our own species. As we suggested in Chapter 3, every living thing is a unique combination of genetic characteris-tics that enable it to adapt to given environmental conditions. The sum total of all the genes of the individuals in a population is termed the *gene pool*. Scientists today are concerned about preserving gene pools in order to preserve the genetically determined characteristics that might prove useful to us for our own survival. For example, a particular fungus produces the lifesaving drug penicillin, and certain species of bacteria produce the important antibiotics streptomycin and tetracycline. Certain flowering plants produce such drugs as morphine (from the opium poppy plant), cocaine (from the coca tree), and quinine (from the cinchona tree). Today, drugs that have been extracted from plants are used to treat leukemia, several forms of tumorous cancer, heart ailments, and hypertension. The discovery of other life-saving drugs depends on the existence of specific plants that some people might consider insignificant and not worth saving. In this context, it is of interest to note that fewer than 2 percent of the world's 250,000 species of flowering plants have been screened for the presence of valuable drugs.

The medicinal value of a seemingly dispensable species is illustrated by the potential role of the nine-banded armadillo in the eventual control of the dreaded disease leprosy. This disease has been a curse to humankind since ancient times, and a cure has been difficult to find because the bacterium that causes the disease grows only in humans and not in laboratory animals. Hence, investigating the nature of the infection and developing an effective vaccine has been nearly impossible. In 1971, however, biochemists discovered that the bacterium flourishes in the nine-banded armadillo, shown in Figure 14.4. It is unlikely that many people would have cared if the lowly armadillo had become extinct as a result of habitat destruction or some other human activity. But because this species is still with us, scientists now have the opportunity to study and perhaps someday to conquer leprosy.

The maintenance of a large gene pool is also of great interest to agriculturalists. All domestic crops and livestock had their origins in native plants and animals, and native species are still needed to provide the new genetic characteristics that will help solve present and future food production problems. For example, the new varieties of wheat and rice that have resulted in increased food production in many countries (see the discussion of the Green Revolu-

Figure 14.4 The nine-banded armadillo, an animal that may be of great importance to research into a cure for leprosy. (Jen and Des Bartlett, Photo Researchers.)

tion, Chapter 17) are products of breeding experiments that utilized thousands of native and domesticated varieties of rice and wheat. Because pests, disease-causing organisms, and technology tend to evolve over time, many presently unknown challenges are bound to present themselves for which we need to be prepared: no one knows, for instance, when a new, perhaps lethal, plant disease (such as the southern corn leaf blight, discussed in Chapter 2) or livestock disease may appear. Also, as agriculture continues to expand into arid regions, more incidents of soil salinity resulting from irrigation are likely to occur, making necessary the development of new crop varieties with a higher salt tolerance. And this trend in expansion, along with the decrease in rainfall being experienced in certain areas, already necessitates the development of new drought-resistant crop varieties. Despite these needs, however, agricultural gene pools are being swept away in the wake of human activities. For example, fifty years ago, 80 percent of the wheat grown in Greece consisted of native strains that were well-adapted to the local environment and resistant to local pests. Today, more than 95 percent of the old breeds have virtually disappeared.

Industry, too, utilizes plant and animal products, and benefits from a rich diversity of species. In 1975, researchers reported that the oil of the jojoba bean could be used by industries requiring a fine grade of oil that was stable under high pressures and temperatures. In just a few years, plantations of this native American desert shrub were started in Arizona, California, and Mexico. Interestingly, because the oil of the jojoba bean is similar to that of the sperm whale, this new source of high-quality industrial oil may save the endangered whale species from extinction.

In addition to its value as a genetic reservoir, each species serves an important function in moving energy and materials within and between ecosystems; hence, each in its way contributes to ecosystem stability. Some species, of course, are more important than others in this respect, and thus their significance is more obvious to us. For example, the alligator is a dominant member of semitropical wetland ecosystems such as the Florida Everglades. Depressions excavated by alligators (gator holes) hold fresh water longest during the relatively dry winters. Hence, these holes are reservoirs that not only preserve aquatic life, but also provide fresh water and food for birds and mammals. If alligators were eliminated, gator holes would quickly fill with sediment and aquatic plants, and wildlife would not be able to depend on these water sources for survival during dry periods. Alligators also affect the stability of the Everglades' ecosystem by their feeding habits: they eat large numbers of gar, a major predatory fish, consequently allowing game fish such as bass and bream to thrive.

Although few species are as important to an ecosystem as the alligator, it is nonetheless true that as more species disappear, diversity diminishes and the number of checks and balances on plant and animal populations decreases. As species are lost, the stabilizing influences of predation, parasitism, and competition are disrupted, and an ecosystem thus becomes more vulnerable to disturbances and in some cases to potential demise (see Chapter 4). Although we usually mourn the passing of an animal species

much more than the loss of a plant species, the extinction of a plant species is often more critical to ecosystem stability. Because plants occupy the base of food webs, a disappearing plant species may take with it 10–20 species of animals that were dependent on it for food or shelter at some time during their lifespans.

Although all arguments for species preservation cited in this section have merit, they also have shortcomings. We actually know very little about the role of most species in the functioning of ecosystems, let alone their potential value to medicine, agriculture, or industry. As far as we can ascertain, few species play all the roles enumerated here, and the value of many species may be only in the minor role they play in ecosystem functions. Without documentation, speculation on the significance of a specific species may only result in a loss of credibility. After all, sixty-two species have become extinct in the United States since the 1600s without any noticable impact on our society. Furthermore, it is easy for us to ignore the possible consequences of species extinction, particularly since they are so nebulous from our point of view. Many people do not even worry about the increased risk of contracting cancer from smoking, even though there is ample evidence that a cause-effect relationship exists. Why then would people be concerned if some obscure species became extinct because we built a dam or highway? Although, as we noted, one can cite long-term economic reasons for species preservation, many people remain unconvinced in view of short-term economic gains.

Some proponents of species preservation are guided by aesthetic rather than economic values. The thrill of watching an eagle soar silently overhead, an afternoon spent listening to the capricious chatter of chipmunks, the taste of wild hickory nuts, the refreshing fragrance of wildflowers, and the softness of a bed of moss: these experiences have no monetary value because they have no equal. Still, those who base their arguments on aesthetics must face the fact that very few people believe that all animals are created equal. While many of us are worried about the potential demise of the great blue whale or the bald eagle, few really care about the fate of the Houston toad (shown in Figure 14.5) or the orange-footed pimpleback clam. Furthermore, when people are hungry, poor, and unemployed, aesthetic concerns tend to rank low on their list of priorities; they are concerned about jobs, not species diversity.

The arguments for species preservation presented thus far have one common element—they are all based on what *we* want; they are people-centered. But many people feel that the inherent value of a species cannot be measured properly by the extent to which human beings can get along without it. They argue that each species has its own rights. If a species exists, it has a fundamental right to continue to exist without being driven to extinction by human activities. Moreover, such noted thinkers as René Dubos, Aldo Leopold, John Muir, and many great religious writers state that we have a responsibility toward other organisms with whom we share the planet to be faithful stewards of the earth and its resources.

Should we preserve other species? It is up to each of us to decide. In making our choice, however, we should remember that when a species becomes extinct, a potentially valuable resource is lost forever. Never again will we have access to its unique and complex genetic reservoir of characteristics. And, in allowing a species to die out, we may fail to fulfill a fundamental obligation to preserve other life forms.

A Historical Overview

The process of becoming extinct is not new. Plants and animals have been disappearing from the face of the earth since life began billions of years ago. Furthermore, extinction is the rule rather than the exception. For example, twenty thousand species of vertebrates (animals with backbones) were alive at the end of the Paleozoic era, approximately 230 million years ago (see Appendix II), but only about two dozen of these species have any living descend-

Figure 14.5 The Houston toad—endangered, but not impressive enough to elicit much human concern. (R. A. Thomas.)

ants today. Interestingly, however, about fifty-thousand species are descended from these two dozen ancestral species.

We have evidence that many species became extinct at the end of geologic periods. One period of significant extinction occurred 65 million years ago, when dinosaurs disappeared. Apparently, dinosaurs were unable to adapt to the changing environment as mountain ranges rose and the swamplands the dinosaurs inhabited dried up. Herbaceous dinosaurs could not eat the new species of plants that evolved under new climatic conditions, and as plant-eating dinosaurs died, the dependent carnivorous dinosaurs also perished.

A more recent mass extinction took place in North America at the end of the last Ice Age. This Pleistocene extinction climaxed about eleven thousand years ago, and affected large mammals: horses, camels, mammoths, mastodons, ground sloths, giant beavers, jaguars, and saber-toothed cats. Figure 14.6 illustrates some of the animals that became extinct during this period. Only 30 percent of the species of big game animals then in existence, including moose and caribou, survived this period.

Figure 14.6 A reconstruction of a Pleistocene scene at Rancho La Brea Tar Pit. Included are vultures, saber-toothed tigers, mammoths, giant ground sloths, and dire wolves. (Charles R. Knight, American Museum of Natural History.)

What caused this mass extinction? Was it a major environmental change? Evidence is not conclusive. Climate certainly underwent a drastic change during the period in which the continental ice sheets receded. As the North American ice sheet melted back, a corridor opened along the MacKenzie River between the Canadian Rockies and the retreating ice front, allowing large masses of extremely cold air to flow out of the Arctic and onto the Great Plains, as shown in Figure 14.7. The cold temperatures may have killed the animals or changed the vegetation that served as their food source. However, a significant criticism of this explanation is that major climatic changes had occurred many times before during the numerous advances and retreats of Ice Age glaciers, but with no notable increase in extinctions. Why was the last glacial retreat so different that it resulted in the mass extinction of large animals?

Professor Paul S. Martin of the University of Arizona points out that the time of the extinctions coincided with the arrival of people in North America. He believes that as these early immigrants crossed the Bering land bridge from Siberia into Alaska and then moved south and east across North America, their hunting activities were largely responsible for the obliteration of large animals. The major objection to this explanation is that the human population was not large enough to bring about so massive an extinction over so great an area as North America. Martin counters this objection by suggesting that upon reaching North America, with its abundant wildlife, the immigrants experienced a population explosion and advanced across the continent as a wave, with the greatest density of hunters at the front, as illustrated in Figure 14.8. He describes these primitive hunters as surprisingly efficient at killing animals in large numbers, setting fires to drive large herds over cliffs or into rivers where they drowned. Hence, large grazing animals, could have been overwhelmed, succumbing in great numbers to the concentrated onslaught of the hunters. Other animals, in turn, would have quickly perished because they lacked the ability to switch to other food sources—the saber-tooth cat, for example, which depended on large grazing animals for food.

Which factor—climate or people—was more important in eliminating so many species of animals? We will probably never know for sure. Both climatic

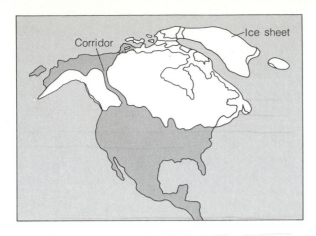

Figure 14.7 The corridor through the continental and mountain glaciers that allowed people from Siberia to enter the southern portion of North America. The lighter area is glaciated and the darker area unglaciated.

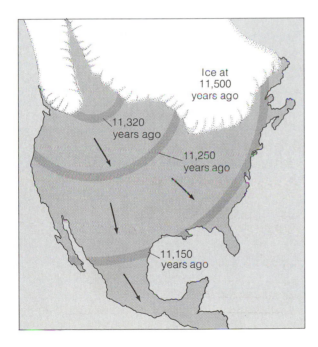

Figure 14.8 The postulated sweep of migrating hunters across North America (represented by the dark bands). As local extinction of large mammals occurred, hunters moved on. (After P. Martin, ''The Discovery of America,'' *Science 179*:969–974. Copyright 1973 by the American Association for the Advancement of Science.)

change and the influx of people probably exerted a stress on wild game populations, but it appears likely that human hunting pressure was the factor that tipped the balance.

Contributing Factors

Relatively few species of plants and animals became extinct during the thousands of years between the Ice Age extinction and the last century or two. But during the last two hundred years, the number of disappearing species rose dramatically, as indicated in Figure 14.9. The total number of extinctions that occur in this century may be ten times greater than that in the seventeenth century. This recent sharp upturn is directly linked to the unprecedented growth of the human population and the increasing stresses created by our exploitation of the earth's resources. We have brought about extinction in several ways: by hunting for commercial products and food, by practicing predator control, by destroying natural habitats, and by introducing alien species. The relative significance of each type of stress varies among species; today, however, the alteration of natural habitat has the greatest adverse impact on native wildlife populations. Figure 14.10 gives the relative signficance of each of the various stresses.

The Destruction of Natural Habitats

Even the most cursory observations indicate that we have grossly modified many ecosystems: complex forest ecosystems have been transformed into farms, where corn and soybeans, hogs and cattle are the dominant species; marshes have been filled in to make way for housing tracts, shopping centers, and industrial parks; and stream channelization, damming, strip mining, and highway construction continue to raise havoc with the remaining natural habitats. Each year, 4750 square kilometers (1900 square miles) of land in the United States are developed intensively.

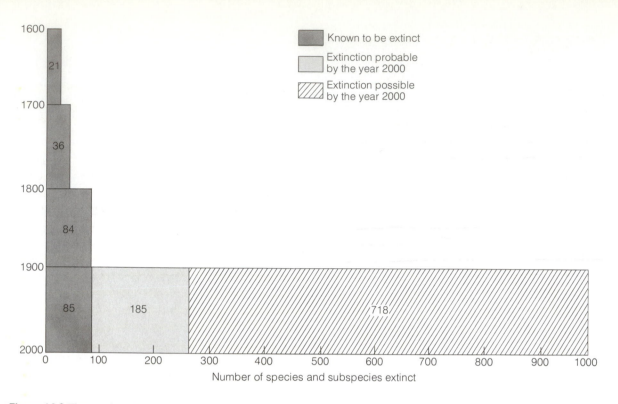

Figure 14.9 The number of extinct species and subspecies of animals by century, with projections for the twentieth century. (After G. Uetz and D. Johnson, ''Breaking the Web,'' *Environment 16*(10):31–39, 1974.)

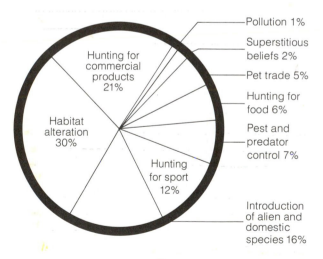

Figure 14.10 The relative significance of human-related pressures for animals presently on the endangered-species list. (After G. Uetz and D. Johnson, ''Breaking the Web,'' *Environment 16*(10):31–39, 1974.)

What are the effects of these massive habitat changes? Each individual has genetically determined tolerance limits, and when we change habitat conditions, we may deprive it of resources essential for its survival. Often, we modify the environment so drastically and quickly that few species are able to adapt to the changing conditions. Thus, when natural habitats for a particular species are destroyed, individuals have fewer places to live, and the very existence of the species is threatened. Frequently, wildlife agencies attempt to relocate animals whose habitat is being destroyed, but such efforts are seldom successful. Many animal species are territorial, and few newly released animals find suitable habitats that are not already occupied and vigorously defended.

Although many human activities, such as lumbering, grazing, and clearing, actually create areas with favorable conditions for successional species (for example, rabbits, deer, elk, and grouse), they also

Figure 14.11 The ivory-billed woodpecker, now thought to be extinct. (James T. Tanner, from National Audubon Society.)

destroy areas for animals that require stable climax conditions (caribou, prairie chickens, wolverines, and ivory-billed woodpeckers). The ivory-billed woodpecker, the largest woodpecker in North America (see Figure 14.11) formerly inhabited lowlands and swamp forests throughout the Southeast. It feeds on wood-boring insects that invade large, dead trees such as decaying cypresses. Each pair of woodpeckers needs about 7.5 square kilometers (3 square miles) of swamp forests to meet its food requirements. Over the years, virgin forests have been cut down and replaced by plantations of uniform-age trees, and as a consequence, the ivory-bill's food source has diminished. Although a few individuals may still survive in remote swamp forests of Louisiana, no recent sightings have been reported.

Kirtland's warbler is another bird with exacting habitat requirements. Approximately two hundred breeding pairs of this bird exist today, all confined to an area of jack pines in central Michigan. The warbler uses the lower branches of the jack pine to conceal the entrance to the nest it builds on the ground. When jack pines reach a height of 2–6 meters (6–18 feet), their lower branches die. Since trees at this height no longer provide cover, the birds refrain from building their nests under them, and thus young jack pines are in continual demand. In former years, when forest fires were common in Michigan, jack pine seedlings often grew in burned-over areas, thereby providing adequate nesting sites for the warblers. But the advent of forest fire prevention and control has resulted in a decrease in the number of burned-over sites. Therefore, the number of nesting sites, and in turn the number of breeding pairs of Kirtland's warbler, has declined.

Damming, draining, and filling has destroyed many wetland areas, which, in turn, has resulted in wildlife losses. In southern Florida, for example, an elaborate system of canals and dikes has transformed wetlands into land suitable for agriculture and commercial development (see Figure 14.12). Formerly, wetland peat soils soaked up the heavy summer rains and slowly released the water to downstream areas during relatively dry winter months. But now "excess" rainwater is diverted to the ocean during summer to prevent flooding. Furthermore, considerable amounts of fresh water are diverted yearround to southern Florida's highly developed east coast for commercial usage. Hence, during particularly dry winters the Everglades National Park and contiguous wetlands have dried up and sometimes have been burned over by intense fires that burn the peaty soil down to bedrock. Reduced water flow to the "Glades" region has dramatically reduced populations of colonial wading birds—in 1935 the census showed 1,500,000 birds; in 1970, 150,000—and has been a contributing factor in the classification of twenty animals in the region as endangered species.

While drainage can destroy wetlands, irrigation can disrupt dryland habitats. For example, the endangered San Joaquin kit fox population now numbers less than three thousand. In 1960, the kit fox roamed 1.2 million hectares (3 million acres) of the dry San Joaquin valley. Today, as a result of irrigation, the valley's desert land has been reduced to less

Figure 14.12 A section of the elaborate system of canals and dikes in southern Florida that provides flood control and water for commercial use. (South Florida Water Management District, West Palm Beach, Florida.)

than a million acres, and the number of kit foxes is dangerously low.

The pollution of natural habitats by oil spills, industrial wastes, and excess organic wastes (see Chapter 7) has also taken its toll on wildlife. Probably the best known examples of pollution-induced losses are those instances in which insecticides, particularly chlorinated hydrocarbons such as DDT, have affected population levels of large birds of prey. Three species, the bald eagle, the peregrine falcon, and the brown pelican, have been affected to the point of becoming endangered species.

In summary, habitat destruction through pollution and changes in land use, particularly the destruction of climax vegetation, is the most important factor threatening the existence of many animal species. But this variable is of equal significance in the loss of plant species as well. Particularly hard hit are species in bogs, marshes, deserts, and virgin prairies and forests. Interestingly, one of the last refuges of plants requiring special or climax habitat conditions is the area along road and railroad right-of-ways and fence rows that were established perhaps decades ago. In the past, these areas were hardly disturbed at all, except perhaps by mowing. But the recent generalized use of herbicides to control vegetation has seriously threatened the continued existence of plant species in these refuges.

Habitat destruction is a by-product of many human activities. Obviously, if we are to slow down the accelerating trend toward species extinction, we must give greater priority to species and ecosystem preservation when we make decisions regarding land use and development.

The Introduction of Alien Species

As long as people have been traveling about the world, they have carried with them, accidentally or intentionally, many species of plants and animals, which they have introduced into new geographical areas. In some instances, an "opening" existed in the new environment, and the alien could "make a living" without seriously interfering with the well-being of native species. But in other instances, the alien was a superior predator, parasite, or competitor, and brought about the extinction or near extinction of native species. For example, some species of birds that were introduced into the United States from Europe, such as starlings and house sparrows, have displaced native songbirds from their favorable habitats. The importation of new parasites has been especially hard on some plant species. As we mentioned in Chapter 4, the American chestnut tree has all but disappeared because of chestnut blight brought from China, and the American elm tree has been eliminated from large portions of its former range in the eastern United States owing to the importation of Dutch elm disease from Europe.

The introduction of alien species is most devastating on islands, where species are often inadequately equipped to deal with invasions of people and their domesticated animals. For example, the dodo bird, pictured in Figure 14.13, lived only on Mauritius, a small island in the Indian Ocean. The dodo possessed two characteristics that were its eventual undoing: it had no fear of people, and therefore could be easily clubbed to death; and it was flightless, and thus had to lay its eggs on the ground. The dodo became extinct by 1681, after pigs brought to the island devoured its eggs. Another illustration comes

Figure 14.13 The dodo, now extinct. (American Museum of Natural History.)

from Abingdon Island in the Galápagos. In 1962, a scientific expedition found that all the tortoises on that island had died. Goats that had been introduced by a fishing party five years earlier had eaten all the plants that the tortoises used for food. Many tortoise shells were found stuck in crevices in the rockier uplands of the island. As the goats ate all the easily accessible vegetation in the flat lowlands, the lumbering tortoises apparently were forced to seek food among the rocky fissures, where many became trapped and died. Today, on other islands in the Galápagos, tortoises still face considerable pressure from alien animals associated with human settlements: rats eat their eggs and young, cats prey on hatchlings, pigs destroy nests and attack adult tortoises, and goats reduce the food supply. On many other islands, including Australia and New Zealand, native species are similarly endangered by domestic animals.

Other factors contribute to the vulnerability of island-dwelling animals. On an island, both the variety of habitats and the area covered by each habitat are physically limited. Such constraints reduce the ability of native animals to adapt to new predators and competitors. For a prey species, the habitat may not be large or diverse enough for it to avoid new predators, and for a species whose habitat is invaded by a new competitor, resource diversity may be inadequate to allow it to coexist with the invading species. Hence, in some instances competitive exclusion leads to the extinction of one of the two species, as was the case for the Galápagos tortoise on Abingdon Island.

Island floras, too, have been severely devastated by the introduction of alien species. In the Hawaiian Islands, nearly one-half of the native vegetation is classified as endangered or threatened. No large herbivorous animals resided on these islands until the 1700s, when Europeans introduced domestic livestock. Because no prior selection pressures had been at work to favor them, thorns or poisons to ward off herbivores were not characteristic of native plants. This situation was aggravated by the fact that the limited habitat size on the islands favored overgrazing. Thus, formerly unthreatened species were devastated by grazing animals.

Although we commonly think of an island as a land mass surrounded by water, ecologically an island is any restricted area of habitat surrounded by dissimilar habitats. Hence, a lake without outflow is an aquatic island. As natural habitats are changed into farms and cities, natural and seminatural areas become increasingly smaller and more isolated, as illustrated in Figure 14.14. Sometimes, these isolated areas are set aside as parks, conservancies, forest preserves, or wildlife refuges. But such preserves are usually widely scattered, and the intervening farms and cities severely retard the migration of many animals. Hence, these natural and seminatural habitats are actually islands as well, and the animals and plants living in them are subject to the same pressures as those living on islands surrounded by water. The smaller the habitat island, the greater is the pressure on inhabitants—a fact to be considered in our land use planning as we continue to invade and carve up natural habitats into smaller and smaller pieces.

Commercial Hunting

Within the last century or so, several species have become extinct because they were hunted out of existence as food. One such species is the great auk—a large, penguinlike, flightless bird that inhabited the North Atlantic from Newfoundland to Scandinavia. Colonies of great auks were decimated in the early 1800s as fishermen exploited them for food and cooking oil. The species became extinct in 1844, when the last two birds were killed to provide skins for a collector. Other species that became extinct because they were overhunted as food include the passenger pigeon, discussed earlier, and the heath hen. The latter, a bird similar to the prairie chicken, was once common from Massachusetts to the Potomac River. But overhunting for the food market so reduced its numbers that by 1880, the heath hen was found only on a single island, Martha's Vineyard, off the Massachusetts coast. The last authentic sighting was in 1932. Today, many whales (blue whale, gray whale, sperm whale, and several others) are on the endangered species list for the same reason, but in their cases the food they represent is often intended for house pets. Other species endangered today because of their value as food include the giant tortoise (for its flesh and liver) found on the Galápagos and Seychelle Islands, the green turtle (for its flesh and eggs), and Steller's albatross (for its eggs).

Many animals are endangered today because they are being overexploited for commercial products other than food—that is, for their furs, hides, tusks, horns, or feathers—or as trophies. Big cats—tigers, leopards, jaguars, and ocelots—have been particularly hard hit by hunters, who sell their skins to

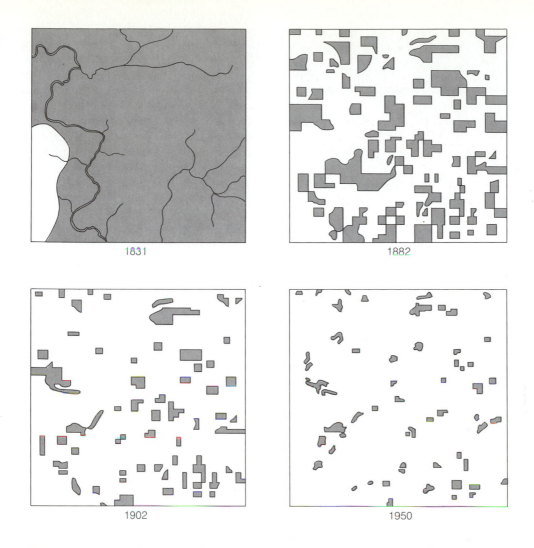

Figure 14.14 The reduction and fragmentation of woodland (shaded areas) in Cadiz Township, Wisconsin, between 1831 and 1950. (After J. Curtis, "The Modification of Midlatitude Grasslands and Forests by Man," in W. L. Thomas, ed., *Man's Role in Changing the Face of the Earth.* Chicago: University of Chicago Press, 1955.)

furriers for the making of fashionable coats and capes. As more and more of these cats are exterminated (fewer than two-thousand Bengal tigers are now left in India), the demand for their skins goes up; hence, the pressure by poachers on remaining populations is tremendous. A pair of snow leopards, members of the species of big cats killed for their skins, is shown in Figure 14.15. Some species of fur

seals, including the monk seal and ribbon seal, are also on the brink of extinction. And some species have been decimated by poorly regulated trophy hunting activities. The bear skin on the floor and the tiger head on the wall represent a shameful waste of endangered wildlife.

The pet trade consumes an astonishing number of wild animals. For example, in 1976, more than 300

Figure 14.15 A pair of snow leopards, members of the cat family that have been intensively hunted. (San Diego Zoological Society.)

million mammals, birds, reptiles, amphibians, and fish were imported into the United States for commercial sale. And this number represents just the tip of the iceberg, since many more animals die in capture or transport. In 1979, fourteen species of monkeys, twelve species of parrots, and nine species of parakeets were on the U.S. Fish and Wildlife Service's endangered species list. These losses are instances of frivolous waste, for most of these animals make very poor pets. And when they lose their novelty, many are flushed away or released to fend for themselves in a foreign habitat.

Superstition plays a role in the demand for some animal products. The three species of Asiatic rhinoceros (Indian, Javan, and Sumatran) are on the brink of extinction because some people believe that powdered rhinocerous horn is an aphrodisiac. And the Chinese tiger population has dwindled because some Chinese believe that the powdered bones of this animal impart vitality and strength.

Many plant species, too, have been severely exploited by collectors. Orchids, lilies, primroses, and palms, for example, are prized for their beauty and are taken from their native habitats to be sold as houseplants or landscape plants. Nearly one-third of all native cacti are thought to be endangered or threatened because they are heavily collected for use as potted plants. Unusual plants (for example, the Venus flytrap) and medicinal plants (ginseng) have also been removed in large numbers from their native grounds.

Species hunted for commercial purposes are frequently caught in a vicious circle. The more they are hunted, the rarer they become, and the rarer they become, the more highly they are prized, and thus the more they are hunted. Figure 14.16 shows somewhat wryly how these pressures can work.

"Had to bag one, Harry—in case this damned Conservation thing doesn't work and they become extinct."

Figure 14.16 An editorial cartoon that alludes to a vicious circle in which rare game animals may be caught. (© 1970 Punch/Rothco.)

Predator Control

The bald eagle, the gray wolf, the red wolf, the eastern cougar, and others have become endangered species partially as a result of *predator control*—that is, efforts made by human beings to protect wild game animals or domestic livestock from attack by natural predators. We have long feared our fellow predators; such notions as the "big bad wolf" are indelible folklore traditions. Underlying this theme is the belief that predators are threats to people, but in reality few authenticated reports have been recorded of unprovoked attacks by wild animals on people. Wild animals attack if they are cornered or if their young appear to be threatened, but under most circumstances, predators are wary of people and avoid direct confrontation. Of course, exceptions are on record. In the United States, five people have been killed by grizzly bears since 1967. All the fatal attacks occurred in national parks or monuments. As tragic as these deaths were, it should be pointed out that given the millions of people who visit these areas each year, the risk of bear attack is quite low. These fatal attacks, however, have provoked considerable controversy over grizzly bear management. For more information on this subject, see Box 14.1.

Not only is the direct threat of predators to humankind negligible, but the effect of predators on the population size of game species under many conditions is also relatively small, as we suggested in Chapter 4. The primary regulatory role of predators appears to be the culling out of weak, sick, injured, and old individuals to maintain a healthier prey population. Hence, predators rarely compete with human hunters for prey, because human hunters usually want to shoot large, healthy animals.

The effects of predators on cattle and sheep is an especially controversial issue. Because ranchers operate on a very narrow profit margin, any loss of livestock is a significant economic problem. Hence, for years, mountain lions, golden and bald eagles, wolves, bears, and coyotes have been shot, trapped, and poisoned because they allegedly kill large numbers of domestic livestock. But obtaining an unbiased estimate of the number of livestock actually killed by predators is difficult. Furthermore, estimating the value of these same predators in controlling pest species, such as mice, rabbits, and ground squirrels, that compete for forage that cattle and sheep eat is perhaps even harder. Nonetheless, hundreds of thousands of dollars are still being spent on predator control every year.

The success of current predator control practices is questionable from both an economic and an environmental perspective. Some methods do exist, however, that could reduce predator losses without further endangering already dwindling wildlife species. For example, research is being conducted to identify chemicals that could be applied to livestock to repel predators by their bad smell or taste. Also, since usually only a few marauding predators cause trouble for ranchers, ranchhands are encouraged to kill only offenders, a sound approach with respect to the well-being of other animals. Furthermore, a few ranchers are raising their sheep in confinement— protected from predators by fencing and buildings. More research and dialogue is needed among people who are willing to consider and implement such alternatives.

Characteristics of Endangered Species

Some species are more vulnerable to human activities than others. The characteristics often associated with endangered or threatened animals are listed in Table 14.1. One feature common among larger animals is a genetically determined low reproductive capacity. The vulnerability of the California condor, shown in Figure 14.17, is due to this factor, for instance. Normally, each female condor lays one egg each year. Once hatched, the chick is usually dependent on its parents for as long as a year. A mother condor with a dependent chick does not lay an egg in the spring following the chick's hatching, but waits until the next spring; hence, a pair of condors may

The Great Grizzly Bear Controversy

wards called the "night of the grizzly," two young women were killed by grizzly bears in apparently unprovoked attacks in separate back-country areas.

A loud public outcry followed the attacks and forced the National Park Service to initiate a program to reduce grizzly-human encounters. The program provided for (1) elimination of sources of unnatural food (mostly garbage) that had altered bear feeding habits and had provoked most of the attacks; (2) temporary closing of campsites and trails in back-country areas where grizzlies are sighted; and (3) relocation or destruction of bears when necessary to protect human life.

This program has been less than a complete success. In August of 1976, in Yellowstone National Park, a 68-year-old man was severely mauled by a grizzly bear when he tried to chase the bear's cubs away from his ice chest. He had failed to heed the warning of park rangers to keep his ice chest inside his camper. About a month later, in Glacier National Park, a grizzly ripped through the tent of a sleeping woman, dragged her away, and killed her. The victim and her two companions had apparently followed all prescribed precautions in setting up camp. Four people were injured in Yellowstone in 1976 and one injured in 1977; in Glacier, five were hurt in 1975 and three in 1976.

Some people fear that the problem of human safety will grow worse. Each year more people visit national parks in the northern Rocky Mountains. Campgrounds are overcrowded, and many backpackers penetrate remote areas. The problem may be aggravated further by a change in the grizzly's behavior. Some biologists believe that the bears have encountered so many humans that they are becoming less people-shy. But they may not be losing their aggressiveness. Hence, more encounters will almost certainly take place and more people will probably be injured and possibly killed.

People with different interests have different views of the grizzly bear situation. The bears threaten livestock as well as people, so ranchers, lumbermen, and some resort owners see the matter one way, while conservationists see it another way. Those who want to save the grizzly point out that other accidents account for far more injuries and deaths in national parks. In Glacier National Park since 1913, only 3 persons have been killed by grizzlies, while 69 have perished from other accidents including drownings (36), falls (16), and being hit by falling rocks and trees (6). Naturally enough, however, the spectre of being mauled by a grizzly bear is particularly frightening to many people, and it has prompted demands that the National Park Service remove all grizzlies from the parks.

How then can the controversy be resolved? First, we must decide whether or not we want to keep a token population of grizzlies in the lower forty-eight states. (It is quite likely that the real future of the grizzly lies in Canada and Alaska, where this species is also found today.) If we decide to maintain a remnant population, then we should set aside a sufficiently large area where the grizzly can reign supreme. All other interests, including lumbering, ranching, and tourism, would have to be rendered secondary in the area, and people who enter it would have to accept full responsibility for their own safety.

Should the National Park Service attempt to save the grizzly bear in the northern Rocky Mountains, or should all grizzlies be eliminated from the region to prevent attacks on people and domestic livestock? What do you think?

Table 14.1 Factors of extinction.

Endangered	Example	Safe	Example
Individuals of large size	Cougar	Individuals of small size	Wildcat
Predator	Hawk	Grazer, scavenger, insectivore, etc.	Vulture
Narrow habitat tolerance	Orangutan	Wide habit tolerance	Chimpanzee
Valuable fur, hide oil, etc.	Chinchilla	Not a source of natural products and not exploited for research or pet purposes	Gray squirrel
Hunted for the market or hunted for sport where there is no effective game management	Passenger pigeon (extinct)	Commonly hunted for sport in game management areas	Mourning dove
Has a restricted distribution: island, desert watercourse, bog, etc.	Bahamas parrot	Has broad distribution	Yellow-headed parrot
Lives largely in international waters, or migrates across international boundaries	Green sea turtle	Has populations that remain largely within the territory(ies)	Loggerhead sea turtle
Intolerant of the presence of man	Grizzly bear	Tolerant of man	Black bear
Species reproduction in one or two vast aggregates	West Indian flamingo	Reproduction by pairs or in many small or medium sized aggregates	Bitterns
Long gestation period: one or two young per litter, and/or maternal care	Giant panda	Short gestation period: more than two young per litter, and/or young become independent early and mature quickly	Raccoon
Has behavioral idiosyncracies that are nonadaptive today	Red-headed woodpecker: flies in front of cars	Has behavior patterns that are particularly adaptive today	Burrowing owl: highly tolerant of noise and low-flying aircraft; lives near the runways of airports

Source: From David W. Ehrenfeld, *Biological Conservation*, p. 129. Copyright © 1970 by Holt, Rinehart and Winston, Inc. Reprinted by permission of Holt, Rinehart and Winston, Inc.

Figure 14.17 A relic of the past, the California condor. (U.S. Department of the Interior, Fish and Wildlife Service.)

produce only one offspring every two years. If disturbed, condors may abandon their nests or chicks. Furthermore, the young birds do not mate until they are six or seven years old. When a species has a low rate of reproduction, as in this example, it can be wiped out by a small decline in the birth rate or an increase in the death rate. In the spring of 1979, a last-ditch effort was made to save the California condor. No more than thirty are believed to inhabit the mountainous areas north of Los Angeles.

Some populations are confined to limited geographical areas either because they have narrow habitat tolerances or because most of their natural habitat has been destroyed. In either case, such a population is particularly susceptible to extinction by devastating storms, earthquakes, or other catastrophic events. A species restricted to a small geographical area may also be more subject to predation, as illustrated by the saying "as easy as shooting fish in a barrel." An example of a particularly vulnerable

species involves the remaining wild whooping cranes, which, until 1976, overwintered only at the Aransas National Wildlife Refuge in Texas, and spent their summers only at the Wood Buffalo National Park in northern Alberta. Since all members of the species stayed together, one severe storm could have wiped out the whooping crane completely.

Conservationists hope to have partially compensated for the whooping crane's vulnerability by establishing a second small whooping crane population at Gray's Lake National Wildlife Refuge in Idaho. In this effort, sandhill cranes were used as foster parents. Fourteen whooping crane eggs were placed in sandhill crane nests in the spring of 1975. That fall, four young whooping cranes followed their foster parents to the overwintering grounds in the Rio Grande Valley of New Mexico. In the spring of 1979, four subadult whooping cranes were counted among the sandhill cranes at Gray's Lake. The concern now is that when the whooping cranes at the new refuge reach sexual maturity (4–5 years), they mate with each other and not with the sandhill cranes, which would create a hybrid species and thus defeat the original effort to preserve the species.

Species that spend most of their time in international waters or whose migratory paths cross international borders may also be particularly vulnerable to extinction because they are subject to the whims of international policies. Several species of endangered whales—especially, the blue whale, gray whale, and sperm whale—exemplify this problem. Because these whales live in international waters, protection of the species requires international cooperation among whaling nations. Although the International Whaling Commission (IWC) was established in 1946 to protect all species of whales from overhunting, it has no policing power and thus has been able to do little to prevent their decimation. Therefore, eight species of whales are now nearly extinct. For many years, nations such as Japan and Russia constantly thwarted any attempts within the IWC to establish catch quotas, and even when quotas were established in the mid-1970s, these countries largely ignored the agreed upon restrictions. During the late 1970s, however, Russia and Japan began to abide by the quotas and regulations, perhaps both because worldwide public pressure to save the whales had increased and because their whaling fleets were aging and becoming unseaworthy.

An important species characteristic not included in Table 14.1 is *critical population size*—the minimum number of individuals needed to ensure the survival of a population. As we noted in Chapter 4, a population that becomes too small is subject to many stresses—a reduced mating stimulus, a tendency toward inbreeding, and an increased vulnerability to predation. We know so little about the factors that control population levels in most species that usually it is essentially impossible to identify critical population size. Still, we can observe the effect when a population drops below its critical size. An example is the plight of the passenger pigeon, described earlier in this chapter. By the 1880s, several thousand passenger pigeons had survived the mass slaughter; they were so widely scattered that hunting them no longer proved profitable. The several thousand pigeons remaining after hunting ceased were below the species' critical level, and the population continued to decline until the last bird died in 1914. In contrast, the eastern cougar, once assumed to be extinct in eastern North America, is apparently recovering population strength due to the survival of a nucleus of only a few breeding individuals in the forests of eastern New Brunswick and western Ontario. The cougar's critical population size is thus demonstrably much lower than that of the passenger pigeon.

Attempts to Save Endangered Species

Because habitat destruction is the major factor in the extinction of plants and animals, the most important strategy to prevent extinction is preserving a representative cross-section of the world's ecosystems. In the United States, the National Wildlife Refuge system has grown since it was begun in 1903 to include 367 sites covering 121,000 square kilometers (50,000 square miles). The distribution of the refuges in the system is shown in Figure 14.18. Of the total refuges, no fewer than 136 protect one or more endangered species of wildlife. Millions of additional hectares have been set aside by private groups and state and other federal agencies. For example, the National Park Service protects the endangered gray wolf on the 2,113-square-kilometer (845-square-mile) Isle Royale National Park in Lake Superior, and protects cacti in Saguaro National Monument and Organ Pipe National Monument (both in Arizona). The United States is by no means the only nation taking such steps toward preservation. In tropical and subtropical nations where the rich diversity of plants and animals is being seriously threatened by human activity, a few sizable, if not yet adequate, natural preserves have been established. Nations now supporting such preserves include Thailand, India, Peru, Costa Rica, Columbia, Kenya, and Tanzania.

Preserves are beneficial in several ways. In most instances, more than one endangered species can be protected in the same area. Furthermore, a naturally functioning ecosystem may not need much management to continue to serve as an effective preserve, particularly when it is a climax ecosystem. Hence, maintenance costs and consumption of resources are minimal.

Establishing preserves, however, does not always ensure adequate protection of endangered species. The poaching of animals and plants on preserves by hunters, collectors, and traders is a significant worldwide problem. The Convention on International Trade in Endangered Species of Wild Fauna and Flora was negotiated in 1973 in an effort to eliminate the motivation for poaching. The convention regulates the export and import of wild specimens and product derivatives of plants and animals that are placed on a list by vote of member nations. Although the list is not identical with the United States Fish and Wildlife list of endangered species (which in-

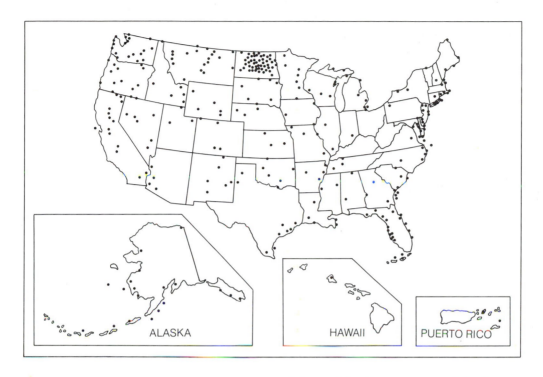

Figure 14.18 The distribution of the National Wildlife Refuges. (After National Wildlife Federation, Special Issue: "Endangered Species," *National Wildlife 12,* 1974.)

cludes nearly seven-hundred species of plants and animals), there is considerable overlap. Thus, in accord with the convention, it is illegal in the United States to import living specimens of endangered species such as the orangutan for sale in pet shops. Likewise, it is illegal to import tiger skin coats for sale or export products made from American crocodiles for sale. Although over fifty nations are now active participants in the convention, many nations (some of them major importers or exporters of live animals or their product derivatives) have not signed the agreement. Nevertheless, with strong enforcement, this treaty should significantly diminish international trade in endangered species and their products.

Some species have been reduced to such small numbers that habitat preservation by itself is not considered sufficient to save them, and more intensive management is needed. The transfer of eggs from an existing population to start a new population, as in the case of the whooping crane, is one such management tool. Captive breeding in zoos, animal breeding parks, or research centers along with subsequent release into the wild has also been tried with some success. The peregrine falcon program at Cornell University has been particularly encouraging. In this program, birds are bred in captivity and the hatchlings are raised until they are large enough to be placed in the wild in uninhabited nests. They are then fed and guarded until they learn to fly and to hunt on their own. However, capture breeding is often unsuccessful. Some species, such as the whooping crane, are difficult to breed, rear, and maintain in captivity in the numbers needed to stock new populations. And animals raised in captivity often exhibit aberrant behavior patterns or fail to develop key behavior patterns, such as those necessary for hunting. Hence, when left to fend for themselves in the wild, many individuals die soon after release. Another significant drawback is that managing species in zoos and breeding parks is more expensive, both in terms of time and resources, than in natural preserves. Still, because the natural habitats of many endangered animals are quickly disappearing, zoos and animal breeding parks may be the last refuges of some wildlife species.

Efforts are also under way to save specific endangered plant species. Special arboretums and gardens are being established where the habitat requirements of endangered plants can be studied so that these plants might be reestablished in the wild eventually. One example is the Waimea Arboretum in Hawaii, where tropical and subtropical plants are being studied in the hope that they can be preserved for future generations. In addition, the International Board for Plant Genetic Resources, which is funded by several United Nations agencies, is establishing regional seed banks for collecting and maintaining samples of traditional crop strains to be made available to crop breeders.

On the legal front, probably the single most important piece of legislation affecting species preservation passed so far is the Endangered Species Act of 1973. This act provides for the identification of threatened as well as endangered species so that preventive action can be taken before a species reaches the critical endangered stage. The act gives consideration to all forms of animal life, and recognizes the importance of plant species as well. One section, section 7, of the act prohibits any federal agency from funding or carrying out a project that threatens the existence of or alters the habitat critical to the survival of an endangered species. Another section implements the Convention on International Trade in Endangered Species of Wild Fauna and Flora, making the United States the first country to ratify this treaty.

Other federal actions that have indirectly helped endangered species include the imposed restrictions on the use of chlorinated hydrocarbon pesticides (DDT, aldrin, dieldrin), which should help to safeguard wildlife, particularly birds of prey. And a 1977 executive order by President Carter strengthens existing controls on the release, escape, or establishment of foreign species into natural ecosystems.

Early in this chapter we raised the question as to whether we should make efforts to preserve other species at all. You should now be in a better position to frame your own views on this issue. For four other viewpoints on our own place in nature, and on how we should deal with other species, see Box 14.2.

What of the Future?

We can take heart from the evidence that some species are making a comeback as a consequence of efforts described in the preceding section. One of the greatest success stories involves the alligator. Twenty years ago, the alligator was heading toward extinction, but in 1977, as a result of conservation efforts, the U.S. Fish and Wildlife Service reclassified about 75 percent of the nation's alligator population from endangered to threatened. The alligators now considered threatened live in Florida and the coastal portions of Georgia, Louisiana, South Carolina, and Texas. Endangered alligators are those in Alabama, Oklahoma, Mississippi, and North Carolina, as well as inland areas of Texas, Louisiana, Georgia, and South Carolina.

The case of the whooping crane, cited above, is also encouraging. The whooping crane population numbered fifteen in 1941; now it numbers more than seventy in the wild and twenty in captivity. And some populations of birds of prey, such as the brown pelicans off the southern California coast, appear to be returning to former levels as a consequence of the virtual cessation of DDT usage in North America.

But the news is not all good. In 1978, a significant

battle over the Endangered Species Act came to a head. It began in 1975, when conservationists filed suit against the Tennessee Valley Authority to stop construction of the Tellico Dam on the Little Tennessee River. The reservoir behind the dam threatened the snail darter, a tiny fish, shown in Figure 14.19, whose only existing breeding habitat is believed to be the shallow, cool, free-flowing waters and gravel substrate present at the dam site. After losing their initial suit, these conservationists took their case to the Federal Circuit Court of Appeals and won. Though the dam was already 90 percent completed, construction was halted in 1977. In 1978, the Federal Supreme Court upheld the Court of Appeals decision. The Supreme Court stated that, as presently written, the Endangered Species Act of 1973 prevented the construction of any federally funded project if it threatened the survival of a species on the United States list of endangered and threatened species.

In succeeding months, a battle raged in Congress over the future of the Endangered Species Act. Developers wanted to weaken the act so that federally funded projects such as dams and highways could be exempted from the act. Conservationists wanted to keep the act as it was to provide strong protection to endangered species. In the end, a compromise was reached, and in an amendment to the act, a cabinet-level committee was established to review requests for exemptions. This special review committee met for the first time in January 1979 to consider two exemption requests. On a motion by the chairman of the Council of Economic Advisers, the committee denied the exemption request for the Tellico Dam. The committee pointed out that the economic justification for the project was dubious even though $100 million had already been spent on it. But the committee did approve an exemption for the Grayrocks Dam project in Wyoming if steps were carried out (these were already agreed to by the project sponsors) to preserve resting sites on the Platte River for the whooping crane. Thus, it would seem that the establishment of a cabinet review committee has not weakened the protection of endangered species that are already listed. (In the fall of 1979, Congress approved an amendment to a public works bill to finish building the Tellico Dam. So the saga of the snail darter continues.)

The amended Endangered Species Act, however, does pose a real danger for those 1850 species that the United States Fish and Wildlife Service was considering for addition to the endangered species list. Now, because of new amendments, before a species can be listed as endangered or threatened, the boundaries of its critical habitat must be designated, an economic impact study prepared, and public

Figure 14.19 The snail darter, a tiny fish (8 centimeters, or 3 inches, in length) that brought into sharp focus the conflicts between development and the protection of endangered species. (U.S. Department of the Interior, Fish and Wildlife Servce.)

Box 14.2

Viewpoints on Saving Endangered Species

Where do you stand on the controversial issue of preserving other species? Many voices have been raised over the often thorny questions involved in human stewardship of life on Earth. Opinions range from living in "complete harmony" with nature to the "complete subjugation" of nature to human material needs. Presented here are four different viewpoints regarding endangered species. Perhaps they will help clarify your own thoughts on the matter.

The first viewpoint is that of René Dubos, a microbiologist at the Rockefeller University who has written much about our species and its relationship to the environment:

> Man is still of the earth, earthy. The earth is literally our mother, not only because we depend on her for nurture and shelter but even more because the human species has been shaped by her in the womb of evolution. Each person, furthermore, is conditioned by the stimuli he receives from nature during his own existence.
>
> If men were to colonize the moon or Mars—even with abundant supplies of oxygen, water, and food, as well as adequate protection against heat, cold, and radiation—they would not long retain their humanness, because they would be deprived of those stimuli which only earth can provide. Similarly, we shall progressively lose our humanness even on earth if we continue to pour filth into the atmosphere; to befoul soil, lakes, and rivers; to disfigure landscapes with junkpiles; to destroy the wild plants and animals that do not contribute to monetary values; and thus to transform the globe into an environment alien to our evolutionary past. The quality of human life is inextricably interwoven with the kinds and variety of stimuli man receives from the earth and the life it harbors, because human nature is shaped biologically and mentally by external nature. (*A God Within,* New York: Scribner's, 1972.)

The second viewpoint is by another biologist, Gardner B. Monent of Goucher College. It appeared in the journal *Bioscience* following the controversy surrounding the killings of two women in Glacier National Park by grizzly bears in 1968.

> The dramatic need of our time is for the intelligent control of our environment, not its personification. A strong sense of stewardship for our wildlife heritage is indeed much needed in our own interests and those of generations yet unborn. But to acknowledge such a need should make it easier not to succumb to a mindless absolutism claiming that every species should be preserved and no discrimination is possible. The Carolinas are less colorful today because of the extinction of their native parakeet, and the world is poorer because of the loss of many species. However, any zoologist could draw up a list of animal and probably plant species that we would be better off without; the common rat, the fire ant, certain species of sharks, the tsetse fly, the malarial organism. Of course, each of these species possesses ecological, physiological, evolutionary, and other aspects of considerable scientific interest. That isn't the point. So does every other species. Removal of any one would result in some change in the intricate web of nature. The upper Mississippi valley is not the same without the passenger pigeon. But this kind of thing has been going on all through the course of evolutionary history. The point is that to abdicate the use of human reason is to throw away the most powerful and unique tool *Homo sapiens* has. Why can't we discriminate? If a

species like the grizzly, which is on the endangered species list, can be saved, all well and good. But there is no scientific basis for the dogmatic assertion that the Park Service or anyone else is obligated to save it or any other species without regard for the human cost in money, lives, or health. (*Bioscience 18:*1105–1108, December 1968.)

In 1978, efforts to amend the Endangered Species Act of 1973 brought political views into the spotlight. The following opinion was expressed by Senator William L. Scott of Virginia, during a floor debate:

> If we look at the act from the political or governmental point of view, we again find that it gives undue consideration to fish, wildlife, and plants over the welfare of people. Not only from the biblical, or evolutionary point of view is mankind superior but also from the political point of view. Our own Government exists to serve people. I do not believe anyone in this body would suggest that the equal protection clause of the 14th amendment applies to fish, wildlife, and plants, but even if it should be so applied, human species would be entitled to the protection now denied them under the Endangered Species Act. It does not appear reasonable that anyone would quarrel with the statement that people should have dominion, as genesis provides, over the fish of the sea, the fowl of the air, and every living thing that moves upon the Earth. Government exists only for the purpose of serving people and apparently this basic fact was not considered fully when we enacted the Endangered Species Act of 1973. (*Congressional Record,* July 17, 1978, S 10903.)

Following the United States Supreme Court decision to uphold lower court rulings that construction of the Tellico Dam be stopped, the following opinions of P. J. Wingate appeared in the *Wall Street Journal:*

> Congress may be in trouble on Election Day, because someone is sure to point out that this 1973 law would, with equal clarity, prevent any federal funds being spent on projects to eradicate such forms of life as the polio virus, the anthrax bacillus, the anopheles mosquito or the diamondback rattlesnake. Now there may indeed be some unexpected problem for people if these four species of life were obliterated, but most voters would be quite happy to run that risk, whatever it is. There may, of course, be some environmentalists who would argue against wiping out the polio virus, the anthrax bacillus, the anopheles mosquito and the rattlesnake, even if we knew how to, but people who have had to deal with these things on a personal basis would be sure to take a dim view of such a position.
>
> The average voter knows that it is necessary to weigh the odds in all such matters and he feels instinctively that the odds are his problems will be less, instead of greater, if such forms of life disappeared. He also feels that it is unlikely that the snail darter and the furbish lousewort [an endangered plant species] will hurt the human race if they continue to have the government spend a few thousand dollars to have them transplanted to other areas.
>
> But not a hundred million dollars, because he also thinks it unlikely that the disappearance of the snail darter and the lousewort would bother him any more than the disappearance of the dinosaur and the dodo bird or perhaps thousands of other species that have vanished during the millions of years in which evolution has been going on. (*Wall Street Journal,* July 6, 1978, p. 12.)

hearings held. All these procedures must be completed within two years of the time the listing is proposed. Given the amount of ecological and economic study now needed for each species and its critical habitat, it is quite probable that very few new species will be proposed for listing and even fewer will gain the protection of the Endangered Species Act.

Now that a few public works projects have been held up, people are becoming aware of the far-reaching implications of the Endangered Species Act. More conflicts are likely to surface in the future as we continue to expand our efforts to meet the perceived needs of a growing society. As in the past, most such conflicts will be resolved through simple modifications in the project plans. In a few instances, impasses may have to be overcome with the aid of a higher authority, such as the cabinet-level committee. But, by and large, affluent countries will be able to give every species the benefit of the doubt without imposing serious economic hardships on its citizens—if only they muster the will to do so. Unfortunately, the same is not true for poorer countries.

On a worldwide basis, the picture for endangered species is bleak. The burgeoning numbers of people in the less developed nations of Africa, Asia, and South America are putting tremendous pressures on wildlife habitats. As we noted in Chapter 5, tropical rain forests and savannas are being destroyed rapidly as demand for land for living space, cultivation, grazing, and forest products increases. Sadly, these biomes provide habitats for myriad species that could not live elsewhere. Before we criticize these developing countries too harshly, however, we must remember that our own consumer demands in the United States have helped to create the markets for such products as veneer, plywood, and paper, which are derived from forested areas. And, to cite another example, in Costa Rica, virgin forests are rapidly being cleared to create cattle-raising grasslands. During the 1960s, Costa Rica's beef production increased by nearly 100 percent, while local beef consumption declined by 25 percent. Nearly all the extra beef produced was imported by fast-food restaurant chains in the United States.

Governments of developing nations face many difficulties in their efforts to save endangered species. Many do not possess the monetary resources or enough properly trained people to set aside and manage large preserves for wildlife. Furthermore, often these countries lack the means to enforce existing poaching laws or to prevent illegal commercial traffic in products derived from endangered species. Undoubtedly, the most difficult problem in these nations is the establishment of priorities. Immediate economic gains have greater political appeal than long-term, unquantifiable values. Given the choice between land for species preservation and land for food and economic development, almost everyone would choose the latter. For these reasons, many species in developing nations appear doomed to certain extinction.

Conclusions

Through its diverse and pervasive activities, our species is probably the single most important force in the natural selection process today. We are determining the course of evolution on the basis of short-term considerations, and we are proceeding in spite of our ignorance as to the full ramifications of our activities. Hence, even without knowing it, we may well be in the process of selecting against ourselves. To reduce species depletion, we must first make species preservation a high priority. It is true that no one cares about the survival of plants and animals until their own survival needs—adequate food, clothing, shelter, and health care—are met. But it is also true that policies based on short-run economic goals could result in serious ecological losses, which in turn could seriously diminish the quality of human life. Thus, our overriding goal should be to make economic and social progress worldwide, enabling all nations to intensify their efforts to preserve the rich species mix still in existence on the planet.

Summary Statements

Hundreds of species of animals and plants throughout the world are in immediate danger of extinction.

Each species may be economically important to us since each species possesses a unique set of adaptive characteristics that might prove useful to us some day in medicine, agriculture, or industry. Each species also contributes to ecosystem stability.

Other arguments for species preservation include the aesthetic value of each species and human responsibility to be faithful stewards of the earth and its resources. In view of short-term economic considerations, however, including the need for jobs, housing, and food, species preservation often ranks low on our list of priorities.

Numerous species have become extinct since life began on earth. Many of these extinctions resulted from the inability of affected species to adapt to a changing environment. More than eleven thousand years ago, many large mammals became extinct in North America as a result of a changing climate and predation by a wave of human hunters immigrating from Siberia.

During the past two centuries, human activities have caused a sharp rise in the number of species to become extinct. Humans have brought about extinction in several ways: destruction of natural habitats, introduction of alien species, commercial hunting, and predator control.

Habitat destruction has had the greatest impact on populations. Most severely affected are plants and animals that occupy climax forests and prairies and unique habitats such as bogs and swamps. The pollution of habitats, particularly with insecticides and herbicides, has also taken a toll.

An introduced species is sometimes a superior predator, parasite, or competitor, and can therefore bring about the extinction or near extinction of native species. Introduced species have been especially devastating to native plants and animals on islands.

Many species are extinct and others are endangered because they have been over-exploited for their furs or hides, or as food, pets, or collectors' items. Several species of predators are also endangered because of their presumed threat to humans, wild game populations, or domestic livestock. Much controversy surrounds the role of predators and the success of current predator control practices.

Certain characteristics make animals particularly vulnerable to human activities. Endangered species often possess one or more of the following characteristics: individuals are large; the animal is a predator; the species is commercially valuable, is intolerant of people, occupies a restrictive habitat or has a limited geographical distribution, and has a low reproductive capacity. Once the population size of a species falls below a critical number, human efforts to save the species from extinction are often fruitless.

The best way to prevent extinction of plants and animals is to preserve their natural habitats. As human pressures increase, zoos, animal breeding parks, and arboretums may prove to be the only safe havens for many endangered species.

The outlook for endangered species is mixed. Some species, such as the American alligator, whooping crane, and peregrine falcon, appear to be making comebacks. In general, affluent countries can afford to save the habitat of endangered species without imposing serious economic hardships—if the people have the will to do so. In developing nations, however, the need for land to provide food and economic development is stronger than the impetus toward species preservation, and many species therefore appear doomed to certain extinction.

Questions and Projects

1. In your own words, write a definition for each of the terms italicized in this chapter. Compare your definitions with those in the text.

2. What are the differences between endangered species and threatened species?

3. List at least five ways in which a plant or animal species may contribute to the well-being of humankind.

4. Describe how human beings may have contributed to the extinction of large animals in North America ten-thousand years ago. How has the nature of human activities changed since that time? What is the significance of these changes for the survival of endangered species?

5. Consider the following argument: "Conservationists are very naive; they are just not living in the real world. When it's a question of 'What's more important, me or an eagle?' nobody is going to choose the eagle." Do you agree or disagree? Explain your answer.

6. Consider the following argument: "Innumerable species of plants and animals have become extinct since life began on earth. People are no different from any other factor in the environment that has acted as a natural selection factor. We are just selecting for those plants and animals that can survive in the environment that we have helped create." Do you agree or disagree? Explain your answer.

7. Explain why the destruction of natural habitats has an adverse impact on native wildlife populations.

8. Go to your city or regional planning office and ask for a current land use map as well as a map thirty to fifty years old. Compare the two maps with respect to residential areas and regions used for commercial development, industry, and

recreation. How much wildlife habitat has been lost to development? Can you detect any habitat islands? If so, are they being managed to preserve valuable habitat? Are they large enough to support the forms of wildlife that are native to the region? What are your community's land use planning policies regarding "green spaces" and habitat preservation?

9. Human-caused pollution is a relatively new threat. Describe several ways in which pollutants can endanger plants and animals.

10. Explain how a small park in the middle of a large metropolitan area is analogous to a small piece of land surrounded by water. What is the importance of habitat islands for the successful management of endangered species?

11. Why are species that live only on islands particularly vulnerable to extinction from the introduction of alien species?

12. Should the United States develop a policy to prevent the importation of all alien species? Discuss the relative merits of importing animals and plants for (1) pet shops, (2) zoos, (3) use in scientific and medical research, and (4) use as a predator or parasite in the control of pests that are already in the United States.

13. Describe the conflicting arguments regarding predator control. Where do you stand? Would you take a different position if the predator in question were an endangered species, such as the bald eagle, than if it were a thriving species, such as the coyote?

14. With reference to the characteristics listed in Table 14.1, describe an imaginary animal that would be extremely vulnerable to extinction at the hands of humankind. Also describe the hypothetical animal that would be least vulnerable.

15. With reference to the characteristics listed in Table 14.1, comment on the survival potential of each of the following hypothetical animals: (a) a mouse-sized omnivore that gives birth to thirty young each year, tolerates the presence of people well, and inhabits weedy fields and lots throughout the temperate portion of the northern hemisphere; (b) a predatory bird the size of an eagle that raises only two young every other year and is found only on a few remote islands in the South Pacific; and (c) a weasel-size predator that has a valuable fur, raises five young every spring, can tolerate the presence of people, and resides in cut-over forests in the Pacific Northwest.

16. Why is critical population size an important consideration in attempts to save endangered species?

17. Describe several measures that are being used to save endangered species. Cite some examples from your region.

18. What method or methods would you advocate as best to save endangered species? Defend your choice.

19. Speculate why the plight of endangered plant species has been largely ignored up to this time while in recent decades considerably more action has been taken to save endangered animals?

20. Keeping in mind that habitat destruction is the major factor contributing to species extinction, what is your position on the application of section 7 of the Endangered Species Act? Should it be changed to allow construction and development, or should it be upheld as in the case of the Tellico Dam in Tennessee? Are compromises possible?

21. Are any plants or animals in your area on the endangered species list? Check with your state Department of Natural Resources or Department of Conservation. What efforts are being made to preserve these species? How can you contribute to these efforts?

Selected Readings

Berwick, S. 1976. "The Gir Forest: An Endangered Ecosystem." *American Scientist 64:*28–40 (January–February). An analysis of the factors affecting the survival of the tropical forest and of the ways to preserve its integrity.

Eckholm, E. 1978. *Disappearing Species: The Social Challenge.* Washington, D.C.: Worldwatch Institute. An examination of the costs and benefits associated with the establishment of priorities to save endangered species.

Ehrenfeld, D. W. 1970. *Biological Conservation.* New York: Holt, Rinehart and Winston. An interesting discussion of factors contributing to endangered species and the destruction of natural ecosystems. Also considers means of species and ecosystem preservation.

Hayes, H. T. P. 1977. *The Last Place on Earth.* New York: Stein and Day. An account of the wild animal herds on the Serengeti Plain in Africa. Critically examines the problems of reconciling wildlife conservation, human population density, and land use.

Laycock, G. 1966. *The Alien Animals.* Garden City, New Jersey: Natural History Press. A popularized account of the introduction and impact of alien animals on native animals and vegetation.

Martin, P. S. 1973. "The Discovery of America," *Science 179:*969–974 (March 9). An examination of the possible role of human beings in the late Pleistocene extinction of large mammals.

National Wildlife Federation. 1974. Special Issue. "Endangered Species." *National Wildlife 12:*3. A beautifully illustrated and well-written overview.

New York Botanical Club. 1977. *Extinction Is Forever.* New York: New York Botanical Club. An in-depth look at endangered plant species around the world and attempts to preserve them.

Smith, R. L. 1976. "Ecological Genesis of Endangered Species: The Philosophy of Preservation." *Annual Reviews of Ecology and Systematics 7:*33–56. A comprehensive examination of reasons for preserving endangered species.

Zimmerman, D. R. 1975. *To Save a Bird in Peril.* New York: Coward, McCann & Geoghegan. An examination of techniques used to save endangered birds. Describes the difficulty of saving endangered species.

Ziswiler, V. 1967. *Extinct and Vanishing Animals.* New York: Springer-Verlag. A good overview of the factors contributing to species extinction and efforts to save species.

A slope of once valuable agricultural land near Concord, California, that has just been contoured by tractors to prepare for construction of a housing development. (Copyright 1978 Barrie Rokeach.)

Chapter **15**

Land Use Management

At this writing, conservationists and resource exploiters are embroiled in heated conflict over land use priorities in Alaska. Alaskan bedrock probably contains vast reserves of oil, natural gas, and metallic minerals. But Alaska also houses our nation's last frontier of pristine wilderness. It is likely that large-scale resource exploitation would cause irreparable damage to Alaska's wilderness and wildlife. Conservationists want to protect and preserve as much of Alaska's wilderness as is possible. But mining and petroleum interests cite our growing resource needs and argue for freedom of exploration. Figure 15.1 shows an example of the wildlife conservationists are seeking to protect.

On September 15, 1977, President Carter cast his lot with the conservationists on this issue. His Alaskan lands proposal to Congress would nearly triple the acreage currently set aside in the state as national parks, wildlife refuges, national forests, and wild and scenic rivers. If this action is approved, mining and oil and natural gas exploitation would be restricted or prohibited on 40 percent of Alaska's total land area.

The Alaskan issue is just one of many conflicts that arise from our multiple demands on the land. We continually modify land, reshaping it for housing sites and for industrial and commercial purposes. We terrace hillsides, drain and fill in wetlands, seal the ground with concrete and asphalt, and whittle out campsites and playgrounds. We grow our food and fiber on land, and we build our great cities on it. We tear up land, extract its rock, mineral, and fuel resources, and use it as a dumping ground for our wastes.

We have carried out these activities with increasing fervor over the years as our population and its demands continued to grow. But the land and its resources are finite. Already the many ways in which we use the land conflict and compete with one another. Prime farmland is gobbled up by urban sprawl, wilderness preserves are eyed by miners, pastureland is grazed by sheep and cattle competing with indigenous wildlife. Meanwhile, our search for living space is pressuring us into building in regions prone to flooding, earthquakes, and other geologic hazards.

In this chapter, we survey our nation's land use patterns and consider the efforts at land use manage-

Figure 15.1 A ram in the proposed Wrangell-St. Elias National Park-Preserve of Alaska. (U.S. Department of the Interior, National Park Service.)

ment currently being made. We then examine geologic hazards, their influence on land use and possible ways of alleviating their impact.

Land Use Conflicts and Control

Figure 15.2 shows the major ways in which we use the land and their comparative significance. In the main, conflicts over land use arise with reference to lands accounted for in the left-hand section of the figure, labeled nonagricultural, although a considerable amount of agricultural land is lost each year to other sorts of activities. Specifically, conflicts in the former category involve decisions on how to use public lands and coastal zones, and on how to control and direct community growth. In this section, we examine these difficulties and the management strategies that have been developed so far to settle them.

Public Lands

Agencies of the federal government currently regulate activities on one-third of the nation's land area through their management of national forests, parks, wildlife refuges, and other *public lands*. This regulatory approach did not always prevail, however. As we noted in earlier chapters, during the first one-hundred years of our nation's history, the unrestricted exploitation of our natural resources was enthusiastically encouraged. The government readily handed over huge parcels of land to anyone willing to exploit its minerals, timber, or water. The result, we know now, was rapid economic development and extensive environmental damage, such as that shown in Figure 15.3. Late in the nineteenth century, more and more people became aware that relentless exploitation exacted a toll in the form of environmental degradation and saw the wisdom of practicing resource conservation. It became apparent that the federal government could use its ownership of public lands for the common good by exercising stewardship over the land. The first gesture in this direction was the founding of Yellowstone National Park in 1872. In the same year, California took steps to protect Yosemite Valley. Twenty years later, the first federal forest reserves, later called national forests, were set aside.

In the early part of this century, the conservation movement was championed by President Theodore Roosevelt. In 1908, concerned over the cost of widespread mismanagement of the nation's natural resources, Roosevelt called a White House Conference on National Resources. One outcome of this meeting was the appointment of a fifty-member National Conservation Commission. Roosevelt also appointed Gifford Pinchot as first chief of the U.S. Forest Service. Together Roosevelt and Pinchot enlarged the nation's forest reserve holdings nearly five-fold. They realized this accomplishment in spite of vigorous objections by ranchers, miners, and others, who were used to having virtual free rein over public lands.

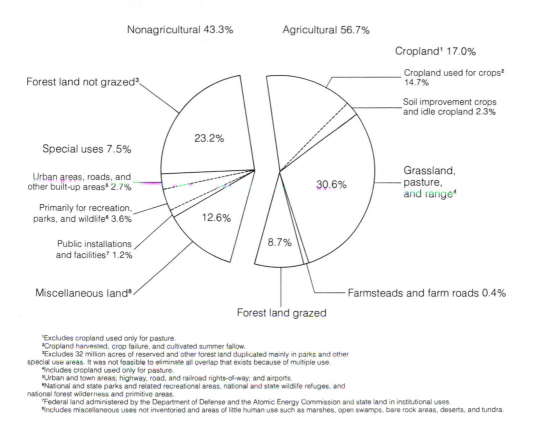

Nonagricultural 43.3%   Agricultural 56.7%

Cropland¹ 17.0%

Forest land not grazed³

Cropland used for crops²
14.7%

Soil improvement crops
and idle cropland 2.3%

Special uses 7.5%

23.2%

Urban areas, roads, and
other built-up areas⁵ 2.7%

Grassland,
pasture,
and range⁴

30.6%

Primarily for recreation,
parks, and wildlife⁶ 3.6%

12.6%

Public installations
and facilities⁷ 1.2%

8.7%

Miscellaneous land⁸

Farmsteads and farm roads 0.4%

Forest land grazed

¹Excludes cropland used only for pasture.
²Cropland harvested, crop failure, and cultivated summer fallow.
³Excludes 32 million acres of reserved and other forest land duplicated mainly in parks and other
special use areas. It was not feasible to eliminate all overlap that exists because of multiple use.
⁴Includes cropland used only for pasture.
⁵Urban and town areas; highway, road, and railroad rights-of-way; and airports.
⁶National and state parks and related recreational areas, national and state wildlife refuges, and
national forest wilderness and primitive areas.
⁷Federal land administered by the Department of Defense and the Atomic Energy Commission and state land in institutional uses.
⁸Includes miscellaneous uses not inventoried and areas of little human use such as marshes, open swamps, bare rock areas, deserts, and tundra.

Figure 15.2 Land use in the United States. (After Council on Environmental Quality, *Sixth Annual Report*, 1975.)

Pinchot's conservation philosophy served as the guiding principle for federal management of public lands through much of this century. Pinchot favored resource development over preservation. He advocated exploitation of the nation's resources for the public good with a minimum of waste. Until the 1960s, preservation of wilderness areas (unexploited lands) was left to the efforts of small, private organizations such as the Sierra Club. Then, spurred by public concern over the nation's diminishing reserves of unspoiled lands, Congress acted to safeguard wilderness areas.

Today, public land falls into two general categories, based on use: regions set aside for the prime purpose of preservation, and areas intended for regulated multiple use. Those areas deemed wilderness areas, wild and scenic rivers, national parks, and wildlife refuges are designed to protect and preserve invaluable natural resources. In national forests and in those regions called national resources lands, a variety of activities are permitted, which are usually subject to regulations. These regulations are intended to keep the multiple uses from interfering with one another, and to ensure the compatibility of

Figure 15.3 Poor timber management, which results in severe soil erosion. (U.S. Department of Agriculture, Soil Conservation Service.)

multiple use with environmental quality.

National forests were established primarily to foster wise forest management that would ensure both a sustained yield of timber and watershed protection. In addition, national forests are used for the grazing of domestic cattle and sheep, recreation, mining, and as preserves for fish and wildlife. The management of these multiple demands in the public interest was first mandated with passage of the Multiple Use—Sustained Yield Act of 1960. More stringent management guidelines became law in 1976, with the passage of the National Forest Management Act. This law called for measures to maintain plant and animal diversity in national forests, minimize soil erosion, and protect water flow and quality.

Although the goal of the U.S. Forest Service is to ensure that the various activities allowed proceed harmoniously, realizing this goal fully is seldom possible. Too many people tramping or driving through national forests can destroy the natural amenities that draw them there in the first place. Livestock can outcompete deer, elk, and other wildlife for available forage. And mining can rapidly counteract the gains in soil conservation accomplished through timber management.

Another major challenge facing our national for-

Figure 15.4 Cattle grazing on national resource lands managed by the Bureau of Land Management. (U.S. Department of the Interior, Bureau of Land Management.)

ests is a consequence of the rising demand for wood products. In 1970, the situation was still favorable, for timber growth rates were then outpacing rates of timber removal. But from that time on, the gap began to close between supply and demand. Now it is clear that by the year 2000, demand will be increasing more rapidly than supply. Since national forests produce less wood per hectare than do private commercial forests, the Forest Service is under pressure to adopt measures to increase per hectare productivity. Techniques adopted to increase production could include application of more pesticides and fertilizers and maintenance of monoculture forests through the controlled burning of competing vegetation. But such practices, if implemented, could jeopardize the wildlife that inhabits the forests, whose protection is a major objective of the U.S. Forest Service.

On our vast holdings of *national resource lands,* use regulation was not always as strict as in national forests. These lands, scattered throughout the nation, have been used primarily for grazing and mining (see Figure 15.4) under the relatively relaxed management of the Bureau of Land Management (BLM). Many such lands have deteriorated considerably because of overgrazing and, especially in the West, the stress of motorcycles and dunebuggies. In an attempt to halt the steady degradation of BLM land, Congress passed the Federal Land Policy and Management Act of 1976 (also known as the BLM Organic Act). This law strengthens the BLM's authority in protecting and managing national resource lands for multiple use and sustained yield. In the spring of 1978, the BLM issued new strict regulations on grazing intended to reverse the trend toward deterioration. And, with similar intent, President Carter signed an executive order in 1977 curbing off-road vehicle use of public lands where such activity is likely to cause environmental damage.

National parks are lands that are regulated in order that a specific natural feature or set of features may be preserved and that the educational and recreational needs of the general public may be met. More than twenty years ago, the National Park Service embarked on a vigorous campaign to attract visitors to national parks. The service built new camping facilities, trails, and access roads. Public response was overwhelming, as evidenced in Figure 15.5. People visited the parks in droves. Campgrounds were overcrowded, traffic was congested, and the noise all but drowned out nature's tranquility. Now, the pendulum is swinging the other way, and the original preservative purpose of our national parks is being emphasized. People are being encouraged to visit in manageable numbers that do not threaten the park's existence. Visitor centers and campgrounds are centralized to minimize recreational sprawl. And in some parks, the traffic problem has been eased by mass transit systems that usher people to nature's wonders, as illustrated in Figure 15.6.

In 1964, Congress established the National Wilderness Preservation System and, by 1978, almost 2 percent of our nation's public land was designated wilderness. *Wilderness areas* are wild or primitive portions of national forests, parks, and wildlife refuges where timbering, most commercial activity, motor vehicles, and human-made structures are prohibited. In 1968, scenic or wild rivers became protected by law. The National Wild and Scenic Rivers System aims to block development on or along selected rivers that have particular aesthetic or recreational value or that are still in a natural, free-flowing state. The rivers in this system are listed in Table 15.1. Also, in addition to wilderness and wild river preserves, our public lands now include national wildlife refuges, which protect the habitats of threatened or endangered species. In the mid-1970s, 367 of these refuges had been established, with at least one in each state except West Virginia (see Figure 14.18).

Certain human activities are permitted in wild and primitive areas as long as they do not infringe on the fundamental objective of preservation. Ranchers may allow their herds to graze in some wildlife refuges and wilderness areas, and public access is provided for recreation. But use of wild areas is rising steadily and, in some regions, overuse by human beings threatens to destroy fragile vegetation, disturb wildlife, and eliminate the possibility of solitude. Unfortunately, traditional methods of controlling use—for instance, paving trails to regulate foot traffic, constructing fences, and planting impact-resistant vegetation—are inappropriate in wild areas. In some wilderness areas, therefore, the number of overnight campers allowed is limited and reservations are sometimes required.

The National Trails System, established by an act of Congress in 1968, provides the public with access to some public lands for hiking and backpacking. Currently, only two trails are included in the system: the Appalachian Trail and the Pacific Crest Trail. The 1968 law also provides that certain local, state, and private trails be designated National Recreation Trails. So far, 116 trails, covering more than 1600 kilometers (1000 miles), have been designated as such.

Although human activities are stringently controlled in wild areas and on other public lands, mining is a glaring exception. In essence, under current mining law, discussed in Chapter 12, the federal government exercises little control over mining sites on most public lands. However, although mining is still allowed in wilderness areas, after 1984, the filing of new claims will no longer be permitted.

Until the 1984 cut-off date, for minerals such as iron, copper, gold, or uranium discovered on public lands, a person need merely file a claim in the local court house. The miner can then maintain the right to that claim indefinitely by demonstrating that at least $100 worth of work is accomplished in working the claim each year. The government exacts no royalty for such claims. But for oil, gas, and oil shale, government control is somewhat more rigid. The

Figure 15.5 A long line of visitors at an entrance to Yosemite National Park. (U.S. Department of the Interior, National Park Service.)

Figure 15.6 Mass transit in Yosemite National Park. (U.S. Department of the Interior, National Park Service.)

Table 15.1 Rivers in the National Wild and Scenic Rivers System.

River	Wild	Scenic	Recreational	Total kilometers
		Kilometers by classification		
Middle Fork Clearwater: Idaho	87	—	211	298
Eleven Point: Missouri	—	72	—	72
Feather: California	53	16	105	174
Rio Grande: New Mexico	83	—	2	85
Rogue: Oregon	53	12	71	136
St. Croix: Minnesota/Wisconsin	—	291	31	322
Middle Fork Salmon: Idaho	166	—	2	168
Wolf: Wisconsin	—	38	1	39
Allagash: Maine	153	—	—	153
Lower St. Croix: Minnesota/Wisconsin	—	19	24	43
Chattanooga: North Carolina/South Carolina/Georgia	64	4	24	92
Little Miami: Ohio	—	29	77	106
Little Beaver: Ohio	—	53	—	53
Snake: Idaho/Oregon	52	55	—	107
Rapid: Idaho	39	—	—	39
New River: North Carolina	—	43	—	43
Missouri: Montana	116	29	95	240
Flathead: Montana	158	66	129	353
Obed: Tennessee	72	—	—	72
Total	1096	727	772	2595

Source: After Council on Environmental Quality, *Eighth Annual Report,* 1977.

federal government leases land holding these resources to developers and shares in the royalties with the states. Despite such control, however, these mining activities can radically interfere with the other purposes of public lands.

Our public lands are coming under increasing pressure from both conservationists and commercial resource exploiters. On the one hand, conservation groups are pressing for the expansion of protected areas and the tightening of use regulations within areas already under federal management. On the other hand, developers, pointing to our needs for

fuels, timber, minerals, and water resources, are arguing for the easing of restrictions. Since both groups address real public needs, it would appear that a compromise is needed whereby we try to meet our resource needs while still actively protecting the quality of the environment.

The Coastal Zone

Currently, public attention is focusing on the use of coastal zones in acknowledgment of the fact that the estuaries, marshes, and other wetland habitats that sustain aquatic life are irreplacable and too fragile to be risked. But the coastline was not always a focus of public concern. Until recently, coastline areas were relentlessly and irreversibily destroyed by uncontrolled and poorly directed development. More than half our population lives on the 160,000 kilometers (100,000 miles) of land that border the Atlantic and Pacific oceans and the Great Lakes, and the sheer density of the population alone has stressed the environment in countless ways. We have long used these coastal areas for waste disposal, recreation, and the siting of industry and power plants, and we have consistently exploited their mineral, natural gas, and oil resources. As a consequence of such human activity, for example, during the period 1955–1964, California lost 67 percent of its estuarine habitats. And in a recent two-year period, so many requests for shoreline modification of Chesapeake Bay were filed with the U.S. Fish and Wildlife Service that, if granted, they would have destroyed 190 kilometers (120 miles) of shoreline.

One example of short-sighted shoreline modification is the construction of artificial barriers that interrupt natural *longshore currents.* Normally, as rivers empty into the sea, they deposit huge quantities of sediment. While a portion of this sediment may build up a delta, another portion is transported along the shoreline by longshore currents. A longshore current is the result of coastal wave action: usually, sea waves approach the shore at an angle, thereby producing a current that flows parallel to shore. This current transports sand along the shore and ensures a continual supply of sand for coastal beaches.

However, this longshore flow of sand can be interrupted by certain artificial structures. After a river is dammed, for instance, the source of sand for a beach is cut off. In this situation, longshore currents transport sand from beaches until nothing but rocky rubble is left behind. In some coastal areas, rock barriers or jetties are constructed to assure boaters calm waters for docking; these structures reduce the force of the waves, thereby decreasing longshore

sand transport. Although these barriers result in calm waters, they do so at the expense of increased sedimentation; harbors become choked with sediment and must be dredged. Also, because most of the sediment is deposited in the harbor, it is unavailable for deposition on down-current beaches. In some regions where breakwaters have been constructed (for example, Santa Barbara, California) it has become necessary to pump sand from the upstream to the downstream sides of the harbor. In effect, energy-consuming pumping does the job that was formerly done by natural longshore currents.

The need to balance use and preservation of coastal zones has spurred shoreline states to develop coastal management measures. Probably the most stringent of these is Delaware's regulation prohibiting the siting of all new industry within 3 kilometers (2 miles) of its shoreline. The federal government is aiding the coastal states' management efforts under the auspices of the 1972 Coastal Zone Management Act. All thirty eligible states are now receiving federal funds intended to help them inventory their coastal resources and set priorities for the orderly use of coastal land and water. By mid-1978, six of the states had received federal approval of their management programs.

In addition, the 1972 law makes federal funds available to assist state efforts to protect estuaries and wetlands because of their biological, scenic, and recreational value. By late 1978, five estuarine sanctuaries had been established and a score of other sites were under consideration for designation as sanctuaries.

Community Growth

A third major category of land use problems involves the control and direction of community growth. Perhaps foremost among these problems is the steady loss of prime agricultural land to development for highways, reservoirs, and housing (see Figure 15.7). Of the half million hectares of cropland taken out of production each year, more than one-half is

Figure 15.7 A housing development on prime farmland. (U.S. Department of Agriculture, Soil Conservation Service.)

taken over by urban and suburban sprawl. If we are to continue to feed our growing population, we must halt or reverse this trend. Our alternative is increasing per hectare food productivity, a course that is likely to entail considerable environmental damage (see Chapter 17) and increased energy consumption.

The amount of land classified as urban in the United States has doubled since 1950, and the growth of freeway systems that facilitate commuting between suburbs and the inner city has contributed significantly to urban expansion. The original goal of the National System of Interstate and Defense Highways, initiated during the Eisenhower administration, was construction of a 66,000-kilometer

(41,000-mile) network of high-speed, limited-access highways linking 90 percent of United States cities with populations over fifty thousand. Although states and cities enthusiastically participated in extensive interstate highway construction during the 1950s and 1960s, today with the interstate system 92 percent completed, many localities are reevaluating the merits of freeways. These highways crisscross the city, as shown in Figure 15.8, breaking up old, established neighborhoods, encouraging industry and population shifts to formerly rural areas, and forcing an increased dependency on motor vehicles. Also, highway construction has had an adverse affect on the landscape, destroying wildlife habitats, disrupting drainage patterns, and distorting scenic panoramas.

Figure 15.8 A freeway cutting a wide swathe through a city. (U.S. Department of Housing and Urban Development.)

In response to these problems, the 1973 Highway Act was passed to give cities an alternative to building more highways. The act allows federal funds earmarked for freeway construction to be used instead for upgrading urban mass transit systems. For example, at the request of city and state officials, federal funds originally designated for construction of a 10.1-kilometer (6.3-mile) segment of expressway in downtown Chicago were reallocated primarily for subway construction. Many obstacles have emerged to block the widespread use of this option, but proposed legislation promises to encourage its widespread adoption in the future.

States are becoming increasingly active in making legislative responses to conflicts in community land use. By the mid-1970s, more than half the states had enacted general land use programs. Dozens of states now have laws regulating the siting of power plants and controlling strip mining. Also, many states have laws that protect wetlands and that identify and protect "critical areas," that is, regions that are particularly sensitive to disturbance or that have historical importance. Furthermore, some states now provide tax incentives to encourage landowners to retain land for agriculture or preserve land for its aesthetic or recreational value.

Traditionally, towns and suburbs have employed zoning ordinances to restrict residential growth. Usually, these regulations specify minimum lot size per home, size of structures, and setback distances.

In addition, some communities have growth control ordinances that limit the number of new dwellings. For example, after its population soared by 25 percent in only two years, Petaluma, California, passed an ordinance allowing the construction of no more than five hundred new subdivision homes each year.

Following the Petaluma example, citizens of Boulder, Colorado, voted in 1976 to curb their city's explosive growth by imposing a quota on new housing. New residential units in Boulder are limited to 415 per year (half the prior rate) in an effort to keep annual population growth at less than 2 percent. Housing permits are awarded on the basis of merit, with consideration given to architectural design, energy system, and accessibility to public services. The Boulder plan also includes provisions for a greenbelt around the city.

Growth control ordinances, however, have drawn considerable fire from critics arguing that the regulations create economic barriers to open housing. The Boulder ordinance, for example, is being challenged by the city's Chamber of Commerce, which claims that the plan has virtually eliminated middle-income housing. No doubt the legality of growth control ordinances will be ultimately determined in the courts.

Land Use and Geologic Hazards

One objective of federal, state, and local land use control strategies is to provide for our resource and recreational needs now and in the future while protecting against environmental disturbance. But another factor is equally significant: planning must account for the fact that some lands are vulnerable to geologic hazards such as floods, earthquakes, and landslides. A comprehensive land use policy, therefore, must regulate habitation of these areas to minimize the risk to life, limb, and property. Such a policy must be based on an informed analysis of the geologic characteristics of specific areas and an un-

derstanding of the nature of the various potential hazards.

Flood Hazard

Floods are the most damaging of our nation's natural hazards. Floods account for the majority of events officially termed disasters by presidential declaration (whereby residents became eligible for federal assistance). Floodwater flattens buildings; erodes valuable topsoil; drowns people and livestock; disrupts municipal water systems; and interferes with communications, transportation, and commerce. Also, as floodwaters recede, they leave behind thick layers of silt and mounds of debris. It is not difficult to imagine the extent of the damage sustained in the flood pictured in Figure 15.9. In 1975, the Federal Insurance Administration estimated the annual loss due to floods in the United States to be $3.8 billion. But property damage aside, the toll of injury and death due to floods is mounting annually nationwide. During the 1970s, flash floods took an average of two-hundred lives per year, twice the flood fatalities of the 1960s and three times those of the 1940s.

The damage and fatalities attributable to flooding are increasing because, simply speaking, the human habitation of floodplains is increasing. People have always been drawn to floodplains by fertile soils, plentiful water for irrigation, and the potential for inexpensive transport. With the shift from an agrarian to industrial-based society, however, the draw has become even stronger. The flat terrains of floodplains have become economically attractive as sites for highways, railroads, buildings, and homes. And with the steady growth of cities, population on floodplains soared. Today, many U.S. cities—for example New Orleans—are built almost entirely on floodplains. About 15 percent of the nation's total urban land and almost 10 percent of the total agricultural land are floodplains and are therefore vulnerable to the phenomenon known as hundred-year flood—that is, a large-scale flood that, as a rule of

Figure 15.9 Flood damage to Sevierville, Tennessee. (Tennessee Valley Authority.)

thumb, can be expected to occur once a century.

Also contributing to the increased flood fatalities in recent years is the fact that remote areas where the flood hazard is unknown and where flood warnings are not readily communicated have become more accessible. It is a growing trend among people in quest of recreation to drive their cars, campers, and mobile homes to flood-prone mountainous areas in search of relative solitude. These vacationers make camp and build cabins in narrow canyons that are

subject to unexpected and rapid rises in stream level. On July 31, 1976, more than 130 lives (mostly campers) were lost in the Big Thompson Canyon in Colorado when heavy rains caused the water level to rise unexpectedly.

Can floods be prevented, or can flood damage at least be reduced? To answer these questions we need to understand the causes of flooding. A flood occurs whenever runoff exceeds the discharge capacity of a river channel, causing water to flow over riverbanks and spread out over the floodplain beyond. Excessive runoff sufficient to cause flooding results from natural events, the activities of people, or some combination of the two. In the former category, hurricanes and intense thunderstorms produce heavy rainfall that exceeds the infiltration capacity of soil. For instance, torrential rains that were remnants of Hurricane Agnes triggered flooding that caused property damage in excess of $3.4 billion in the northeastern states during the summer of 1972. In the same year, 237 lives were lost in Rapid City, South Dakota, when exceptionally heavy rains caused flash flooding. An example of flash flood damage is shown in Figure 15.10. Rapid spring snow-melt also causes a sharp rise in river levels, a problem that is sometimes compounded by the break-up of river ice. Huge slabs of ice pile up at bridges or channel narrows, causing water to back up and flood.

Human-caused flooding arises from development. As cities grow and expand, new roads and buildings render more and more land impervious to water. Sewer systems in many cities are unable to handle the huge volumes of water that can accompany a summer downpour. Consequently, basements, viaducts, and low-lying areas are subject to rapid inundation. Another human factor is the removal of the land's vegetative cover. Vegetation slows the flow of runoff and thereby reduces flood threat. Logging, overgrazing, and mining are particularly notorious for disturbing the land and contributing to the flood hazard. In addition, the failure of dams, levees, and other structures designed, ironically, to prevent flooding, is a major cause of flood calamities.

Until recently, our nation's flood control efforts consisted exclusively of a variety of engineering projects aimed at excluding floodwaters from populated areas. But in spite of billions of dollars spent on this structural approach, flood-related death, injury, and property damage continue to rise. Today, floodplain management is receiving more emphasis as a flood control strategy. Basically, the shift has come about because structural efforts at flood control failed to contain floodwater and resulted in significant environmental damage. And, in fact, this trend in flood control reflects the pattern of land use trends generally.

Structural flood control measures are matched to specific characteristics of a flood-prone area. Where possible, land along river valleys is shaped into terraces, beaches, or levees to confine floodwaters. In some cases, the discharge capacity of a river is enlarged by dredging the channel, or channeling (straightening) the stream. *Stream channelization,* a controversial flood control strategy, is discussed in detail in Box 15.1. In other instances, earthen or concrete dams, dikes, or floodwalls are built to detain or divert floodwaters. These flood protection measures are enhanced by erosion control measures that protect the soil and vegetative cover of the watershed.

Most flood control structures are actually designed to accommodate only moderate flooding. Nonetheless, the presence of any flood protection device often engenders a false sense of security in local citizens and inspires land development and home construction. The stage is then set for a potentially catastrophic flood.

Several recent disastrous dam failures have demonstrated the vulnerability of many of these flood control structures. In June, 1976, the Teton Dam in eastern Idaho failed while its reservoir was being filled for the first time. The gap in the dam is shown in Figure 15.11. Fourteen lives were lost and property damage approached $1 billion. Subsequent investigations revealed that a combination of geological factors (the most significant was a highly permeable foundation rock) and inadequate design was responsible for the dam's failure. And on November 6, 1977, an earthen dam failed during heavy rains above the campus of Toccoa Falls Bible College in Georgia, resulting in thirty-nine deaths. The Teton and Toccoa Falls disasters heightened national concern over dam safety.

One consequence of the Teton disaster was the delay (leading, perhaps, to the eventual scrapping) of the $1 billion Auburn Dam project in California. Designed as the largest concrete dam of its kind in

Figure 15.10 Flash-flood damage to homes in Red Butte Village, west of Casper, Wyoming. (U.S. Department of Agriculture, Soil Conservation Service.)

Figure 15.11 The failure of the Teton Dam in eastern Idaho. This view from downstream shows the gap in the dam. (U.S. Department of the Interior, Bureau of Reclamation.)

the world, the Auburn Dam would impound the North Fork of the scenic American River above Sacramento. Following the Teton failure, scientists raised concerns that earthquakes might cause the dam to burst, imperiling the lives of 750,000 people in metropolitan Sacramento. The magnitude of the risk involved prompted the White House (with Congressional support) in 1977 to block the project until the dam is demonstrated to be safe. Consequently, the Bureau of Reclamation, which built the Teton Dam, embarked on a $25 million investigation of the geologic hazards at the proposed Auburn Dam site.

The Toccoa Falls disaster strengthened governmental resolve to inspect the nation's fifty-thousand dams for potential hazards. The U.S. Army Corps of Engineers has been charged with conducting the inspection program. Although this plan was approved by Congress in July 1977, little action was taken until a presidential directive following the Toccoa Dam failure.

Congress responded to the catastrophic floods of 1972 by enacting the Federal Flood Disaster Protection Act of 1973, which reflects the new emphasis on floodplain management as opposed to structural control measures. This law requires that in order to be eligible for federal flood insurance, local governments must adopt floodplain development regulations. Construction projects within regions identified as flood hazard areas that do not qualify for flood insurance are denied federal funding altogether. Floodplain development regulations call for the elevation or floodproofing of buildings and provisions for a floodway that will allow floodwaters to pass through a community with a minimum of damage.

The goal of the 1973 law is to foster the wise use of floodplains and other flood-prone areas and thereby to reduce the toll of floods. This law encourages communities to use floodplains in ways that are compatible with periodic flooding: for agriculture, forestry, some recreational activities, parking lots, and wildlife refuges, rather than as construction sites for homes and industries.

Box 15.1

One of the more controversial flood control measures is stream channelization, the straightening and ditching of meandering stream channels. Channelization alleviates the danger of upstream flooding by allowing water to flow rapidly downstream. Projects to this effect are sponsored by such federal agencies as the Soil Conservation Service, the Bureau of Land Management, and the U.S. Army Corps of Engineers. Although stream channelization may successfully control upstream flood waters, it can merely displace the flood problem downstream. And it sometimes triggers a chain of environmental problems.

In some areas, stream channelization has disrupted fish spawning grounds and destabilized stream banks, leading to accelerated erosion. And it has increased the sediment load of streams, thus reducing aquatic life—in some areas the recreational fishing potential has been virtually destroyed. The Sport Fishing Institute reports that game fish in some North Carolina streams were reduced by 90 percent following channelization. On the other hand, in heavily silted areas, channelization can actually restore the aquatic habitat necessary for some species of fish and other aquatic life.

In 1962, the Kissimmee River, which connects Lake Okeechobee and Lake Kissimmee in south central Florida, was a gently meandering stream some 160 kilometers (100 miles) long. By 1971, it

Earthquake Hazard

In August 1976, one of the greatest natural disasters of recorded history struck T'ang-shan in northeast China. More than seven-hundred-thousand people

Channelization of the Kissimmee River

had been drastically transformed by stream channelization into a straight, swiftly flowing canal only 93 kilometers (58 miles) long. Now citizen action groups, conservationists, and some government officials are asking that the channelization be reversed.

The straightening and ditching of the Kissimmee River was one component of a comprehensive flood control plan for southern Florida. Engineers reasoned that a channelized Kissimmee would more effectively regulate the water levels in the two lakes it connected and also provide a drainageway for developing areas, including housing tracts, in the lake watersheds. The project was authorized in 1954. Construction began in 1962, and the job was completed nine years later at a cost of almost $24 million.

Even before the project was finished, voices were raised in opposition to channelization. In 1972, critics charged at a public hearing that channelization was contributing to the eutrophication of Lake Okeechobee. The next year, the Florida legislature appointed an investigative body to look into the problem. This study group concluded that channelization had been followed by intensified land use—especially agriculture and housing—and had thereby accelerated the runoff of fertilizers and other nutrients into Lake Okeechobee.

Critics argued that channelization was causing other problems besides increased eutrophication. They noted that it had drained wetlands, causing waterfowl populations to decline precipitously. They argued that the potential for flood damage was actually increased because the newly drained land lured more people onto floodplains. And they pointed out that channelization was not designed to accommodate runoff from the torrential downpours that accompany tropical storms.

In 1976, in response to mounting evidence that channelization of the Kissimmee River was causing serious ecological harm, the Florida legislature unanimously adopted the Kissimmee River Restoration Act. This law called for measures, including a reduction in the river's nutrient content, to restore the water quality of Lake Okeechobee. And it established a coordinating council to propose specific remedies. After a lengthy review of the problem, the coordinating council recommended that the best way to improve water quality was to control runoff from agricultural lands.

Many critics were not satisfied with the coordinating council's recommendation. They argued that the original intent of the Restoration Act was to dechannelize the Kissimmee River. This would be a $30 million endeavor—more than channelization had cost. But public and political support for such a restoration effort is strong, and dechannelization may well begin soon.

---

perished when a strong *earthquake* destroyed homes and businesses in the congested city. Local buildings had not been constructed to withstand earthquake tremors, and the city was situated on unstable river sediment that shifted in the violent quake.

A multitude of earthquakes have occurred during the course of human history—some catastrophic, some very minor. Most major earthquakes occur along boundaries of gigantic crustal plates (refer back to Figure 12.5). Adjacent plates grind against one another along boundary zones, and occasionally huge slivers of rock become stuck in these zones, causing energy to accumulate within the rock. Then, abruptly, the rock slivers break free, releasing the trapped energy in the form of waves, which travel great distances across the globe. In large earthquakes, the energy released may be fifty times greater than that generated in a nuclear explosion.

The West Coast is the primary site of earthquake activity in the United States. The sliding of the Pacific plate past the North American plate produced the San Andreas fault—a 430-kilometer (270-mile) rift that has been the location of numerous severe earthquakes (see Figure 12.7). One of the more infamous of these was a quake that leveled much of San Francisco in 1906. Figure 15.12 shows some interior damage sustained in that quake.

The violent shaking of the earth is not the only cause of quake damage: earthquakes can trigger landslides and avalanches of rock, mud, and snow; they can disrupt the flow of rivers and groundwater, causing rivers to flood and wells to run dry; and, in some coastal areas, major earthquakes can generate the very dangerous *tsunamis,* seismic sea waves.

At sea, the energy in a tsunami is dispersed within a large volume of water, and the wave therefore poses no threat to navigation. It travels along at tremendous velocities (more than 800 kilometers per hour), but at the inconspicuous height of less than 2 meters. When a tsunami approaches certain shorelines (particularly around the Pacific), however, its energy gradually becomes focused into a wall of water that can be taller than a three-story building. During the Alaskan earthquake of March 1964, for example, a tsunami more than 10 meters (32 feet) above high tide level crashed into Kodiak harbor, causing extensive damage. The loss of life accompanying a tsunami can be staggering. In August 1976, a very strong earthquake centered off Mindanao in the Philippines triggered a series of tsunamis that claimed five-thousand lives. These lives were lost in spite of a tsunami warning system that has operated in the Pacific since 1948.

In view of the death and destruction attributable directly or indirectly to earthquakes, it is appropriate to ask whether anticipating them is possible. In recent years, with the growth of population in seismically active regions, earthquake prediction has become a high-priority objective of scientific research. Several forecasting schemes are currently under study. One technique relies on the computation of recurrence rates, that is, the determination of earthquake frequency in a specific area from past records. This method is not satisfactory, however, since historically earthquakes are random events. Also, several violent earthquakes have occurred in locations with no prior record of significant earthquake activity.

Another earthquake prediction technique is based on the identification of *seismic gaps,* that is, regions of relative quiescence within a seismically active belt. The existence of seismic gaps may indicate that energy is gradually accumulating in anticipation of a major quake. Although predicting the timing of an earthquake is not possible using this method alone, seismic gaps may pinpoint areas that are not appropriate for new housing developments.

The primary objective of most earthquake prediction research today is the identification of premonitory events, that is, signals that precede potentially destructive earthquakes. Extremely sensitive instruments monitor the buildup of earthquake energy in rocks along fault zones. Other instruments detect slight changes in land elevation or changes in very small earthquake waves that may occur before a major quake. Behavior changes in animals prior to an earthquake are also being studied.

Once prediction methodology has been perfected and refined, we will be able to take precautions to minimize the damage and injury sustained in major earthquakes. But even before prediction methods are perfected, we can institute certain safety procedures in earthquake-prone regions. (Figure 15.13 shows the zones of earthquake risk in the United States.) We can write building codes and zoning regulations to limit the height of buildings and otherwise mini-

Figure 15.12 The interior of a building wrecked by the San Francisco earthquake of 1906. (U.S. Geological Survey photo by W. C. Mendenhall.)

mize earthquake damage. Also, we can limit building on terrain prone to sliding during earthquakes or in coastal areas where tsunami threat exists.

Landslide Hazard

Landslides are the downslope movements of rock, soil, mud, or a mixture of these materials. As with floods, earthquakes, and volcanic eruptions, landsliding is a natural phenomenon that contributes to the evolution of the landscape. Some landslides are gradual, and others are abrupt and catastrophic. In remote regions, landslides may pass unnoticed, but in populated areas they cause considerable property damage, wrecking homes and in some cases burying entire towns.

Land slips along inclined fracture planes within

rock layers and soil, or when the entire mass of rock or soil becomes saturated with water. The actual slide is usually triggered by an earthquake or by the infiltration of huge quantities of water during periods of prolonged rainfall. But human activities can also trigger landsliding. The terracing of mountains and hillslopes for home sites or highways can create dangerously steep slopes that fail under heavy rains. The removal of vegetation through overgrazing, timber cutting, or the preparation of construction sites may allow water to more readily infiltrate soil, transforming it into a mud slurry that rapidly flows downslope. Irrigation and other activities that alter groundwater flow may also saturate sloped land and trigger sliding.

The nation's West Coast is particularly prone to landsliding, as Figure 15.14 indicates. This region is

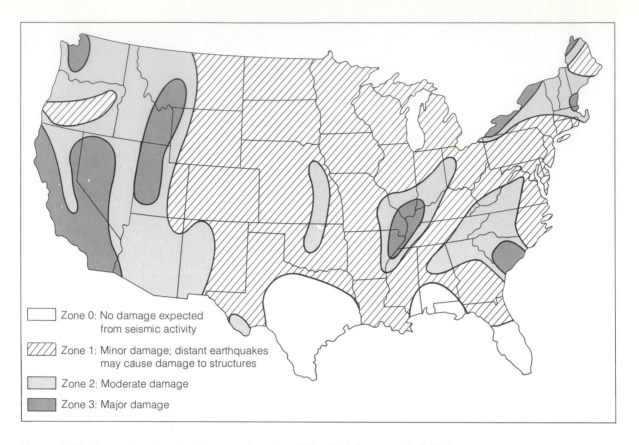

Figure 15.13 Zones of earthquake risk across the nation. (After D. N. Cargo and B. F. Mallory, *Man and His Geologic Environment.* Reading, Massachusetts: Addison-Wesley, 1977. Reprinted with permission.)

Legend:

☐ Zone 0: No damage expected from seismic activity

▨ Zone 1: Minor damage; distant earthquakes may cause damage to structures

▤ Zone 2: Moderate damage

▩ Zone 3: Major damage

hilly or mountainous, seismically active, and subject to intense rainfall, all characteristics of landslide-prone areas. The hazard to people in this region is increasing, for as the demand for living space soars, new housing subdivisions are built on dangerous hillsides and mountain slopes. Figure 15.15 shows the result of landsliding on a construction site.

Although we are able to map areas of landslide susceptibility, our ability to control sliding is limited. We can stabilize hillslopes denuded of vegetation by revegetation, construct retaining walls at the base of hillslopes, and install elaborate drainage systems to divert surface and subsurface water away from steep hillslopes. But these strategies are not always successful, because many variables contribute

simultaneously to landsliding. The most reasonable strategy at the present time is to discourage construction in landslide-prone regions.

Conclusions

As our population grows, increased competition for our finite lands becomes inevitable. Conflicts over land use where a choice exists are bound to become more frequent, and, in the near future, as we step up our efforts to find new sources of fossil fuels, they will doubtless grow more severe. Our public lands will become the target of increased pressures from housing developers and resource exploiters, and, as

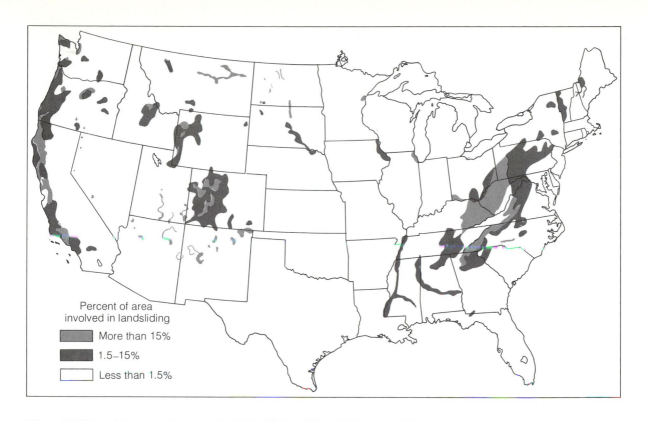

Figure 15.14 Landslide potential across the United States. (After U.S. Geological Survey, *Annual Report,* 1976.)

Figure 15.15 Landslide damage in Orinda, California. (U.S. Department of Agriculture, Soil Conservation Service.)

Box 15.2

Resolving Land Use Conflicts

Conflicts about land use can sometimes—although not always—be resolved through compromise. Consider the example of New Hampshire's Franconia Notch State Park, the site of some of nature's most striking sculpturing. Craggy granite cliffs, pristine lakes and streams, and pine forested mountain peaks provide spectacular vistas. Unfortunately, the park also lies in the path of Interstate 93, a major highway that stretches from the outskirts of Boston to the Canadian border.

Work on the Notch segment of the highway was held up for two decades as conservationists fought to keep the roadbuilders out of the park. Highway interests claimed that the new road was essential to the well-being of the state's economy and pressed for its construction through the park. Conservationists argued that the highway would destroy some of the park's natural amenities and threaten a famous landmark, the Old Man of the Mountains (a 15-meter-tall natural rock sculpture, perched on a mountainside, that resembles a craggy human profile).

Compromise resolved the controversy late in 1977. Interstate 93 will be routed through Franconia Notch, but construction will be limited to upgrading the two-lane road that presently winds through the park. This will minimize disruption of parkland, and the segment will be the second of only two places along the 66,000-kilometer (41,000-mile) interstate system that has two instead of four or more traffic lanes.

In other cases compromise is not feasible and other strategies must be employed to settle land use conflicts. An example is the expansion of the Redwood National Park in California. Conservationists feared that continued logging activity on land near the park would disturb the park's resources. They therefore pressed to protect the park by expanding its boundaries to include an adjacent 19,500 hectares (48,000 acres) of timberland. But opposition came from local loggers and sawmill workers who argued that closing the land to logging would cause job layoffs as well as disruption of the local economy.

Congress agreed with the conservationists and passed the Redwood National Park Act in March 1978. But Congress also recognized the need to alleviate the local economic impact. Hence, the law not only approved the addition to the park but also provided federal funds to offset potential unemployment. Concessions to loggers and sawmill workers included retraining and relocation benefits, preferential hiring for park jobs, and unemployment compensation at rates comparable to or higher than the worker's maximum income prior to being laid off.

---

the cities expand, more people may be forced into regions prone to floods, earthquakes, and other geologic hazards.

Only two general responses to these trends are open to us: either we curb our growing demands for resources or we institute more stringent land use regulations. Controls will have to be designed to balance environmental quality and resource exploitation. Inevitably, such a balance will call for compromise—for instance, that described in Box 15.2—by both conservationists and resource exploiters.

Summary Statements

Agencies of the federal government regulate activities on one-third of the nation's lands through management of national forests, parks, wildlife refuges, and other public lands.

Some public land is set aside for the prime purpose of preservation and other lands are intended for regulated multiple use.

Mining is the one human activity not subject to stringent controls in wild areas and in other public lands.

Conservationists are increasingly pressing for expansion of protected areas, while resource exploiters argue for the easing of use restrictions.

The need to balance use and preservation of coastal zones has spurred shoreline states to adopt management measures.

New highways, reservoirs, and housing developments are contributing to a steady loss of the nation's prime farmland.

States control land use through laws regulating construction sites and tax incentives. Communities traditionally control growth through zoning ordinances.

Land use control strategies must account for the fact that certain lands are vulnerable to geologic hazards.

Damage attributable to flooding is on the rise because floodplains are more densely inhabited than ever and remote flood-prone areas are more accessible to more people.

Floods may be caused by natural events (for example, excessive rains and rapid spring snow-melt) and by human activities that alter the vegetative cover. In the past, our nation's flood control efforts focused almost exclusively on structural alternatives (dams and levees, for example). Today, the emphasis is increasingly on floodplain management as a flood-avoidance strategy.

An earthquake is a violent shaking of the land that can trigger landslides, avalanches, floods, and tsunamis.

Earthquake forecasting research primarily focuses on recurrence rates, seismic gaps, and premonitory events.

The terracing of mountains or hillslopes for home sites or highways may create dangerously steep slopes of rock and soil that slide under heavy and prolonged rainfall.

The future is likely to see increasing competition for lands as our demands continue to grow. On multiple-use lands particularly, more conflicts among users appear inevitable.

Questions and Projects

1. In your own words, write definitions for the terms italicized in this chapter. Compare your definitions with those in the text.

2. How is land use regulated in your community? Your local planning agency can provide you with valuable information.

3. Is your region prone to floods, earthquakes, or landslides? What precautions are taken to minimize the impact of these hazards? Are approaches merely structural or do they include management efforts?

4. Some critics argue that community growth control ordinances are discriminatory. What is your view? Support your opinion.

5. Strong arguments support our need for more mining and the preservation of wildlife habitats. However, most mining activities are not compatible with the preservation of sensitive wildlife habitats. Devise a plan to resolve this conflict in land use.

6. Someday scientists may be able to pinpoint locations where future earthquake activity is likely. Suppose such a prediction were made for your community. How would you and your fellow residents react? What would be the economic impact?

7. List activities that are appropriate along a river floodplain. Order these activities in terms of their importance. What criteria do you use to set priorities?

8. Identify means other than zoning and growth control ordinances that a community could use to curb its growth.

9. Comment on the notion that some conflict in land use is an inevitable consequence of our way of life.

10. Prepare a list of the demands that each individual places on the land. List these demands in order of importance.

11. Enumerate those environmental factors that must be considered in selecting a suitable site for a new (a) shopping center, (b) copper smelter, and (c) hospital.

12. What is the primary purpose of national forests?

13. How strictly are mining activities regulated on public lands?

14. Summarize the various strategies employed by communities to regulate their growth.

15. Outline the advantages and disadvantages of urban freeway systems.

16. Why is the toll of fatalities and property damage due to floods on the rise?

17. Explain the recent shift away from structural control of floods and toward floodplain management.

18. What human activities contribute to flood threat?

19. What measures can be taken in earthquake-prone areas to reduce the toll of death, injury, and property damage?

20. Enumerate methods of earthquake prediction currently under study.

Selected Readings

Cargo, D. N., and B. F. Mallony. 1977. *Man and His Geologic Environment.* Reading, Massachusetts: Addison-Wesley. An introductory text that describes causes of geologic hazards.

Carter, L. J. 1976. "Wetlands: Denial of Marco Permits Fails to Resolve the Dilemma," *Science 192:*641–644 (May 14). A summary of conflicts between real estate developers and conservationists concerning the mangrove swamps of southwest Florida.

————. 1977. "Auburn Dam: Earthquake Hazards Imperil $1 Billion Project," *Science 197:*643–649 (August 12). An account of the concerns that led to delays in construction of the Auburn Dam in California.

Council on Environmental Quality. 1975. *Sixth Annual Report.* Washington, D.C.: U.S. Government Printing Office. Includes a summary of land use trends and critical issues related to the nation's land use policies.

Griggs, G. B., and J. A. Gilchrist. 1977. *The Earth and Land Use Planning.* North Scituate, Massachusetts: Duxbury Press. A basic text that introduces the geologic information needed for land use decisions.

Mark, R. K., and D. E. Stuart-Alexander. 1977. "Disasters as a Necessary Part of Benefit-Cost Analysis," *Science 197:*1160–1162 (September 16). An article advocating the inclusion of disaster costs in the economics of water control projects.

Soil Conservation Society of America. 1976. "Managing Floodplains to Reduce the Flood Hazard," *Journal of Soil and Water Conservation 31:*44–62 (March/April). A collection of papers on floodplain dynamics, the nation's increasing vulnerability to floods, and approaches to floodplain management.

Tank, R. W. 1976. *Focus on Environmental Geology.* New York: Oxford University Press. Includes case histories and selected readings on geologic hazards.

Trefethen, J. B. 1976. *The American Landscape: 1776–1976, Two Centuries of Change.* Washington, D.C.: The Wildlife Management Institute. A well-illustrated historical account of the growth in our land needs.

United States Geological Survey. 1978. "Nature To Be Commanded . . .," *Earth-Science Maps Applied to Land and Water Management,* Geological Survey Professional Paper 950. Washington, D.C.: U.S. Government Printing Office. An exceptionally well-illustrated demonstration of how the geological sciences provide valuable information for land use analyses.

Part III

Fundamental Problems: Population, Food, and Energy

In previous chapters, we studied the ways in which the environment works, and the ways in which human activities have disrupted the environment's normal functions. We also considered human attempts to rectify past mistakes in resource management, and evaluated strategies for improving environmental quality and for utilizing natural resources more efficiently. Throughout these discussions, we repeatedly pointed out that the success of any strategy is affected by two critical variables: future population growth and future supplies of energy. Furthermore, because agriculture has a significant impact on environmental quality, because often it is the central issue in land use controversies, and because we need a continual supply of food to stay alive, the quality of our lives in the future will depend greatly on world food supplies.

In this section, we consider these three variables, representing the fundamental problems facing humankind: the size of our own population, the sources of our energy supply, and our patterns of food production. In Chapter 16, we first examine the impact of continued population growth on environmental quality and resource management, and then consider the principles governing population growth. In the context of these principles, we evaluate strategies to slow human population growth. In Chapter 17, we evaluate the promises and environmental limitations of various efforts to enhance food production. In Chapters 18 and 19, we consider present and future energy supplies and energy consumption in the various sectors of our society. Chapter 18 deals mainly with the role of conventional energy sources (coal, oil, natural gas, and uranium) in meeting our energy needs, and Chapter 19 examines the promise of alternative sources of energy, including solar and wind power. We hope that in examining these critical problems, each reader will consider the potential consequences in global as well as personal terms and begin adopting ways that lead to solutions.

A table at a railway station in Bombay, India, where birth control devices are distributed free to the public. (Agency for International Development.)

Chapter 16

Human Population: Growth and Control

Crowding, inflation, unemployment, pollution, and dwindling resources—all these circumstances are related to the fact that too many people are competing for a finite number of resources and opportunities. Today, more than 4.3 billion people inhabit this planet; tomorrow 160,000 more, and next year 60 million more will be added to our population. Although the U.S. population is growing more slowly, our numbers nationwide are expected to increase by at least another 15 percent, to 252 million, within the next fifty years.

The pressures of population growth have already forced us to make some changes in our way of life. In the United States, communities faced with dwindling resources are formulating zoning ordinances to limit their growth. In preceding chapters, we identified the kinds of conflicts that arise when too many individuals compete for the same resource: for instance, should water be used for municipalities or irrigation, forest lands for timber or recreation, land for housing tracts or agriculture? Our high level of affluence determines to a great extent the nature and frequency of these conflicts and has implications not only for ourselves but also for other people of the world.

Although Americans make up less than 6 percent of the world's population, we consume five times our fair share of the resources exploited in the world each year. Hence, each American actually has a much greater impact on the environment than does the average citizen of a less developed nation such as India or Kenya.

In turn, we in the United States are influenced in many ways by population growth in other countries. Gainful employment is a growing problem worldwide. The promise of work in this country, where the level of affluence has been consistently high, has drawn 3–5 million illegal immigrants, as many as two-thirds citizens of Mexico, thereby swelling the number of job seekers competing with each other in this country. Each year, an estimated eight-hundred-thousand additional people enter the country illegally.

As we have seen, the fact that world population is growing also means that the demand for goods is rising. Unfortunately, enhancing the available supply of valuable resources (water, minerals, timber, and food) to meet these demands has become increasingly difficult. Hence, as demand continues to exceed supply, prices go up and scarcity-induced inflation becomes a significant worldwide problem. These trends are reflected in the price rises for many products marketed in the United States (see Figure 16.1). Furthermore, in our attempts to enhance the supply, we have begun to overexploit some resources (ocean fisheries, discussed in Chapter 17, are a good example). Consequently, the supply of these resources is actually declining while the demand continues to rise, further fueling inflation and perhaps continued overexploitation.

On the agricultural front, a growing population means growing demand for food. The resulting expansion of agriculture leads in turn to more soil erosion. And a larger demand for fuel leads to the burning of more fossil fuels and wood, which raises carbon dioxide and particulate levels in air. The growth of industries and commerce to provide needed jobs further contaminates air and waterways and will continue to do so unless pollution control efforts are stepped up accordingly. If unchecked, these exploitive activities could contribute to climatic change and further aggravate the problems of water supply and pollution.

Many of the conditions that result from population pressures—overcrowding, unemployment, pov-

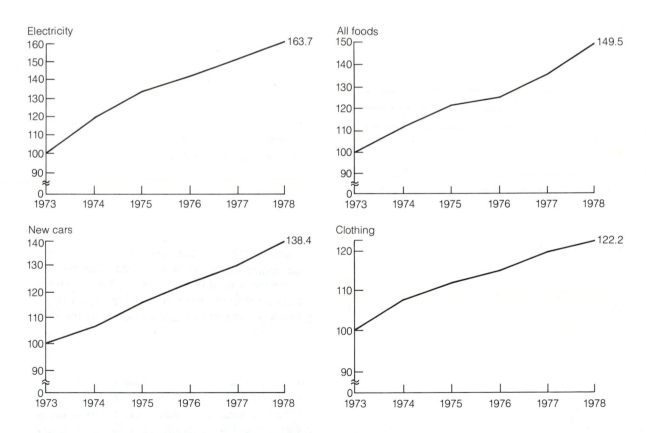

Figure 16.1 Price trends for some key items in family budgets in recent years. Prices for 1973 are set equal to 100. (After U.S. Department of Labor.)

erty, hunger, and illness—lead to social dissatisfaction. Such unrest in less developed countries is already evident today in the form of ever more frequent political confrontations and an intensification of terrorist activities. Often, extreme dissatisfaction erupts into war, usually with neighboring countries, as a nation seeks to relieve its population pressures. Today, our well-being in the United States depends on a continual supply of many resources from other countries (see Table 16.1), many of which are themselves suffering from the burdens of supporting a burgeoning population. Political upheaval in these countries can cut off our access to essential resources. The Iranian situation in 1979 and the resulting reductions in oil imports constitute a prime example.

At this writing, many people in the world still feel immune from the problems associated with population pressures. But population growth affects us all. Each new baby places additional pressure on the earth's food, energy, water, mineral, and space resources, and the more affluent the parents, the stronger will be the pressures of the individual child on these resources. Eventually, even the wealthiest people will feel these effects. Pollution, for example, touches everyone; no ecological islands exist where one can go to escape it. Inflation affects all economic classes, though at this point what may be just an inconvenience to the affluent can seriously threaten the survival of the poor. But eventually, as shortages become more apparent, the wealthy will begin to suffer the lack of necessities, too.

Population growth is not the only factor responsible for our problems. Rising affluence, inappropriate technologies, and mismanagement of resources also contribute directly or indirectly to our difficulties. But unless population growth is slowed significantly, few countries will be able to meet the needs of future generations, let alone provide adequate food, water, shelter, health care, and education to those people already living. In this chapter, we explore some of the causes of the recent human population explosion and evaluate existing strategies for population control.

A Historical Overview

Throughout most of human history, the human population was quite small, grew quite slowly, and even experienced occasional declines. Figure 16.2 shows the general trend of population growth over the last ten-thousand years. Our ancestors were vulnerable to a hostile environment: food often was scarce; protection against disease nonexistent, and clothing and shelter were primitive. About ten-thousand years ago, humankind began a transition from a hunting-and-gathering existence to an agricultural way of life. With the cultivation of crops and domestication of animals, the food supply became more reliable and population size began to grow somewhat faster. But, throughout, famines took their toll in human health and human lives. Figure 16.3 is a classic illustration of the effects of famine.

Historically, epidemics of disease have also severely limited population growth. In the Middle Ages, bubonic plagues reduced Europe's population substantially in the fourteenth and again in the seventeenth century. Figure 16.4 shows the impact of plague on the population of Europe over time. These

Table 16.1 United States imports of some leading commodities for 1977.

Commodity	Value (in millions of dollars)
Petroleum and related materials	41,526
Metals and metal products	12,246
Chemicals	5,432
Ores and metal scrap	2,234
Lumber	2,098
Natural gas	1,945
Wood pulp	1,215
Rubber, including latex	645

Source: U.S. Department of Commerce

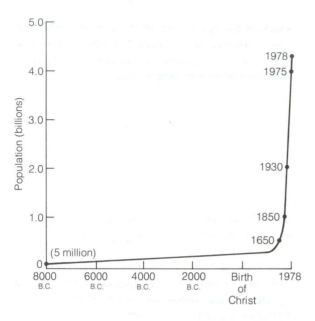

Figure 16.2 The growth of the human population from ten-thousand years ago to the present.

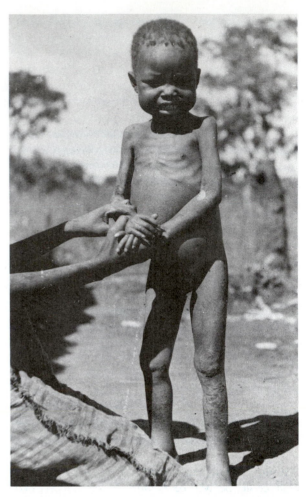

Figure 16.3 A child in East Africa suffering from a form of protein-calorie malnutrition called kwashiorkor. (FAO photo.)

epidemics were frequently accompanied by severe outbreaks of influenza and typhus. Whole populations were terror-stricken by these outbreaks, since no one knew the causes of the diseases and no one could prevent the epidemics.

By the early 1800s, however, the medical profession had begun to control many diseases by means that we now take for granted. One of the first great advances occurred in 1796, when Edward Jenner demonstrated that smallpox could be prevented by inoculating human beings with material from cowpox lesions. Today, smallpox, which once killed hundreds of thousands of people each year, has been virtually eliminated from the face of the earth. Other diseases were shown by Louis Pasteur and Robert Koch to be caused by bacteria, and these findings also led to cures.

As empirical knowledge increased, the ancient beliefs that demons and noxious vapors caused ill-ness were slowly dispelled, and helplessness in the face of disease was gradually reduced by improved sanitation and medical care. As we saw in Chapter 7, acceptance of the germ theory of disease inspired the chlorination of drinking water and the development of sewage treatment systems. Antiseptic techniques were developed for use in doctors' offices and hospitals. (Surprising as it may seem today, as recently as

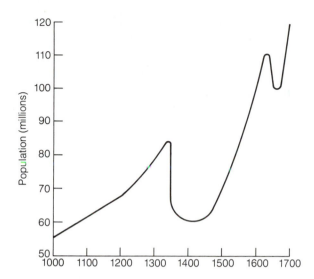

Figure 16.4 The impact of recurrent plagues on the population of Europe. (After W. Langer, "The Black Death," *Scientific American 210:*14–19, February 1964. Copyright © 1964 by Scientific American, Inc. All rights reserved.)

one-hundred years ago, enlightened doctors who advised their colleagues to wash their hands between examinations were often ridiculed and banished from their profession.) Personal hygiene also improved: bathing became more frequent, and cotton clothing that could be washed frequently became popular.

Another major breakthrough occurred when Walter Reed discovered that the *Anopheles* mosquito was the transmitter of yellow fever. With this discovery it became possible to limit some diseases by controlling their vectors. And Alexander Fleming's accidental discovery that a culture of a mold named *Penicillium* prevented the growth of bacteria represented still another breakthrough. Since that discovery, many other antibiotics (streptomycin and tetracycline are examples) have been isolated from microorganisms.

Improved sanitation and victory over smallpox, malaria, cholera, polio, and tuberculosis were major

factors in the recent increase in human population growth. Another significant factor was the continual improvement of agricultural technology, which reduced the chances of crop failure. Improvements in the storage and transportation of food and advances in communications softened the blow of local famines by providing access to distant food resources.

Still, the rapid growth of the human population in modern times is unprecedented, and no single factor explains it. You might gain an appreciation of how unique this period of explosive growth really is by considering some major population mileposts. It took several million years (about 99 percent of human history) for the human population to reach 1 billion, which occurred some time around 1850 (refer back to Figure 16.2). In marked contrast, humankind doubled its population to 2 billion in only eighty years more, and doubled it again, to 4 billion people in forty-five years. Hence, we are living in a period of population growth unparalleled within such a short time period. We turn now to a consideration of the principles of human population growth, and then examine the implications of this period of growth for the future.

Exponential Growth, Age Structure, and Population Momentum

We are currently in the *exponential phase* of human population growth. *Exponential,* or *geometric, growth* occurs when a factor increases by a constant percentage of the whole in a constant time period. For example, our savings account may grow (collect interest) at the rate of 6 percent per year. The next year we will collect interest not only on the principal, but also on last year's interest. Most of us are more familiar with *linear,* or *arithmetic, growth,* which is an increase by a constant amount in a constant time period. For example, a youth growing in height by a constant rate of two centimeters per year is experiencing arithmetic growth. Exponential and linear growth differ in some surprising ways. To properly assess the impact of current population growth on our personal well-being, you need to understand these differences.

An old legend helps illustrate one of the differences between exponential and linear growth. Once upon a time, a clever citizen presented a beautiful chess set to his king. The king was so pleased that he asked how he could reward his subject. In return, the man asked the king to give him one grain of wheat for the first square on the board, 2 grains of wheat for the second square, 4 grains for the third square, 8 grains for the fourth square, and so on. That is, with each square the numbers of grains would be doubled. Thinking that this was a small price, the king readily agreed. Table 16.2 shows what happened. The tenth square required 512 grains, the fifteenth required 16,384 grains, and the fortieth more than 549 billion. Needless to say, the king ran out of wheat long before the sixty-fourth square was reached. Obviously, the king had been thinking about the more familiar linear growth when he agreed to the price. In the instance of linear growth, he would have only had to add one extra grain for each square. If the total request had been met by linear growth, the total cost to the king would have been less than a bushel of grain. But meeting the request for all sixty-four squares in terms of exponential growth would have required the world's annual production of wheat for the next two-thousand years.

This story illustrates just how quickly huge numbers can be generated by exponential growth. But how does exponential growth affect human numbers? The human population is currently doubling (experiencing an 100 percent increase) about every forty years. If we assume for illustrative purposes that this doubling time will continue, the world population in 2090 would be 32 billion. That is, shortly after the third centennial of our country, the world would need to support nearly eight times the number of people it does today. At this rate, forty years later, in 2130, the world's population would be sixteen times greater than it is today.

Another significant aspect of exponential growth can be illustrated by a modern-day version of an old French riddle. Assume that you are the proprietor of a plush resort overlooking a beautiful lake. One day your groundskeeper informs you that an algal bloom is growing on the far side of the lake. The groundskeeper has determined that the algal bloom is doubling in size each day and that it will take thirty days for the bloom to cover the entire lake. He asks you what he should do about it. You answer that you are very busy with other matters and that anyway the bloom is too small to worry about now. You tell him that you will decide what to do when the pond is half covered. That should give you plenty of time to stop the algal growth. The riddle is, on what day will you have to decide? The answer? The twenty-ninth day. Since the bloom is growing exponentially, once it covers half the pond, you will have just one day to consider the alternatives, make a decision, obtain the needed supplies and equipment, and apply them to the lake. (It is hoped that the treatment would require less than a day to be effective.) If you cannot accomplish all of this in one day, the lake will be completely covered by the algal bloom the next morning, and your unhappy patrons may pack their bags and leave. Hence, exponential growth has another deceptive characteristic. It is explosive, and can

Table 16.2 An illustration of exponential growth.

Selected squares on chessboard	Number of grains of wheat required
1	1
2	2
3	4
4	8
5	16
10	512
20	524,288
40	549,755,809,568
64	9,223,372,036,854,775,807

reach a fixed upper limit within a much shorter period than linear growth. Because we are lulled into a false sense of security by our familiarity with linear growth, we may fail to appreciate the dire consequences of exponential growth.

If we disregard for the time being the effect of migration on population growth, a country's population grows exponentially if its *crude birth rate* exceeds its *crude death rate*. For example, in the United States in 1978, our crude birth rate was 15 per 1000 people, and our crude death rate was 9 per 1000 people. The difference is 6 per 1000, which translates into an annual exponential growth rate of 0.6 percent. *Percent annual growth rates* are calculated by use of the following equation:

$$\text{Percent annual growth rate} = \left(\text{crude birth rate} - \text{crude death rate} \right) \times 100$$

The significance of a growth rate is easier to visualize when we consider its corresponding *doubling time,* that is, the length of time needed for a population at a fixed growth rate to double (see Table 16.3). At the 0.6 percent growth rate, the population of the United States would double in 115 years. In marked contrast, Kenya's population is growing at the rate of about 3.5 percent per year. If this rate does not change, Kenya's population will double in fewer than twenty years.

These birth and death rates are termed crude because they do not tell us anything about the *age structure* within a population. Age structure influences the number of deaths and births in a population, and therefore affects population growth. With regard to reproductive and death potential, not all individuals making up a population are equal. First, only women of intermediate age can give birth. The reproductive age of most women is between the ages of fifteen and forty-four. Second, the very young and the old are more likely to die than those individuals of intermediate age. Hence, a country's population can be subdivided into three major age groups: prereproductive, reproductive, and postreproductive. By classifying the individuals in a country in one of

Table 16.3 Doubling time associated with some growth rates.

Annual population growth rate (in percent)	Number of years in which population will double	Example
0.5	140	Sweden
1.0	70	Spain
1.5	47	Chile
2.0	35	India
2.5	28	Egypt
3.0	24	Ecuador
3.5	20	Rhodesia
4.0	18	Libya

these three categories, we can generate age structure diagrams. Such diagrams can help us to determine the current growth status of the population and predict what may happen in the future.

One type of age structure diagram, a broad-based pyramid, is illustrated in Figure 16.5 for Mexico. A broad-based pyramid, that is, an age structure in which a large percentage of the population is quite young, indicates a recent history of high birth rates. Such a broad base also indicates that a large number of individuals will be entering reproductive age and greatly increasing the size of that category. As a consequence, a significant increase in the total number of births can be expected. In contrast, the top of the pyramid is narrow, indicating that a relatively small percentage of the population is entering old age. Hence, the total number of deaths can be expected to remain relatively small. With a large increase in the number of births and little change in number of deaths, a population with such an age structure can be expected to grow rapidly in coming years.

A contrasting age structure diagram is illustrated for Sweden in Figure 16.5. In Sweden, birth rates have been low for some time, and, as a consequence, the prereproductive group is about equal in size to

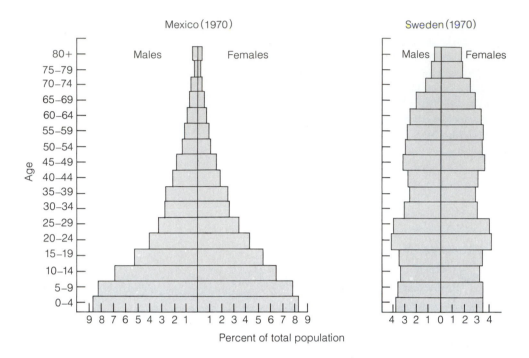

Figure 16.5 The age structure of Mexico, which has a recent history of high fertility, compared with that of Sweden, which has a recent history of low fertility. (After R. Freedman and B. Berelson, "The Human Population," *Scientific American 231*:31–39, September 1974. Copyright © 1974 by Scientific American, Inc. All rights reserved.)

the reproductive group, and the postreproductive group remains the smallest. Because the prereproductive and reproductive groups are about equal in size, we can expect little change in the number of births as the prereproductive group enters reproductive age (assuming that the preferred family size of this generation remains the same as that of the parent generation). And as the present reproductive group moves into old age, the number of deaths should increase. Hence, the age structure of Sweden denotes a population that is expected to grow little in coming years.

Crude birth and death rates are the numbers quoted the most widely for describing population growth. But because of the influence of age structure

on a population, we must be careful how we interpret these rates. For example, in the United States, the birth rate had become so low by 1974 that the crude birth rate nearly equaled the crude death rate and the country was nearly at the *replacement rate.* That is, parents were having just enough children to replace themselves. Because some children die before they can marry and raise a family, and because some women are physically unable to bear children, the average family size in the United States must be 2.1 children, rather than 2.0 to ensure replacement.

The fact that fertility in the United States dropped to the replacement rate, however, cannot be interpreted to mean that our nation's population will not continue to grow. Our population age structure, as

Fundamental Problems: Population, Food, and Energy

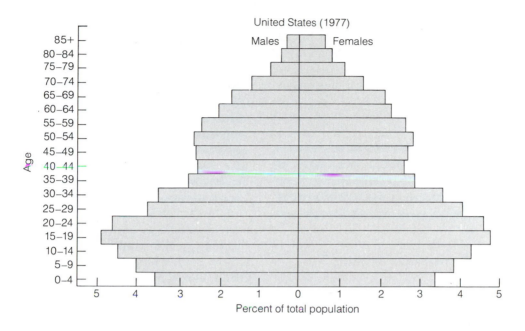

United States (1977)

Males | Females

Figure 16.6 The age structure of the United States, which shows that the baby-boom generation is now entering reproductive age. (U.S. Bureau of the Census.)

illustrated in Figure 16.6, indicates population growth will continue. Because of the post-World War II baby boom, a significant proportion of people in the United States are either in or about to enter the twenty-to-thirty-year age bracket, the prime child-producing years. In contrast, the number of people over sixty is smaller than the number in the fifteen-to-thirty age group. Even if these young people average only 2.1 children per family, the population will continue to increase for another fifty to sixty years, because so many new families are forming. Hence, the total number of births will continue to be greater than the total number of deaths. At the present birth rate, our population is expected to be at least 15 percent greater (30 million more people) by 2025 than it is today. At that time, the sizes of all age classes should be relatively equal and the population more nearly *stationary,* that is, its size is not chang-

ing. This state of no growth is often referred to as *zero population growth* (ZPG). But if the crude birth rate increases again at any time during the next fifty years, the population will continue to grow beyond projections. Hence, we see that both growth rates and age structures must be employed in predictions of a nation's future population growth.

An analysis of the U.S. age structure in Figure 16.6 also illustrates the concept of *population momentum.* Once a population begins to grow, it tends to continue to grow. Because it includes a large proportion of individuals in reproductive ages, a growing population continues to grow for some time, even if each couple, on the average, just replaces itself. Thus, the fact that the country experienced a baby boom in the late 1940s and early 1950s, indicated by the large number of fifteen- to thirty-year-olds, means that population growth will continue for at least another fifty years. Population momentum is particularly crucial for countries that are in the midst of an unprecedented population boom. In these countries, 40–50 percent of the people are younger than fifteen years old and will soon be raising families. Hence, a long time is needed to achieve a stationary population, even if the birth and death rates approach each other. We can use Mexico as an example. Even assuming that replacement reproduction was achieved immediately and then maintained (not a safe assumption), the population would still double within fifty years; it would take over a century for Mexico to achieve a stationary population.

Variations Among Nations

We can infer from the preceding discussion that the pattern of world population growth illustrated in Figure 16.2 is not the entire picture. Actually, population growth rates vary considerably among nations. Although exceptions do exist, countries can be classified roughly in one of two groups according to their growth rates. One group, commonly called the developed countries, includes the nations of Europe and North America, the USSR, Japan, Australia, and New Zealand. The other group, usually referred to as the less developed countries, includes the nations of Central and South America, Africa, and Asia. The developed nations generally have a low rate of growth—less than 1 percent per year. A few countries (East Germany, West Germany, Luxembourg, Austria, and Belgium) are essentially not growing at all. In marked contrast, less developed countries have an average annual growth rate of more than 2 percent, which means that their populations double every thirty-five years or less (refer back to Table 16.3). Some emerging African nations (Kenya, Ghana, and Uganda are examples) have growth rates of more than 3 percent. Their populations are doubling in fewer than twenty-four years.

What accounts for these dramatic differences in population growth rates? Let us begin to answer this question by examining the conditions that have contributed to low growth rates in developed countries. In an earlier section of this chapter, we saw that advances in medicine, sanitation, and agriculture contributed to a gradual decline in death rates. In addition, social changes associated with industrialization have been occurring during the past one-hundred years or so that have lowered birth rates. Here we consider these changes as they affected the United States in order to demonstrate their relation to level of development.

With the advent of the Industrial Revolution, many people began to move to the cities and, since that time, rural populations have steadily dwindled. On the farm, a large family was needed to do the chores, but in the city, with the abolition of child labor, children no longer contributed to the economic welfare of the family. Furthermore, the rearing of children became much more expensive because of inflation and rising expectations (braces on teeth, ten-speed bicycles, and a college education). In 1977, $62,000 was required to raise a child in an average income family from cradle through college.

In recent years in the United States and other developed nations, an increasing trend toward greater personal freedom has come into conflict with the responsibilities of caring for a large family. This trend has been particularly evident for women, many of whom are seeking alternatives to motherhood as a means of personal fulfillment. Many people wish to have greater mobility and to do what they want, when they want. In addition, people are paying less attention to traditional religious teachings that might encourage larger families. The net effect of these social changes of the past century is that the birth rates now nearly equal the death rates in many developed countries. This decline in birth rate following the lowering of death rate since industrialization is referred to as the *demographic transition.*

While the demographic transition was taking place in Europe and North America, however, most of the

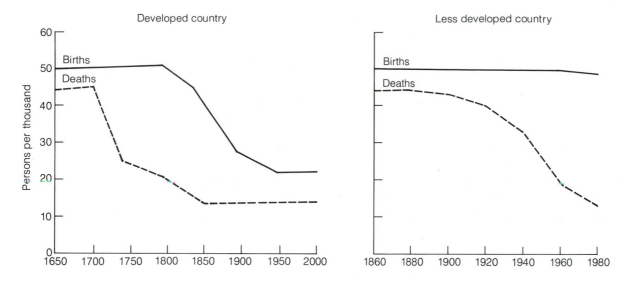

Figure 16.7 Contrasting trends for birth and death rates for a typical developed country that has experienced the demographic transition and a less developed country. (After J. R. Echols, "Population Versus the Environment: A Crisis of Too Many People," *American Scientist 64*:165–173, 1976.)

rest of the world was still suffering high mortality rates. (Figure 16.7 shows the contrasting trends in birth and death rates in a developed country and less developed country.) Not until World War II did improvements in health care and sanitation reach the less developed countries. Once they did, death rates dropped precipitously almost immediately. But the social and economic conditions that contributed to the demographic transition in the developed countries did not—and still do not—prevail in many less developed countries. Hence, though mortality rates have dropped dramatically in many less developed countries, birth rates remain high, and thus populations in these nations continue to swell. Some population experts are concerned that the birth rates of many less developed countries will not fall fast enough or soon enough to limit growth to the point at which development can occur and relative prosperity can be attained.

Population Control

Three factors determine a nation's population size: its birth rate, death rate, and migration rate. As we have seen, high mortality once controlled human population size significantly, but many technological advances have greatly reduced the number of human deaths resulting from famines and epidemics, and today even natural disasters have only a minor influence on world population growth. One of the worst human disasters in recent years was a typhoon that hit Bangladesh in 1971, killing an estimated 1 million people. In 1976, about seven-hundred-thousand people perished in a severe earthquake that rocked northern China. As devastating as these events are to local populations, the world population growth is now so high that these losses were made up in about three to five days.

Migration was once one of the great safety valves

for relieving population pressures. Many countries, including the United States, once welcomed newcomers by the hundreds of thousands. Unless we are native Americans, each of us can trace our roots back to another country. To escape famines, unemployment, and religious and political oppression, hundreds of thousands of people immigrated to the promised land—the United States.

Today, many countries are limiting immigration because they do not possess adequate resources to readily assimilate large numbers of immigrants. Unoccupied land well suited for settlement is no longer available in any significant amounts. Furthermore, migration from one country to another does little to relieve population pressures on the earth, which must now sustain more than 4.3 billion people. Yet as long as economic inequalities exist among countries, people from poor countries will strive to enter richer countries. With immigration becoming more restrictive, many people will immigrate illegally. For more information on the issues of illegal immigration in the United States, see Box 16.1.

Since migration is becoming more restrictive, and few people would suggest raising mortality rates, what are the options for relieving population pressures? A country has only one rational means of population control open to it: developing a program that somehow limits the number of children born into each family. However, creating and implementing such policies are likely to be matters of controversy. In the following section, we consider some important social factors that must be accounted for if population control is to be carried out successfully.

Socioeconomic Factors

A growing body of evidence, as well as common sense, indicates that birth rates decline when parents find it desirable to have smaller families. If a country is to achieve population control, the people must want fewer children. But in many countries, most people still want to raise large families. What, then,

Box 16.1

The United States has the world's biggest illegal immigration problem. Nobody knows how many illegal aliens are in this country, but most estimates are in the 3–5 million range. Thousands of "illegals" come from several Latin American countries including Guatemala, Colombia, and Ecuador, but the vast majority are from Mexico. Most simply walk across the 3200-kilometer-long (2000-mile-long) border between Mexico and the United States. The hard-pressed U.S. Border Patrol can do little to stem the flow; the border is simply too long to be effectively patrolled. The Border Patrol, an arm of the Immigration and Naturalization Service, catches about 165,000 illegal aliens each year just in the vicinity of El Paso, Texas, a major crossing point. And it estimates that for every person it apprehends there, as many as ten make it across. As one Border Patrolman put it, "If a guy is willing to walk out of town, he can pretty well get by us." In 1977, the Immigration and Naturalization Service apprehended and deported nearly one million illegal immigrants, most of them Mexicans.

It is not hard to understand why many Mexicans want to leave their country. Mexico's annual rate of economic growth is 2 percent, while its population growth rate is at least 3.2 percent. Half of Mexico's 18-million-member labor force is unemployed. This sorry situation results in dire poverty for millions on the Mexican side of the border. The result: thousands of "illegals" crossing the border each day in search of jobs.

Where do the illegal migrants go after they get to the United States? In the past most stayed fairly close to the border, usually in rural areas. But now

Illegal Aliens in the United States

many move inland to larger cities throughout the nation. New York, Los Angeles, Houston, Chicago, and Miami have especially large populations of "illegals."

The influx of these migrants has caused much worry and resentment among Americans who believe they have a large impact on the job market and the cost of social services here. Some labor unions have argued that because illegal aliens are willing to work for low wages and rarely complain about their treatment for fear of being deported, they get jobs that would otherwise go—at higher wages—to legal residents, particularly minority-group members, women, and youths. Other groups and individuals dispute this assertion, contending that while "illegals" may take some jobs away from legal residents and depress wages in some localities, the effects are not widespread or serious. They point out that in the United States there has always been a group at the bottom of the economic ladder forced to accept menial jobs. Perhaps as most Americans move up the economic ladder they spurn the low-wage, low-status jobs, causing many employers to tolerate and even encourage "illegals" to fill these lowly slots in the economic system.

What of the charge that illegal aliens greatly overburden schools and welfare systems? Many seem to believe this contention, perhaps with some justification in certain localities. In Texas, several local school boards have refused to provide free public schooling to children who cannot prove that they are legal residents. Other people dispute this position, however. They argue that nearly 75 percent of the illegal aliens pay taxes. They also contend that because most "illegals" want to avoid any contact with governmental authorities, they are afraid to claim social services. Recent experience in Los Angeles lends credence to this argument. That city has a large population of illegal aliens, but welfare officials there say that relatively few are on welfare. By one estimate, fewer than 10 percent of the illegal immigrants in this country receive welfare payments or have children in school.

Several aspects of the illegal-alien controversy seem clear. One is that as long as great economic inequalities exist between this country and others, people will come here even if they have to enter illegally. Another is that as long as large-scale illegal immigration continues, the United States will not achieve a stationary population. Still another is that if the massive illegal influx continues, government agencies will come under increasing pressure to stop it.

Those who oppose cracking down on illegal aliens include some agricultural interests who want their supply of cheap labor to continue, certain Hispanic rights groups, and U.S. Catholic bishops who fear a growing discrimination against all Latin Americans. Those who want stricter immigration policies because they fear increased competition for jobs include the AFL-CIO, certain other unions, and the Urban League. The Zero Population Growth organization, which sees "illegals" as thwarting the chances of achieving a stationary U.S. population, also favors tighter immigration regulations.

What is your position on illegal immigration?

are the factors that motivate people to limit the size of their families?

The answer to this question is not at all clear and remains the subject of sharp debate. Some population experts believe that birth rates decline most rapidly when people's basic social needs have been met. They cite an equitable distribution of income and basic education for all children as factors in the reduction of fertility. For example, recent dramatic declines in birth rates in Taiwan, South Korea, Sri Lanka, and Malaysia have been attributed to a more equal income distribution among the people. But other population experts point out dramatic exceptions. Kuwait, for example, has the highest per capita income in the world, and the distribution of income, goods, and services among the people is so equitable that little poverty exists in the country. However, Kuwait's population growth rate (5.9 percent per year) and crude birth rate (43 per 1000 per year) are among the highest in the world.

Reduced infant mortality has also been cited as a factor leading to a reduction in birth rate. In less developed countries, children are a form of social security for their parents' old age. But in nations where infant mortality is high, a large number of births is necessary to ensure that children will survive to take care of their parents in their declining years. For example, in India, a couple needs to have at least six children to ensure that one male offspring will survive to the father's sixty-fifth birthday. Given these circumstances, we can appreciate the motivation for having large families. Nevertheless, reducing infant mortality does not necessarily slow down population growth. Trends in Pakistan and Bangladesh indicate that birth rates dip slightly as child mortality declines, but that increased rates of survival of children may later result in more population growth if parents continue to have large families because of the children's economic value or because of cultural or religious traditions.

Hence, the relationships between social and economic conditions and family size are unclear. In some nations, at least, a decline in family size is associated with improved socioeconomic conditions. Perhaps with their basic needs satisfied, citizens are more receptive to education campaigns that portray the impacts of large families on their personal well-being and on society. People who are reasonably comfortable economically are probably also more amenable to governmental strategies designed to encourage small families. Such policies might include raising the minimum allowable age of marriage, creating tax laws to limit the number of "tax deductible" children, and instituting housing benefits and employment benefits that favor small families.

Once citizens begin to perceive the benefits of small families, birth control education and devices must be made readily available to them. In 1974, a United Nations Conference on Population at Bucharest, Romania, resolved that each government must accept responsibility for ensuring that every couple within its borders has the means to plan their families. By 1977, 96 percent of the people in Latin America and Asia and 72 percent of those in Africa lived in countries that had adopted or supported family planning programs, but many problems remained in actually disseminating information and devices to the people. Even the birth control programs in the United States, which cost $120 million annually, do not reach all of the people.

Subtle but crucial differences exist between the concepts of *family planning* and *population control*. Family planning emphasizes a couple's ability to control their fertility so that they can start their family and space their children as they wish. It does not imply that people must limit the number of children they have. Population control measures, however, are specifically intended to affect the size of a population as a whole. Since people nearly everywhere want more children than would result in a stationary population, successful population control requires that the nation as a whole limit the number of children per family to the replacement rate.

Some population experts are concerned either that improving basic social conditions and providing family planning services will not lower the birth rate quickly enough or that these procedures will not work at all. Hence, they believe that more drastic measures will probably be needed in some countries. But recent experiences indicate that attempts to implement compulsory population control can backfire. Reactions in India are a case in point. Recognizing that the growth of its population was stressing the country's resources to their limits, in 1976, Indira Gandhi's government began to put strong punitive pressures on civil servants who had more than two children. Subsidized housing, maternity leave, and medical care were denied to couples if one member did not get sterilized after the couple had two children. In the state of Maharashtra, sterilization of all men with three living children was made compulsory. Furthermore, abortion of any pregnancy that would result in a fourth child was mandatory. A year after these measures were implemented, Indira Gandhi was overwhelmingly defeated in a national election. Many factors contributed to her loss, but her government's unpopular population control policies were an important factor. The press had reported instances of overzealous attempts to enforce the law. Allegedly, extreme force had been used to bring in men for sterilization. To meet quotas, health service personnel were reported to have indiscriminately picked men off the streets, and reputedly some sterilizations had been performed under unsanitary conditions. Such alleged abuse of human rights led to heated public protest, and the coercive sterilization program was phased out just before the elections.

Birth Control Methods

Once an individual or a couple is motivated to avoid having children, either temporarily or permanently, a variety of choices exists as to the method of birth control to be used. The various alternatives are listed in Table 16.4. Because circumstances vary among different societies and people, and even during different phases of a couple's lifespan, no single birth control method ever has been, or probably ever will

Table 16.4 Methods of birth control.

Methods that prevent entry of sperm
Abstention
Coitus interruptus (withdrawal)
Condom
Spermicides
Diaphragm
Sterilization
Methods that avoid or suppress the release of the egg
Rhythm
Oral contraceptive (the pill)
Methods that prevent implantation
Intrauterine device (IUD)
Methods that prevent birth in case of pregnancy
Abortion

Source: After S. J. Segal and O. S. Nordberg, "Fertility Regulation Technology: Status and Prospects," *Population Bulletin 31*, 1977.

be, universally accepted. But we can enumerate the key characteristics of an ideal method: it should be safe, effective, inexpensive, convenient, free of side-effects, and compatible with local cultural, religious, and sexual attitudes.

The safety of birth control methods is of utmost importance. Figure 16.8 shows the number of annual deaths associated with several methods of birth control along with the age of the women using them. These data are based on women in the United States and Great Britain. To properly assess the safety of birth control methods, we must keep in mind that pregnancy itself is not without risks to the mother. Complications during pregnancy and childbirth result in about 23 deaths per 100,000 expectant mothers per year in the United States and Great Britain. The birth control method that may pose the most serious health risk is *"the pill,"* essentially a combination of female hormones that controls the reproductive cycle. British studies indicate that an increased tendency toward blood clotting, resulting in higher risk of strokes, heart attacks, and lung

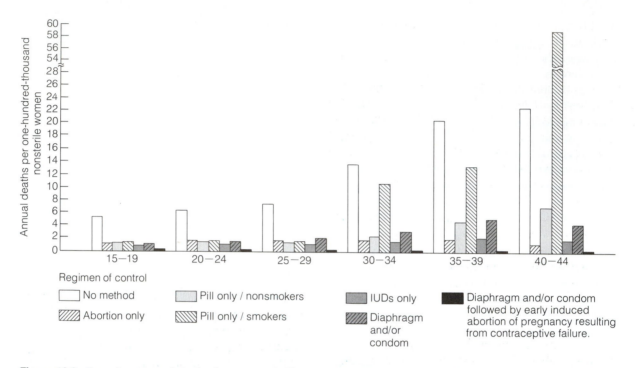

Figure 16.8 Annual number of deaths in women of different ages using various methods of birth control or no birth-control method. (After C. Tietze, "New Estimates of Mortality Associated with Fertility Control," *Family Planning Perspectives* 9:74–76, 1977.)

damage, is associated with the use of the birth control pill. The rate of death from these three causes in Britain is higher for users of the pill than it is for nonusers. Among nonpregnant women under age forty, the rate is about 1 per 100,000 for women who do not use the pill; for users of the pill the rate is 3 per 100,000. For women older than forty, the risks are considerably higher—12 deaths per 100,000 for those who do not use the pill, and 36 per 100,000 for those who do.

These interactions are complicated by other factors. Heart disease, for example, is more frequent among women who are diabetic or obese, have high blood pressure, or who smoke. Actually, the risk of pill-taking is only slightly greater than the risk of smoking, but to smoke and use the pill together sharply increases the hazard. Because of these risks,

the United States Food and Drug Administration has advised doctors against prescribing the pill for older women and those women with histories of high blood pressure and cardiac disorders. For younger women without these health problems, the risk of death from clotting while taking the pill is much smaller than that from complications during pregnancy and childbirth.

A common fear today is that the properties of chemicals that we take into our bodies may cause cancer. Are birth control pills carcinogenic? Evidence so far is inconclusive. Recent studies in England and in Connecticut suggest that pill users actually run a lower risk of benign breast tumors, and hence the pill may act to prevent breast cancer. In contrast, a California study indicates that women who use the pill before they bear children may run a higher risk of contracting breast cancer later in life. At least another decade of research is needed before the relationship, if any, between cancer and the pill is established.

Little serious health risk is associated with the use of the *intrauterine device* (IUD)—a small plastic or metal object that is placed inside the uterus and left there as long as contraception is desired. Some bleeding and discomfort are usual for a short time after insertion, and if these symptoms continue or are excessive, the IUD should be removed. However, there is no evidence of any relationship between the IUD and increased cancer. Also, the *diaphragm* (a rubber cap that covers the cervix and prevents entry of the sperm into the uterus) is nearly free from side-effects and complications. As a matter of fact, the higher death rates associated with the use of diaphragms and condoms as illustrated in Figure 16.8 actually relate to deaths of mothers at time of birth in pregnancies due to contraceptive failure.

Following the discovery of the castastrophic effects of thalidomide on abnormal fetal development (see Chapter 3), birth control methods have been regulated and monitored with increasing strictness. The regulatory requirements for contraceptive drugs are now more stringent than for other classes of medication. New legislation also requires more safety regulations for nondrug methods of birth control such as the intrauterine device. More tests for effectiveness and safety are required, including tests for side-effects, reversibility of the procedure, and potential carcinogenic effects. Up to fifteen years of research, development, and testing may now ensue before the federal government approves a new contraceptive procedure. The pill and other birth control methods in use today are under constant scrutiny for possible adverse health effects. Yet the long-term risks remain unknown, particularly for the pill. Thus, in choosing a method of birth control, each couple must weigh the known health risks of a particular method against its advantages.

The effectiveness of birth control devices varies considerably, as Table 16.5 indicates. *Abortions* and

Table 16.5 Percent of couples who experience contraceptive failure within the first year of use of contraception.

Method	Percent failing
Pill	2.0
IUD	4.2
Condom	10.1
Spermicides	14.9
Diaphragm	13.1
Rhythm	19.1
Others	10.8

Source: After B. Vaughan, J. Trussell, J. Menken, and E. F. Jones, "Contraceptive Failure Among Married Women in the United States, 1970–1973," *Family Planning Perspectives 9,* 1977.

sterilization are essentially 100 percent effective. The pill is also very effective, but a woman must remember to take it each day; each forgotten pill increases the chances of pregnancy. The *rhythm method* generally is less successful, for two reasons. The time period during which a woman is capable of conception is often difficult to determine, particularly in women with irregular menstrual cycles. Furthermore, to allow an adequate margin of safety, the period of abstention from coitus should amount to a considerable fraction of the month. Since hormonal changes during this time make some women particularly inclined to have sexual relations, a high degree of self-restraint is required if the rhythm method is to be successful.

Cultural, religious, and sexual attitudes greatly influence the choice of birth control measures. Undoubtedly, the most controversial method in this regard is abortion. Actually, recent efforts to legalize abortions are not based primarily on its use for birth control, but rather on a woman's right to privacy, and concern over the mortality rates associated with illegal abortions. Prior to the 1973 decision by the Supreme Court to legalize abortion, an estimated 250 women died yearly in the United States from infections and internal hemorrhaging resulting from self-induced abortions or unsterile help from un-trained people. Today, fewer than 40 women die each year from legal or spontaneous abortions. Abortions performed by trained personnel under sanitary conditions during the first three months of pregnancy pose a much smaller risk than does pregnancy itself, as Figure 16.9 shows. Few people would claim that abortion is preferable to contraception, but each year more than 1 million women in the United States obtain abortions. Either they are unaware of other alternative forms of birth control, other methods of birth control are not available to them, or they find abortion acceptable.

In 1977, Worldwatch Institute estimated that more than 280 million couples around the world used some form of contraception. Voluntary sterilization was the most popular form of birth control—an estimated 75 million couples used this method. The use of voluntary sterilization is expected to continue to grow because surgical procedures have been simplified and people now realize that sterilization does not hinder their sexual activities. With 55 million users, the pill is the second most preferred contraceptive. The *condom,* one of the oldest and simplest means of contraception, is now used by an estimated 30 million couples.

Research continues for better means of birth control. Of high priority is the development of long-acting forms of contraception, which would make the daily chore of taking a pill or frequent checks for expulsion of an IUD unnecessary. Injections of female hormones, given at intervals of up to three months, are now being tested. Trials are also being conducted on contraceptive implants that may last as long as six years. These implants are tubes made of rubberlike compounds that slowly release synthetic hormones to prevent ovulation. The tubes are implanted under the skin usually in the arm or buttock.

Until the early 1970s, little research was done on improving male contraceptive techniques. Now, a major recent strategy under study involves employing synthetic male hormones to modify the normal functioning of the reproductive system. The focus of

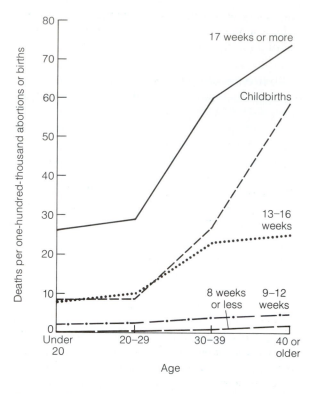

Figure 16.9 Mortality associated with childbirth compared with that associated with legal abortion at different times after conception.

most research in this area has been the suppression of sperm formation. But progress is slow, since many combinations of hormones have caused side-effects such as impotency and incompatibility with alcohol consumption. A major hazard to be overcome is the danger that a partially damaged sperm could fertilize an egg and lead to defects in a fetus. If a male contraceptive is ever successfully developed, it will probably take the form of a pill, an implant, or an injection.

Any new advances in birth control technology will be slow in reaching the people. Research to find new procedures is complex and expensive and requires many years of testing. Even after a particular device has been tested for years, the United States Food and Drug Administration may still consider its associated health risks too high and rule it unacceptable. For the views of one authority on the issues of birth control, abortion, and other population-related subjects, see Box 16.2.

The Transition to a Stationary Population

In the late 1970s, the population of the United States reached a significant milepost—population size began to level off. We can begin to sense the implications of moving toward a stationary population by comparing the age structure for an ultimately stationary population in the United States, as shown in Figure 16.10, with the present age structure, illustrated in Figure 16.6. The basic difference is that the stationary population will have proportionally more older people and fewer young people. This changing age structure is already forcing many adjustments. As the number of children declines, the impact on the elementary and secondary education system is substantial. Across the nation, schools are closing and teachers are being laid off. By the early 1980s, a similar impact will be felt by colleges and universities. A major unresolved question is whether this decline will be viewed as an opportunity to improve the quality of education or simply as an opportunity to cut taxes.

As the baby boom generation grows up, it is likely that the demand for such goods and services as food and clothing for children, toys, sporting facilities, and nursery schools will decline. Advertising campaigns are already beginning to focus on older people, as Figure 16.11 illustrates. At the same time, severe burdens will be placed on the social security system and private pension plans. Simply put, fewer working people will be supporting more retired people. Hence, a greater percentage of working people's income will be needed to support the elderly.

In the job market, the labor force of the future will

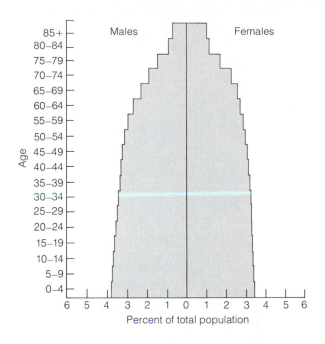

Figure 16.10 The ultimate age structure of the United States if the country achieves a stationary population during the next century. (After C. Westoff, "The Population of Developed Countries," *Scientific American 231*:108–121, September 1974. Copyright © 1974 by Scientific American, Inc. All rights reserved.)

be older, more predominantly female, and more heavily white collar in orientation. The number of jobs, such as teaching, associated with young people will decline, and career opportunities in health professions and social services for the elderly will proliferate. Eventually, the total work force will become smaller. To compensate, people may be encouraged to retire later in life. Such a development would ease the economic pressure on pensions and the social security system.

A significant factor complicating the transition to a stationary population is that the total number of births is very likely to fluctuate rather than remain constant. As we mentioned in our discussion of age structure, the total number of births in the United

Figure 16.11 As long as the birth rate remains low, there are relatively more older people in the population. This may account for why, the "Pepsi Generation" has expanded to include all age groups rather than the young exclusively. (Reproduced with the permission of PepsiCo, Inc.)

States is likely to increase significantly in the next decade as the baby boom generation begins to settle down and have children. As a result, as many children may be in elementary schools in 1995 as were enrolled in the peak year of 1970, when the current decline began. Hence, communities laying off teachers and closing schools today, will need more schools and many more teachers within a decade. These alternating periods of bust and boom will cause considerable difficulties in the maintenance of a high-quality educational system, and will force cyclic adjustments in all other facets of our social, economic, and political system.

Our society has clung tightly to its youth culture, and Americans have generally abhorred the thought of growing old. Fortunately, the shift in age structure will occur gradually and we will have the time to plan ahead for a smoother transition. Some may fear this transition, but it is inevitable. Clearly it is better to provide for an orderly transition now than to wait until the future, when the world will be more crowded, resource-poor, and environmentally degraded.

What of the Future?

Some of our insights into the future give us hope; others are cause for despair. On the positive side, world population growth has begun to slow down. Annual world growth peaked at 69 million additional humans in 1970, and since that time has slowly declined. In no region of the world is the population growth rate increasing significantly at this time. Birth rates are now either edging downward or holding steady in all but a few countries. Some governments are even adopting programs to bring their population growth to a halt. These countries include Bangladesh, China, Mexico, and India.

Given these events, many population experts are predicting a continued decline in annual population growth rates. But, as you can see in Table 16.6, the degree of predicted decline varies considerably, according to the agency making the predictions. Researchers in the Community and Family Study Center at the University of Chicago are obviously more optimistic about future declines in population growth than is the United States Census Bureau. (One reason for the large variation in estimates is that each agency uses different combinations of fac-

Table 16.6 Projected annual population growth rates for 1980–1985 and 1995–2000 under medium assumptions for the world.

Area	CFSC*		World Bank**		United Nations†		U.S. Census Bureau‡	
	1980–1985	1995–2000	1980–1985	1995–2000	1980–1985	1995–2000	1980–1985	1995–2000
World	1.73	1.34	1.80	1.54	1.95	1.64	1.8	1.7
More developed countries	0.71	0.45	0.78	0.46	0.83	0.60	0.7	0.5
Less developed countries	2.11	1.62	2.13	1.65	2.32	1.94	2.1	2.0
Africa	2.86	2.51	2.75	2.83	2.86	2.77	2.9	2.8
Latin America	2.68	1.92	2.94	2.38	2.71	2.37	2.9	2.4
Asia	1.84	1.36	1.91	1.48	2.11	1.64	1.9	1.8

*University of Chicago Community and Family Study Center. Estimates made in 1978.
**Estimates made in 1977.
†Estimates made in 1973.
‡Estimates made in 1977.
Source: After A. O. Tsui and D. J. Bogue, "Declining World Fertility: Trends, Causes, Implications," *Population Bulletin 33,* 1978. Courtesy of the Population Reference Bureau, Washington, D.C.

tors such as fertility level, social and economic development, and effectiveness of family planning programs to arrive at its estimates.)

The table also shows that the great disparity in growth rates between developed and less developed nations will continue, and that growth rates will remain quite high in many African nations while more moderate rates will be experienced in most nations in Latin America and Asia. If we accept them, the more optimistic estimates by the Community and Family Study Center provide hope that the demographic transition has begun in some less developed nations and that efforts to slow population growth are beginning to prove successful. In this light, the world population crisis appears resolvable.

But any optimism we feel must be guarded. The Community and Family Study Center warns that their optimistic projects are based upon a big *if*—*if* organized family planning continues in less developed countries. Hence, investments in family planning must be maintained and probably intensified if the predicted declines in population growth are to be realized. Moreover, economic and industrial development programs should also be continued in the less developed countries.

Another consideration should serve to balance our optimism. Although the growth rate may be declining, we need to realize that each of the four estimates points toward a much larger world population by the year 2000. The study group from the University of Chicago projects a world population of 5.8 billion in the year 2000. The World Bank estimate is 6 billion, and the United Nations estimate is 6.3 billion. All these estimates represent a doubling of the world's population in approximately forty years. Can our food, water, and energy resources keep pace with the added demand such a population will make on them?

We have evidence that the population race is already being lost in some poor countries. In the 1970s, death rates rose in several impoverished nations, notably Bangladesh, India, Sri Lanka, and the Sahelian countries of Africa. The added deaths were not primarily attributable to war and epidemics, but rather to hunger and nutritional stress. Large-scale droughts and floods reduced the food supply, forcing a decline in food stocks and sharp rises in food prices. Often, available food was insufficient and, in any case, it was beyond the economic reach of the poor. Children and elderly people were severely affected: millions died from starvation and famine-induced diseases. And, for each person who died, hundreds more suffered. Millions of children who somehow survived semistarvation have inherited the grim legacy of irreparable brain damage.

But even this is not the whole story. Stressed by overpopulation, many of these regions are evolving into ecological disaster areas. Deforestation and overgrazing, and their consequence, severe soil erosion, are reducing the land's productivity. Many subsistence farmers and nomads are losing their means of earning a living and producing an adequate food supply. Tragically, the land's carrying capacity for people is diminishing while the human population continues to grow.

Conclusions

In the global perspective, we are approaching the earth's capacity to sustain us. In addition, the peoples of the world do not function as one happy family, but rather are separated by cultural, social, political, and geographical barriers. Yet few, if any, countries are self-sufficient; all rely on trade with other nations. When the demands of a country's population outweigh its capacity to produce or trade for needed resources, the country is overpopulated. Some countries have already exceeded their carrying capacity and others are rapidly closing in on it. A nation that exceeds its carrying capacity is faced with three alternatives: it can rely on the good will of more prosperous nations, raid its neighbors for needed resources, or allow death to reduce its population to below carrying capacity. In any event, international conflicts are the inevitable consequence of intensified competition for finite resources and opportunities.

The human population must cease to grow. Nature

Box 16.2

The following excerpts are taken from a speech delivered by John D. Rockefeller III, an authority on population and family planning, before the House of Representatives' Select Committee on Population in 1978. Rockefeller founded the Population Council in 1952 and served as its chairman until his death at the age of seventy-two in July 1978.

On the Size of the U.S. Population
In the long run no substantial benefits will result from further growth of the nation's population. Rather, population growth is an intensifier and multiplier of many problems—environmental, social, political, economic. The nation has nothing to fear from a gradual approach to population stabilization.

On the Moral Aspects of Population Control
Fundamental to any consideration of the population problem . . . is the strong moral consensus which has been emerging over the past several decades. The realization is increasingly widespread that motivations in the population field are not negative and restrictive, but positive and constructive. The concern is not merely about numbers and fertility, but about human values and the quality of human life.

On the Roles of Women
I speak of the "roles" of women very deliberately. There has been too much of a tendency to see women only in their *reproductive* role as wives and mothers. It is time more attention was paid to their *productive* roles, as important contributors to the social and economic life of every country. . . . As long as the social status and economic security of women depend largely on the number of children

The Population Problem: One Man's Perspective

they have, as long as development programs that do reach women deal with them largely or solely in their roles as mothers, they will have good reason to continue having many children. . . . Clearly, we are dealing not only with the roles of women, but with the prevailing attitudes of men. Based on long-standing custom and tradition, these dominant attitudes will not be easily changed. But . . . we should make sure that our programs abroad are based on a new sensitivity to the key significance of the roles of women.

On Contraceptive Research

Even if it were possible to develop the perfect contraceptive, that by itself would not solve the population problem. There would remain such questions as its availability, the motivations for using it, the context within which it is available, to say nothing of the deeper developmental and social problems I have mentioned.

Yet, indisputably, contraceptive research is of critical importance. A better contraceptive is an essential element for solving the population problem.

There seems to be a widespread notion that we are doing about all we can in the field of contraceptive research. Nothing could be farther from the truth. . . . Only about $120 million is being spent worldwide each year. That funding should triple.

On Teenage Pregnancy

Teenage pregnancy has recently approached crisis proportions in this country. One million teenage girls become pregnant each year. Two-thirds of those pregnancies are unintended, and two-thirds occur out of wedlock. . . . In the Alan Guttmacher Institute's report on the subject, one authority is quoted as saying: "The girl who has an illegitimate child at the age of 16 suddenly has 90 percent of her life's script written for her. Her life choices are few, and most of them are bad."

There are two causative factors to consider in respect to teenage pregnancy. One is a biological fact, the gradual lowering of puberty. In 1840 the average young woman in the western world menstruated for the first time at the age of 17; her modern counterpart at about 12. The second factor is the prevailing attitude of adults in our society, the inhibitions and taboos which have kept our young people in relative ignorance concerning human sexuality. In combination, these two factors have resulted in too many pregnancies, broken lives, abortions, unwanted children, and child abuse.

The answer to the problem of teenage pregnancy is to be found more in helping our young people than in blaming them. The truth is that we know next to nothing about the state of sexual learning in our society . . . I mean not just the biology and mechanics of sex, but the moral values and responsibility, the sense of concern and caring for others as well as one's self, that should permeate this most natural and important aspect of human life.

On Abortion

There is only one question really at issue in regard to abortions, namely, whether they will be safe or unsafe. Abortions will not go away if they are illegal. . . . There are three elements in each decision concerning abortion: the woman, the unwanted child, and the fetus. All attention is currently focused on the third, the fetus. I submit that the others are important, too; that is, what happens to the life of the mother and what happens to the life of the unwanted child.

dictates that our numbers, like those of all species, must come into balance with the supply of needed resources. Will we achieve this balance by rational and orderly efforts to lower the birth rate, or will we reach it through chaos and a grim rise in the death rate?

Summary Statements

Unemployment, inflation, crowding, dwindling resources, and pollution are all related to the fact that too many people are competing for a finite number of resources and opportunities. Population pressures are felt both in this country and abroad. The more affluent the population, the greater is their impact.

For thousands of years, because of the vulnerability of human beings to a hostile environment, the human population remained quite small and grew slowly. The development of agriculture, advances in disease prevention and cure, and improved sanitation were major factors contributing to the recent unparalleled population growth.

The human population is currently growing exponentially. Exponential growth differs from linear growth in that it can more quickly generate large numbers and reach a fixed upper limit.

A population's growth rate is a function of its birth rate, death rate, and age structure. If a population contains a large proportion in the prereproductive group, then the population should grow. Even a population at the replacement rate will grow if the reproductive and prereproductive groups within it are large. A stationary population would have nearly equal numbers in the prereproductive and postreproductive groups.

Developed nations have low birth and death rates, and the combination of these factors result in low population growth. Less developed nations have high birth rates and low death rates, which together result in high population growth.

Limiting the number of children per family appears to be the most rational means of population control. However, many people still want large families. The improvement of basic social conditions may function in the reduction of population growth, but the exact role of this factor, if any, remains unclear. Nevertheless, once couples are motivated to have small families, birth control education and devices must be made readily available to them.

A variety of means exist for limiting family size. Although no perfect method is available, key characteristics include safety, effectiveness, low cost, convenience, and compatibility with cultural, religious, and sexual attitudes.

Pregnancy results in more deaths per year than does any means of birth control except illegal or self-induced abortions. The pill may pose the most serious health risk, particularly when combined with smoking. Abortion and sterilization are essentially 100 percent effective. The rhythm method is much less successful. Abortion is the most controversial method with respect to cultural and religious attitudes.

Research on improved birth control methods continues. But extensive testing is required, and any new advances in birth control technology will be slow in reaching the public.

As our population becomes stationary, it will contain more older people and fewer younger people. This shift in age structure will eventually have many ramifications for our society, but because the shift will be gradual, we will have the time to plan ahead for a smoother transition.

Predictions of future population trends give us cause for both hope and despair. The world growth rate has dropped slightly in recent years, and family planning facilities are becoming ever more widely available in many countries. But many people, particularly in less developed countries, still want large families. In some countries, the death rate appears to be rising because of starvation and famine-induced diseases. Ample evidence exists to suggest that we are approaching the earth's capacity to sustain us. Just how the human population will come into balance with the supply of needed resources remains unclear.

Questions and Projects

1. In your own words, write a definition for each of the terms italicized in this chapter. Compare your definitions with those in the text.

2. How is the well-being of the United States tied to the population control policies of other nations?

3. List five factors that contributed to the growth of the human population in the last two-hundred years.

4. Describe the differences between exponential and linear growth. In the context of population growth and resource consumption, why are these differences important?

5. Describe why age structure must be considered along with birth rates and death rates in predicting the population growth of a nation. In view of the most recent trend, a small decline in world birth rates, what is the significance of the interactions of these factors for resource management during the next fifty years?

6. Why is it necessary to understand the concept of population momentum before one can realistically predict when a country will achieve a stationary population?

7. Describe the process known as the demographic transition.

8. A desire for greater personal freedom has contributed to the decision of many couples to have few or no children. But some people are concerned that these couples will take on less responsibility in their communities; that is, couples with few or no children may not be concerned about the quality of schools, youth opportunities, or the kind of world that will be left for the next generation. How would you respond to someone who voiced this concern?

9. Name three major population factors that determine a nation's population size.

10. What factors could lower the birth rate in many less developed countries? Can birth rates be lowered significantly without peer pressure or government coercion?

11. What are the characteristics of an ideal birth control method? In your opinion, which currently available method best meets these criteria? Defend your choice.

12. The transition in the United States to a stationary population will require many adjustments. Describe some of the expected social, environmental, and economic changes. How will your life be influenced by these adjustments?

13. What factors would tend to raise the American birth rate? What factors would lower it? Which set of influences do you think will be more important during the next decade?

14. Comment on the following statement: Because the entire population of the United States could stand on an area roughly half the size of New York's Manhattan Island, we really do not have a population problem.

15. When is a country overpopulated? What countries meet these criteria? What are these countries doing to solve their population problems?

16. Although we strongly encourage other countries to develop and implement population policies, the United States does not have an official policy on population. Suggest some probable reasons for this apparent paradox.

17. Evaluate the following statement: Population control is necessary, but by itself it is not sufficient to solve many of society's problems.

18. Visit a Planned Parenthood Office in your community. What are the major problems that the agency is attempting to solve? What are the prevailing attitudes of the agency's clients about their desired family size? Do staff members perceive a difference between family planning and population control?

19. Design a survey to assess attitudes about family size. Administer the survey to several diverse groups, such as your class, a religious organization, and your neighbors. Compare the responses and identify the reasons for similarities and significant differences among the several groups.

Selected Readings

Brown, L. R. 1979. *Resource Trends and Population Policy: A Time for Reassessment.* Washington, D.C.: Worldwatch Institute. A consideration of the trends in per capita production of essential resources and their impact on future population growth.

Brown, L. R., P. L. McGrath, and B. Stokes. 1976. *Twenty-Two Dimensions of the Population Problem.* Washington, D.C.: Worldwatch Institute. An overview of twenty-two dimensions of society that are adversely affected to some degree by human population growth.

Ehrlich, P., A. Ehrlich, and J. Holdren. 1977. *Ecoscience: Population, Resources, Environment.* San Francisco: W. H. Freeman and Company. A sourcebook containing several chapters that comprehensively survey the world population situation and existing policies to slow down population growth.

Population Reference Bureau. *Annual Population Data Sheet.* Washington, D.C.: Population Reference Bureau. A concise summary of population data listed by country.

Ridker, R. G., ed. 1976. *Population and Development.* Baltimore: Johns Hopkins University Press. A book in a series by the Resources for the Future organization that identifies and describes many socioeconomic factors as determinants of fertility. Makes recommendations for stimulating economic growth and family planning programs while retarding population increase.

Scientific American. 1974. "The Human Population," *Scientific American 231:*3 (September). A special issue that provides an excellent overview of the varied aspects of human population growth.

Segal, S. J., and O. S. Nordberg. 1977. *Fertility Regulation Technology: Status and Prospects, Population Bulletin 31* (March). Washington, D.C.: Population Reference Bureau. A comprehensive report evaluating current methods of fertility regulation, including recently developed methods, methods under investigation, and new areas of research. Easily understood by the nonexpert.

Stokes, B. 1977. *Filling the Family Planning Gap.* Washington, D.C.: Worldwatch Institute. Describes accomplishments on a global scale in providing family planning services and identifies existing problems.

Teitelbaum, M. 1975. "Relevance of Demographic Transition Theory for Developing Countries." *Science 188:*420–425 (May 2). Suggests that the demographic transition theory for Europe may not be applicable to the less developed countries.

Tietze, C., and S. Lewit. 1977. "Legal Abortion." *Scientific American 236:*21–27 (January). Examines the effects of legalized abortion in many parts of the world.

Tsui, A. O., and D. J. Bogue. 1978. "Declining World Fertility: Trends, Causes, Implications." *Population Bulletin 33* (October). Washington, DC.: Population Reference Bureau. An optimistic evaluation of recent declines in world fertility, its causes, and its implications for future world population growth.

Westoff, C. F. 1978. "Marriage and Fertility in the Developed Countries." *Scientific American 239:*51–58 (December). An examination of current social changes that appear likely to further the decline of fertility and of some of the ways that nations may respond to a population decrease.

The two extremes of the world food situation—starvation in the Sahel and surplus wheat in the United States. (*Top:* FAO photo. *Bottom:* Carl Davaz/Topeka Capital-Journal.)

Chapter 17

Food Resources and Hunger

During the late 1960s and early 1970s, a severe drought held the Sahel region of Africa in a deadly grip. The Sahel zone encompasses six nations—Mauritania, Senegal, Mali, Upper Volta, Niger, and Chad—and has a population of 27 million people. As the drought progressed, hundreds of thousands of cattle, goats, and camels began to die, and many nomadic people lost their means of making a living. The frontispiece to this chapter shows the effects of the drought on one nomadic family. As the drought intensified, the people of the region, with no food and no means of support left, were forced to enter refugee camps. Many arrived at these camps in such a weakened condition that they soon died. Worse still, many never even made it to the camps, having perished en route.

The outside world failed to notice the desperate plight of the Sahelian people until mid-1973. Then, nations around the world sent food to the stricken area, saving hundreds of thousands of lives. But by that time 100,000–250,000 people had perished from starvation. What caused this tragedy? The answers to this question illustrate some of the complexities involved in feeding not only the inhabitants of a single,

isolated area, but also the rapidly growing world population.

The Sahel zone is situated in the savanna transition zone, between the hot, steamy equatorial region to the south and the arid Saharan desert to the north. Normally, the Sahel experiences a rainy summer and a dry winter. But during the late 1960s, the monsoon winds that normally bring rain to the region arrived later than usual. When they did appear, they lasted for shorter periods than in normal years, and hence produced less rain. Average annual precipitation declined by 45 percent—a particularly disastrous turn of events for a region with only marginal rainfall in the best of times. As a result, soil dried up and blew away, livestock perished, and people starved.

The roots of this tragedy were not anchored in the climate alone. From the 1920s to the 1960s, worldwide temperatures had been mild and monsoons generally reliable. During that time, the Sahel received aid from developed nations, and social conditions improved: increases in medical services lowered death rates, and imported technology spurred food production. As a consequence, human and livestock populations experienced a burst of growth. Prolonged drought had occurred before in the Sahel, but when it returned in the late 1960s, its effects were intensified due to this new growth in the human and livestock populations.

Restricted migration compounded the plight of the Sahelian people during the drought. In earlier times, these nomadic herdsmen had been able to flee when drought struck and travel southward to localities where water and food were more plentiful. But now rigid political boundaries exist between countries, and this option is no longer available. And even if these people had been able to move southward, they may not have fared much better. The nations bordering the Sahel, such as Nigeria and Ghana, were already having difficulty providing adequate resources for their own populations. An influx of an additional 10–20 million people from the Sahelian countries would have overwhelmed the resources of neighboring nations.

The monsoon rains returned in 1974, but conditions in the Sahel zone have improved little since then. The legacy of overpopulation remains. During the drought, large herds of livestock had overgrazed the land, stripping it bare of vegetation. The dry, thin layer of topsoil had vanished with the wind, leaving only rock and sand behind. It is in this way that the Sahara Desert continues its slow, unrelenting march southward, engulfing land that once supported the nomads and their herds. The people of the Sahel are still "ecological refugees"; they continue to live in camps, having no livestock and hence no means of self-support. Their former way of life has been completely destroyed. These people are the victims of overpopulation in an area that was too fragile to sustain their resource demands.

The basic problems associated with world food supply can be summed up succinctly. Never before have so many people existed on earth. Few places have the capacity to raise the additional food necessary to feed more people. And present efforts to feed the world's peoples are resulting in widespread, intensive degradation of the earth's ecosystems to an unprecedented degree. During the past decade, world food production has at times been unable to keep pace with population growth, and it has even declined in some years. Quite simply, we may be approaching the limits of the earth's capacity to sustain us. However, as we saw in the preceding chapter, despite this possibility, the human population is continuing to double every forty years.

In this chapter, we explore the viable alternatives that exist for feeding the growing human population. We also survey the constraints on food production with which all the nations of the world will have to wrestle. Finally, we analyze several contrasting projections as to how long we can realistically expect to feed the human population at its current rate of growth.

Minimizing Food Losses

About half the world's food production is consumed or destroyed by pests such as insects, rodents, and fungi. Obviously, one major way to feed more people is to reduce the amount of food consumed by these competitors. Experience in developed countries indicates that this task can be partially accomplished with the use of pesticides. In the United States, due to the widespread use of pesticides and well-made storage facilities, the rate of food losses to pests is about a third less than the rate of food loss worldwide. Largely due to this relatively low U.S. food-loss rate, food prices in our country are 30–50 percent lower than they otherwise would be. Unfortunately, these savings in food have brought costs of their own.

Ironically, pesticides not only save food, but in some instances they also diminish food availability. In addition to killing target pests, many pesticides kill nontarget species that benefit food production. One such beneficial organism is the honey bee. As honey bees forage for pollen and nectar, they also pick up pesticides that may kill them. Pesticide poisoning contributed to a 20-percent decline in the honey bee population over the past decade in the United States. This decline has serious ramifications, since nearly one-hundred crops depend upon honey bees for pollination. These crops include apples, cherries, cucumbers, melons, strawberries, and such forage crops as alfalfa and clover. Over $18 million have been paid to beekeepers by the federal government as compensation for monetary losses resulting from pesticide poisoning of their bee colonies.

Another risk associated with pesticides is their potential to disrupt natural pest control mechanisms such as parasitism and predation (discussed in Chapter 4). An example of this sort of disruption involves cyclamen mites, which can be serious pests to strawberries. Attempts to control these mites with parathion, a synthetic insecticide, only served to increase a particular cyclamen mite population by fifteen to thirty-five times its original size. What caused this population explosion? Ironically, parathion proved to be more toxic to another mite species that preyed on the cyclamen mite. The death of the predator mites allowed the population of the pest mites to explode despite the spraying. Hence, continued spraying resulted in more crop losses, not less.

Some pesticides, particularly the chlorinated hydrocarbons, break down slowly and therefore persist in the environment for many years. Since these chemicals are soluble in the fatty tissues of organisms, they accumulate in food chains. The accumulation of these substances in aquatic food chains reduces natural reproduction in some fish populations and, as we saw in Chapter 7, makes their flesh unfit for human consumption. Pesticide accumulation has also contributed to the population declines of several endangered species (see Chapter 14).

Even if no risks were associated with pesticides, we would find that resistance among pest populations inhibits the effectiveness of these chemicals. In nearly 225 species of the pests influencing agricultural production, genetic strains exist that are resistant to one or more pesticides. How does such resistance develop? We can begin to answer this question by noting that tens of thousands of individuals of a single species are commonly present in a hectare of land, and that within such a large population, considerable genetic variability exists. Some members of this varied population may be naturally resistant to a particular insecticide, and thus are not killed when exposed to it. After several applications of the insecticide, most nonresistant members are eliminated and a small resistant population is left. In the absence of intraspecific competition (between resistant and nonresistant members), the numbers of resistant insects explode. Figure 17.1 is a schematic representation of the selection for a pesticide-resistant population.

When growers apply pesticides to their crops, in effect they are taking natural selection into their own hands. However inadvertently, they are selecting for those individuals in the pest population that are most resistant to the pesticide. Thus, instead of eliminating the target population, they contribute to the growth of a pest population composed primarily of resistant members. Usually, in an attempt to overcome this greater resistance on the part of the pest population, growers increase the frequency of application or the concentration of the pesticide, or both.

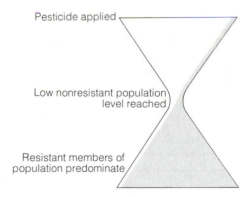

Pesticide applied

Low nonresistant population level reached

Resistant members of population predominate

Figure 17.1 A schematic representation of the selection for a pesticide-resistant population. The shaded area represents pesticide-resistant members of the population; the unshaded area represents pesticide-sensitive members. The passage of time goes from the top to the bottom of the figure.

Or they turn to a more toxic chemical. In any case, they usually continue the selection process, and in so doing breed "super-resistant pests," whose destruction of crops is increasingly more difficult to control with pesticides.

In analyzing pesticides from a cost-benefit standpoint, we find ourselves faced with a dilemma. On the one hand, pesticides are effective in combating our food competitors, and they therefore make possible the distribution of more food at cheaper prices. On the other hand, pesticides themselves contribute to losses of food (particularly aquatic), create super-resistant pests, reduce the populations of some wildlife species, and otherwise disturb the world's ecosystems. And our assessment of the value of pesticides is made more difficult by the enormous variety of poisons on the market. More than a thousand different chemicals are used as pesticides, and these occur in over thirty-thousand different formulations. Most pesticides are formulated to kill specific types of pest species. *Insecticides* are directed at insects, *rodenticides* at rodents, *fungicides* at fungi such as blight and rust, and *herbicides* at weeds (that is, any plants growing where they are not wanted). Each pesticide formulation has different properties and hence behaves differently in the environment.

To evaluate each pesticide we would have to weigh the risks and benefits of each independently. Basically, we would assess the value of a particular substance by determining the tradeoffs involved in using it. The chlorinated hydrocarbons have already been deemed the most hazardous to the environment through this process. This judgment has led the EPA to ban the widespread use of several insecticides in this group—DDT in 1972, aldrin and dieldrin in 1974, and heptachlor and chlorodane in 1975—though the agency still permits their use in an emergency, such as an epidemic of a disease transmitted by an insect vector. When chlorinated hydrocarbons were removed from general use, other insecticides—for example, the organophosphates—came into greater use. As Table 17.1 indicates, the major properties of organophosphates are more compatible with the environment than those of chlorinated hydrocarbons; the former are generally less toxic and less persistent than the latter.

Alternative Pest Control Methods

The surest way to avoid the negative effects of pesticides, of course, is to use alternative methods of pest control. Ideally, such alternatives should affect only the target pests, should control pests permanently at harmless densities, and should not select for resistance in target pests. Nature is one source of such alternatives. We know that natural means of pest control are functioning all the time, since much of the plant material produced each year survives the onslaught of such pests as insects and rodents. And it is true that we have been able to devise some effective pest control strategies by adapting ecological principles to particular situations. For example, we saw in Chapter 4 that natural predators and parasites can be used to control pest populations in the method known as biological control. Recall the example cited there of the importation of ladybugs to control the cottony-cushion scale, which threatened the citrus industry in California. By maintaining populations of ladybugs, citrus growers were able to eliminate the scale problem. Biological control has several advantages over pesticides: it is nontoxic, nonpolluting, and relatively inexpensive. Furthermore, the method is usually self-perpetuating; that is, once the predator or parasite population is established, it does not have to be reintroduced. And normally the possibility that the pest might develop a resistance to the control is minimal in comparison with pesticides. As usually occurs in nature, both the pest and its control agents coevolve to maintain a stable interaction.

Biological control does have some shortcomings, however. The control population needs time to effectively reduce the pest population, and farmers usually want to kill pests immediately. Most farmers would be unable to absorb the economic losses incurred before pests were brought under control by

Table 17.1 Relative toxicities of selected chlorinated hydrocarbons and organophosphates. (The maximum value for each for the four categories is 4 and the minimum value is 1.)

Common name	Toxicity to rats	Toxicity to fish	Longevity	Food chain accumulation	Total
Chlorinated hydrocarbons					
DDT	2.7	3.7	4.0	3.1	14.2
Dieldrin	3.1	3.9	4.0	3.0	14.0
Organophosphorus chemicals					
Malathion	1.8	3.2	1.1	1.0	7.1
Parathion	3.6	3.3	1.3	1.0	9.2

Source: After J. Weber, "The Pesticide Scorecard," *Environmental Science and Technology* 11:756–761, 1977.

predators or parasites. And a significant investment in time and money may be necessary before the control program is even instituted, since careful study of the pest and its interactions with potential predators and parasites is required to determine the best control agent.

Even after a control agent has been identified, a chance always exists that the control agent will undergo a change in lifestyle and become a pest itself. An example of such a turnabout occurred in Jamaica, where the Indian mongoose was introduced to control rats in sugar cane fields. Although the mongooses did reduce the rat populations initially, the rats subsequently took to trees where the mongooses could not reach them. Now rats are causing considerable damage to tree-nesting birds, and the moongooses' diet has expanded to include poultry and ground-nesting birds. Furthermore, the mongooses carry rabies. Thus, the mongoose, introduced as a control agent, is now a pest in Jamaica.

Another alternative pest control method is the use of naturally occurring pest-killing substances. Many natural pesticides exist; the difficulty lies in identifying them and making them readily available. One example of a natural pesticide is pyrethrum, produced when the flowers of a daisylike plant are ground into a powder. This substance is a compo-

nent of several commercial insecticide aerosols. Figure 17.2 shows pyrethrum flower heads being harvested.

Other naturally occurring chemicals are currently being examined for their potential role in pest control. One such group of chemicals consists of the *juvenile hormones*. These chemicals are produced by many insect species. Changes in the concentration of its juvenile hormone initiate various stages in an insect's life cycle. Figure 17.3 illustrates the role of juvenile hormone in the life cycle of a moth. If we apply a juvenile hormone in the proper concentration at the right time, we can disrupt a pest insect's life cycle, rendering the organism harmless to crops. Juvenile hormones have several advantages over many pesticides; their toxicity is low, they are nonpersistent, and they are specific to certain insect species, leaving other species unaffected.

Another natural approach to pest control is the use of cultivation processes that discourage or inhibit the growth of pest populations, thereby reducing the need for pesticides. One such practice is crop rotation, which prevents pest populations from building up in soil and crop residue. For example, by alternating corn with wheat or soybeans from year to year, farmers can inhibit the growth of corn rootworm populations in the soil. Also, diverse crops

Figure 17.2 Harvesting pyrethrum flowers in Kenya. The flower heads will be ground into a powder and used as a powerful insecticide. (FAO photo.)

and fencerows act as barriers to pest migration and serve as refuges for the natural enemies of pests.

The development of disease-resistant crops is another effective tool in the struggle to reduce pest damage. For example, at one time the Hessian fly caused several hundred million dollars of damage to wheat annually. By breeding strains of wheat resistant to this pest however, crop geneticists have relegated the Hessian fly to a minor status. Other pests, such as the corn borer, aphid, and loopworm, have also been controlled through the development of resistant crop varieties. However, obtaining resistant varieties that also produce high yields is a major difficulty. Another problem is that pest populations eventually overcome the defense mechanisms of existing crop varieties through natural selection, and geneticists must therefore continue to develop new crop strains. In the wheat-growing regions of northwestern United States, for example, a new variety of wheat can be expected to maintain its resistance for only about five years before evolving pests begin to reduce its yield.

Beginning in the early 1960s, the concept of *integrated pest management* began to receive considerable attention. Integrated pest management is based on the notion that nature controls population size by a variety of mechanisms—including predators, parasites, and juvenile hormones—and that agriculturalists can control pests by making maximum use of these natural controls. Growers using integrated pest management supplement these natural controls by using such techniques as crop rotation and by raising pest-resistant crops and livestock varieties. Pesticides, too, are included in integrated pest manage-

Fundamental Problems: Population, Food, and Energy

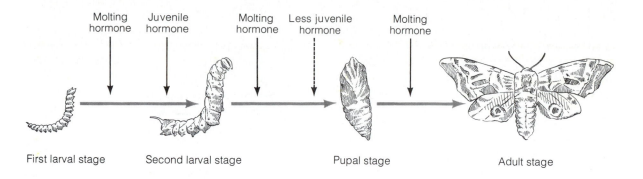

Molting   Juvenile   Molting   Less juvenile   Molting
hormone   hormone   hormone   hormone   hormone

First larval stage   Second larval stage   Pupal stage   Adult stage

Figure 17.3 The role of juvenile hormone in the development of a moth. When the molting hormone plus large amounts of juvenile hormone are present, the organism will remain in the larval stage. When the juvenile hormone is present in very small concentrations or is absent, the larval form will develop into the pupal stage. When no juvenile hormone is present, the adult stage develops.

ment, but are used only when absolutely necessary and in the smallest quantities possible to do the job effectively. The actual combination of control techniques depends on the crop and the pest. Integrated pest management has been used successfully to control pests on several crops, including alfalfa, cotton, cabbage, apples, and peaches. Many researchers believe that integrated pest management holds great promise for economic, environmentally sound pest control. However, since controlling pests by this means is complex, the development of each individual management program requires a great deal of research and testing.

In 1975, in recognition of the problems associated with pesticide usage, the National Academy of Sciences strongly recommended that agriculturalists make better use of new insights into population biology and ecosystem functions in their pest control efforts. Progress since then has been slow. In 1976, the EPA licensed the first two juvenile hormone insecticides. Also, the use of integrated pest management is still under study. In general, however, most farmers are paying little attention to alternatives to pesticides. In fact, given the direction that modern agriculture is taking, pesticide usage in the United States will probably continue to increase. (Pesticide

production in the United States increased from 45 million kilograms, or 100 million pounds, in 1947 to more than 726 million kilograms, or 1600 million pounds, in 1976.) Fencerows are disappearing as farms become larger and farmers employ bigger more efficient machinery. And the tendency toward monocultures, discussed in earlier chapters, is creating a greater need for pest control, and for most farmers this need translates into a heavier dependence on pesticides. Farmers prefer to use pesticides because they are relatively easy to apply, and because these substances represent only about 2 percent of total food production costs. Hence, farmers view pesticides as inexpensive insurance to protect their large investment. Therefore, unless they are made economically competitive with pesticides, alternative pest management techniques are likely to be largely ignored by U.S. growers.

Unfortunately, pest problems are particularly severe in less developed countries where people can least afford to lose food. In these areas, farmers are often working at the subsistence level, and are unable to afford pesticides even when they are available. Hence, these impoverished, unsophisticated farmers are in great need of inexpensive, simple-to-use alternatives. Pest control programs in these areas should center on the development of pest-resistant crop varieties and the improvement of local cultivation practices. Also, crops must be made more secure from pests after harvesting in many less developed countries. Specifically, rat-proof structures are needed that can be fumigated with nonpersistent pesticides. In many areas today, even bumper crops do little to alleviate hunger, since excess crops must be left outside to rot or be devoured by rats and birds because storage and transportation facilities are inadequate.

Cultivating New Lands

From the beginnings of agriculture, some ten-thousand years ago, until approximately 1950, the major means for increasing world food production was to cultivate new lands. Today, however, no readily available reserves of arable land are left in the world. Even in the United States, the large reserves of untilled cropland that lay fallow in the 1950s and 1960s are now back in production. Furthermore, little suitable space is left for the expansion of croplands in the temperate regions of Asia and Europe. The potential for expansion that does exist lies in two types of regions: humid tropical and semiarid. In this section, we examine the possibilities of and limitations on expanding agriculture into these areas.

Expansion in the Humid Tropics

The Amazon Basin of South America and the rain forests of Africa are the only remaining large tracts of arable land with rainfall sufficient to allow for intensive agriculture. These areas are particularly desirable for agriculture because they house a large segment of the world's hungry people and their human populations are growing rapidly. You can see in Figure 17.4 that this biome lies within the regions of the world in which diet is at best barely sufficient. The lush vegetation of the tropical rain forests seems to suggest a high soil fertility and an excellent potential for crop-growing, but in this case appearances are deceiving. Most rain forest soils are actually quite low in nutrient and humus content. High temperatures and heavy precipitation accelerate rates of decomposer activity. Dead plants and animals break down rapidly, but few nutrients are released directly into soil. Rather, for the most part decomposition is carried out by fungi that grow in live plant roots and extend into soil. As these fungi break down dead plants and animals, the mineral nutrients they release are taken up and transported directly into plant roots. Those few nutrients that are released directly into soil are quickly carried away by heavy rains. Thus, essentially all mineral nutrients in tropical ecosystems are tied up in vegetation, not in soil.

The ramifications of this mode of nutrient distribution on agriculture have long been known to the inhabitants of tropical rain forests. To compensate, they practice *slash-and-burn agriculture*. In this method, farmers cut down and burn vegetation on a small plot, usually less than 1 hectare (2.5 acres) in area (see Figure 17.5). The ashes of the burned vegetation contain some of the nutrients that had been contained in the plants, and are thus used as fertilizer. Within three to five years, harvesting of crops (along with their incorporated nutrients) coupled with heavy rainfall depletes the soil of its nutrients. Farmers then move on to another area and begin anew. After twenty to twenty-five years, the first plots are once more overgrown with vegetation, so the people can return to them and start the cycle again. Slash-and-burn agriculture is the best farming method yet devised for these tropical soils. Unfortunately, the yield of this technique is small.

Nutrient-poor soils are not the only factor limiting agriculture in tropical areas. Some tropical soils, though low in most nutrients, are high in iron and aluminum. When vegetation is removed, the tropical sun bakes these soils into a hard material called *laterite*, which can be used as a building material. If the cleared area covers only a hectare or so, as with the slash-and-burn technique, it will be revegetated

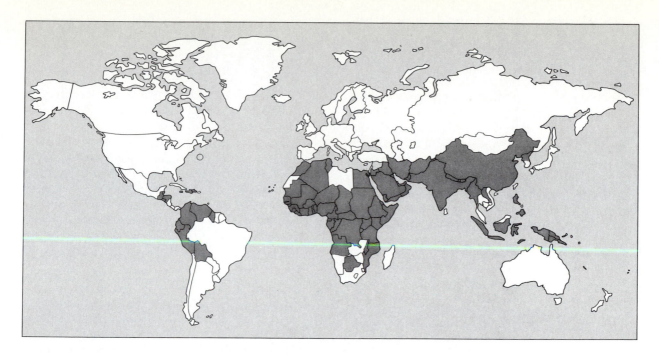

Figure 17.4 The geography of hunger, 1969–1971. Shaded areas are the regions of the world in which either human diets are deficient in calories or protein or dietary needs are barely being met. (After "Assessment of the World Food Situation, Present and Future." United Nations World Food Conference, 1974.)

relatively quickly. If the plot is larger, however, it will become a vast rock pavement that can support little vegetation.

The tropical climate contributes to other problems that make intensive agriculture difficult. Warm temperatures, high moisture, and a year-round growing season encourage the proliferation of insects, fungi, weeds, birds, and mammals, necessitating intensive pest control. However, frequent heavy rains quickly wash pesticides away so that farmers—if they can afford such substances or can gain access to them at all—must make frequent applications. Also, soil erosion is accelerated in tropical areas due to the high rainfall rate.

Assessing the potential of these areas for increasing the world's food supply is difficult. One significant advantage is that the climate allows for multiple cropping—that is, the growing of more than one crop annually in the same fields, where water is available year-round. Also, evidence is beginning to indicate that proper farming practices can overcome the limitations inherent in tropical soils. But soil researchers have much to learn about managing tropical soils before food can be produced on new lands in the humid tropics.

Unfortunately, the time necessary for learning how to farm these lands productively and for transmitting this knowledge to local farmers may be running out. Due to increasing population pressures in these areas, more land each year is being subjected to the slash-and-burn technique. Clearings are being made larger and closer together, and the time allowed for a clearing to regenerate is being shortened. Other efforts are being directed toward large-scale, energy-intensive agriculture—that is, cultivation practices that require large amounts of fossil fuel

Figure 17.5 A site of slash-and-burn agriculture in the Amazon Basin, northern Brazil. (FAO photo.)

energy, fertilizers, and pesticides, and the use of large machinery—a technology that is ill-suited to the tropical environment. And more than 2 percent of the Amazon Basin's forests are cleared for agriculture and forest products annually. At this rate, the great Amazonian rain forest will essentially vanish in forty years. Some may view this destruction as necessary for providing food and forest products to impoverished people. Others, however, are greatly concerned that the loss of the vast genetic reservoir lying within the disappearing forest will have serious long-term ramifications.

Expansion in Arid and Semiarid Regions

The other major areas of potential expansion are semiarid and arid regions, for example, the southwestern United States, the Sahara region of Africa, the Middle East countries, and the Australian interior. Conditions in these regions are the direct opposite of those found in tropical rain forests. In many

of these arid and semiarid areas, soils are fertile because precipitation is low and very little nutrient-rich water runs off the soil or seeps away. On the other hand, because natural precipitation is too low to support the growth of crops, water must be brought in via irrigation projects.

Irrigation is already a significant factor in world food production. The United Nations estimates that more than 27.6 million hectares (69 million acres) are now being irrigated for crops, 15 million hectares in the United States. Rice, which is the food staple for more than half the world's population, is almost always grown as an irrigated crop. Without irrigation, some fruit and vegetables would be scarce. In the United States, more than 70 percent of the nation's lettuce, carrots, celery, cantaloupe, and radishes are grown on irrigated land.

But, like pesticides, irrigation involves tradeoffs. One significant drawback is that only one out of every four liters of water drawn for irrigation is actually taken up by plants. Figure 17.6 shows the

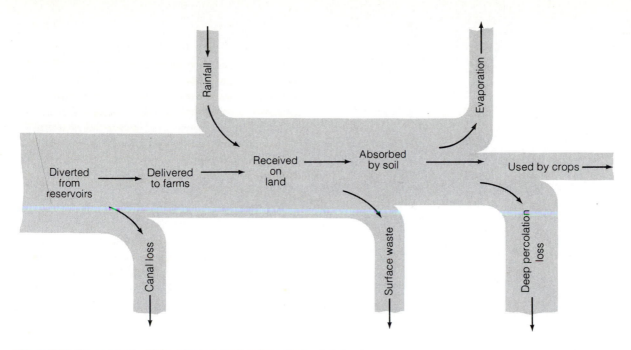

Figure 17.6 The points at which water is gained and may be lost in irrigation operations.

sources and losses of water in an irrigation project. Seepage from irrigation canals can be so substantial that it raises the water table (see Figure 17.7), creates a marsh, and makes cultivation impossible. Methods do exist to prevent this waterlogging, but they are often sophisticated and expensive.

Another problem associated with irrigation is the buildup of salts at the soil surface due to water evaporation, as shown in Figure 17.8. Crop yields decline as soil salinity increases. Methods are available to prevent salt buildup, but flushing away the excess salts requires the use of large quantities of additional water that could otherwise be used directly in irrigation to increase crop production.

In 1977, the United Nations estimated that 20 percent of the world's irrigated land was adversely affected by salt buildup and waterlogging, and that

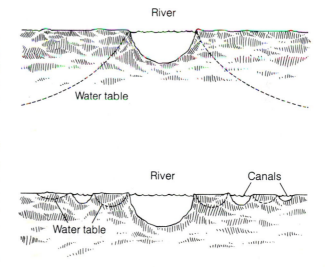

Figure 17.7 Seepage from irrigation canals, resulting in a rise in the water table.

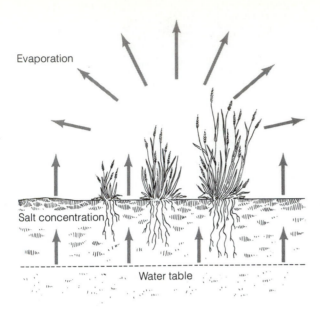

Evaporation

Salt concentration

Water table

Figure 17.8 Salt accumulating in the upper soil as a result of water evaporation.

productivity on this land had consequently dropped by 20 percent. In the United States, more than 400,000 hectares (1 million acres) have suffered at least some loss of fertility because of increased salt concentration. Figure 17.9 shows the effect of salt buildup on cultivated land.

Despite the problems associated with irrigation, the United Nations projects that an additional 23 million hectares (58 million acres) will be irrigated by the end of the century. Many of the rivers and watersheds, however, that lend themselves to damming and irrigation have already been developed. Therefore, any newly cultivated regions would have to be drier or farther away from water sources than those now irrigated. Water requirements for irrigation in arid and semiarid lands would be greater than in more temperate zones, since water loss from plants and soil in arid climates is greater. Also, soil salinity problems are generally more intense in arid regions. Hence, further expansion of irrigation into arid lands would be costly, from both an economic and an environmental standpoint.

Given the costs, the UN projection is probably overly optimistic. Agriculture is unlikely to be expanded into arid and semiarid regions to any significant degree until rainfall patterns can be altered to increase precipitation over desert areas. But in this area, too, we face obstacles. Precipitation enhancement could have serious adverse effects on the environment, as we saw in Chapter 8, and therefore we must proceed cautiously in our experimentation with climate modification.

Of greater importance than irrigating new lands, however, is preserving soil fertility in presently irrigated lands. Actually, for the same monetary investment, more food could be produced by rehabilitating existing irrigation systems than by developing new lands. With proper management, problems of water-logging and salinity could be controlled in many instances. But management is costly: irrigation systems must be continually maintained and drainage, groundwater, and salinity control closely monitored. Still, allowing faulty systems to continue functioning will produce millions of hectares of marshes and salt flats.

Management is not the only cost involved in maintaining existing irrigation systems. Irrigation is energy intensive: crop production involving irrigation requires three times the amount of energy used in normal rainfed production. Thus, as energy costs continue to climb, irrigation costs may become prohibitive for many farmers. Analogously, as our sources of fresh water become increasingly stressed by growing demands, farmers may have difficulty affording the very water they need to irrigate their crops. Employing water in food production increasingly conflicts with other water uses. In some instances, irrigation may be seen as a poor investment of water compared with other enterprises. For example, a given volume of water in Arizona generates about a thousand times more personal income when used for manufacturing than for agriculture. Furthermore, since irrigation is a consumptive use of water, and accounts for up to 80 percent of water usage in water-poor western states, it may well

Figure 17.9 Salt damage to a carrot crop. This scene is common where arid lands are irrigated, but in the Coachella Valley of California, where this photograph was taken, land usually can be reclaimed through leaching to flush the salts from the soil. (U.S. Department of the Interior, Bureau of Reclamation.)

be sacrificed to uses in which water is recoverable as the competition grows more intense. As we noted in Chapter 8, conflicts between agricultural and urban-industrial uses are likely to intensify as future demands for the fixed water supply increase.

Much of the arable land of the world is already under cultivation. Any large-scale expansion into new lands will be costly and in some instances will have to await development of new agricultural technology. A 1977 report by the National Academy of Sciences estimates that, worldwide, the area under cultivation can be expanded by no more than 1 percent. A similar study in 1978 by the United States Department of Agriculture projects only a 6-percent increase in land under cultivation. It is clear from these figures that any significant increases in future food production will have to originate on lands already under cultivation.

Increasing Production on Cultivated Land

We have seen that the feasibility of bringing new lands under cultivation is limited. An alternative to expanding agricultural efforts, however, is to increase the per hectare production of the land already under cultivation. Since 1950, more than 70 percent of the increase in world grain production has stemmed from improved yields on land already in production. Higher yields have been achieved not only in nations with advanced agricultural technologies, but also in many of the less developed countries. These nations have imported American and European agricultural technology to improve their food production.

Initial attempts at increasing yields in developing nations did not meet with much success. Often, varieties of corn and wheat from the United States fared poorly under local soil and climatic conditions. In addition, these varieties were often particularly susceptible to native pests. When fertilizers were applied to wheat and rice varieties traditionally grown in these countries, grain heads often became too heavy for the tall, spindly stalks, causing the plants to fall over. Thus, these grains were difficult to harvest and vulnerable to spoilage and pests. To overcome these difficulties, crop geneticists spent many years of careful work developing new, high-yielding varieties of grain, mainly rice and wheat. These new varieties were genetically engineered to make more efficient use of fertilizer application: they were shorter and had thicker stems that could support the heavier heads of grain (see Figure 17.10). Some of the early varieties were highly susceptible to pests, but strains currently in use are more pest-resistant, at least temporarily. Many of these improved varieties also mature in a shorter period than native strains. Hence, if sufficient water is available, it is possible for more than one crop to be raised during a single growing season.

In the late 1960s, when these high-yielding varieties were introduced and the utilization of fertilizers and pesticides was increased, crop production in some less developed countries rose dramatically. Mexico and Pakistan became exporters, rather than importers, of wheat. India became temporarily self-sufficient in meeting its cereal needs in 1972. To varying degrees, yields improved in many countries in South America, Asia, and North Africa. And in some nations, the rate of food production exceeded the population growth rate; thus people in these countries enjoyed a greater per capita food supply. The term *Green Revolution* was coined to refer to these remarkable events. Some scientists believed that the Green Revolution would solve most of the world's food problems.

Unfortunately for many people, however, the Green Revolution has proved to be less than a total success. Many countries, particularly those in Africa, have failed to experience a significant increase in their per capita food supply. For the world as a whole, grain production has been erratic. In 1972 and 1974, world production in cereals actually declined, both in developed and less developed countries. Encouragingly, from 1975 through 1978, significant gains were made in world food production. During this time, however, per-capita cereal production continued to fluctuate, as shown in Figure 17.11.

What are the reasons for this erratic pattern of food production? Why hasn't the Green Revolution lived up to expectations? Many factors contribute to successful agriculture: favorable weather; improved seeds; effective pest, water, and soil fertility management; and healthy economic incentives to farmers. Limitations in any of these areas will reduce agricultural production. In Chapter 2, we saw that the trend toward genetic uniformity in crops increases the chances for widespread losses if a new disease or pest species develops. And in this chapter, we discussed the limitations on crop yield imposed by pests and problems associated with irrigation. We turn now to an exploration of several other factors that influence food production on land: weather, soil erosion, fertilizer supply, and energy supply.

Climatic Change

Climate is one of the primary controls of food production on land. Climate regulates soil development, the length of the growing season, and the amount of water and sunshine available for crops. Thus, climate governs the types of crops that can be grown in particular regions and determines crop yields. Even in this age of advanced technology, our food supply is largely at the mercy of the weather. This vulnerability became quite evident in the

Figure 17.10 A new dwarf strain of high-yielding wheat (the wheat on the left, with shorter stalks) compared with a native strain. The development of this strain has contributed significantly to increased crop production in many countries. (FAO photo.)

United States during the 1974 growing season, when the country was hit by a series of weather extremes: heavy spring rains, a midsummer drought, and an early autumn freeze combined to reduce corn harvests by 11 percent and soybean harvests by 16 percent from 1973 levels. In subsequent years, however, rains generally came at the right time (notable exceptions occurred in California and portions of the upper Midwest). Other weather conditions were also favorable across the rest of the nation. Farmers produced bumper crops in subsequent years so that large grain surpluses developed, as evidenced in the frontispiece to this chapter.

New crop varieties including those produced by the Green Revolution, are particularly vulnerable to changes in climate. Many new varieties of wheat require high soil moisture, and therefore they must be grown in areas where rainfall is dependable or irrigation adequate. New rice varieties do not yield well when they are submerged for too long, and thus must be grown where natural flooding can be controlled. In fact, many crop-breeding programs have been based on the assumption that the weather would be near optimal. Not until recently have some seed growers begun to select for crop varieties with an increased tolerance to more stressful weather.

While agriculture in all parts of the globe is influenced by climatic change, cultivated regions that

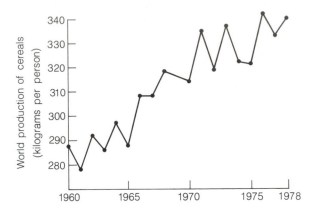

Figure 17.11 World per-capita production of cereals, 1960–1978.

experience marginal climatic conditions are most vulnerable to this variable. Such localities include northern latitudes, where growing seasons are perilously short, and portions of the subtropics characterized by a drought-prone monsoon climate. The Sahel drought, discussed earlier in this chapter, brought on as it was by the failure of the monsoon rains, illustrates what can happen in marginal lands when weather patterns shift.

The effects of short-term climate changes can be disastrous on a local level, but a permanent change in world climate, believed possible by many climatologists, could have serious effects worldwide. Throughout the first half of this century, weather conditions were generally favorable for successful crop production. Temperatures were generally moderate, rainfall was sufficient, and growing seasons were long. But, as we noted in Chapter 9, recent experience indicates that weather conditions may be deteriorating in areas that now have weather that is favorable for agriculture. Some parts of the country have been experiencing varying combinations of record winter cold, summer heat, droughts or floods, and shortened growing seasons. The weather seems

to be getting more changeable—that is, it is destabilizing—and hence less reliable for agriculture.

How can we cope with these changing conditions? One suggestion is that we implement weather modification projects. But, as noted above, we know too little, particularly about the environmental impact of such efforts, to attempt large-scale weather modification—if indeed we are capable of making such alterations at all. Our only practical alternative is to learn to live with climatic changes and be prepared to modify our agricultural practices accordingly. The response of growers to the 1976–1977 drought in California illustrates the kinds of modifications that are possible. When California farmers learned that they would obtain only a small fraction of their normal allocation of surface irrigation water, they turned to groundwater supplies. Many farmers drilled new wells or reconditioned old wells that had not been in use. They substituted crops that required less water (oats and barley, for example) for rice, which has a high water requirement. Some farmers stopped using their flood irrigation systems and installed more costly trickle irrigation systems (see Figure 17.12), thereby cutting water use by 30–50 percent. Many farmers installed pumps to recirculate runoff irrigation water from the foot of fields back to the heads. The combined effect of these procedures was successful: though the drought was severe, production in 1977 in California was about equal to that of former years when the weather was good. But the costs of maintaining crop yields were high. Although gross farm receipts nearly equaled those of earlier years, the California Department of Food and Agriculture estimated that net farm income dropped by 17 percent due to the added expenses of installing trickle irrigation equipment and new pumps and of drilling new wells.

Although California crop growers were able to adapt to the drought, cattle producers were not so fortunate. Because supplies of grass and water were inadequate during the drought, ranchers had to truck in hay and water. But these measures were very expensive, and many ranchers were forced to sell all

Figure 17.12 Flood irrigation of a mature almond orchard (*top*), and trickle irrigation of a young almond orchard (*bottom*). Both are in California's Central Valley. (U.S. Department of the Interior, Bureau of Reclamation.)

their cattle except their breeding stock. As a consequence, beef prices dropped, and California ranchers suffered losses of about $500 million in 1977.

The California drought ended as suddenly as it had begun, further illustrating the variability of weather. Heavy rainfall began in mid-December of 1977 and continued through March 1978. In just three months, river valleys were flooded and reservoirs nearly filled, and more than enough water had become available for irrigation.

The success of California's growers in adapting to climatic change during the drought (at least in the short run) could not be repeated in many less developed countries. California farmers are highly skilled and have access to the financial credit, machinery, energy, and information necessary for adapting quickly to change. In many less developed countries, however, such services are not available, and farmers must rely on their own meager resources to cope with adversity. Hence, inclement weather in these regions often leads to local famines.

If, in fact, the world's climate is shifting, human beings around the world may find it necessary to adapt by changing not only agricultural practices but also attitudes toward food storage and reserves. Some people have suggested that we should seriously consider the ancient biblical strategy of storing grain in the good years to provide food for the lean. In fact, the United Nations has proposed that a world food bank be established to serve this purpose. However, given the inevitable international disagreements over where reserves would be stored and who would control them and pay for them, a world food bank is a long way from becoming a reality.

Soil Erosion

Although fertile soil is absolutely essential for good crop and livestock production, this priceless resource continues to be eroded at an alarming rate. The erosion of cropland by runoff water is particularly severe. Each year, nearly 4 billion metric tons of sediments are carried into the waterways of the forty-eight contiguous states, three-quarters of the total originating in farmlands. Winds annually blow away another billion metric tons of soil. The U.S. Department of Agriculture estimates that more than a third of our nation's cropland is suffering soil losses at a rate too high to sustain current levels of productivity. And, according to another estimate, 7 percent—93,000 square kilometers (36,000 square miles)—of the nation's total cropland is being so severely damaged by soil erosion that further degradation could only be prevented if the land were allowed to revert to forests or grasslands.

Erosion problems, however, are not unique to this country. Soil erosion in less developed countries is at least twice as severe as it is in the United States. These problems will intensify as demand for food increases. Already people are moving into marginal lands where cultivation induces severe soil erosion. In both developed and less developed countries, hillsides are washing away and deserts are expanding.

Many methods of erosion control are available to farmers. For example, when farmers plow furrows up and down rolling hills, runoff flows rapidly downhill, accelerating soil erosion (see Figure 17.13). But plowing furrows parallel to land contours (*contour farming*) and flattening local slopes into terraces inhibits erosion. Such procedures do require an increase, albeit small, in farming time and fuel use, but the savings in valuable soil make up for the added cost.

Crops differ in their ability to anchor the soil against erosion, and the planting technique called *strip cropping,* illustrated in Figure 17.14, takes advantage of this feature. In strip cropping, a crop planted in widely spaced rows, such as corn or soybeans, is alternated with a crop such as alfalfa that forms a complete cover and thus reduces soil erosion. During the winter months, crops such as annual rye or clover can be sown to protect the land. Most farmers do not favor this latter procedure, however, since they usually plow the land in the fall to give themselves a head start on spring planting. Another soil-conserving cultivation method involves no tillage, or *minimum tillage,* techniques. In this strategy,

Figure 17.13 Severe soil erosion on a hillside resulting from improper cultivation practices. (U.S. Department of Agriculture, Soil Conservation Service.)

Figure 17.14 Strip cropping and contour cultivation, which reduce soil erosion. (U.S. Department of Agriculture, Soil Conservation Service.)

plowing, planting, and weed control are combined in one operation through the use of specialized machinery that plants, seeds, and applies herbicides and fertilizers to untilled soil. Combining these activities minimizes the disturbance to the soil surface, and the potential for erosion is thereby greatly reduced. Finally, a soil management practice specifically designed to reduce wind erosion is the planting of trees to form shelterbelts (see Figure 17.15). Following the Dust Bowl days, many shelterbelts were established in the Upper Great Plains.

Despite the fact that soil erosion control measures do exist, few farmers in the United States currently employ them. Although nearly $15 billion has been spent on soil conservation since the Dust Bowl in the 1930s, soil erosion is still one of our most significant environmental problems. But the ramifications of this problem are not limited to agriculture. A failure to conserve soil fertility could have a chain-reaction effect on the nation's economy. A loss in soil fertility would reduce crop production. This loss in turn would leave us with fewer agricultural exports, which we need to offset at least partially, our very costly imports of foreign oil. (For an analysis of the relationship between petroleum consumption and food production in this country, see Box 17.1.) On a larger scale, a loss in valuable topsoil worldwide would defeat efforts to feed even the 4.3 billion people on earth today, let alone the additional billions to come in the future.

The Fertilizer Supply

Today's high crop yields are totally dependent on soil fertilizers. These substances contain the major plant nutrients—nitrogen, phosphorus, and potassium. World commercial fertilizer usage has jumped almost 500 percent in the last quarter century—from 18 million metric tons in the early 1950s to more than 98 million metric tons in 1977. The United Nations estimates that if growing food demands are to be met, fertilizer use must increase another 300

percent by the end of this century. Where will farmers obtain the nutrients to meet this increased demand for fertilizer?

In the past, the application of *organic fertilizers* in the form of crop and animal residues was a major means of improving soil fertility. Many farmers also helped restore soil fertility by growing a leguminous crop—peas, beans, alfalfa, or clover—as one of their rotated crops. As we noted in Chapter 2, bacteria that live in nodules on roots of these plants increase the nitrogen (ammonia) content of soils.

With the development of new technologies, new supplies of plant nutrients were tapped. Phosphorus in the form of phosphates, and potassium in the form of potash, were extracted from rich deposits in the earth. These inorganic forms could be granulated and sold commercially. And the Haber process, described in Chapter 2, provided a means of synthesizing ammonia from atmospheric nitrogen to be used as a fertilizer. Because they were easy to handle, relatively low in price, and effective at producing high yields, these new forms of fertilizers were widely adopted. They also allowed farmers to obtain higher crop yields without having to raise livestock to produce manure.

Given the United Nations projection cited above, worldwide supplies of commercial *inorganic fertilizers* will be adequate for at least the next five years. But a major limitation to fertilizer production—and thus to improving world food production—is the unequal geographical distribution of potash and phosphate deposits and natural gas reserves (needed in the Haber process for ammonia production). The largest reserves of phosphate are in Morocco, the United States, the Soviet Union, and Tunisia. Germany, Canada, the Soviet Union, and the United States have large potash deposits, and Mexico and the Middle East nations, with their large petroleum reserves, have the greatest potential for manufacturing ammonia. In contrast, the deposits of phosphate and potash and reserves of natural gas are quite limited in many less developed countries. Further-

Figure 17.15 Shelterbelts in North Dakota. (U.S. Department of the Interior, Bureau of Reclamation.)

more, these nations usually lack the financial resources both to develop the deposits they already do have and to import sufficient quantities of fertilizers.

As a consequence of these regional differences, fertilizer usage contrasts markedly among nations. In 1976, fertilizer was used most heavily in Europe, where 200 kilograms were used per hectare (180 pounds per acre) of arable land. In marked contrast, only 11 kilograms were applied per hectare (10 pounds per acre) in Africa, and the application rates of other nations fell somewhere in between. For the most part, the less developed nations had the lowest rates, which is ironic, since increased fertilizer usage would have been more beneficial in those countries than in the developed countries. Applying additional fertilizers to already heavily fertilized land generally brings relatively small improvements in yields, whereas fertilizing nutrient-depleted soil, as in the less developed nations, results in dramatic increases in yield. But before fertilizer usage in impoverished nations increases significantly, many difficult problems of international aid and trade will have to be solved.

The unequal distribution of reserves is not the only limitation on inorganic fertilizer production. Extracting and processing inorganic fertilizers requires considerable quantities of fossil fuel energy. In the United States, the manufacture of fertilizers consumes more energy than any other sector of farming operations, nearly 28 percent of the total used in agriculture. Hence alternatives to commercial inorganic fertilizers would not only allow impoverished farmers to improve soil fertility, but they would also permit farmers in developed countries to cut accelerating production costs.

At first glance, favoring animal manure over commercial inorganic fertilizer might seem to be a real alternative. However, all but about 10 percent of the total manure produced annually in the United States is already being returned to the land. This unused manure would only be economically competitive with commercial, inorganic fertilizers where crops and livestock were raised very close to each other. Because of labor and energy costs sustained in loading, hauling, and spreading, manure costs exceed those of conventional fertilizers if the manure has to be transported more than 2 kilometers (1.2 miles) from its place of origin. And in some less developed countries, manure is the only available fuel for cooking and space heating, uses that are given higher priorities than fertilizing the soil.

Crop rotation using leguminous plants such as soybeans and alfalfa is beneficial on two counts: the nitrogen-fixing bacteria in the nodules of these plants increase soil fertility for future crops, and the soybeans and alfalfa themselves require less nitrogen to yield well than crops such as corn. Because cereal grains provide most of the world's food, and because these grains (especially corn) require large quantities of nitrogen fertilizer, major research efforts are currently being devoted to developing strains of nitrogen-fixing bacteria that will live in nodules on the roots of grain plants as they do on the roots of legumes. But success in this area is not likely to occur in the near future, if ever. Two other research efforts have better chances of success: improving the nitrogen-fixing capabilities of the legumes currently used, and inoculating the soil with species of free-living, nitrogen-fixing blue-green algae.

The Supply of Fossil Fuels

The significance of fossil fuel energy in raising crop yields is illustrated in Table 17.2, in which energy usage in rice production is compared for modern, transitional, and traditional production methods. In

Soil and Oil

But can the current levels of productivity in the corn, wheat, and soybean heartlands be sustained? In 1971, it was estimated that in the North Central United States, 67 percent of all cropland needed conservation treatment. Since then, highly erosive and sloping soils have been placed in production of export crops, replacing forage crops.

The seriousness of the erosion problem is further indicated by a more recent analysis showing that unrestricted land use would result in a national soil loss figure of 20 metric tons per hectare per year, twice as high as the maximum tolerable rate, according to expert opinion. This could imply that for each ton of grain going to Europe or Japan, we export several tons of topsoil to the Gulf of Mexico!

Soil is a crucial element in the farm production equation. How shall we live, if both soil and oil are depleted? Perhaps we need a negative severance tax on sediment—that is, payments for keeping soil in place. This idea was basic to the national soil conservation policy that has succeeded in breaking the back of the erosion problem, but not in reducing it to a tolerable level. Meanwhile, the programs implementing the policy have been allowed to wither over the past two decades.

Ironically, this neglect is in part attributable to the phenomenal success of another national policy of even longer standing, namely federal-state cooperation in the use of public funds for farm production research, development, and demonstration.

Historical trends suggest that soil losses are not necessarily caused by high yields: good conservation and high productivity are compatible. But it is equally clear that some soils are being mined. The implication is that the freedom to use any land for any purpose is to be tempered with a judgment as to how the private and the common enduring interests are best served.

Who is responsible for this? Soil conservation practices often appear not to be good business over the short haul. We should not depend on ethically inspired voluntarism any more than we can in other conservation issues. The stewardship challenge is one for the nation and its institutions, to be met through a voluntary partnership based on material interests. But a mere revival of the old system and adequate funding of existing programs will not be sufficient.

Farm operations can have a significant environmental impact, and undue loss of soil is classified as a nonpoint pollution source. Granting blanket exemptions for farm operations or regimentation through permits and fines are nonsolutions. But much can be said for an amalgam of short-term risk sharing in the production and marketing of crops with long-term risk sharing in the conservation of soils, as long as participation is voluntary.

Such a policy may not be popular. But it is fair to ask whether protection against the vagaries of weather and markets should be extended without assured conservation of the soil resource. Without such a provision, our now profitable solar energy enterprise may well decline through a bad trade of soil for oil.

Table 17.2 Fossil fuel energy required for rice production by modern, transitional, and traditional methods (in thousands of kilocalories per hectare).

Input	Modern (United States)	Transitional (Philippines)	Traditional (Philippines)
Machinery	4,184	335	172
Fuel	8,983	1,602	—
Fertilizer	11,071	2,447	—
Seeds	3,410	1,724	—
Irrigation	27,330	—	—
Pesticides	1,138	142	—
Handling and Transportation	8,531	29	—
Total Energy Consumption	64,648	6,279	172
Crop Yield (kg/ha)	5,800	2,700	1,250

Source: After *F.A.O. Monthly Bulletin of Agricultural Economics and Statistics* 25:3, February 1976.

traditional rice cultivation, fossil fuels are used only for the manufacture of tools. Humans and animals supply power for sowing and harvesting, and manure serves as fertilizer. Where the Green Revolution has gained a foothold—that is, where transitional farming methods are used—additional fossil fuels are required for the manufacture and operation of small machinery, and for the production and application of small amounts of fertilizers and pesticides. In modern rice production, much larger quantities of pesticides and fertilizers are applied and larger farm equipment is employed. Also, in modern methods of rice farming fossil fuels are consumed in the pumping of irrigation water and the grain-drying process. Although rice yields from modern agriculture are double those from transitional farming methods, fossil fuel consumption is nine times greater. For other crops, the energy expenditure necessary to double crop yields is not as great. Nonetheless, many observers are troubled by the nagging suspicion that fossil fuel supplies and costs will soon limit food

production. For example, it is estimated that to double the world food supply on today's cropland would require roughly a threefold increase in energy usage. Some observers wonder if fossil fuel supplies will be adequate to fulfill the energy needs required to double food production.

Barring oil embargos, fossil fuel supplies should be adequate in developed countries in the coming years, but costs are likely to continue to rise (see Chapter 18). Strategies are available that could significantly reduce agricultural energy consumption in developed countries if widely adopted. One strategy is to aim for achieving the best food energy return for the fossil fuel energy investment. In Figure 17.16, the energy gain per energy input is compared for various crops. Sorghum, sugar cane, and corn offer the best food energy return per unit of invested fossil fuel. Although the growing of crops such as wheat, oats, and soybeans requires comparable fossil fuel investments, these crops yield only about half as much food energy. In marked contrast, such crops as

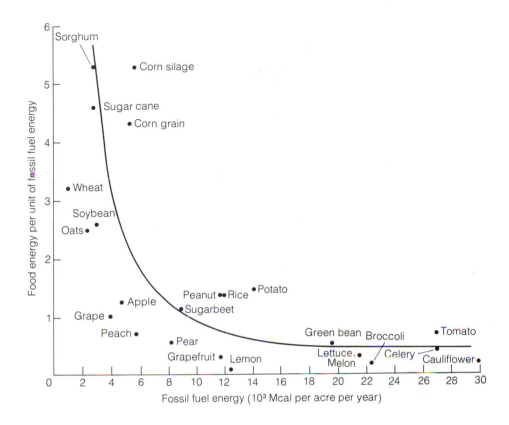

Figure 17.16 Food energy per unit of fossil-fuel energy invested for various crops. Cereals produce the most food energy per unit of fossil-fuel energy invested. Many vegetables produce less food energy than the fossil-fuel energy invested. (After G. Heichel, "Agricultural Production and Energy Resources," *American Scientist* 64:64–72, January/February 1976.)

tomatoes, lettuce, green beans, peaches, and pears yield less food energy than the fossil fuel energy required for their cultivation.

Although crops such as corn and sorghum return the most food energy per unit of invested fossil fuel energy, many people prefer meat over these food-stuffs. Consequently, these grains are fed to live-stock. However, as we saw in Chapter 2, the longer the food chain, the more energy is lost between trophic levels. The total food energy in livestock is approximately 20 percent of the energy input. This percentage is comparable to energy returns from vegetables and fruit crops that require considerable energy for their production but that are consumed directly by people. The significant point here is that livestock production successfully competes on an energy-efficiency basis with the cultivation of many vegetables and fruits. Although livestock lose ap-proximately 80 percent of the energy fed to them, this loss is partially compensated for when they are fed crops whose production requires less fossil fuel.

To obtain the best possible food energy return on our fossil fuel energy investment we would have to turn away altogether from grain-fed livestock and vegetables and fruits with high energy requirements. Instead, we would have to eat corn, wheat, oats, and soybeans. But such a drastic change in eating habits, while advantageous from the standpoint of fuel conservation, would eliminate from our diets the variety of foods required to meet our minimal nutritional needs for protein, vitamins, minerals, and fiber.

We can save some fossil fuel energy by modifying crop production practices. We have already identified certain alternatives that would reduce our dependence on energy-intensive agriculture: substituting manure for commercial fertilizer (where manure is available nearby), using minimum tillage techniques to reduce fuel spent in cultivation, and rotating crops to decrease the use of pesticides and commercial fertilizer. But the effectiveness of these alternatives is limited by significant constraints. As we have seen, unused manure is scarce. Minimum tillage requires more pesticides than standard cultivation techniques, since, with this procedure, crop residues are allowed to accumulate in the fields, and pest problems are thus aggravated. And systematic crop rotation restricts the choice of crops to be planted, whereas farmers prefer to plant the crops predicted to give the best monetary returns in a given year. All these constraints are economic in nature. It seems clear that economically hard-pressed farmers are unlikely to choose energy-efficient practices that would actually cost them more money than the fuel required in energy-intensive procedures.

Prospects for Increased Production

Clearly, we must increase food production on cultivated land in order to feed the additional 60 million people who will be added to the world population each year. However, as we have seen in this section, many obstacles stand in the way of this goal. The largest gains can be realized in less developed countries, but most small farmers in these countries can afford few, if any, of the resources needed to significantly improve their yields. Hence, in the absence of financial resources, the only means open to these farmers for increasing production are the planting of mixtures of high-yield crop varieties that are pest- and disease-resistant, the planting of legumes that support highly active nitrogen-fixing bacteria, and the use of appropriate hand tools. Even by optimizing their chances of success through these means, however, farmers in less developed countries will need technological and financial assistance from developed countries.

Although farmers in developed countries are often able to employ advanced agricultural technologies and gain access to needed resources, they too face difficulties in enhancing their future yields. Because of rising costs and declining net incomes, farmers are finding it increasingly more difficult to purchase necessary production resources. The alternative, switching to less expensive agricultural practices, would probably result in a diminished yield, and is certain to involve economic and environmental tradeoffs that may prove too costly. Hence, some agriculturalists believe that farmers in developed nations are approaching the upper limits on per hectare yield.

Enhancing the World Protein Supply

Protein is essential in our diet for normal physical and mental development. If deprived of sufficient protein, a baby or young child may experience stunted growth and may be mentally retarded. Tragically, these effects are not reversible; no amount of remedial education can correct the mental damage sustained through early protein deficiency. The United Nations estimates that as many as 500 million people may be suffering from protein malnutrition today. In the future, when several billion more of us inhabit the globe, the extent of protein deficiency could be much greater. Clearly, the efforts now under way to improve the world's supply of protein—in effect, to increase the production of high-quality food—are of great importance. In this section, we explore ways in which individuals can improve the protein content in their diets, and we describe the on-going efforts in several areas of research.

Our bodies build proteins by assembling twenty component substances, *amino acids,* in innumerable combinations. Of these twenty amino acids, eight are not synthesized by the body and must be supplied by food. These eight are referred to as the *essential amino acids.* We assess the nutritional value of protein by determining the amount of essential amino acids it contains. The higher this amount, the greater is the value of the protein.

Foods derived from animals have greater nutritional value than those derived from plants. Animal protein contains the greatest quantities of all eight essential amino acids. Of all food protein, protein from eggs has the best mix of amino acids for humans; the most valuable protein sources following eggs are fish, milk, cheese, and meat. These animal products also supply valuable minerals and vitamins.

Plant protein is poorer in quality than animal protein, since it is usually low in one or two of the essential amino acids. Proteins provided by cereals (rice, wheat, and corn) are particularly low in the essential amino acid lysine. And soybeans, while high in lysine, are low in another essential amino acid, methionine.

Presently, the world's population obtains about 70 percent of its dietary protein from cereals, vegetables, and legumes. A significant way to improve the protein content of the human diet would be to educate people to eat more high-protein plant materials and to develop means of incorporating these materials into their regular diets. Soybeans are the best source of plant protein available today. After the oil is extracted from soybean seeds, the residual soybean meal contains about 50 percent protein. The protein in soybean meal has the highest nutritive value of any plant protein.

Much soybean meal is fed to livestock as a protein supplement, but recently human beings have begun to eat more of this material in various forms. For example, soy flour made from soybean meal can be added to wheat flour, and the bread and pastries made from the mixture, though their texture and appearance remain relatively unchanged, have a significantly higher protein value than those made from pure wheat flour. Soybean protein is also used as a *meat extender.* By a special process, soy flour is turned into a material similar in texture to meat. This textured soy flour can be used to replace substantial amounts of meats in prepared foods calling for ground beef. By federal standards, up to 30 percent of the meat in meals prepared for school lunch programs may be replaced by textured soybean protein.

In another form, fibers spun from soybean protein are pressed together, flavored, and colored to simulate meats. Although it can be used as a meat extender, spun soy protein is usually used alone and sold as a *meat substitute.* Thus far, spun soy flour has been used to produce analogs of bacon, ham, beef, chicken, and seafood. Soybean protein is less expensive than meat protein, and could probably be produced in much greater quantities. But many people object to the strong "beany" flavor of soybean foods, and this taste problem limits its use. Research is currently being channeled into efforts to simulate beef, chicken, and seafood flavors more closely.

Food made with cereal grains (bread and breakfast cereals are examples) can be fortified through the addition of essential amino acids such as lysine. These amino acids can be synthesized or taken from natural high-protein sources such as soy flour. In the United States today, white bread is fortified with replacements for the nutrients lost in the processing of grain.

The meat from cattle, sheep, goats, and buffalo accounts for another 25 percent of the world's protein supply. Hence, the world's protein supply could be significantly enhanced if livestock management were improved, particularly in less developed nations where livestock grow slowly and suffer high mortality rates. For example, about 67 percent of the world's milk cows reside in less developed countries, but these cows produce only 20 percent of the world's milk supply (see Figure 17.17). Much could be done to improve meat and milk production without adding to the land area used to support livestock. For example, livestock could be bred for improved meat and milk production and disease resistance; pests reduced through vector control; and pasture and rangeland enhanced through increased fertilizer application, more careful selection of forage plants, and prevention of overgrazing. These measures, if accepted by farmers globally, could increase animal protein production by 30–50 percent worldwide. Given the high nutritional value of animal protein, any increase in production would be significant.

Some people, citing the inefficiencies involved in raising livestock, believe that human beings should eat plants only. Because 7 kilograms of grain are needed to produce 1 kilogram of meat, they argue, many more people could be fed if we all became vegetarians. Proponents of this approach also point out that the production of animal protein puts stress on fossil fuel reserves. As Table 17.3 indicates, producing animal protein generally requires five to ten times more fossil fuel energy than producing an equivalent amount of plant protein. In this regard, the food production method that consumes the greatest amount of fossil fuels is the raising of cattle in feedlots, where all feed is brought in to animals and all wastes are hauled away. As a result of these energy expenditures, the production of a kilogram of feedlot cattle protein requires twenty times more fossil energy than is needed to produce a kilogram of plant protein.

Nevertheless, a worldwide shift to vegetarianism could actually increase protein deficiency rather than

Table 17.3 Comparison of protein yield per fossil fuel input for various foods.

Food product	Fossil fuel input/ protein yield
Soybeans	2.06
Wheat	3.44
Corn	3.63
Rice	10.01
Beef (rangeland)	10.10
Eggs	13.10
Pork	35.40
Milk	35.90
Beef (feedlot)	77.70

Source: After D. Pimentel et al., "Energy and Land Constraints in Food Protein Production," *Science 190:*754–761, 1975.

reduce it. In order to design a well-balanced diet containing the necessary vitamins and minerals as well as adequate protein, vegetarians must have a sophisticated understanding of principles of nutrition. Most people are unwilling to make the effort required to gain such knowledge. Besides, global vegetarianism would waste the forage plants that grow on grazing lands. These grasses and shrubs are eaten by livestock and converted by them into animal protein but are of little food value to humans directly. Thus, in the interests of using our fuel and food resources in the most efficient ways possible, it appears that most of us will continue in our traditional role as omnivores.

To summarize, enhancing the protein supply—for example, through increasing the use of soy protein and improving livestock management—could substantially improve human nutrition in many less developed nations. But considerable research is needed before problems of development and distribution—and, in the case of soy products, palatability—are overcome and the full potential of these methods can be realized.

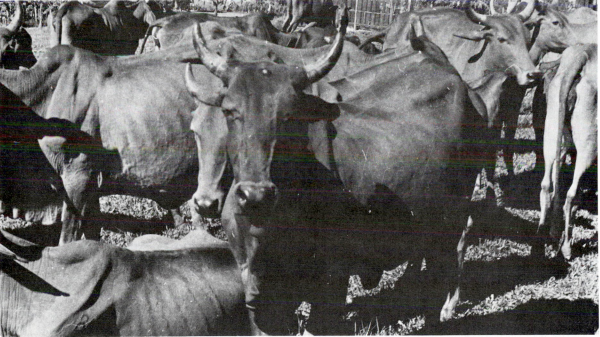

Figure 17.17 The contrasts between cattle in developed and less developed countries. *Top:* Dairy cattle from Wisconsin, high in general vigor and milk production. (Photo by Jack Markham.) *Bottom:* Cattle from the Philippines, showing the effects of feeding on grasses of poor nutritional content. (FAO photo.)

Harvesting the Oceans

So far in this chapter, we have focused on terrestrial food production. We turn our attention now to the oceans as sources of food. Although the world fish harvest contributes only about 5 percent of the total annual protein supply, this figure belies the importance of fish protein in many countries. The United Nations lists more than thirty countries in which fish protein represents 40 percent of total animal protein supply. In many of these nations, people rely so heavily on a diet of rice or starchy roots that without fish protein many citizens would suffer from severe protein deficiencies.

At this time, the future of the world's fisheries is in question. One reason for this uncertainty is that fish harvests have been leveling off recently. During the 1950s and 1960s, harvests increased from 21 million metric tons to 70 million metric tons annually. This rise represented a 5 percent annual growth rate. The world's per capita fish supply was thereby enhanced and the supply of much needed protein increased. But since 1970, the annual fish catch has leveled off, fluctuating between 66 and 74 million metric tons.

Despite the lack of growth in fishing harvests over the last decade, some fishery experts believe that a yield of 100 million metric tons of fish per year is a realistic expectation. To achieve this goal we would have to increase the world catch by a third. How could we do it? One possible strategy would be to step up fishing efforts in areas considered to be underfished. Two such regions are the Indian Ocean and the southwest Atlantic. Also, it is possible that a number of species could support a heavier harvest. These include squid in the eastern Pacific, capelin in the northwest Atlantic, and anchovies off the coast of West Africa. And observers expect future technological developments such as satellites to aid fishermen in finding and harvesting large schools of fish.

All these possible methods of increasing world fish harvests are still merely sophisticated means of hunting and gathering. Another approach to improving the fish harvest, however, is to treat the sea as a vast resource to be cultivated through the technology of *aquaculture,* analogous to the technology of agriculture on land. Current annual production from aquaculture is about 4 million metric tons, but United Nations experts believe that this yield could be tripled if aquaculture efforts were intensified.

One form of aquaculture is fish farming. An example of the cultivation of fish is found in the Far East, where for many years fish have been raised in the canals that bring water to the rice paddies, such as that shown in Figure 17.18. In another fish-farming method, part of a lake or bay is blocked off with an earthen dam and the resulting ponds are stocked with fish. And in still another aquaculture technique, successfully practiced in Japan, Norway, New Zealand, and the United States, trout and salmon are grown in floating cages in estuaries and coastal waters.

More complex forms of aquaculture are currently under study. Two possible techniques are the fencing off an entire bay for fish raising, and the growing of shrimp or turtles in large tanks. Although such efforts could yield valuable protein, they are still just possibilities. We have a lot to learn about the nutritional needs of the species to be farmed and many technological problems to overcome, for example, waste removal and disease prevention. The huge capital investment required by such enterprises will doubtless prove to be an obstacle as well. Thus, aquaculture is not yet practiced on a large scale.

Encouraging people to eat the less desirable rough fish, such as skate and dogfish, is one sure way to increase, at least to some degree, the amount of available protein. Rough species are now processed into fish meal, which, for the most part, is fed to livestock, particularly chickens and hogs. As a consequence, about 1 million metric tons of livestock are produced from an annual catch of about 20 million metric tons of rough fish. If fish meal were made more palatable, this protein source would become directly available to people.

In direct contrast to scientists who believe that we can increase the world fish harvest, others are con-

Figure 17.18 Tilapia fish being farmed in the canals that irrigate rice paddies in Thailand. (FAO photo.)

vinced that we are already at or near the sustainable level of harvest. They point out that despite the development of sophisticated technology and intensified fishing efforts, the level of harvest has remained stable over the last decade. These skeptical observers believe that any increase we might manage would probably be small. To justify this outlook, they cite continuous overfishing and the on-going pollution and destruction of aquatic habitats.

To assure a sustained catch in future years, fishermen must leave behind sufficient numbers of fish to maintain an adequate reproduction rate. But *overfishing*—the harvesting of more fish than are produced in a given time period—is currently resulting in the decline in takes of the Peruvian anchovy, East Asian sardine, California sardine, Northwest Pacific salmon, Atlantic salmon, Atlantic herring, Atlantic cod, and Atlantic haddock. Furthermore, annual harvests of such species as the bluefin tuna, yellowfin tuna, cod, ocean perch, and yellowtail flounder have failed to grow despite increased efforts.

The repercussions of overfishing will intensify unless fishermen begin practicing conservation. But fishery conservation is limited in several ways. Because relatively little is known about the population dynamics of marine fish, appropriate fish harvest quotas—limits on allowable takes—are usually established on a trial and error basis. If a quota is set too high, the fishery will continue to decline. And commercial fishermen will suffer either way: if the quota is too low, their take will be limited; if too high, stocks will be depleted.

A more severe limitation on conservation is the lack of cooperation among nations in establishing and enforcing quotas. Many fish species migrate over large areas of oceans, and until recently many fishing areas lay beyond national territorial limits. Conflicts over fishing rights have led to serious confrontations between nations. In an attempt to calm such anger, the United Nations has sponsored a series of Law of the Sea Conferences to develop an international treaty providing for the orderly development of ocean resources, living and nonliving. But progress on the treaty has been slow, and some nations have taken unilateral action, extending their territorial limits and setting their own catch quotas for both domestic and foreign fishermen.

As an example of this sort of unilateral action, in March 1977 the United States extended its territorial waters from 19 to 320 kilometers (12 to 200 miles). Within this 320 kilometer zone, catch quotas were established to permit a sustainable yield. Under the new quota system, American fishermen are given preference, and are allowed to catch all they can within the quota limits. Foreign countries, after receiving permits setting catch quotas, may take the remainder, if any. To receive a fishing permit, foreign nations must agree to let American ships fish in their territorial waters. In addition, foreign ships must allow United States Coast Guard personnel to come on board and inspect their cargo and records. Other nations, including Canada and Mexico, have also expanded their territorial limits.

The other threat to sustained yield from the oceans is the pollution and destruction of aquatic habitats. This danger results from the fact that the most productive areas of the oceans—estuaries, reefs, and regions of upwelling—are adjacent to continents. These zones are becoming seriously polluted by eroded sediments, dredge spoils, pesticides, and industrial wastes, as described in Chapter 7. Furthermore, commercial, industrial, and recreational development is threatening to eliminate fish and shellfish habitats altogether. Finally, coastal waters are being exploited with increasing frequency for petroleum and mineral deposits. Hence, if coastal zones are not properly managed, world fish production might actually decline.

Although much expert attention focused on enhancing the fish harvest, the surest way to increase the amount of available fish is to minimize wastage and spoilage of harvested fish. A United Nations study suggests that as much as 40 percent of the catch in some less developed nations is lost or spoiled before it can be consumed. Hence, if adequate refrigeration and processing facilities were provided to preserve fish already caught, protein malnutrition would be significantly reduced.

If recent history is a reliable indicator, the world fish catch may have already reached its maximum. Moreover, even the current harvest level may not be sustained unless better fishery and coastal zone management is implemented soon around the world. The contribution of aquaculture is limited by technological and economic constraints and quite probably will never produce anywhere near the quantity of fish produced in the oceans. Although protein provided by fish will continue to be important in the human diet, it is unlikely that the world fish harvest will provide more than the 5 percent of the total annual protein supply it represents today.

Conclusions

In our attempt to predict whether world food supplies will be adequate to meet future needs, we can be certain of one thing: barring a holocaust such as nuclear war, nearly 6 billion people will inhabit the planet by the year 2000. The amount of food that will be available at that time is much less clear. Further developments in the Green Revolution could greatly improve the world food situation, but significant obstacles to increasing production remain. Efforts to increase yield hinge not only on the availability of such resources as seed, fertilizers, pesticides, water, and energy, but also on such environmental factors as favorable climatic conditions and the control of soil erosion, salination, and pollution. Even if we can overcome these limitations, however, the social and political climate will also influence production greatly: effective land use policies, economic incentives for conservation and increased production, and international cooperation are all necessary to an improved yield. Because of the complex nature of these variables, estimates of future food supplies vary widely.

Some predictions are optimistic. A 1969 report by the National Academy of Sciences titled *Resources and Man* estimates that food production could be increased tenfold. And Dr. Roger Revelle of the Center of Population Studies at Harvard University estimates that, if agricultural technology continues to develop at its present rate, the earth's resources will be sufficient to provide enough food to sustain a world population of 40-50 billion. Such estimates, however, rest on two tenuous assumptions: that the resources necessary to implement Western agricultural technology are available in sufficient quantities around the world; and that environmental degradation will not limit food production. Unfortunately, these assumptions appear to be invalid.

Regarding the first assumption, the needed resources are not available in sufficient amounts throughout the world. In fact, the less developed countries with the lowest rates of food production are those that have the fewest resources for improving their yields. And even in developed countries the costs of these resources relative to the monetary return on crops are so high that they may soon begin to limit production significantly.

The second assumption—that environmental degradation will not limit food production—is equally invalid. We have cited numerous examples in this book of environmental degradation. Much agricultural land is actually declining in fertility because of soil erosion, salination, and waterlogging. And pollution and overfishing are endangering our ability to improve the harvest from the oceans. Clearly, environmental damage is already a significant constraint on productivity.

In view of these limitations, such researchers as Lester Brown of the Worldwatch Institute suggest that a more realistic expectation regarding potential food production is one more doubling, though even meeting this limited goal will require considerable effort and involve many environmental risks. Dr. Norman Borlaug, who won the Nobel Peace Prize in 1970 for his work on improving wheat varieties, estimates that we managed to gain thirty years through the Green Revolution in the race against mass starvation. Before this period ends, we must bring demand in line with supply by slowing population growth significantly.

Undoubtedly, we will continue to increase our food yield for some time. But history indicates that we should expect to experience years in which harvests will be poor and production will decline. In the long run, whether or not the population will level off before we exceed earth's capacity to feed us remains an open question.

Summary Statements

About half of the world's food production is consumed by pests. In developed countries, the use of pesticides has reduced crop loss significantly.

Some pesticides kill beneficial insects and disrupt natural pest control mechanisms. Through selection for resistance, populations of more than two-hundred species of pests are resistant to at least one pesticide. Because some pesticides are persistent, they accumulate in food chains in toxic concentrations.

Each pesticide must be evaluated on its own merits. The pesticides most dangerous to the environment are chlorinated hydrocarbons.

Alternatives to pesticides include biological control, the use of juvenile hormones, and agricultural practices such as crop rotation and the breeding of pest-resistant crop varieties. All these alternatives have both advantages and disadvantages.

Integrated pest management is based on a combination of several pest control procedures selected to provide the best control of a pest under given environmental conditions.

No decline in pesticide usage is expected until alternative pest management techniques that are economically acceptable to farmers are perfected.

A potential exists for expanding agriculture into humid tropical and semiarid regions. But many tropical soils presently limit agricultural expansion because of their low fertility and tendency to bake into bricklike laterite.

Constraints on expansion into semiarid regions are the limited availability of water for irrigation and the associated problems of seepage, salination, increased costs, and competition for water. Because of rising costs and limited resources, areas under cultivation worldwide are unlikely to be increased by more than a few percent.

The development of dwarf varieties of wheat and rice, an effort referred to as the Green Revolution, averted the threat of massive starvation in the late 1960s. During the 1970s, however, grain production became erratic and total world production declined.

Climate is one of the primary controls of food productivity. Crop failures in recent years in both developed and less developed countries illustrate the vulnerability of our food supply to weather. Because of support services, farmers in developed countries are often better able to adapt to climatic change (at least in the short run) than farmers in less developed countries. History illustrates that we should expect the weather to become less dependable for agriculture.

Though it is essential for food production, we are allowing the soil to erode at a rapid rate. Many procedures are available to inhibit soil erosion, but few farmers around the globe are employing these techniques.

Today's high yields depend on the improvement of soil fertility through fertilizer application. Although supplies of commercial fertilizers seem adequate for the near future, the unequal distribution of fertilizer resources around the globe could severely limit the growth of world food production. Also, the costs of commercial fertilizers are high. Inexpensive alternatives include the increased planting of nitrogen-fixing legumes and inoculation of the soil with free-living, nitrogen-fixing blue-green algae.

Modern agricultural practices consume large amounts of fossil fuels. Some fossil fuel energy can be saved by reducing the consumption of grain-fed livestock and the vegetables and fruits that require energy-intensive cultivation procedures.

Many people now suffer from protein deficiency. People can improve the protein content of their diets by eating plant materials with high protein content (soybeans) and fortified foods. Protein deficiency would also be reduced through the improved management of livestock, particularly in less developed countries.

Fish are an important protein source for people of many countries. The future of the world's fisheries is unclear. Some experts predict that fish harvests will be significantly improved through advances in harvesting technology, greater utilization of less desirable fish, and developments in aquaculture. Other experts believe that continued overfishing and destruction of aquatic habitats will preclude significant increases in ocean harvests.

Because of the complex nature of growing, processing, and distributing food, estimates of future world food supplies vary greatly. Given expected resource and economic limitations, one more doubling of total world food production is a realistic estimate. Meanwhile, world population continues to grow and climatic and economic uncertainties cloud the future.

Questions and Projects

1. In your own words, write out a definition for each of the terms italicized in this chapter. Compare your definitions with those in the text.

2. Describe how an insect population develops a resistance to a particular insecticide.

3. List the characteristics of an "environmentally sound" insecticide.

4. List the advantages and disadvantages of biological control.

5. List and discuss the factors that you would consider in deciding how to control a pest insect in your garden. Would you base your decision on other considerations if you owned a large farm?

6. People do not like to find insects in their food or to eat "blemished" fruit. But it has been estimated that 10-20 percent more insecticides than would otherwise be required are used to reduce the incidence of insects in food and to improve the appearance of fruits and vegetables. Considering the energy costs and health and environmental problems associated with insecticides, is this additional usage justified? What factors must be taken into consideration?

7. Although the forest vegetation in tropical areas is usually very lush, the soils are often quite infertile. Explain this apparent paradox.

8. Discuss the pros and cons of irrigating cropland. If possible, visit an irrigated farm.

9. What is the Green Revolution?

10. Why is climate a critical influence on crop production? In recent years, has severe weather reduced crop production in your area? If so, how have the reductions affected your community's economy?

11. How do strip cropping and minimum tillage help to reduce soil erosion?

12. Vegetables are important sources of vitamins and minerals, but vegetable farming is energy intensive and often requires irrigation. As a society, can we afford to continue to grow vegetables this way? Do alternative methods of growing vegetables exist?

13. Compare the level of energy consumption of United States agriculture with that of less developed countries. Speculate on how agriculture in different countries might be affected in the near future by dwindling supplies of fossil fuels.

14. Describe the factors that currently limit crop production on land already under cultivation. How might these limitations be overcome? Might attempts to overcome these limitations result in adverse consequences? If so, what are they?

15. Why are animal products better sources of protein than plant products? How does the amount of animal protein in the American diet compare with that of other countries? What are some of the health implications of a diet high in animal products?

16. Why did world fish harvests level off in the 1970s? Discuss the potential of the oceans to continue to serve as an important protein source.

17. The trend in both agriculture and aquaculture is toward greater utilization of fossil fuels. At the same time, we are destroying many of nature's free-food-producing areas such as estuaries. Comment on the long-term advisibility of such priorities.

18. What is the most important thing that you could do as an individual to help solve the world food problem?

19. The United States is one of a handful of countries that regularly has extra food to export. What should we do with this food? Store it for the lean years? Sell it to the highest bidder? Give it away to the needy? Cut back production to meet our own needs alone, thereby reducing the deterioration of American farmland? Name some other alternatives. How would your answers differ if you were a farmer, a conservationist, a stockholder of a grain-exporting company, or a laborer?

20. Have land use conflicts arisen between agriculture and other interests in your community? What is your community doing to resolve these conflicts?

Selected Readings

Brill, W. 1977. "Biological Nitrogen Fixation," *Scientific American 236*:68–81 (March). Examines the bacteria and blue-green algae that are the major contributors to the world nitrogen cycle and explains how they can be used to improve soil fertility.

Brown, L. R. 1978. "Vanishing Cropland." *Environment 20*:6–15 (December). An examination of the many ways in which croplands are being lost.

Bryson, R., and T. Murray. 1977. *Climates of Hunger: Mankind and the World's Changing Weather*. Madison: University of Wisconsin Press. Reviews the climatic causes of drought and famine in the past and creates a perspective for examining current climatic trends.

Carson, R. 1962. *Silent Spring.* Boston: Houghton Mifflin. A classic book that first brought the environmental problems of pesticides to public attention.

Eckholm, E. P. 1976. *Losing Ground: Environmental Stress and World Food Prospects.* New York: Norton. A discussion of the environmental costs of attempts to improve food production.

Eckholm, E. P., and L. Brown, 1977. *Spreading Deserts—The Hand of Man.* Washington, D.C.: Worldwatch Institute. Describes the role of human activities and climate in the expansion of the world's deserts.

Gulland, J. 1975. "The Harvest of the Sea," in *Environment: Resources, Pollution, and Society,* W. W. Murdock, ed. Sunderland, Mass.: Sinauer. Examines the limits to growth of world's fisheries and points up the values of international cooperation and sound management.

Heichel, G. H. 1976. "Agricultural Production and Energy Resources," *American Scientist 64:*64–72 (January/February). An analysis of expenditures of fossil fuel in current farming practices along with suggestions for improving energy efficiency in agriculture in the future.

McEwen, F. L. 1978. "Food Production—The Challenge for Pesticides," *Bioscience 28:*773–779 (December). A look at the beneficial role of pesticides in agriculture.

National Academy of Sciences. 1969. *Resources and Man.* Washington, D.C.: U.S. Government Printing Office. An examination of the potential food production on land and sea.

———. 1975. *Contemporary Pest Control Practices and Prospects.* Washington D.C.: U.S. Government Printing Office. A comprehensive overview of the achievements and problems of contemporary pest control. Considers prospects for pest management and makes recommendations for future pest control strategies.

———. 1977. *World Food and Nutrition Study.* Washington, D.C.: U.S. Government Printing Office. Assesses the problems of world nutrition and food supply and makes recommendations for research and development.

Pimentel, D., W. Dritschilo, J. Krummer, and J. Kutzman. 1975. "Energy and Land Constraints in Food Protein Production," *Science 190:*754–761 (November). An analysis of probable future constraints on food production.

Pimentel, D., J. Krummel, D. Gallahan, J. Hough, A. Merrill, I. Schreiner, P. Vittum, F. Koziol, E. Back, D. Yen, and S. Fiance. 1978. "Benefits and Costs of Pesticides Used in U.S. Food Production," *Bioscience 28:*772, 778–784 (December). A critical analysis of the role of pesticides in American agriculture.

Scientific American. 1976. *Food and Agriculture 235:*3 (September). A special issue examining how food is produced and how it nourishes, and identifying the well-fed and the hungry. Describes how the food situation may be improved in the future.

Triplett, G. B., and D. M. Van Doren. 1977. "Agriculture Without Tillage," *Scientific American 236:*28–33 (January). Explores how planting without plowing may save energy, water, and soil.

Weatherly, A. H., and B. M. G. Cogger. 1977. "Fish Culture: Problems and Prospects," *Science 197:*427–430 (July). Analyzes the potential role of aquaculture in contributing to world protein needs.

Weber, J. B. 1977. "The Pesticide Scorecard," *Environmental Science and Technology 11:*756–761 (August). Describes measures to test toxicity of pesticides and presents a rating scheme to compare their environmental safety.

Power lines stretch across rice farms in the fertile Sacramento Delta area north of Davis, California. (Copyright 1978 Barrie Rokeach.)

Chapter 18

Assessing the Energy Crisis

In the early evening of July 13, 1977, one by one and in rapid succession, the boroughs of New York City went dark. In the midst of a summer heat wave, the "Big Apple" came to a grinding halt, suffering the ultimate of insults. Relief-giving air conditioners stopped functioning; elevators became motionless, serving as temporary jail cells for their riders; traffic lights went out, leaving traffic hopelessly snarled; and the contents of refrigerators and freezers quickly warmed in the summer heat and began to spoil. Worst of all, under the cover of darkness, hundreds of stores were plundered, and the entire life investments of many small businesspeople were wiped out. In Figure 18.1, satellite photographs show the area during the blackout and following the restoration of full power.

The plight of New Yorkers during this experience still serves as a grim reminder of our utter dependence on an uninterrupted flow of electricity. What must we do to make sure that when we flick on a switch, we will get power? And, since we know that tapping energy sources affects environmental quality, what effects are we having on the environment by maintaining the energy flow? To begin answering these questions, we must examine the myriad ways in which we use energy and the extent to which our energy reserves are sufficient to fulfill these purposes. In this chapter, we address these topics and then turn to the crucial questions of how quickly we are using fuel reserves, how long we can expect them to last, and in what ways we might supplement conventional energy sources by developing alternative technologies.

A Historical Overview

For centuries, people have been fascinated by devices that save human labor, provide entertainment, make life more comfortable, and amplify living muscle power. In modern times, this love affair has led to the development of a parade of energy-consuming machines and devices on which we now depend so heavily that life is nearly unimaginable without them. But our overreliance on fuel-driven mechanisms originated in the innocent desire to improve the quality of life. Less than a century ago, to most people, horsepower still meant just that—horse-

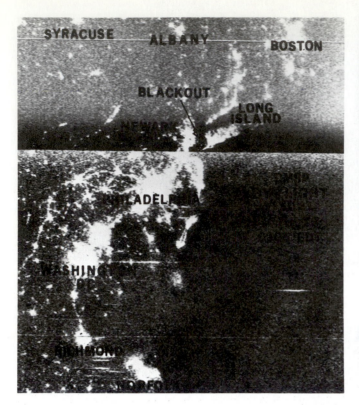

Figure 18.1 Nighttime satellite photos of the New York City area (*left*) during the blackout on July 14, 1977, and (*right*) on September 15, 1977, with full power. (U.S. Air Force.)

power. Stagecoaches and surreys were pulled by horses, and teams of horses augmented a farmer's own sweat and aching muscles.

The development of the steam engine marked the beginning of our dependence on fuel. Initially, steam engines were used to pump water from mines. By 1817, steamboats were providing regular service on the Ohio and Mississippi rivers. After 1825, steam locomotives—iron horses—were pulling their loads over rapidly expanding rail networks. And by the 1850s, clumsy steam engines were providing power to drive factory machinery and farm equipment such as threshing machines (see Figure 18.2). Although the first steam engines burned wood, coal was soon the preferred fuel, because it provided more heat per kilogram and was easier to handle.

In 1859, in Titusville, Pennsylvania, the first commercial oil well gushed forth (see Figure 18.3),

and by the 1870s, small engines fueled by natural gas and gasoline had been developed. These inventions were soon followed by larger, more powerful petroleum-burning engines. The first horseless carriages, handbuilt in backyard garages, appeared around the turn of the century, and were the harbingers of America's most voracious energy-consuming machines.

During the following years, invention after invention came into common use. In 1882, the first centralized electricity-generating systems in the United States began operating in Appleton, Wisconsin and New York City. Electric appliances did not begin appearing until around the turn of the century—the electric stove was introduced in 1896 and the electric vacuum cleaner and clothes washer in 1907.

In succeeding years, as the cities swelled with people, demand for energy grew. Around 1930, Chi-

Figure 18.2 (*Above*) A steam-powered threshing machine. This photograph was taken in the 1920s. (U.S. Department of Agriculture.)

Figure 18.3 (*Left*) The Drake oil well near Titusville, Pennsylvania, the first oil well in the United States. (Drake Well Museum.)

cago was linked to Texas gas fields by a 60-centimeter (24-inch) natural gas line. The electrical energy supply for Los Angeles received a boost in 1936, when the city was connected to hydroelectric facilities at Hoover Dam by the first long-distance high-voltage transmission lines. Elsewhere, electricity became more readily available as larger, more efficient coal-fired electricity-generating plants replaced antiquated power plants built in the 1890s. These new plants operated at higher steam pressures and were more than ten times more efficient than the old variety.

Between 1935 and 1955, high-compression diesel engines, which could operate on less expensive fuel, replaced gasoline and steam engines as industry's work horses. During this same period, electric motors came into widespread use. After World War II, new pipeline systems made possible an increase in the use of natural gas in gas appliances such as stoves, clothes dryers, and refrigerators. The appearance of petroleum-based plastics (polystyrene in 1937, nylon in 1938, Teflon in 1944, and Orlon in

1948) ushered in an era of synthetic fibers and inexpensive manufacturing materials. In the years that followed, more energy-consuming devices appeared: air conditioners, electric clothes dryers, color televisions, to mention but a few. Fuel is required not only to operate these appliances, but also, in significant amounts, to produce, distribute, and market them.

It was the seemingly endless supply of inexpensive energy that spurred this great surge of inventions, the result of which was a great increase in human productivity. Assembly line workers were able to accomplish more in a day because the energy that ran their machines in effect supplemented their own muscle power. Teamsters shifted from horsedrawn wagons to eighteen-wheelers. Construction crews and miners went from using picks and shovels to driving huge power shovels and bulldozers literally capable of moving mountains. (Figure 18.4 shows an example of such heavy power equipment.) And, as illustrated in Figure 18.5, farmers went from guiding single-bottom horse-drawn plows to employing multibottomed plows pulled by four-wheel-drive,

Figure 18.4 The world's largest truck, with a carrying capacity of approximately 320 metric tons. Each tire weighs about 3.6 metric tons. (Terex Division, General Motors.)

Figure 18.5 The effect of fossil fuels on agriculture over the past seventy years. *Top:* A farmer plowing in Montana in 1908. (U.S. Department of Agriculture.) *Bottom:* A modern farm plow. (Deere & Company.)

225-horsepower tractors. The era of large-scale landscape modification had arrived.

The ready availability of inexpensive fuel engendered complex changes in lifestyles and social structure too. It enabled large masses of people to flee rural America and move to cities and suburbs in search of a better life. And it led to improved communications, another development that changed the very fabric of our society. Today people are more mobile and better informed than they were a hundred years ago: instead of the pony express we have a complex system involving television, telephone, and telecommunication satellites; instead of stagecoaches we have buses, trains, and jet planes.

Throughout this dramatic transition period, few people concerned themselves with the possibility that we might run out of fuel. The first clear signal that energy supplies were finite came to the American people in the fall of 1973, when the Arab oil countries shut off all shipments of oil to the United States to press for higher prices. The immediate impact on most citizens was personal inconvenience and discomfort. Many of us had to adjust to gas shortages at local service stations, cooler homes in winter, carpools, and an 88-kilometer- (55-mile) per-hour speed limit. But the long-term significance of the Arab oil embargo was the growing awareness it initiated in us of our dependency on an uninterrupted and copious energy flow. This new consciousness inspired us to reevaluate the state of our domestic fuel reserves and our energy-consumption practices.

In response to the situation dramatized by the 1973 oil embargo—generally referred to as "the energy crisis"—President Nixon announced the creation of Project Independence. The goals of this project were to develop energy reserves in the United States, primarily oil, so that we would never again find ourselves dependent on foreign oil. In 1972, the year before Project Independence was created, the United States had imported about 29 percent of its oil supply from oil-exporting nations, at a cost of $4.5 billion. In mid-1975, two years after

Nixon's project was announced, imported oil accounted for 35 percent of the total consumed, for which we paid $25 billion. In 1978, we were importing 48 percent, and government projections indicate that by 1980 we are likely to be importing over half of our oil.

On October 15, 1978, five years after the Arab oil embargo and following nearly a year and a half of deliberation, Congress passed the National Energy Act. Provisions of this act are expected to reduce our oil import needs by 2.5–3 million barrels a day by 1985. The act is designed to encourage a shift toward the use of more coal and toward a more efficient and equitable use of U.S. energy resources. After the act was passed, President Carter stated, "We have declared to ourselves and the world our intent to control our use of energy, and thereby to control our own destiny as a nation."

Even following the passage of the National Energy Act of 1978, however, we still find ourselves in a state of turmoil over energy. Arguments abound over which measures we should employ to conserve existing energy supplies, and to which new energy sources we should give special attention. Because the government has appeared indecisive in the past, many people do not believe that an energy crisis exists at all. (Nonetheless, long lines reappeared at gasoline stations during the summer of 1979.) This attitude is compounded by the public's now prevalent distrust of people in authority, be they government officials, scientists, or oil company executives. Yet, in truth, the government's energy policy decisions will influence the quality of our environment, economy, national security, and international relations for decades to come. All of us must come to grips with the seriousness of the energy-supply situation in order that we will be willing to make the necessary sacrifices in our habits of fuel consumption. Therefore, we begin our examination of the energy question by identifying current trends in energy use and assessing the prospects for future energy supplies.

Trends in Energy Use

Today, our nation consumes about fifteen times more energy than it did a century ago, as indicated in Figure 18.6. This dramatic rise in energy consumption is attributable in roughly equal parts to the growth in population and the soaring per capita energy demand, and it occurred in all sectors of our economy—residential-commercial, industrial, transportation, and electricity generation—as Figure 18.7 suggests. When we divide the total energy consumed by the number of people in the United States, the results are 28 barrels of oil, 2,600 cubic meters (91,000 cubic feet) of natural gas, and 2.6 metric tons of coal per individual per year. (See Appendix I for conversion factors for units of energy.) We can gain perspective on just how these vast amounts of energy are used by examining consumption trends within each of our society's economic sectors.

Energy is used in both homes and commercial

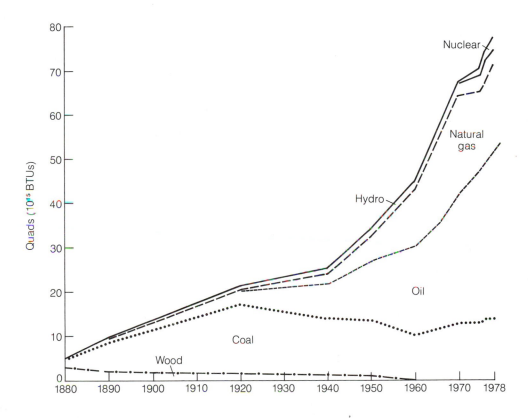

Figure 18.6 Energy use in the United States over the last century. Approximately 1.8 quads of biofuel energy is not accounted for in recent data. (Data for 1880–1976: Federal Energy Administration, *Energy in Focus—Basic Data,* Washington, D.C., 1977; data for 1977–1978: U.S. Department of Energy, *Monthly Energy Review,* January 1979.)

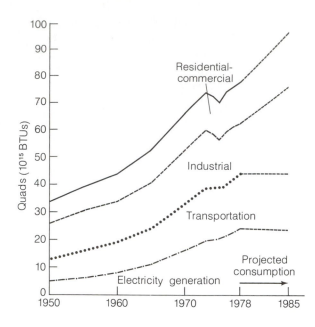

Figure 18.7 Energy consumption in the United States by economic sector. Projected consumption assumes that 1976 policies have not been changed. (Projected data: Federal Energy Administration, *Annual Report, 1976/1977,* Washington, D.C.: U.S. Government Printing Office, 1977.)

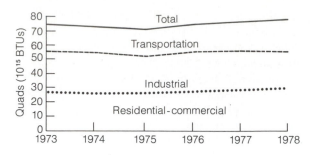

Figure 18.8 Energy consumption for various sectors of the U.S. economy, with electrical consumption included in each. (After U.S. Department of Energy, *Monthly Energy Review,* January 1979.)

buildings for basically the same purposes: heating, air conditioning, lighting, ventilation, and the powering of small electrical appliances. In 1978, the residential-commercial sector, as indicated in Figure 18.8, accounted for 37.8 percent of the total energy used in the United States. This figure includes energy used to generate electricity consumed within this sector. Residential and commercial usages were about equally divided: homes accounted for about 21 percent and business establishments about 17 percent of our total consumption. Between 1950 and 1972, the rate of energy use in the residential sector increased by 4 percent per year, about double the rate of population growth.

Space heating is the single most important energy-consuming process in homes and businesses. In addition, air conditioning was a major contributing factor to the recent rise in energy consumption. For example in 1950, fewer than 1 percent of all U.S. homes had any type of air conditioning, but today about half are so equipped. Other appliances found in homes with increasing frequency include clothes dryers, self-defrosting refrigerators, color televisions, and improved but often excessive lighting. Given the fact that our population is increasing, it is clear that total energy consumption in the residential-commercial sector will only drop if we learn to use energy more wisely.

In 1978, the transporting of people and industrial goods accounted for 26.4 percent of our nation's total energy consumption. Gasoline is by far the most important fuel for transportation; 77 percent of the fuel consumption in the transportation sector was gasoline. Much of this fuel is wasted through unnecessary travel, idling engines, wasteful driving habits, unnecessarily large vehicles that require excessive amounts of gas, and losses through evaporation and spillage. Of the total horsepower dedicated to transportation, 95 percent is utilized on our highways; trains, airplanes, and ships use the remaining 5 percent. Government forecasters predict that in 1985 we will be using about the same amount of gasoline as we did in 1975 despite continual population growth. This prediction is based on the assumption that by 1985 new automobiles will meet the mileage requirement of 11 kilometers per liter (26 miles per gallon) set for the auto industry by Congress.

The industrial sector of our society is the largest user of energy, accounting for about 35.8 percent of the 1978 total. The chemical industry is the largest single consumer within this sector, since it not only requires energy for manufacturing, but also uses petroleum as the raw material for many of its products, such as pharmaceuticals, plastics, and pesticides. The steel-producing, petroleum-refining, and paper and aluminum industries are also heavy users of energy. Most industrial enterprises have the means to bring about some immediate reductions in energy waste.

The Arab oil embargo in 1973 helped call attention to energy prices, which had begun to increase a few years earlier. These price rises sent tremors through every segment of our society. Skyrocketing electric and fuel bills dealt stiff blows to people on fixed incomes, and energy-intensive industries experienced sharp profit declines when they were unable to pass these added expenses on to consumers in the form of higher prices. Even those individuals who succeeded in conserving energy became disenchanted with making sacrifices, since utility bills generally climbed no matter what they did. And in 1979, the slowdown of exports by Iran began the second round of sharp price increases for energy products.

Our breakdown of total energy use by sector indicates where energy is being consumed and for what purposes. But if we are to assess the status of energy supplies for the future, we must also be able to predict, to some degree, the extent of future demands. As was indicated in Figure 18.7, the Department of Energy predicts that total energy demands will increase by about 38 percent between 1975 and 1985. Furthermore, increases are projected for all consuming sectors. Where will this energy come from?

Contemporary Energy Sources and Supply

For the last twenty-five years, the United States has relied almost entirely on the fossil fuels—oil, natural gas, and coal—for its necessary energy. Nuclear fuel is a relatively new energy source; only since 1973 has it met more than 1 percent of our nation's total energy demand. We have just seen that our usage of these fuels has steadily increased and is projected to continue increasing. In light of these increases, a question fundamental to our whole way of life is, How long can these nonrenewable resources last? Answering this question is difficult, since the expected lifetime for nonrenewable resources depends on future energy consumption trends, economic growth, import levels, and the success of future exploration efforts. Keeping these uncertainties in mind, we turn now to an examination of the present and projected supply for each of our major energy sources.

Natural Gas

Natural gas constitutes only 4 percent of our energy reserves but, in 1978, it furnished 29 percent of our total energy (see Figure 18.9). Nearly all our natural gas comes from wells within the lower forty-eight states, including offshore regions. Only 5 percent is imported—mostly from Canada—and even smaller amounts of synthetic natural gas are produced from coal. As indicated in Figure 18.10, the production and consumption of this resource increased rapidly for years, but the flow of natural gas peaked in 1973. By 1976, production had already dropped by 12 percent.

Natural gas is an ideal fuel. It is clean-burning, producing only carbon dioxide and water as endproducts. If the dirtier fuels, oil and coal, were used exclusively in home heating systems, the resultant air pollution problem would necessitate the installation

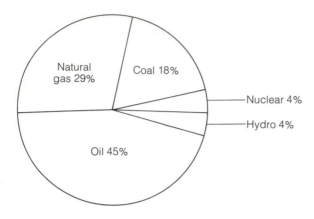

Figure 18.9 Energy supplied by various fuels in the United States in 1978. (After U.S. Department of Energy, *Monthly Energy Review*, January 1979.)

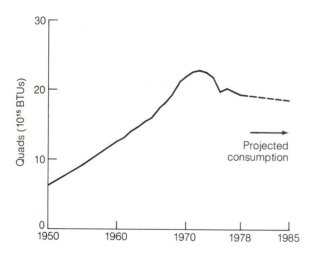

Figure 18.10 Annual domestic consumption of natural gas in the United States. (Data for 1950–1976; Federal Energy Administration, *Energy in Focus—Basic Data*, Washington, D.C., May 1977; data for 1977–1978: U.S. Department of Energy, *Monthly Energy Review*, January 1979; data for 1985 (projected): Federal Energy Administration, *Annual Report 1976/1977*, Washington, D.C.: U.S. Government Printing Office, 1977.)

of an air pollution control device on each individual system. But if natural gas were used to heat homes, control devices would be unnecessary. Also, natural gas is easily transported by pipeline directly to where it is needed and can be utilized more efficiently than other fuels.

To ensure that residences will have an uninterrupted energy supply, the federal government is trying to force industries and electric utilities to change from natural gas to more expensive and dirtier but more abundant fuels, primarily coal, though few have done so. In addition, the Department of Energy has ordered most gas utilities not to hook up any new residential or commercial customers. The government is also encouraging home and business owners to insulate, weatherstrip, and employ other energy-conserving techniques. Undoubtedly, increased efficiency and further curtailment of natural gas use will be necessary as gas reserves continue to shrink.

Up to now, natural gas has been sold at a fraction of its replacement cost. In a move to match the price of natural gas with its replacement costs, the government will lift price controls on new natural gas on January 1, 1985. This action will force industries and utilities that currently consume 60 percent of natural gas flow to switch to alternative fuels.

How long will natural gas supplies last? No one can definitely say, because the exact size of our reserves is not and probably will never be known precisely. Furthermore, uncertainties regarding the severity of winter weather conditions and restrictions on use blur predictions. We do know, however, that the outlook for natural gas is the bleakest of those for all our fossil fuels. Over the past ten years, although we have been using about 560 million cubic meters (20 trillion cubic feet) of natural gas a year, we have discovered only 280 million cubic meters (10 trillion cubic feet) annually. Known reserves will last only ten to eleven years at 1978 consumption rates. And, more discouragingly, the amount of existing undiscovered gas is estimated to be 14 trillion cubic meters (500 trillion cubic feet)—about a thirty-five-year supply at current consumption rates. In total, then, we may have a forty-five-year supply, providing that we do not increase our present rate of usage. But long before the forty-five years have passed, many people will have had to reduce their reliance on natural gas; some communities have already begun to do so.

Exploration for new natural gas supplies will require considerable money and time, as will the extraction and delivery of new discoveries. For example, although Alaskan gas supplies were discovered in the mid-1960s, this natural gas is not expected to reach the lower forty-eight states until 1985, when the Alcan Gas Pipeline is completed. (Figure 18.11 shows the route of the pipeline.) Even at the peak of delivery, this highly publicized source will furnish only 10 percent of our 1990 demand. Furthermore, construction of the pipeline will cost $10 billion, about one-quarter of the estimated value of the gas.

The diminishing of natural gas supplies in the future means that we will have to increase greatly our use of coal. In Europe, insufficient natural gas supplies have been supplemented for several decades by *coal gasification*. In this process, also used in the United States during the 1920s, coal is heated in the absence of oxygen to drive off methane gas, a fuel. The remaining coal is made to react with steam to produce carbon monoxide and hydrogen, which also yield energy when burned. The three gases—methane, carbon monoxide, and hydrogen—are mixed together to form a fuel source with a relatively low energy content (3.3–5.0 megajoules per cubic meter, or 100–150 BTUs per cubic foot). This fuel can be burned in nearby power plants or industries. To make coal gas with a high energy content (10–20 megajoules per cubic meter, 300–600 BTUs per cubic foot) that can be distributed to distant sources by pipeline, producers must increase the methane content. During the coal gasification process, 10–50 percent of the energy content of the coal is sacrificed to produce the gas.

Coal gasification is fairly simple to describe, but in reality it requires a costly, complex network of pipes and reaction vessels resembling an oil refinery. Furthermore, if the huge demands of this country are to be met through coal gasification, the process would have to be scaled up much beyond the size of existing plants. Part of the present U.S. research effort in this area is an attempt to produce coal gas directly within coal seams (*in situ* mining) located in Wyoming. Despite the difficulties, coal gasification has the advantage of converting a dirty fuel into a cleaner fuel, since sulfur can be removed much more easily from gas than from coal. But synthetic natural gas derived from coal is now very expensive and therefore is not expected to contribute significantly to gas supplies in the near future.

A short-term means of increasing natural gas supplies is to import more natural gas by employing *liquefied natural gas* (LNG) tankers, in which natural gas is liquefied by being chilled to $-162°C$ ($-259°F$). Figure 18.12 shows an LNG tanker being unloaded at a regasification terminal. In the liquefied state, natural gas occupies about $1/600$ of its normal volume; hence, more LNG than gas can be transported per tanker. But liquefied gas is extremely hazardous, and elaborate precautions must be taken to prevent explosions during transport and loading and unloading. In addition, imported liquefied natural gas is considerably more expensive than gas supplied by pipelines.

Crude Oil

Of the total energy consumed in the United States, 45 percent is derived from crude oil. In 1978, about 48 percent of our oil came from foreign wells. The historic trends since 1950 are indicated in Figure 18.13. Since the Arab oil embargo, the member nations of the Organization of Petroleum Exporting Countries (OPEC) have raised the price of crude oil about fivefold. The rapid rise in our payments to foreign countries is charted in Figure 18.14. As a consequence, in our balance of trade we have gone from the black (as net exporter of goods) to more than $30 billion in the red in 1977 (as net importer).

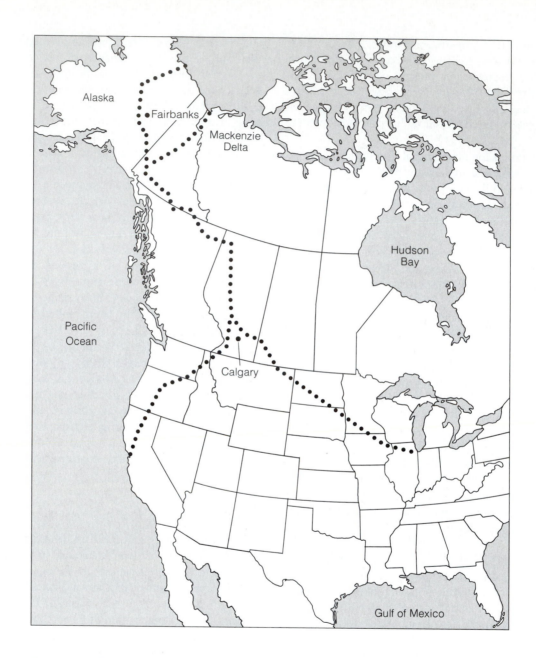

Figure 18.11 Route of the Alcan pipeline, which will supply natural gas to the Midwest and the West Coast by about 1985.

Figure 18.12 A liquefied natural gas (LNG) tanker unloading at a regasification terminal on Chesapeake Bay. This natural gas comes from Algeria and is used in the middle Atlantic states to augment dwindling U.S. supplies. (El Paso Natural Gas Company.)

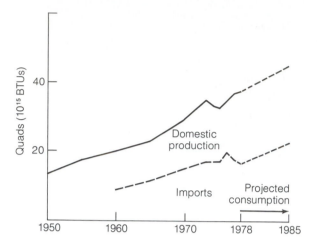

Figure 18.13 Past and projected U.S. oil consumption. Note the projected increase in dependence on foreign oil despite Alaskan oil coming on line in 1977. (Federal Energy Administration.)

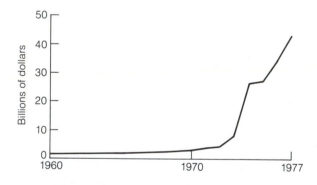

Figure 18.14 U.S. expenditures for oil imports, 1960–1976. (After Federal Energy Administration, *Energy in Focus—Basic Data,* Washington, D.C., May 1977.)

Predictions are that this deficit will continue to grow at least over the next few years. This cash outflow enlarges our national debt and fuels worldwide inflation. But because so much of our oil comes from the international market, our government can play only a

relatively minor role in determining world oil prices. Thus, the significant question is whether any means of reducing our dependency on foreign oil is open to us.

Some potential to increase oil production lies in rejuvenating old, idle wells. In most cases, more oil remains in the ground than was originally extracted. Whether production from these wells is continued is partially controlled by economics: if oil prices rise, putting some old wells back into production may be profitable. Thus, we could employ costly secondary and tertiary recovery techniques to squeeze more oil from a well (see Chapter 12). The amounts of oil recovered from individual wells range widely—from as little as 10 to nearly 100 percent of the oil originally present. About 40 percent recovery, however, is thought to be the overall upper recovery limit for crude oil.

To increase production of oil at home, U.S. geologists have focused their oil exploration activities on frontier areas on the outer continental shelf and in Alaska. In 1975, 97 percent of the oil produced within the United States originated in the lower forty-eight states. But by 1990, Alaskan oil and new offshore oil sources are expected to account for about 37 percent of the domestic oil flow. In making the shift, we will have to construct new transportation facilities—for example, deep-water ports, supertankers, storage facilities, pipelines, and coastal refineries. Oil from frontier regions is much more expensive than that from already producing areas because extraction costs in these hostile environments are high. Figure 18.15 shows a costly offshore drilling rig. Also, many observers are concerned over the environmental degradation that results from the development of new sources and associated oil-handling facilities. (Recall the environmental consequences of oil spills discussed in Chapters 7 and 8.)

In addition to obtaining more oil from frontier regions, we may also be able to increase domestic oil production by exploiting *oil shale*—rock containing organic material that can produce oil when heated to high temperatures. Although its potential as an oil

Figure 18.15 An offshore oil-drilling rig, used to find and extract oil reserves on the continental shelves. (U.S. Department of Energy.)

Although the technology for processing oil shale is available (Figure 18.16 illustrates a method), the present economic climate is unfavorable for large-scale exploitation. At the present time, the Middle East oil available to the United States is less expensive. In addition, serious environmental considerations inhibit exploitation. For every liter of oil produced, 1.2–2 liters of water are consumed (this water is nonrecyclable), and an additional liter of water may be required for waste disposal. Ironically, regions where high-grade oil shale deposits are found are the semiarid states of Wyoming, Colorado, and Utah. Water shortages in these areas would probably limit oil shale production to about 170 million liters (1 million barrels) per day—about 5 percent of the nation's current oil consumption.

A second—almost overwhelming—obstacle to oil shale development is the sheer volume of material that has to be processed and the associated waste disposal problem. For each 170 liters (1 barrel) of oil produced, 1.4 metric tons of shale would have to be processed. And in the production of 170 million liters (1 million barrels) of oil per day, 1.4 million metric tons of shale would have to be processed, resulting in 1.2 million metric tons of waste. This residue is highly alkaline and poor in nutrients, and, especially in semiarid climates, it resists revegetation. In addition, the volume of the processed shale is greater than the raw material itself, since retorting (heating) causes the rock to expand. Thus, additional costs would be entailed in spreading topsoil over spent shale dumps to effect land reclamation.

Because most rich oil shale deposits are under federal lands (80 percent of the total), oil companies must obtain leases to exploit these sources. Under the authority of the Department of the Interior, six small tracts of land in the Central Rockies were recently leased for development of small-scale oil shale extraction operations. Although these efforts involve conventional mining techniques prior to processing, some research has been done there on methods of in-place processing. Government scientists anticipate that this approach eventually will

source has been known for decades, oil shale is just beginning to receive serious attention. The quantity of oil that may be generated by the processing of oil shale deposits is enormous. In fact, some high-grade oil shale in the United States may yield about 9 percent oil, or 105 liters (28 gallons) of oil per metric ton of shale. Recoverable oil shale deposits in the United States may eventually yield 600 billion barrels of oil—roughly a one-hundred year supply at present consumption rates. Also, some scientists have speculated that lower grade shales (not economically extractable by current technology) may house as much as 300 trillion liters (2 trillion barrels) of oil.

Figure 18.16 An artist's conception of Union Oil's planned experimental oil-shale project near Grand Valley, Colorado. Shale is mined by conventional underground techniques and treated to extract oil. Wastes will be piled and revegetated in the bottom of the valley. (Union Oil Company.)

alleviate somewhat the problems of water use and waste disposal, but the technology for large-scale implementation of this method is at least a decade away. By 1990, oil from shale and coal (synthetic oil) may account for about 180 million barrels per year, or, at most, 2 percent of consumption.

In the future, coal will undoubtably serve as the raw material for the manufacture of petroleum liquids. *Coal liquefaction* is a complicated process—whereby coal is heated in the absence of oxygen to produce liquid fuels—and more difficult than coal gasification. Complex and costly chemical plants are required to produce major changes in the chemical structure of coal. Here, too, economics will dictate when coal liquefaction will begin to supplement the flow of petroleum liquids from oil refineries. Exactly when coal liquefaction will become important is open to debate. In 1978, imported crude oil cost about $14–15 per barrel, while the production of petroleum liquids with an equivalent heating value through coal liquefaction cost in the range of $17–23 per barrel to produce. When this price gap narrows, coal liquefaction may become an important process for producing liquid fuels.

Given the uncertainty as to when oil shale processing will contribute to our domestic oil supplies, how long will our domestic oil reserves last? Again, as with natural gas, we do not know the exact size of our oil reserves. But assuming that we continue to consume oil produced in the United States at the 1978 rate while using imports to meet oil supply deficits, estimates are that we have a ten-year reserve. Furthermore, using the same assumptions, scientists speculate that another eight-year supply is likely to be found. Yet another undiscovered twenty-four-year supply is also tacked onto government estimates. Included in these projections is the highly publicized 10-billion-barrel Alaskan oil reserve, which would last only 1.6 years if it were our sole source of oil. Now that Alaskan oil is flowing, domestic production is expected to increase until 1985, and then to resume its decline. Regarding new discoveries, in the late 1970s we found only about a half barrel of oil in the time it took us to consume a whole barrel.

These estimates dramatically illustrate the crisis associated with our oil supply. Oil produced within the United States will never again match our demands. In fact, if we increase our demands, as is projected, we will have to import even more oil than we do now. And some scientists predict that world oil production will peak around 1990. Thus, all countries will experience a reduction in the available oil supply in the near future and will be in competition for shrinking world reserves.

Coal

Today, the United States derives 18 percent of its energy from coal. In 1920, coal was the major fossil fuel; at that time it accounted for 73 percent of the nation's total fossil fuel consumption. The relative importance of coal as an energy source diminished in subsequent decades with the discovery of vast oil reserves around the globe. Oil was cheap and burned easily in diesel engines, which replaced coal-fired steam engines. Until recently, natural gas and oil were the preferred fuels because they are cleaner burning than coal, but, as we have seen, the supply of these fuels is rapidly dwindling and interest is being renewed in using coal more extensively. In steam production, coal is about half as expensive as the oil required to generate an equivalent amount of steam. Thus, economic considerations constitute a strong argument for an increase in the use of coal.

Coal is used primarily to generate electric power and as a heat source for various manufacturing processes. The metallurgical industry uses coal to produce steel from iron ore. The coals used in these metallurgical processes are high grade (low in volatile substances), and typically they command a price about two to three times greater than the price of coal used to generate steam. These high-grade coals account for about one-quarter of the country's coal consumption and about 50 percent of the coal industry's profits.

Coal is the most abundant of the fossil fuels, and U.S. deposits are large enough to meet our nation's needs for several hundred years, even at consumption rates several times greater than contemporary rates. As shown in Figure 18.17, coal deposits are scattered throughout the lower forty-eight states. Hence, as we switch from natural gas and petroleum to coal, these regions will experience more environmental and economic impacts related to coal mining. The most important deposits in the East are located in the Appalachian region. These deposits consist primarily of bituminous coal, which has a relatively high heating value but moderate to high levels of sulfur (see Chapter 12). Bituminous coal from Illinois, western Kentucky, and Indiana are important centrally located deposits. Farther west, important coal fields are located in the Dakotas, Montana, and Wyoming. Coal from these lignite and subbituminous deposits is lower in heating value and produces more ash, but it is also much lower in sulfur content. Coal production from these various areas is summarized in Table 18.1.

Certain economic and regulatory policies will affect future coal consumption and the environmental

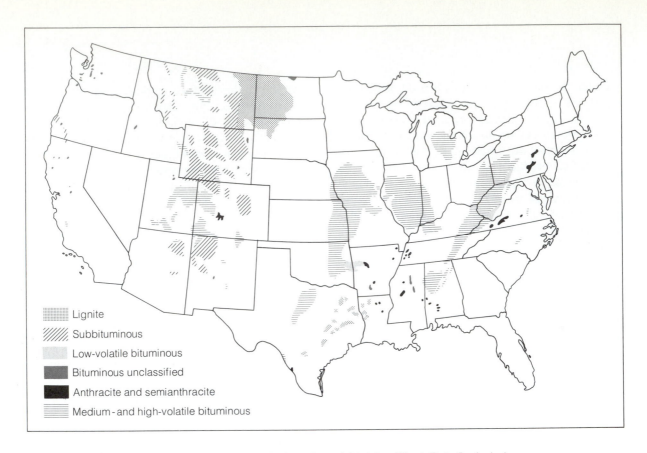

	Lignite
	Subbituminous
	Low-volatile bituminous
	Bituminous unclassified
	Anthracite and semianthracite
	Medium- and high-volatile bituminous

Figure 18.17 Location and types of coal deposits in the lower forty-eight states. (Illinois State Geological Survey, Circular 499, *Trace Elements in Coal,* 1977.)

impact of increased coal mining, processing, and burning. The air pollution requirements of the Clean Air Act of 1970 and subsequent amendments forced hundreds of electric utilities to switch from coal to low-sulfur oil by the early 1970s. However, now that oil is more expensive and in short supply, the government would like to see these plants switch back to coal. Until now, the government's preference on this matter has generally been ignored. New industrial and electricity-generating plants, however, are being designed to burn coal or nuclear fuel rather than oil.

Controversies are now arising over how coal burning should be regulated. The Environmental Protection Agency has decided that all new coal-burning plants must meet emission standards for sulfur dioxide. Presently, new power plants can meet these standards either by burning low-sulfur coal (coal that contains less than 0.3 kilograms of sulfur per 1 billion joules, or 0.6 pounds per 1 million BTUs) or by installing systems that remove sulfur from stack effluents (see Chapter 11). Installing and operating desulfurization air pollution equipment adds about 30 percent ($6–7) to the cost of a metric ton of coal. Thus, plants have an economic incentive to choose the first alternative—burning low-sulfur western coals. But coal companies that handle high-sulfur coal argue that sulfur-removal equipment

Table 18.1 Annual coal production by region (in millions of tons).

	1976	1985
East of Mississippi		
Middle Atlantic	232	———
South Atlantic	170	———
Midwest	128	———
Total	530	705
West of Mississippi		
Southwest	28	———
Central	7	———
North Central	85	———
West	10	———
Northwest	5	———
Total	135	360
National total	665	1,065

Source: Federal Energy Administration, *Annual Report, 1976/1977*, Washington, D.C.: U.S. Government Printing Office, 1977.

should be required on all new coal-burning electricity-generating facilities. Such a requirement would allow these companies to sell more high-sulfur coal. Projections show that if the Clean Air Act is amended to include this provision, western coal production would decline, since burning eastern coal, with its higher energy content, would be more economical. To complicate the issue even further, western strip mines are ten times more labor efficient than eastern underground mines, and therefore stepped-up mining in the East would mean more jobs. However, underground mining is much more dangerous than strip mining. Thus, the political decision as to how air pollution from coal is to be controlled will affect the cost of the coal itself, and this factor in turn will determine which geographical area experiences the greatest increase in mining activities, and, hence, the more economic and environmental impact.

In addition, changes in the geographical distribution of coal-mining activities will create a measure of social and cultural turmoil. Western and Appalachian communities adjacent to coal mines are currently experiencing sudden, substantial economic growth. Most of these communities are located in sparsely populated regions in which administrators lack the expertise and financial resources to plan carefully for expansion—for example, with respect to providing sewers, water supplies, roads, housing, schools, and hospitals. Also, many people in these regions (especially westerners) prefer the quiet, stable lifestyle to which they have become accustomed. They are skeptical, to say the least, about the desirability of allowing mining and commercial enterprises to invade their old stomping grounds (see Box 18.1). Concerned residents are right to be skeptical. The more coal production is intensified, the greater will be the associated environmental problems: strip mine damage, acid mine wastes, air pollution, and fly ash accumulation (see Chapters 12 and 13). Still, coal-mining areas always experience environmental problems. But some of the problems of coal burning can be avoided through utilizing coal gasification and liquefaction processes.

World Fossil Fuel Resources

In the preceding sections, we saw that U.S. natural gas production peaked in 1973 and domestic oil production has been steadily declining since 1970, although Alaskan oil is expected to stabilize domestic production for about a decade. How long can we expect U.S. and world fossil fuel resources to last? The data in Table 18.2 shed some light on this frequently asked question.

These data indicate that U.S. oil and natural gas reserves are already precariously low. Furthermore, our demands are so high that we are rapidly burning up the small reserve we have. Global natural gas and oil reserves are expected to last four to six times longer than U.S. reserves. But even these reserves

Box 18.1

Coal, Water, and Controversy:

The Yellowstone River enchants the eye as it flows through the Paradise Valley, just south of Livingston [Montana]. It is the nation's last major undammed river outside Alaska, and its upper reaches here are brimming with trout that thrive in the icy water swirling down the mountains through Yellowstone National Park.

Now, however, the Yellowstone's waters are swirling in controversy too. The Montana Department of Fish and Game wants to limit water withdrawals from the Yellowstone and its 70-odd tributaries to about the present level.

The department and its supporters say that without the limitation, the Yellowstone's fish and wildlife populations will be threatened. Some of the river's tributaries are already dammed, and those who want strict water-use limits fear the same eventual fate for the Yellowstone. Indeed, a Yellowstone dam near Livingston has been an on-again, off-again proposition for 75 years. "A dam would destroy one of the best trout streams in the world," says John Bailey, whose family owns a fly-fishing shop in Livingston.

But many ranchers, businessmen and city officials along the river, who often compete among themselves for water, say the Fish and Game Department wants to go too far. "We don't want a dam on the Yellowstone, but we do want to use more water for growth," says Robert L. Jovick, city attorney for Livingston. "The people who live here are entitled to enjoy the area without having a substandard economy."

The dispute might be a routine struggle between environmentalists and developers except for one thing—43 billion tons of coal. The Yellowstone flows right through eastern Montana's Fort Union coal field, containing some 40% of the country's coal that's suitable for strip mining. The river's water would be a valuable tool for developing the low-sulfur coal, which could help fuel an energy-hungry nation without the pollution problems of coal from the eastern U.S.

Tenneco Inc. has acquired major coal reserves in the Fort Union field and is eyeing water-use rights. Tenneco's Intake Water subsidiary already has permission to divert some water from the lower Yellowstone, and now it wants to build a dam on the Powder River, a Yellowstone tributary. "Our objective is to acquire enough water and coal rights to build a coal-gasification plant," says Gary Cheatham, Tenneco's director of national resources.

Other energy companies want to dip their buckets into the Yellowstone and its tributaries too. General Electric Co.'s Utah International unit hopes to build a Powder River project that would compete for water with Tenneco's proposed dam (the two companies are in court over the matter). The line for water rights also includes such energy giants as Mobil, Getty and Gulf Oil.

• • •

Environment vs. Development in Montana

If Montana were 1,000 miles east of where it is the Yellowstone wouldn't be much of an issue. But like much of the Western U.S. Montana is semi-arid, and when flying low over the Yellowstone River Valley it's easy to see how precious water is in the state. Most of the autumn landscape is a soft brown color. But along the river and its tributaries lie bright-green squares and rectangles, showing the reach of irrigation ditches.

Green circles spot the landscape like huge green polka dots, the product of "center pivot" irrigation systems able to pump water to higher ground. "There's a lot more land that could be irrigated around here if only they'd let us have more water," says Charles Pierson, who raises cattle and hogs east of Livingston.

Also visible from the air are reddish streaks of earth called scoria rock, indicating the rich coal veins just beneath the surface. Some half-dozen strip mines have opened in eastern Montana in the past decade. Most of the 27 million tons mined last year was shipped out by rail to be burned elsewhere.

"Stripping and shipping," as it's called, is fine with the environmental interests because Montana has tough land-reclamation laws and because the process doesn't use water. "We know that the coal is going to be dug, but that can be done without tapping the Yellowstone for water," says James A. Posewitz of the Montana Fish and Game Department. "There isn't any reason to have a coal-gasification plant here."

But the energy companies see a big reason—money. "It makes much more sense economically to convert the coal to synthetic gas and then to ship it elsewhere," says Tenneco's Mr. Cheatham. "It's twice as efficient to ship energy in the gaseous state as it is to ship coal."

∙ ∙ ∙

Caught between the energy and environmental interests are townspeople and ranchers in the valley, and they don't trust either side. "The Fish and Game Department is using the idea of a dam on the Yellowstone as a scare tactic so it can grab all the water," says Mr. Pierson, the Livingston rancher. "We shouldn't get paranoid about a dam and do some stupid things."

Mr. Jovick, the Livingston city attorney, says that while his town wants to grow it also "doesn't want to be a boom town. We know the problems that can bring."

∙ ∙ ∙

But the lure of economic growth and its benefits may prove irresistible to an area that hasn't seen much of it. Even environmentalists concede they're unlikely to get the strict water-use limits they want for the Yellowstone River. "We're shouting temperance," says the Fish and Game Department's Mr. Posewitz. "That never did sell very well."

Table 18.2 U.S. and world fossil fuel reserves, resources, and annual rates of consumption (in quads, or 10^{15} BTUs).*

	United States			World		
	Natural gas	Oil	Coal	Natural gas	Oil	Coal
Rate of consumption	22	38	15	17	116	79
Proved reserves	235	191	10,700	2,200	3,900	130,000
Ultimately recoverable resources remaining	735–1,730	630–3,560	45,000	11,100	6,900–10,400	233,000
Lifetime of proved reserves**	11	5	730	47	34	1,600

* Consumption rates and proved reserves for the United States are for 1976; for the world, 1972.
** Assumes stated consumption rate, that is, no growth.
Sources: Federal Energy Administration, *Energy in Focus—Basic Data,* Washington, D.C., May 1977; and after P. R. Ehrlich, A. H. Ehrlich, and J. P. Holdren, *Ecoscience: Population, Resources, Environment.* San Francisco: W. H. Freeman and Company, 1977, p. 402.

will be largely depleted by the early part of the next century. To keep up with present worldwide consumption rates, we would have to discover oil reserves equivalent to those in Iran or Kuwait every three years, or equivalent to those in Texas or Alaska every six months. The oil-exporting countries of the world produce about 40 million barrels of oil per day. About 27 percent of the exported oil comes from a single country—Saudi Arabia. Until 1979, when it curtailed its exporting activities sharply, Iran was the second largest oil-exporting nation. Whether the United States will continue to have access to world oil and natural gas supplies is also an open question. The answer depends on political decisions made in foreign countries and the world economy in general. Nevertheless, since world oil production is expected to peak in the next fifteen years, every nation will have to reduce its dependency on oil.

Both U.S. and world reserves of coal are much larger than gas and oil deposits, and they will undoubtedly be relied on more heavily in the future. In fact, coal reserves are large enough to last for several centuries, even if consumption rates were to increase several times over. Dr. M. King Hubbert, an expert on fossil fuel resources, predicts that coal production may not reach its peak until the year 2000. By increasing our utilization of domestic coal reserves, the

United States can become less vulnerable to economic pressures in the world energy market.

Nuclear Energy

Despite the fact that nuclear energy has been providing electrical power for over two decades, the health effects and safety of nuclear reactors are still controversial issues. Referenda held in several states in the mid-1970s favored the continued development of nuclear power. But on March 28, 1979 a new chapter in the controversy surrounding nuclear power began when the Three Mile Island nuclear power plant at Middletown, Pennsylvania, experienced the worst accident in the history of the industry. Through a series of mechanical failures and human error, radioactive water spilled into the nuclear reactor building. Procedures used to shut down the reactor resulted in the release of "puffs" of radioactive gases from the plant. Because of the potential hazard posed by escaping radioactivity, pregnant women and young children within 8 kilometers (5 miles) were advised to evacuate the area. Thousands of other people in the area joined in the exodus. A week was required to cool the reactor so that people could move back into their homes and reopen their businesses. While no one was killed or injured in the

Three Mile Island accident, an additional 1–10 cancer deaths are projected in the future as well as an equal number of nonlethal cancer disorders because of low-level radiation exposure.

The accident is also leaving an enormous wake of legal and political problems. Suits have been filed against the owners and designers of the plant for reimbursement for lost income, devalued property, endangered health, and psychological stress. Furthermore, because of flaws in the laws controling nuclear liability, it is unclear "who is to pay." Consumers argue that the plant owners should pay, while the company argues that it cannot continue to operate without additional revenue from its customers. No doubt it will take at least several years to clear up the legal and financial tangles created by the accident. Meanwhile, political and popular support for nuclear power has sharply declined.

To clarify the issues associated with nuclear power, we begin our investigation by considering briefly the process of nuclear fission, which is the basic principle governing the operation of nuclear power plants.

Nuclear Fission. The generation of power in all commercial nuclear reactors in service today is based on a nuclear chain reaction called *nuclear fission,* which involves the splitting of the *nuclei* (the extremely small, dense centers of atoms) of uranium atoms that have a mass (weight) of 235 atomic mass units. The process, illustrated in Figure 18.18, begins when a uranium-235 nucleus captures a *neutron* (a fundamental atomic particle found in the nuclei of elements). The introduction of this extra neutron into the nucleus destabilizes the nucleus, causing it to split into two large fragments (hence the term nuclear fission). Furthermore, each uranium-235 nucleus that undergoes fission ejects two or three neutrons. If other uranium-235 atoms capture the ejected neutrons, these atoms also undergo fission, thereby releasing more neutrons and thus sustaining a *nuclear chain reaction.* The fissioning of the uranium nucleus releases tremendous amounts of energy, which, of course, is the reason the reaction is carried out. The fissioning of a sample of nuclear fuel about the size of a golf ball would release energy equivalent to that generated by 650,000 liters (4,000 barrels) of oil. In a nuclear reactor, nuclear fission reactions are run under water so that released neutrons are slowed (moderated) and thereby captured by uranium-235 nuclei more efficiently.

The fissioning of uranium-235 is a random process; that is, the nucleus never splits in a predictable manner, and the size of the fragments varies with each fission. Uranium-235 atoms are known to fission in more than thirty different ways, forming over 450 different types of fragments, called *isotopes.* Isotopes are variations of atomic mass of a particular element. Isotopes of a particular element have similar chemical properties but widely varying nuclear properties. Almost all isotopes produced in the fission process are *radioactive,* meaning that they emit dangerous radiation. As the fission process continues, radioactive isotopes accumulate in high concentrations within the reactor. Also, as fission proceeds, small amounts of plutonium-239 build up in the reactor fuel elements. This material, if separated out—a very hazardous and technically difficult task—can be used as a component of nuclear bombs.

Radiation Hazards to Humans. We have long known that large doses of radiation are harmful to living tissue. The effects of massive radiation exposure to organisms are given in Table 18.3. The tissues of the human body most sensitive to radiation damage are the bone marrow (the site of red blood cell formation), the lymph system (which acts as a filter for tissue fluids and produces white blood cells), and the lining of the intestinal tract. Acute doses, such as those given in Table 18.3, occur only following the detonation of a nuclear bomb or, possibly, following accidental exposure to *high-level nuclear wastes* from a nuclear reactor.

Of greater concern is the effect on living tissue of chronic exposure to low levels of radioactivity. Firm evidence suggests that people who have been ex-

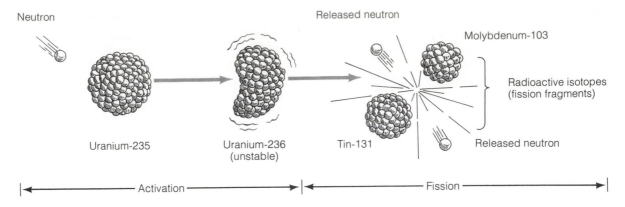

Neutron

Released neutron

Molybdenum-103

Radioactive isotopes
(fission fragments)

Uranium-235

Uranium-236
(unstable)

Tin-131

Released neutron

|◄—————————— Activation ——————————►|◄————————— Fission —————————►|

Figure 18.18 The process of nuclear fission. The sequence shown here is called a chain reaction.

Table 18.3 Effects of exposure on people to high radiation doses.

Dose (rem)*	Effect
100,000	Death in minutes
10,000	Death in hours
1,000	Death in days
700	Death for 90 percent within months (10 percent survive)
200	Death for 10 percent within months (90 percent survive)
100	No short-term deaths, but the chances of cancer and other life-shortening diseases are greatly increased. Permanent sterility in females; 2–3 year sterility in males.

* Rem = a radiation dose unit that measures damaging effect in mammals.
Source: After W. W. Murdoch, ed. *Environment*. Sunderland, Massachusetts: Sinauer Associates, 1975.

posed to low-level radiation over extended periods of time show an increased incidence of leukemia (cancer of the white blood cells), skin cancer, thyroid cancer, and lung cancer. For example, studies of children who have received x-ray treatments in the neck re-

gion have shown an increased incidence of thyroid cancer. Generally, a five- to twenty-year latency period elapses following exposure before these diseases are diagnosed.

Low-level exposures also accelerate the aging process, thus shortening lifespans, and can cause chromosome damage (mutations). Human fetuses are particularly susceptible to radiation damage. One study has showed a 30-percent increase over normal in incidence of leukemia and cancer of the nervous system in children whose mothers were irradiated with x-rays during pregnancy. Another study showed a higher than normal incidence of Down's syndrome (a genetic defect that results in mental retardation) in children whose mothers were exposed to x-rays.

People concerned about the possibility of exposure—simply through proximity—to nuclear power plant radioactive wastes cite these known effects of radiation when they ask what effects radioactive wastes will have on human health. This question is at the root of the controversy surrounding nuclear power, and the answer is open to debate. Proponents of nuclear power point out that we are exposed continually to natural background radiation from space and from radioactive isotopes such as uranium-238, uranium-235, thorium-232, and potassium-40 that occur naturally in the earth's crustal

Table 18.4 Average annual radiation dose per person in the United States.

Source	Dose (millirems*)
Natural	
Rocks, soil, sun	130
Man-made	
Medical and dental x-rays	72
Fallout from past weapons testing	4
Television sets	0.1
Commercial products	1.9
Air travel	1.3
Commercial nuclear power	0.003
Total	209.303

* Millirem = radiation dose unit that measures damaging effect in mammals; one-thousandth of a rem.
Source: Energy Research and Development Administration, "Nuclear Energy" (pamphlet), Washington, D.C.: Office of Public Affairs, 1976.

rocks. (Table 18.4 shows the various sources of radiation and their relative doses.)

Exposure to radiation has an additive, or cumulative, effect. That is, radiation from medical and dental x-rays are added to the natural background radiation, thus resulting in greater total exposure. It is therefore important that exposure be minimized wherever possible. Furthermore, some scientists believe that no threshold value exists below which radiation is harmless. They claim that tolerance ranges vary among people, and that even small increases in radiation exposure will be injurious to some members of a population. Other experts argue that a definite threshold value does exist, and that as long as exposure levels are below this threshold, radiation is harmless. Resolution of this controversy awaits further research.

The health hazards of radiation are a consequence of *alpha* and *beta particles* and *gamma rays* emitted by radioactive materials. These particles and rays can disrupt the molecular architecture of living things, thereby inducing cancer and mutations. Each of the hundreds of radioactive waste products (isotopes) of nuclear fission reactors emit at least one of these three forms of radiation. The fact that concrete several feet thick is required to stop gamma rays attests to their highly energetic nature. Gamma rays can readily penetrate an entire organism, resulting in what is known as whole-body exposure. Figure 18.19 shows the penetrating power of alpha and beta particles and gamma rays.

Beta particles have less penetrating power than gamma rays and are stopped more readily by body tissue. Although a beta particle may possess the same amount of energy as a gamma ray, it may cause more localized damage in an exposed organism because the energy is absorbed over a shorter distance. Also, if an element that emits beta rays becomes concentrated in one part of the body, that area will receive a higher dose of radiation. For example, strontium concentrates in bone; therefore, if a person ingests strontium-90, a beta-emitting isotope, bone marrow will be the primary site of radiation damage. To protect yourself from beta particles from an external source, you would have to place several centimeters of solid material (this book for example) between yourself and the source.

Alpha particles have the weakest penetrating power of the three types of radiation, but they damage exposed surfaces significantly. For example, if we were to inhale air containing radioactive radon, a gaseous nuclear waste, the lining of our lungs would suffer most of the damage. Exposure to alpha particles can be prevented by a thin protective covering, such as a sheet of aluminum foil, placed over the source. Gaseous alpha emitters such as radon, however, must be kept in sealed containers.

Each isotope disintegrates at its own characteristic rate, spontaneously emitting radioactive rays as it gradually becomes transformed into a more stable form. The change is called decay, and we indicate the rate of decay of a particular isotope by a measurement known as *half-life*—the time it takes half the nuclei of a radioactive isotope to decay to a more stable form. Figure 18.20 illustrates the decay of the

Exposure to external radiation source

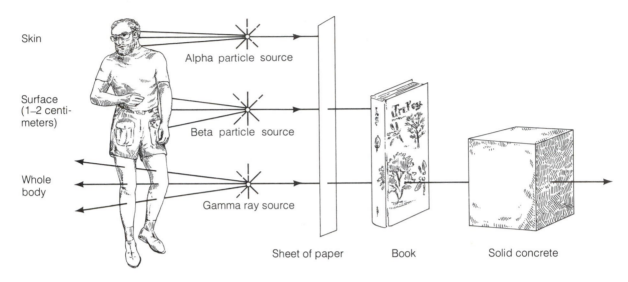

Figure 18.19 Relative penetrating power of particles and rays emitted by radioactive sources.

isotope phosphorus-32 (^{32}P). The half-life of phosphorus-32 is fourteen days. Phosphorus-32 decays through changes in its nucleus to form sulfur-32 (^{32}S), a nonradioactive isotope. If we start with one gram of phosphorus-32, after fourteen days, 0.5 grams remain (the other 0.5 grams is sulfur-32). After another fourteen days have passed, 0.25 grams of phosphorus-32 remain, and so on.

Half-lives of radioactive wastes produced by fission reactors vary widely, from 0.96 seconds for xenon-143, to as long as 24,400 years for plutonium-239. Usually, at least ten half-lives must elapse before a source decays to the point at which it no longer constitutes a serious radiation threat. Thus, the half-life of a radioactive isotope determines the length of time in which these hazardous materials must be kept in isolation to prevent human exposure. The first several hundred years are the most critical, because nuclear wastes emit the highest radiation levels during this period. Because some isotopes in

nuclear waste have long half-lives, future generations will have to watch over these wastes for centuries, making sure they are stored away securely. Radioactive waste could only be strictly monitored in this way as long as the society remains stable—an unlikely possibility in view of past history. Truly, we leave to our children a unique legacy in the form of radioactive wastes.

Food Chain Accumulation. We have learned from our past experience with aboveground nuclear weapons testing that radioactive forms of some elements accumulate via food chains. For example, strontium-90—a by-product of nuclear weapons testing—threatens human health because it moves from soil to plants to cows' milk, and is eventually deposited in human bones. Because most strontium-90 remains in the bones of exposed individuals, and because this isotope has a half-life of twenty-eight years, strontium-90 represents a life-long radi-

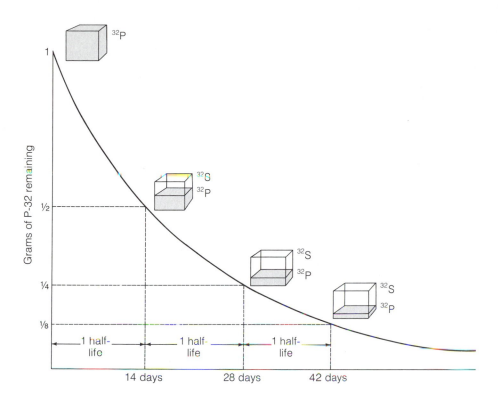

Figure 18.20 The amount of radioactivity from phosphorus-32 (^{32}P) decreasing by one-half every half-life (fourteen days). The decay of ^{32}P results in the formation of sulfur-32 (^{32}S).

ation hazard. Because of such potential effects, it is imperative that radioactive nuclear contaminants be prevented from entering food chains insofar as this is technically feasible. Other radioactive isotopes that are of concern in food chains are cesium-137 and iodine-131.

The Nuclear Fuel Cycle. What are the potential sources of exposure to radiation in nuclear power plants? We can identify critical points at which exposure may be hazardous by examining the cycle that fuels nuclear plants. This cycle, illustrated in Figure 18.21, begins in either open-pit or underground uranium mines. On these sites, radioactive radon gas poses the major radiation health hazard to uranium miners. If inhaled in sufficient concentrations over extended periods, this gas induces lung cancer. Proper ventilation in mines greatly reduces this hazard. Radon gas emitted from uranium mine waste heaps may also pose a threat to people downwind.

Uranium ores are converted into fuel elements for reactors through a complex process that includes the enrichment of the fissionable uranium content from 0.7 percent to about 3 percent uranium-235. Once enriched, individual fuel pellets (shown in Figure 18.22) are encapsulated in an expensive alloy (zirconium alloy). This cladding acts as a container to retain 99.99 percent of the radioactive waste products that build up within the fuel elements as ura-

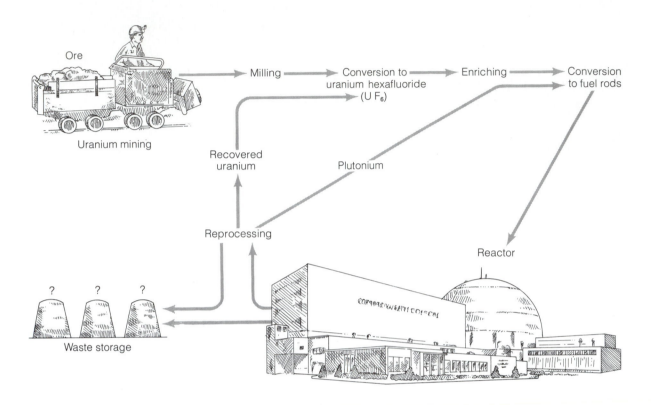

Ore

Uranium mining

Milling → Conversion to uranium hexafluoride (UF_6) → Enriching → Conversion to fuel rods

Recovered uranium

Plutonium

Reprocessing

Reactor

? ? ?

Waste storage

COMMONWEALTH EDISON

Figure 18.21 The nuclear fuel cycle. The cycle is not complete, since no permanent storage sites or reprocessing plants are in operation.

nium-235 is fissioned. These enrichment and encapsulation processes present few health hazards.

Encapsulated fuel pellets are placed into fuel rods, which are inserted into the reactor core. Each rod can produce the energy equivalent of 180 metric tons (about three railroad cars) of coal. Nuclear fission reactions take place within the reactor core, and the heat released by these reactions is absorbed by water surrounding the core. This heat is removed from the core area when the water is cycled through a closed, high-pressure primary loop (Figure 18.23) into a steam generator. Pressurized steam formed here is fed via a secondary loop to steam turbines. Because the primary loop is isolated from the secondary loop,

any radioactive materials that leak from fuel elements of a pressurized water reactor are retained in the primary loop. From this point on, a nuclear-powered electricity-generating plant is no different from a coal-fired electricity plant—the steam produced in the secondary loop is used to drive steam turbines that turn electric generators.

Operating nuclear power plants release only a minute amount of radioactivity into the environment through cooling water and into the atmosphere as gases. The Nuclear Regulatory Commission (NRC) requires that these releases be kept "as low as practicable," and that all releases be monitored and reported. Even most critics of nuclear reactors concede

Figure 18.22 *Above:* Encapsulated nuclear fuel pellets. (Hanford Engineering Developmental Laboratory.) *Right:* Fuel rods containing pellets. (Westinghouse photo by Jack Merhaut.)

that these levels of released radioactivity are too low to pose a health threat. Table 18.4 shows the level of radiation exposure from nuclear power plant emissions relative to other sources.

However, the public could be exposed to dangerous levels of radiation in the event of a catastrophic accident. Such a release could take many lives and permanently contaminate the environment. To prevent accidental release, nuclear reactors are constructed so that the reaction is carried out within a sealed reaction vessel (see Figure 18.23). Additional containment is provided by a thick concrete shell. But even these precautions do not guarantee that radioactive material will not be released. The possibility still exists that a reactor's cooling systems, along with its emergency backup cooling system, might fail. In this event, a reactor could grow so hot

that it could melt its way through the containment vessel into the soil and rock under the reactor. Since only about 3-percent fissionable material is incorporated into nuclear fuels, no possibility exists that a reactor will detonate like a nuclear bomb and create a nuclear holocaust. Nevertheless, in case of a complete *meltdown*, high levels of radioactivity would be released into the environment. Even though elaborate backup devices are installed in nuclear reactors to prevent such a catastrophe, no one can say with absolute certainty that such a reactor accident will never occur.

After fissionable uranium-235 has decreased from 3 percent to about 1 percent (through fissioning), spent nuclear fuel assemblies are removed from the reactor core (Figure 18.24). These assemblies contain high-level radioactive wastes and are stored

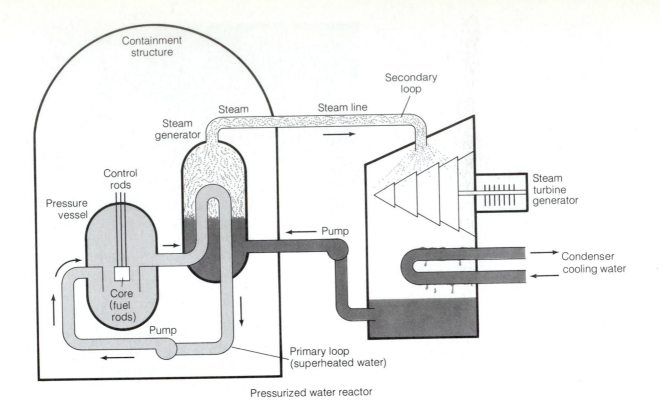

Pressurized water reactor

Figure 18.23 A schematic drawing of a pressurized nuclear reactor. (Science and Public Policy Program, University of Oklahoma, 1952.)

under water in pools at nuclear power plant sites. This method of storage is used because no nuclear fuel reprocessing centers or permanent storage sites exist for these radioactive wastes. In late 1977, in response to this urgent problem, the President's Council on Environmental Quality assigned to the NRC the responsibility of determining an acceptable method of permanently disposing of these dangerous wastes. The council took the position that the NRC should either find a solution to the nuclear waste disposal problem or cease licensing nuclear power plants.

At present, about 2700 metric tons of spent nuclear fuel are stored at sixty-five nuclear power plants throughout the country. At many of these plants, the storage capacity is being approached. Temporary government storage sites for radioactive

wastes are not expected to be available before 1985. Thus, either the storage facilities at some plants will have to be enlarged or the plants will have to be shut down.

Compounding the storage problem are about 240 million liters (60 million gallons) of military nuclear weapons wastes, which have accumulated over the past thirty years. These high-level liquid radioactive wastes are stored in underground tanks, such as those shown in Figure 18.25 (photographed prior to burial), located at four sites in the United States. The volume of these military wastes is considerably larger than spent nuclear power plant wastes, but the exact procedures for permanent disposal are yet to be determined. Rough estimates of the costs for final disposal of military wastes alone are about $20 billion.

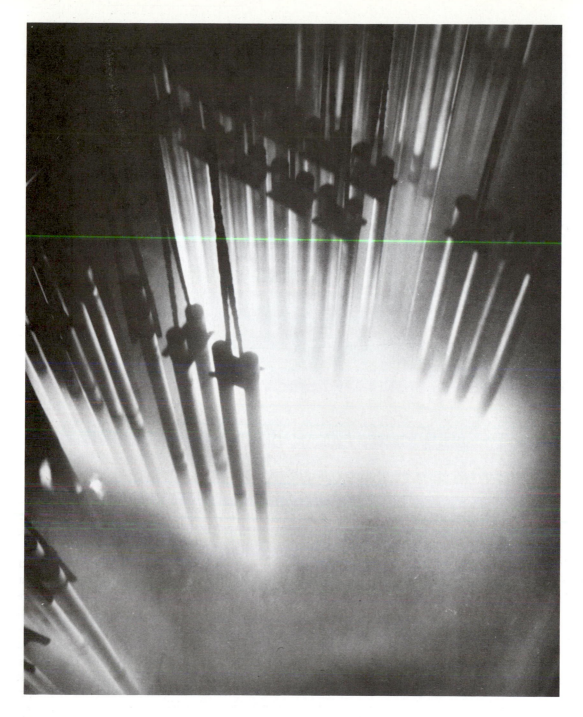

Figure 18.24 Radiation emanating from spent fuel-rod assemblies under water. The brightest assemblies (top and center) were just removed from a reactor. In front of them are assemblies that were removed a month earlier. At the extreme lower left are assemblies that have cooled for three-and-a-half months. (E. I. du Pont de Nemours & Company.)

Figure 18.25 Storage tanks for high-level liquid radioactive wastes under construction at the U.S. Department of Energy's Savannah River Plant in South Carolina. The double-walled steel tanks are shown here prior to encasement in concrete. (E. I. du Pont de Nemours & Company.)

The lack of permanent storage facilities for these military-related liquid wastes has led to several major spills. A total of nearly 2 million liters (500,000 gallons) of these wastes has leaked from storage tanks at sites in Hanford, Washington, and Savannah River, South Carolina. The worst spill occurred in 1973, when 450,000 liters (115,000 gallons) leaked from a tank at the Hanford site into the surrounding soil. Fortunately, no human casualties have resulted from any of these spills.

Currently, plans are under way by the NRC to convert high-level wastes from nuclear power plants into a suitable solid form within five years after they are removed from a nuclear reactor. The Commission has not yet specified how this conversion is to be

accomplished. It is most likely that the wastes will be solidified into a glassy material, probably in the form of cylinders. Smaller pellets illustrating what the disposed material would look like are shown in Figure 18.26. These solidified wastes are to be delivered within ten years to a government repository. A site for the repository has not yet been chosen, but the most likely selection is a stable geologic formation, such as unfractured granite or an abandoned salt mine, where wastes can be buried at least 600 meters (1800 feet) underground. Figure 18.27 shows one possible design for such a disposal system, and Figure 18.28 shows possible locations of such disposal sites across the nation. Fortunately, the volume of high-level radioactive wastes going to permanent

Figure 18.26 One proposal for handling high-level radioactive wastes—immobilizing them by incorporating them into glassy materials. They would be subsequently permanently stored in this form at a yet-to-be-approved permanent storage site. (U.S. Department of Energy photo by Dick Peabody.)

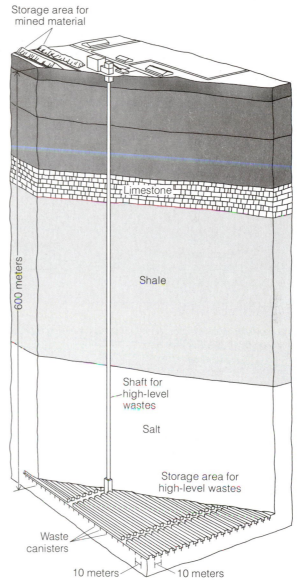

Figure 18.27 A plan for a permanent storage method in deep underground salt deposits. (B. L. Cohen, "The Disposal of Radioactive Wastes from Fission Reactors," *Scientific American 236:*25, June 1977. Copyright © 1977 by Scientific American, Inc. All rights reserved.)

disposal sites will be small, about 2 cubic meters (70 cubic feet) per year for each nuclear reactor. The Department of Energy is to have such a repository in operation by 1984, not only for commercial nuclear power plant wastes, but also for military nuclear wastes. However, at this writing it seems highly unlikely that this deadline will be met. In fact, the government is now proposing that additional facilities be built for the temporary storage of spent fuel from nuclear reactors. Possible sites for such a temporary facility are West Valley, New York; Barnwell, South Carolina; and Morris, Illinois. Electric utilities would pay for the cost of such facilities, which are estimated to raise the cost of nuclear power by about 3 percent.

The reprocessing of nuclear fuels is another potential means of handling nuclear wastes. Through reprocessing, valuable unfissioned uranium-235

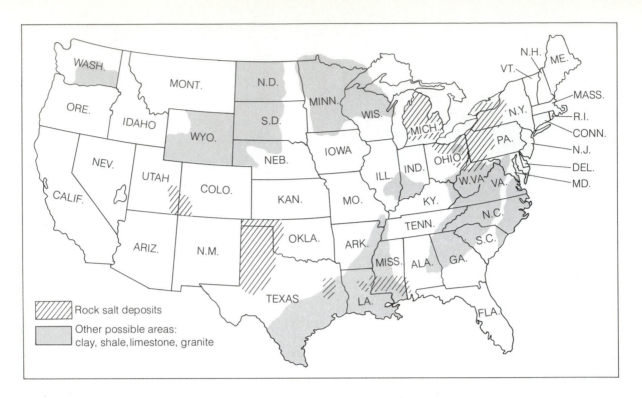

Figure 18.28 Geologically favorable areas for the possible permanent disposal of radioactive wastes. (General Accounting Office Report, September 9, 1977.)

would be retrieved, though the disposal of the remaining radioactive wastes would still be a problem. When a nuclear reactor is shut down for refueling, spent fuel rods not only contain about 1 percent valuable unfissioned uranium-235, but also a similar amount of fissionable plutonium-239, which is formed during operation. The fissionable material recovered from spent fuel rods from three reactors would be sufficient to fuel a fourth. In essence, the operation of a reactor actually creates additional new fissionable fuel. Currently, no reprocessing plants are licensed to recover fissionable materials from spent nuclear fuel. In the past, private companies that carried out the complex reprocessing of nuclear wastes encountered serious economic and technical difficulties, and have shut down.

One major obstacle to reprocessing is the lack of permanent repositories for radioactive reprocessing wastes. Another is the fact that reprocessing plants would isolate plutonium-239, the primary ingredient of nuclear bombs. Means of ensuring that this material will not fall into the hands of terrorist groups or deranged members of society must be established before reprocessing plants can be licensed. In view of the complexities of these issues and the ponderous pace of governmental regulatory agencies, it may be some time, if ever, before a reprocessing plant is operational. Given the high capital costs involved plus uncertainties as to future governmental policy regarding nuclear wastes, private industry is finding the financial risks of operating reprocessing plants too great. In this regard, the United States may find it necessary to follow the lead of Great Britain and France, where nuclear wastes are reprocessed at government-operated facilities. Meanwhile, nearly 2700 metric tons of nuclear reactor fuel wastes await reprocessing or final disposal, and the nuclear fuel cycle thus remains incomplete.

A problem related to waste disposal is how to dispose of nuclear reactors themselves after they wear out. Reactors are designed to be in service for only about thirty to forty years, since after that time the metal containment vessels lose some of their strength. During this period, the reactor containment vessels become highly radioactive, making dismantling and disposal a problem. Dismantling will have to be accomplished by expensive remote-controlled techniques. Such costs are estimated at $30-100 million—3-10 percent—of the cost of constructing each reactor. How these high disposal costs will be passed on to customers has not yet been determined by the NRC. Also, the dismantling of the reactors themselves will add to the volume of radioactive wastes requiring permanent storage. On the other hand, the alternative to dismantling—filling the reactor with concrete, fencing it off, and guarding the site for decades—is unacceptable, since under these circumstances radioactive materials would not be isolated sufficiently from the public. Hence, in the absence of procedures to dispose of nuclear wastes, we are left with a clear and present health hazard. As we shall note in the following section, nuclear power vies with coal as the primary replacement for our diminishing supplies of oil and natural gas. But our inability to dispose of nuclear wastes will weigh heavily in the choice between nuclear power and coal as the means of generating electricity.

Nuclear Fuel Supplies. Estimates of the size of U.S. uranium resources for the nuclear power industry indicate that at our present level of technology there is only limited room for growth. The size of uranium resources can be expressed as the number of 1000-megawatt—an average-size plant—power plants that can be supplied over their expected thirty- to forty-year lifetimes. Existing reserves are estimated by the Federal Energy Administration to be sufficient to supply 125-150 1000-megawatt nuclear reactors. In 1980, the equivalent of 66 plants of this size will be in operation. Probable resources still undiscovered are estimated to be large enough to

supply an additional 190-250 1000-megawatt plants. If the Federal Energy Administration's estimate that 110 nuclear reactors will be in operation by 1986 proves accurate, it will be necessary to mine more expensive reserves as well as to locate and develop new uranium resources. Some scientists, however, feel that the government estimates are overly optimistic, and that known reserves and probable reserves combined will only be sufficient to fuel 100-200 nuclear reactors. However, if new nuclear technology such as the breeder reactor (discussed in Chapter 19) is perfected and socially accepted, uranium resources will be increased at least seventyfold.

Energy Resource Utilization Policies

The decline in oil and natural gas reserves and, to some extent, nuclear fuel, is forcing us to consider alternate means for meeting energy demands. It is becoming clear that we need to adopt explicit public policies to aid us in making the wisest use of our remaining resources. Because each fuel has unique advantages, specific and limited purposes should be designated for each. For example, natural gas is a clean-burning fuel that, from the public health viewpoint, should be used in situations where installing pollution abatement equipment would be too costly—for example, to heat homes. Also, natural gas is a key ingredient for the manufacture of fertilizers, and therefore burning it in large power plants would seem unwise.

Gasoline and other liquid petroleum fuels such as diesel fuels are essential for private and public transportation and for powering agricultural equipment. Because no readily available substitutes have been discovered as yet, reserves should be carefully managed so that oil will be available to perform these essential services in the future. Because of the essential role of oil in transportation, this fuel is too valuable to be used to generate electricity. Hence, to make the wisest use of natural gas and oil, we must phase these fuels out as the means of meeting future electric power demands.

We cannot develop wise fuel utilization strategies without considering the total *energy efficiency* of the system under study. For example, moving toward a technology more dependent on electrical energy would actually reduce total efficiency. For a host of technical reasons, modern coal-fired and nuclear-powered electricity-generating plants are only about 33 percent efficient. In other words, generating one unit of electrical energy that is ultimately used requires three units of energy. Hence, if a consumer reduces his or her electrical energy demand by one unit, three units of energy will actually be saved. Consider, for example, two identical homes that require 200,000 kilojoules (200,000 BTUs) of energy for one month's space heating needs. One home is heated electrically, whereas the second home is heated with natural gas. The electrically heated home will require 200,000 kilojoules (200,000 BTUs) of electrical energy, because electrical heating devices are 100 percent efficient. But a total of 600,000 kilojoules (600,000 BTUs) of energy was required to generate the electricity used to heat the home. The gas-heated home, on the other hand, will require a total of 250,000 kilojoules (250,000 BTUs), since about 20 percent of the potential energy in the gas goes up the chimney. Thus, the burning of natural gas in homes for space heating is a more efficient use of this resource than using the natural gas to generate electric power to be used in space heating.

Because electric power plants are inefficient energy converters, the federal government is urging utilities to switch to coal or nuclear fuels. Such a switch would force us to use natural gas more efficiently, and in so doing we would be extending the lifetime of this precious resource. But in some parts of the country, power plants are designed to burn only natural gas. In Texas, for example, in 1975, 90 percent of the state's electricity was produced by the burning of natural gas, and, if existing plans are implemented, 60 percent of the 1985 demand in Texas will still be met by natural gas. Some states already generate all of their electricity by using either coal or nuclear fuel.

Making the Choice: Coal or Nuclear Power

The dwindling of our oil and natural gas supplies is forcing us to decide between coal and nuclear power as the main source of power in the production of electricity. The decision is not an easy one, and much controversy has been generated by the debate. In this section, we compare the tradeoffs involved in deciding between coal and nuclear fuel.

Nuclear reactors pose some unique and formidable problems that justifiably evoke the sort of public concern expressed by public officials and private citizens after the Three Mile Island nuclear reactor accident. To determine the relative risks associated with nuclear plants, the government had earlier funded a $30-million study, conducted by Professor Norman Rasmussen of the Massachusetts Institute of Technology. The final report of this enormous effort is known as the Rasmussen Report, which estimates that the worst possible nuclear reactor accident—complete meltdown—could claim thirty-five-hundred lives immediately from acute radiation poisoning and an additional forty-five-thousand lives in years following the accident from latent radiation-induced cancers. Furthermore, the report estimates, such an accident would cause about $14 billion in property damage. But the report states that the chance of an individual being killed by a nuclear accident is extremely remote. In fact, as Table 18.5 indicates, being killed in an automobile accident is about a million times more likely than being killed by a nuclear accident. Furthermore, according to the report, the chance of a given reactor experiencing complete meltdown during the course of a year is about one in twenty thousand. The possibility that such an event could occur as a result of sabotage, however, is not accounted for in these estimates. Review of this report in 1978 has brought these probabilities into question and indicated that the chances of being killed by a nuclear reactor

Table 18.5 The probability of being killed by various accidents in the United States (1969).

Accident type	Total number	Individual chance per year
Motor vehicle	55,791	1 in 4,000
Falls	17,827	1 in 10,000
Fires and hot substances	7,451	1 in 25,000
Drowning	6,181	1 in 30,000
Firearms	2,309	1 in 100,000
Air travel	1,778	1 in 100,000
Falling objects	1,271	1 in 160,000
Electrocution	1,148	1 in 160,000
Lightning	160	1 in 2,000,000
Tornadoes	91	1 in 2,500,000
Hurricanes	93	1 in 2,500,000
All accidents	111,992	1 in 1,600
Nuclear reactor accidents (100 plants)		1 in 5,000,000,000 (estimated)

Source: Executive Summary of the "Reactor Safety Study: An Assessment of Accident Risks in U.S. Commercial Nuclear Power Plants. "WASH-1400(NUREG-75/014), U.S. Nuclear Regulatory Commission, October 1975.

accident may be somewhat higher.

Scientists remain divided as to the importance of the risks of nuclear power. Those who oppose nuclear power as an option cite the ominous statistics associated with the total meltdown of a reactor, suggesting that though the probability of such an occurrence may be small, the risk, given the consequences, is too great to take. They also cite our inability to dispose of waste and dismantled power plants as sufficient reasons for not employing nuclear technology. Proponents of nuclear power, on the other hand, argue that the hazards associated with nuclear power are minimal, especially when compared with other risks that we casually accept. They also point to the excellent safety record of the nuclear power industry. Not a single member of the public has been killed in the United States by the accidental emission of radiation from a nuclear power plant, nor have any fatal accidents from radiation occurred at commercial nuclear power plants, although some close calls have been experienced at smaller experimental reactors. For further discussion of the possible danger of nuclear power plants, see Box 18.2.

To support their position, proponents of nuclear power plants also cite the deleterious effect of relying solely on coal as an energy source. They note that mining itself is dangerous, and that the health hazard posed by air pollutants emitted from coal-fired power plants is considerable. Since these factors are quantifiable, evaluating both sides of the issue may be possible by comparing projections regarding the negative effects of nuclear-powered and coal-fired power plants. In making this comparison, we have drawn on the research of D. J. Rose, P. W. Walsh, and L. L. Leskovjan.

In this projected scenario, four-hundred 1000-megawatt power plants would be operating across the nation, the number of plants required to meet our current electrical energy demands. If all four-hundred plants were nuclear-powered, the estimated number of deaths associated with electric power generation would be about two hundred per year (see Table 18.6 for a breakdown of the projections for both nuclear-powered and coal-fired plants). Because many assumptions go into such estimates, critics of the nuclear industry claim that the number of deaths would be as high as six hundred per year.

If all four-hundred plants used to generate electricity were coal-fired, our concern would be with the health effects of sulfur dioxide, nitrogen oxides, and the small particulates emitted from power plant stacks. Unfortunately, the health effects of these substances have not been studied nearly as intensively as those of radiation hazards. Nevertheless, the EPA estimates that if 1975 sulfur dioxide pollution standards were met, the number of deaths exceeding expected overall mortality figures would be about one death per year per 1000-megawatt plant. Hence, the estimated annual toll would be about four-hundred deaths. In fact, since many unknown factors were assumed in the formulation of the mortality figures for coal and nuclear fuel, the risks may be comparable. But if air pollution standards are not met or if they are relaxed substantially, the number of excess deaths associated with coal-burning plants would climb to about twenty deaths per plant per year or more. Furthermore, coal-fired power plant emissions aggravate chronic respiratory disease. If air pollution standards were met, the number of power-plant-related cases of such diseases would be insignificant. But if sulfur restrictions were lifted, about twenty-five-thousand additional cases of chronic lung disease per 1000-megawatt plant would be expected. Currently, nuclear power plants meet their emission standards, which is more than can be said for most coal-fired plants, and future compliance of coal-fired plants is by no means assured.

Other risks are associated with the extraction of

Box 18.2

Nuclear power plants are now generating electricity in some twenty countries around the world. Presently there are at least seventy-two licensed in this country, and more are under construction or on order. Advocates of nuclear power argue that the industry's safety record is impressive; they point out that so far not a single fatality has been caused by the operation of a commercially run nuclear plant anywhere in the world.

But does this mean that nuclear plants are really "safe"? Can we trust the assertions of some nuclear-power advocates that the danger of a catastrophic accident is negligibly small?

Various risk factors can cause or contribute to nuclear accidents, and some of them already have:

Inadequate testing and design. Rapid shutdown of the Three Mile Island nuclear power plant in 1979 was not possible because of a large hydrogen gas bubble that formed in the top of the reactor vessel. This troublesome bubble could not be removed because the valve had to be opened manually; no one could get near the valve because the area was highly radioactive, and a person would have received a lethal dose of radiation in less than a minute. Proper design would have provided a motor operable by remote control to alleviate such a problem when it occurred. The possibility of hydrogen bubble formation had not been forseen by designers.

Equipment failure. The Three Mile Island accident was triggered by a minor equipment failure—a valve closed that should have remained open. This

How Accident-Proof Are Nuclear Power Plants?

failure was followed by a least one other equipment failure—a pressure relief valve that opened failed to reclose. These equipment failures set off in rapid succession warnings to the plant operators. With a number of simultaneous equipment failures and warning signals, the operators were undoubtedly confused as to the exact reasons for the problems, and therefore were confused as to which emergency procedures should have been implemented.

Human error: carelessness, indecision, etc. At the Three Mile Island accident, an auxiliary system that should have taken over when the initial breakdown occurred was unavailable because it was undergoing maintenance repairs. The system should not have been operated without the backup system in a ready state. Also, human error contributed to the severity of the Three Mile Island accident because an emergency cooling-water circulation system was shut off for a short period of time. This system should have been left on but was incorrectly turned off in the confusion.

Questionable or faulty risk assessment. In January 1979, the NRC repudiated the 1975 Rasmussen report, which had concluded that the chances of a major nuclear accident were extremely remote. The NRC's rejection of the report did not mean that it thought the report's estimates were too high or too low, but rather that those estimates could err significantly in either direction—in other words, that they were less reliable than reported. About two months later, the NRC ordered five nuclear plants shut down because the engineering firm that de-signed all five had made a faulty assessment of their ability to withstand earthquakes. (The firm discovered the error itself and reported it to the NRC.)

That same year, two scientists at Columbia University's Lamont-Doherty Geological Observatory reported that the three reactors of a nuclear power plant at Indian Point, New York, were within a kilometer (0.6 mile) of an active geological fault. The presence of the fault had long been known, but the fault had been presumed inactive, that is, not earthquake-prone. The Lamont-Doherty scientists found, however, that "Since 1976, several shocks [weak local earthquakes] have occurred on the fault . . . both to the southwest and northeast of the plant as well as almost directly beneath it." And after analyzing the history of seismic activity in the area, they estimated that there was a 5- to 11-percent chance that the nuclear plant would, sometime in the next forty years, experience an earthquake of a strength equal to or greater than what it was designed to withstand.

These and other considerations suggest that the chances of a major nuclear accident may not be vanishingly small, as some supporters of nuclear power have claimed. But such data do not prove that nuclear power is "unsafe." No technology is perfectly safe, and in the real world safety must always be a relative matter. The nuclear-safety question is really three questions, all of which need further study: What are the risks? What are the benefits? and Are we willing to accept the risks to enjoy the benefits?

Table 18.6 Comparison of estimated number of annual fatalities associated with four-hundred 1000-megawatt nuclear and coal-fired electric power plants.

	Nuclear		
Activity	Accidental	Radiation related	Total
Uranium mining and milling	69	0.4	69.4
Fuel processing and reprocessing	19	16	35
Reactor manufacture	16	—	16
Reactor operation	15	47	62
Waste disposal	—	0.1	0.1
Transport of nuclear fuel	14	4	18
Total	133	67.5	200.5

	Coal	
Activity	Accidental	Air pollution
Mining	200	4000?
Coal transport	200?	
Coal burning—1975 sulfur dioxide	—	
If emission standards are not met	—	8,000–40,000
If emission standards are met		400
	400	4,400–44,000
Total		4,800–44,400?

Source: After D. J. Rose, P. W. Walsh, and L. L. Leskovjan, "Nuclear Power—Compared to What?" *American Scientist 64*:291, 1976.

both fuels, and these must be accounted for in our overall evaluation. A typical nuclear power reactor requires only about 27 metric tons of uranium per year, while a coal-fired plant of comparable size requires 1.8 million metric tons of coal. Mining sufficient coal for the four-hundred coal-fired plants would claim about seven-hundred-and-fifty lives annually, while seventy lives would be lost in mining and milling the necessary uranium. Transporting enough coal for four-hundred plants would be expected to take about five-hundred lives, compared with about twenty for the transport of the much smaller volume of nuclear fuel. And in the context of environmental quality, the strip mining of coal disrupts eighty times the land surface area disrupted by the mining of the amount of uranium necessary to generate an equivalent amount of power. This land-disruption figure does not account for damage associated with coal slag, which creates the problem of acid runoff that eliminates aquatic life in streams.

One final consideration to be weighed in the waste disposal problem. The burning of coal does present a waste disposal problem, albeit much different in nature from that concerning nuclear waste. Burning the 1.8 million metric tons of coal required by a 1000-megawatt plant produces about 180,000 metric tons of ash per year. In fact, in the thirty-five-year lifetime of a 1000-megawatt plant, enough ash would accrue to cover a football field to a depth of 1.2 kilometers (4000 feet).

Clearly, the optimism that was once associated with nuclear power has diminished substantially. Nuclear power has been received less than enthusiastically by society. The size of our nuclear fuel resources is smaller than earlier estimates indicated. But coal, the alternative to nuclear power, also poses substantial hazards to the people working in the industry and to the public affected by its pollutants. We have to decide which risks we want to take and what price we can afford to pay.

Conclusions

In this chapter, we have identified many of the negative effects and many of the risks associated with our society's attempts to satisfy its seemingly unquenchable thirst for more energy. At the same time, we have recorded many of the benefits that have contributed to a steady rise in our standard of living. But a continued reliance on conventional fuels is creating problems of limited supply, negative health effects, and substantial increases in consumer costs. Not only the United States, but the whole industrial world faces these problems. And, at the same time, the developing nations hope to improve their lot by moving toward a more modern, more energy-intensive society.

We are at a critical stage in our nation's history with respect to the utilization of energy resources. We are faced with three alternatives: (1) importing additional quantities of expensive petroleum; (2) switching from natural gas and oil to more abundant sources such as coal and nuclear fuels; and (3) cutting back on our reliance on traditional energy sources. Each of these alternatives involves consequences that would affect the life of every individual in the nation. Importing more oil would be politically unwise and would represent a continual financial drain of our nation's monetary resources. Burning more coal or using more nuclear fuels would have serious effects on the quality of our health and the environment. And fossil fuel conservation measures —though they would not inevitably restrict lifestyles or trigger unemployment, as is widely believed— would by no means solve the continuing supply problems posed by the demands of our population. Furthermore, all alternatives would require national economic adjustments.

In the next chapter, we evaluate alternative energy sources as supplements to conventional energy sources. In addition, we consider ways to curb the appetite of our energy-consuming technology—the hungry giant.

Summary Statements

During the past century, we have seen the rapid development of energy-consuming technology. This factor, along with the growth of our population, has resulted in a fifteenfold increase in our nation's energy use over the last hundred years.

Domestic production of natural gas, a preferred fossil fuel, peaked in 1973. At 1978 consumptions rates, known domestic reserves are predicted to last ten to eleven years; unidentified resources may provide another thirty-five-year supply at these same consumption rates. Natural gas supplies can be enhanced slightly over the short term by the importation of liquefied natural gas, and over the long term by coal gasification.

Crude oil is the raw material for a vast array of petroleum-based products. Only 52 percent of the nation's oil consumption in 1978 was met by domestic sources of production. At 1978 consumption and importation rates, known domestic oil reserves are estimated to last ten years, and undiscovered resources perhaps another twenty-four years. Oil shale deposits in the United States are large, and our exploitation of them could contribute in a small way to future supplies if oil prices continue to climb. Coal liquefaction, another means of producing petroleum liquids, would only become economically feasible if oil prices rose still further.

Coal reserves are scattered throughout the United States, and are large enough to last for several centuries. A greater reliance on coal for energy will result in more landscape disruption, air pollution, acid mine wastes, and fly ash disposal problems. Furthermore, mining communities will experience rapid growth and concomitant social change.

World reserves of natural gas and oil are estimated to be four to six times larger than U.S. reserves. Worldwide oil production is expected to begin dropping in the 1990s. World coal reserves are huge.

Nuclear energy can be used to augment the production of electrical energy. High levels of radiation are emitted from nuclear waste products, and therefore these wastes must be permanently isolated to protect the public against radiation-related diseases such as cancer. Permanent storage methods, however, have not been developed, and some temporary storage facilities at some nuclear power plants are nearly filled to capacity.

Reprocessing plants for waste from nuclear fuels would recover valuable unfissioned uranium-235 and plutonium-239. No reprocessing plants are currently operating, however, because safety measures have not been established for handling the purified nuclear weapons ingredients, such as plutonium-239, that would be isolated at these plants.

Nuclear fuel reserves are at least sufficient to meet projected demand for the next thirty to forty years.

To evaluate various energy utilization strategies we must consider total energy efficiency. Given our present level of technology, using electricity exclusively as an energy system would result in much fossil fuel waste, since power plants have inherent inefficiencies.

Over the next several decades, coal and nuclear fuels will be the major fuels used to meet electrical energy demands. Both alternatives pose formidable problems.

Although nuclear power plants do not contribute significantly to environmental pollution, a real, albeit slight, possibility of a catastrophic accident exists. In contrast to nuclear-powered plants, coal-fired power plants cause greater air pollution. Also, the mining of coal is more disruptive of the landscape than the mining of uranium, and, overall, coal mining is more hazardous to humans. Both coal-fired and nuclear-powered plants produce wastes, but nuclear wastes require special handling and disposal techniques.

The alternatives for extending domestic supplies of natural gas and oil include importing more oil, increasing our reliance on coal and nuclear fuel, and reducing demand. All of these alternatives would require economic adjustments.

Questions and Projects

1. Write out a definition for each of the terms italicized in this chapter. Compare your definitions with those in the text.

2. Discuss with your parents and/or grandparents how energy consumption patterns have changed in your family over the past half century. Submit a written report and also make interfamily comparisons in your class.

3. Would an oil embargo today have a greater impact on the nation than the 1973 oil embargo did? Explain your answer.

4. Is it possible for the United States to fulfill its projected energy demands without experiencing conflicts of interest with other countries?

5. Try to trace down the points of origin of the coal, natural gas, and oil burned in your region.

6. What are the important variables in assessing the quality of coal? What type(s) are burned in your community?

7. We are accustomed to think of petroleum solely in terms of gasoline, but actually petroleum is used in a variety of common household products. List some of these products and identify those you could do without.

8. We often encounter statements that say in effect that we are going to run out of oil in X years. State reasons why such statements are oversimplifications.

9. Describe the rate at which you use toothpaste from a tube. Is your rate of usage dependent on the size of the reserve in the tube? Does an analogous situation exist with respect to our oil supplies?

10. The exploitation of energy sources often creates conflicts with the management of other resources. Cite examples of such conflicts in your community and give possible ways of resolving them.

11. Find out what fraction of the various types of fuels are used to generate electrical power in your state. Contact your utility for state energy planning reports.

12. If your electric utility burns significant quantities of coal, where does it dispose of its fly ash? What are the environmental impacts of fly ash disposal in your area?

13. If electrical consumption was to increase in your community, where would you locate the power plant to minimize negative environmental impacts? What type of plant would you choose?

14. Contact your local electric utility and determine how officials plan to meet projected electrical demands. What assumptions have been made in the preparation of these projections?

15. How do alpha, beta, and gamma rays differ?

16. If a radioactive source of strontium-90 (the half-life is twenty-eight years) is emitting 10 billion radioactive particles per second, at what rate would the source be emitting particles after eighty-four years?

17. Which segments of the nuclear fuel cycle are the most hazardous?

18. Divide your class into two groups and debate the pros and cons of nuclear power.

19. Develop a table that compares the risks associated with using natural gas, oil, coal, and nuclear fuels.

Selected Readings

Bebbington, W. P. 1976. "The Reprocessing of Nuclear Fuels," *Scientific American 235*:30–41 (December). A discussion of technology for recovering valuable nuclear fuels from nuclear wastes.

Christiansen, B., and T. H. Clack, Jr. 1976. "A Western Perspective on Energy: A Plea for Rational Energy Planning," *Science 194*:578–584 (November). A look at the impact of increased coal mining in the western states.

Cohen, B. L. 1977. "The Disposal of Radioactive Waste From Fission Reactors," *Scientific American 236*:21–23 (June). A discussion of the characteristics of radioactive wastes and alternatives for their disposal.

de Marsily, G., E. Ledoux, A. Barbreau, and J. Margat. 1977. "Nuclear Waste Disposal: Can the Geologist Guarantee Isolation?" *Science 197*:519 (August). A discussion of the relative stability of various geological formations as sites for the permanent storage of high-level nuclear wastes.

Dorf, R.C. 1978. *Energy, Resources and Policy.* Reading, Massachusetts: Addison-Wesley. A textbook that examines conventional and future energy technologies in light of resource availability and related policy decisions. Some sections are more technical than others.

Drake, E., and C. Reid. 1977. "The Importation of Liquefied Natural Gas," *Scientific American 236*:22–29 (April). Examination of technology associated with importing natural gas from foreign sources.

Ehrlich, P. R., A. H. Ehrlich, and J. P. Holdren. 1977. *Ecoscience: Population, Resources, Environment.* San Francisco: W. H. Freeman and Company. Chapter 8 gives an overview of the problems associated with the exploitation of energy resources.

Flower, A. R. 1978. "World Oil Production," *Scientific American 238*:42–49 (March). A discussion of world oil reserves, their locations, and the limits to oil recovery rates.

Fowler, J. M. 1975. *Energy and the Environment.* New York: McGraw-Hill. A text written for the nonspecialist. The book is an excellent source for students seeking further insight into energy resources, economics, and conversion technology.

Gamble, D. J. 1978. "The Berger Inquiry: An Impact Assessment Process," *Science 199:*946–952 (March). Describes the Mackenzie Valley gas pipeline and explains how the Canadian government considered the concerns of Canadians regarding energy policies, resource allocations, industrial development, unique cultural matters, and self-determination.

Griffith, E. D., and A. W. Clarke. 1979. "World Coal Production," *Scientific American 240:*38–47 (January). Survey of world coal resources and consumption rates.

Hammond, A. L. 1976. "Coal Research (III): Liquefaction Has Far to Go," *Science 193* (September). A summary of the progress made toward the liquefaction of coal.

Hammond, O. H., and R. E. Baron. 1976. "Synthetic Fuels: Prices, Prospects, and Prior Art," *American Scientist 64:*407. A discussion of various technologies available for coal gasification and liquefaction, including economic and social impacts.

Hammond, R. P. 1979. "Nuclear Wastes and Public Acceptance," *American Scientist 67:*146–150 (March–April). Discussion of disposal methods for high-level radioactive wastes and procedures for gaining public acceptance of disposal methodology.

Hayes, E. T. 1979. "Energy Resources Available to the United States, 1985 to 2000," *Science 203:*233–239 (January). Overview of fossil and nuclear fuel resources and consumption rates for the remainder of this century.

Healy, T. J. 1974. *Energy, Electric Power, and Man.* San Francisco: Boyd and Fraser. A text focusing primarily on conventional and future technologies for producing electricity. Includes a discussion of energy resources.

Marshall, E. 1979. "NAS Study on Radiation Takes the Middle Road," *Science 204:*711–714 (May). A summary of the third National Academy of Sciences report on the biological effects of low levels of ionizing radiation and the controversy surrounding the report.

Maugh II, T. H. 1977. "Underground Gasification: An Alternate Way to Exploit Coal," *Science 198:*1132 (December). A review of projects assessing technology for *in situ* coal gasification.

Nef, J. U. 1977. "An Early Energy Crisis and Its Consequences," *Scientific American 236:*140–142 (November). An historical account of what happened in previous societies that faced energy crises.

Revelle, R. 1976. "Energy Use in Rural India," *Science 192:*969–975 (June). Discussion of how energy is used and wasted in a poor country.

Rockefeller, N. A., 1977. *Vital Resources,* Lexington, Massachusetts: Lexington Books. Report of the Commission on Critical Choices for Americans. Overview of energy and its relationship to environment, economics, and international stability.

Schipper, L., and A. J. Lichtenberg, 1976. "Efficient Energy Use and Well-Being: The Swedish Example," *Science 194:*1001–1013 (December). An examination of economic and social factors important to comparisons of energy use among nations.

Stoker, H. S., S. L. Seager, and R. L. Carpenter. 1976. *Energy: From Source to Use.* Glenview, Illinois: Scott Foresman. A textbook dealing with the science of providing energy to society.

Van Tassel, A. J., ed. 1975. *The Environmental Price of Energy.* Lexington, Massachusetts: Lexington Books. Major sections of this book address energy conservation and problems associated with implementing alternate sources of energy.

Walsh, J. 1977. "Texas Power Companies Converting from Natural Gas to Coal, Lignite," *Science 198:*471 (November). An account of environmental, economic, and technological issues that Texas utilities must face in switching from natural gas to coal in generating electricity.

Toyotas, fresh from Japan, being unloaded at the import dock at Benicia, California, for distribution to Toyota dealers in Northern California. (Copyright © 1978 Barrie Rokeach.)

Chapter 19

Averting the Energy Crisis

"Wood stoves in stock!" "Build your own wind-powered electric generator!" "Cut fuel bills by installing a heat-recirculating fireplace!" "Heat your home with free solar energy!" Sales slogans such as these are appearing with increasing frequency in newspapers and magazines and on radio and television. They reflect the public's growing realization that a continuous flow of energy is the lifeblood of our society. As more and more people learn about the conditions affecting our existing fuel supplies, they are becoming convinced that the energy crisis is real and will grow worse if we continue to rely on fossil fuels. Consequently, some concerned citizens are turning to alternative energy sources.

In the preceding chapter, we identified three alternative strategies for facing the fossil fuel crisis: increasing our imports of foreign oil; switching from natural gas and oil to other energy sources, such as coal and nuclear fuels; or cutting back on our reliance on traditional energy sources. In this chapter, we pursue the latter possibility by surveying alternative energy sources and conservation techniques intended to supplement or stretch conventional energy sources. Some of the alternative options currently under study will be operable in the near future—for example, solar and wind power. For others, such as nuclear fusion, we will have to invest at least several decades of experimentation and development before they are capable of channeling significant quantities of energy for public use.

We turn now to a consideration of the costs, benefits, and environmental effects of the more promising alternative energy sources, assessing these options in terms of their potential importance in the overall energy-supply picture. We end the chapter with an examination of some specific examples of conservation techniques designed to save conventional fuels.

Nuclear Power: Breeder Fission Reactors

In Chapter 18, we saw that one of the products formed in nuclear fission reactors was plutonium-239 (^{239}Pu). This readily fissionable material is formed from the nonfissionable isotope of uranium, uranium-238 (^{238}U). In conventional reactors, the amount of ^{239}Pu formed is limited; thus, the opera-

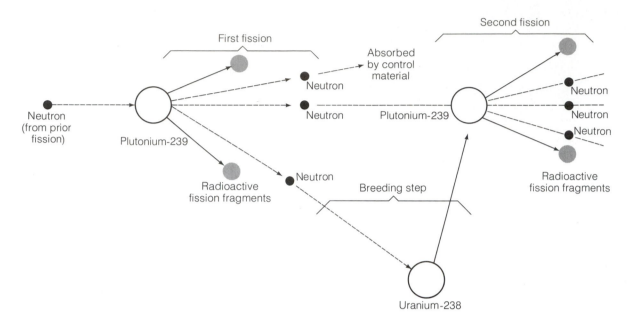

Figure 19.1 The sequence of nuclear reactions in the core of a breeder reactor.

tion of about three reactors is required to produce enough fissionable material from ^{238}U to fuel another nuclear reactor. However, by increasing the amount of plutonium or ^{235}U in the fuel and making modifications in the nuclear reactor, we can make the reactor create, or breed, more ^{239}Pu than it uses up. Reactors with the capacity to breed ^{239}Pu are therefore called *breeder fission reactors.*

The strategy employed in a breeder reactor (see Figure 19.1) is the conversion of the much more abundant but unfissionable ^{238}U into fissionable ^{239}Pu. Basically, the sequence of reactions in the core of a breeder reactor consists of three processes: (1) fission of the fuel ^{239}Pu, which releases energy and two to three neutrons per fission; (2) a breeding step, in which ^{239}Pu is formed from ^{238}U; and (3) a control system that absorbs the excess neutrons so the reactor does not overheat and explode. All three

processes go on simultaneously, and the continuous flow of energy released by the nuclear reactions is carried away from the core by hot liquid sodium, as shown in Figure 19.2. The hot molten sodium generates the steam necessary to drive electric turbine generators.

The principal advantage of breeder reactors is that they can create fissionable fuel from the more abundant isotope of uranium, ^{238}U. Conventional reactors fission primarily ^{235}U, which constitutes only 0.7 percent of the total amount of uranium mined. Breeder reactors, however, can utilize both ^{235}U and ^{238}U. And mined ^{238}U for breeder reactors is already in plentiful supply, because it is a by-product of nuclear weapons production. In fact, current supplies of ^{238}U could supply breeder reactors for four-hundred 1000-megawatt electricity-generating plants (the approximate number of plants necessary to meet

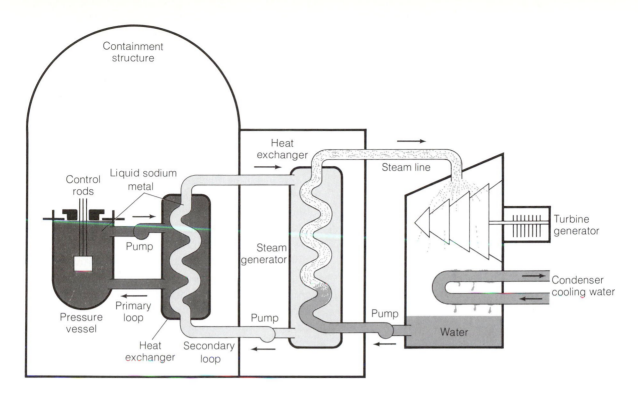

Figure 19.2 A breeder reactor. (Atomic Industrial Forum.)

our current rate of electrical consumption) for about five-hundred years. Thus, if breeder reactor technology were implemented, our total uranium reserves would easily last for tens of centuries.

Breeder reactor technology, however, has some potentially catastrophic aspects. First, plutonium, the fissionable material produced, when ingested is one of the most toxic substances known. Furthermore, in sublethal internal dosages it is a carcinogen. Also, as we mentioned in Chapter 18, if breeder technology is implemented, terrorist groups could acquire ^{239}Pu and use it to manufacture nuclear devices. Only 4.5 kilograms (10 pounds) of ^{239}Pu are required to produce a nuclear bomb the size of the one dropped on Hiroshima, and breeder reactors will produce tons of the material. Although stringent security measures might be taken to prevent theft, the chance remains that a conspiracy to obtain the material could be

perpetrated. Finally, even aside from concerns over exposure or theft, breeder reactors pose nuclear waste disposal problems similar to those of conventional nuclear power plants.

The role of the breeder reactor in the energy picture of the future is as yet unclear. At this time, the first full-scale breeder reactor in the United States has yet to be built. U.S. nuclear scientists have tested small breeder reactors for years, and have proposed to build the first full-scale breeder reactor at Clinch River, Tennessee. But President Carter vetoed most of the funding for the project, citing the serious unresolved problems of safeguarding ^{239}Pu from terrorists and disposing of nuclear waste. In Europe, however, a consortium of electric utilities has agreed to construct a 1200-megawatt breeder reactor electricity-generating facility (about the size of most conventional nuclear reactors) in the

Creys-Malville farming area of France. This project, the Superphénix plant (named after the mythical bird that rose out of its own ashes), is being modeled after a relatively successful smaller demonstration plant.

At least one or two decades of operational experience with a few breeder reactors are needed before full-scale implementation of this technology becomes feasible. And even then, full-scale operation would require that we establish fuel reprocessing centers to support the breeder reactors. As noted in Chapter 18, the implementation of reprocessing technology, too, has been inhibited by the dangers inherent in the materials, and so far no such centers are in operation. Even though breeder reactors have an almost limitless potential to furnish energy, we may decide that building and operating them are not worth the risks of possible sabotage and exposure to nuclear wastes.

Nuclear Power: Fusion Reactors

Another type of nuclear reaction that releases enormous quantities of energy is *nuclear fusion*. In this process, the nuclei of light elements (elements of low atomic mass), principally isotopes of hydrogen, are fused together to form heavier elements, and in the process tremendous amounts of energy are released. The energy-releasing reactions that occur within the sun and other stars are nuclear fusion processes.

Physicists are trying to duplicate these processes on a small scale, but in the laboratory the fusing of two nuclei together is a difficult process, requiring the heating of fusion fuels to temperatures approaching 100,000°C (180,000°F). At these temperatures, containment vessels melt, and therefore reacting substances—called plasmas—must be contained by a complicated array of large magnets. One of the most promising fusion reactions now under study is the fusion of deuterium and tritium to produce helium. This reaction is diagrammed in Figure 19.3. At our present level of fusion technology,

the amount of energy derived from the fusion process is considerably smaller than that required to carry out the process. Thus, a great deal of research is needed before fusion will become a practical source of energy (if it ever does). For this reason, the first commercial fusion reactor is not expected to be operating before the turn of the century. A portion of an experimental fusion-powered reactor is shown in Figure 19.4.

If the technology of nuclear fusion is perfected, however, fusion reactors promise to offer significant advantages over both fission and fossil fuel power-generating systems. For one thing, the ocean contains vast amounts of deuterium, and the environmental and economic costs of extraction would be minimal. The fusion of the deuterium contained in a cubic kilometer of sea water would yield an amount of energy nearly equivalent to that contained in all the earth's oil reserves. Also, fusion does not pose the hazards of nuclear fission, because reaction products are inert, though some of the equipment associated with fusion reactors would become radioactive. The construction of fusion reactors, however, would require the use of exotic (and therefore scarce) materials, and this factor alone could limit the implementation of nuclear fusion technology, even if it were perfected.

Solar Power

Solar energy is the driving force behind the water cycle, the wind, and the weather (see Chapters 6 and 9). Some of the sun's energy is also trapped by plants, thereby becoming the energy source available to all living things. Direct solar energy has been used by human beings for centuries, but only on a very small scale. Since Roman times, people have used the sun's rays to evaporate reservoirs of sea water in the recovery of salt.

Using solar power to meet our energy needs is a particularly attractive option, since solar energy is

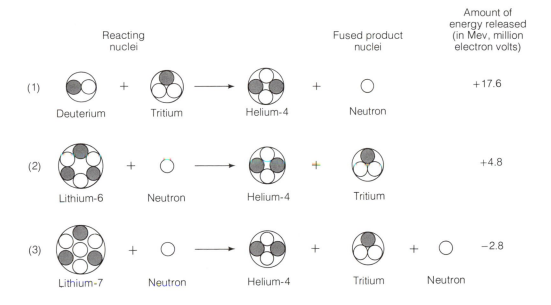

	Reacting nuclei		Fused product nuclei		Amount of energy released (in Mev, million electron volts)

(1) Deuterium + Tritium → Helium-4 + Neutron   +17.6

(2) Lithium-6 + Neutron → Helium-4 + Tritium   +4.8

(3) Lithium-7 + Neutron → Helium-4 + Tritium + Neutron   −2.8

Figure 19.3 The fusing of deuterium and tritium, a promising energy-releasing nuclear fusion reaction (reaction 1). Deuterium is abundant in sea water. The tritium required for reaction 1 would be produced from two common isotopes of lithium, lithium-6 and lithium-7, via reactions 2 and 3.

free and will last as long as the sun itself—at least several billion more years. However, not until late 1974 did Congress provide funds to encourage large-scale research and development, and the building of demonstration projects concerning solar-powered systems for heating, cooling, and electric power generation. This action, when it did come, was spurred by the recognition that fossil fuel reserves were dwindling and that nuclear energy sources involved hazards. Solar energy is manifested in many forms: radiation, photosynthesis, wind, running water, and ocean currents. The President's Council on Environmental Quality now optimistically predicts that, with a strong national commitment to using solar power, as much as one-quarter of our energy needs could be met by solar sources in the year 2000. The estimates for the various solar technologies are summarized in Table 19.1.

Radiation Collectors

Each day, sunlight delivers an average of 16 megajoules per square meter (1400 BTUs per square foot) to the earth's surface. The total amount of solar energy falling on U.S. lands yearly is about six-hundred times greater than the amount of energy our nation presently consumes during the same period. Another way of expressing the value of solar energy is to say that, over the course of a year, $5000 worth of energy shines on the roof of a small house (93 square meters, or 1000 square feet), assuming that this energy could be collected and sold at average electricity rates.

The collection of solar radiation, however, presents some problems, since, compared with fossil or nuclear fuels, sunlight is diffuse in nature. Hence, to

Figure 19.4 A portion of an experimental nuclear fusion apparatus. This facility, at Oak Ridge, Tennessee, is part of the effort to demonstrate that a fusion-powered reactor could be practical in the future. (U.S. Department of Energy, Frank Hoffman.)

use solar energy, we need *solar collectors*—panels designed specifically to collect and concentrate the sun's rays. At present, the few existing solar energy systems are generally used for space or water heating in small buildings such as homes, apartments, schools, or small businesses. In the near future, solar-powered air conditioners may cool these same places. Solar panels do not produce high temperatures, and therefore these devices are not adequate for most industrial purposes.

Basically, solar collectors are framed panels of glass that trap solar energy. The sunlight is usually allowed to pass through two layers of glass before it is absorbed on a blackened metal plate. The absorbed heat energy is transferred from the absorbing plate to either air or a liquid, which is conveyed by fans or pumps to wherever it is needed. Figure 19.5 shows two examples of solar panels currently in use, and Figure 19.6 shows schematic diagrams of two solar heating systems. Typically, solar collectors

Figure 19.5 Examples of solar panels currently in use. *Top:* A modular home heated primarily by solar energy. (U.S. Department of Energy—Los Alamos Scientific Laboratory, photo by Johnnie Martinez.) *Bottom:* The installation that heats the wash water and the milking parlor at the U.S. Department of Agriculture's Animal Genetics and Management Laboratory in Beltsville, Maryland. (U.S. Department of Agriculture.)

Table 19.1 Council on Environmental Quality's estimates of maximum solar contribution to U.S. energy supply under conditions of accelerated development (in quads per year of displaced fuel).*

Solar-related technology	1977	2000**	2020**
Heating and cooling (active and passive)	Small	2–4	5–10
Thermal electric	None	0–2	5–10
Intermediate temperature systems	None	2–5	5–15
Photovoltaic	Small	2–8	10–30
Biofuels	1.3	3–5	5–10
Wind	Small	4–8	8–12
Hydropower	3	4–6	4–6
Ocean thermal energy conversion	None	1–3	5–10

*Total U.S. energy demand in 1977 was 76 quads. [The figure in 1978 was 78 quads.] Estimated total U.S. energy demand is from 80–120 quads for the year 2000 and from 70–140 quads for the year 2020.

**The estimates in these columns are not strictly additive. The various solar electric technologies will be competing with one another, and their actual total contributions will be less than the sum of their individual contributions.

Source: Council on Environmental Quality, *Solar Energy*, Washington D.C.: U.S. Government Printing Office, April 1978.

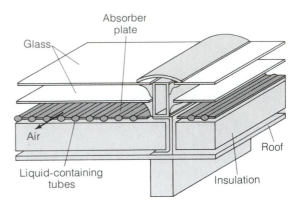

Figure 19.6 *Above:* A cutaway view of a solar panel. *Right:* Schematic diagrams of two solar heating systems, one that uses water to transfer heat (*top*) and one that uses air (*bottom*). (Federal Energy Administration.)

capture 30–50 percent of the sunlight reaching the collector.

All technologies based on the collection of solar radiation are limited by the fact that the source is not continuous. Not only does the sun cease to shine at night, but its intensity varies both seasonally and geographically. For example, in the northern states, the variation in the sun's intensity from winter to summer is 6–23 megajoules per square meter (500–2000 BTUs per square foot). Figure 19.7 shows the mean daily solar radiation across the nation for December (when solar radiation is least intense) and June (when it is most intense).

The most formidable obstacle to widespread and immediate use of solar energy for heating is cost. Although the fuel itself is "free," the initial cost of the equipment necessary to utilize it is high. At this writing, available collector designs range in price from $7–12 per square foot. A typical house requires about 46–65 square meters (500–700 square feet) of collectors, which translates into about $3500–$8000. In addition, solar heating systems require heat-storage facilities, so that heat will be available at night and during overcast periods. Excess heat collected when the sun is shining brightly is usually stored in insulated water tanks or in compartments filled with rock. The size of the storage facilities required in a particular system depends on the variability of local weather conditions. The construction costs for storage facilities add approximately $1000 to the cost of a system. Finally, conventional heating systems are needed as a backup to withstand prolonged periods of cloudiness, which increases the capital cost still further.

Although the initial costs for solar heating (or cooling) systems are much higher than for conventional systems, over the long term, solar systems promise to be more economical than systems that rely on the increasingly more expensive conventional fuels. But because many variables influence the eco-

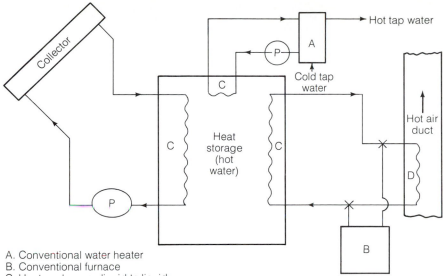

A. Conventional water heater
B. Conventional furnace
C. Heat exchange–liquid to liquid
D. Fan coil heat exchange–liquid to air
P. Pump

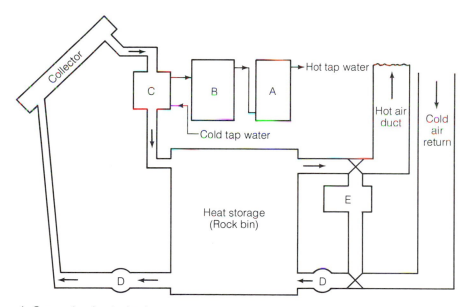

A. Conventional water heater
B. Solar hot water storage
C. Air to liquid heat exchanger
D. Blower
E. Conventional furnace

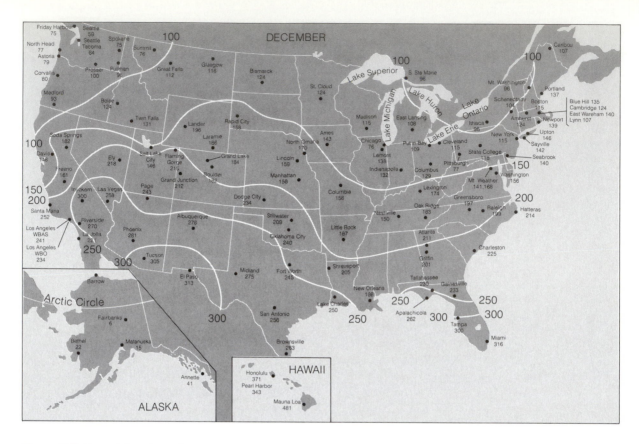

Figure 19.7 Mean daily solar radiation at the earth's surface in the United States. The smallest amount of radiation is received in December and the largest in June. (Units are in langleys; 1 langley = 1 calorie per square centimeter, or 3.69 BTU per square foot.) (Federal Energy Administration.)

nomic feasibility of solar heating or cooling, people contemplating such a venture should analyze their particular situations carefully. For example, unless a home is well insulated and has small windows on the north side, the investment in solar collectors sufficient to heat the house will be high. Other variables to consider are the degree of cloudiness at the site, the number of windows in the house, and the size and efficiency of the system, including its storage component. Swimming pool heating costs are another factor, (solar collectors can be used to heat pools in summer). Once the total cost of the system, along with interest rates, has been computed, it should be weighed against the projected costs of more conventional systems.

Using solar energy to cool buildings is more complex than using it to heat them. Nevertheless, solar cooling is particularly attractive, since the peak electrical demand for most locations occurs in summer. Although solar cooling systems are not economically competitive at present, they are expected to become so in the near future.

Another option in the heating and cooling of buildings is utilizing architectural design in capturing some of the sun's energy in cool periods and in protecting against excessive heating in summer. Design features that contribute to optimizing internal temperatures include south-facing windows to capture winter sunshine and shade-producing overhangs to keep out hot summer rays. Systems that are de-

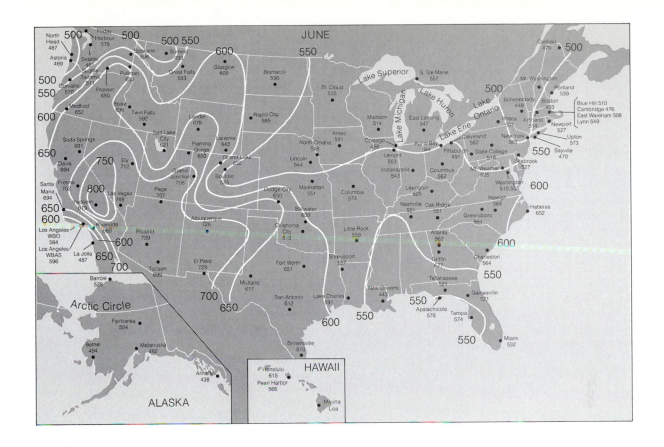

signed to be highly energy-efficient architecturally are called *passive solar collectors,* whereas systems that utilize solar panels are termed *active systems.* The relative importance that a passive solar system can play is illustrated in a new experimental classroom building at the Massachusetts Institute of Technology. This structure will derive about 85 percent of its heating requirements from passive design features. Such design features save fossil fuels over the entire lifetime of a building and add little to its initial cost. Unfortunately, adding passive design features to old buildings is very expensive.

If the costs of solar hardware drop as prices of fossil fuel energy continue to rise, owners of homes and commercial properties might well convert to solar energy in the 1980s and 1990s. The President's Council on Environmental Quality projects that by the year 2000 we might expect a maximum of 2-4 quads to be harnessed by active and passive solar systems. For reference purposes, total U.S. energy consumption in 1978 was 78 quads. But conversion on a large scale would require billions of dollars of capital and millions of metric tons of copper, steel, glass, and petroleum for manufacture of the necessary equipment. Thus, conversion to solar energy systems would not be without its significant costs.

Scientists and engineers are currently studying ways of converting solar energy into electricity. In one conversion system—called a power tower system—curved mirrors that track the sun would be used to collect and concentrate diffuse sunlight, focusing the sun's energy on a single heat-collection point. Concentrated sunlight in these systems can produce high temperatures—up to 480°C (900°F), high enough to convert water into high-pressure steam for driving turbine generators. A large solar-power system of this sort would require a field filled with tracking mirrors (see Figure 19.8). In fact, the generation of 50 megawatts of electricity (enough for fifty-thousand people) would require about 2.6 square kilometers (1 square mile) of land, making the land unusable for other purposes. At least another decade of experimentation is required before any commercial plants using this technology become a reality. And even then, such plants would only become possible if the cost of solar collectors were greatly reduced. The Council on Environmental

Figure 19.8 An artist's conception of a solar collection system for the production of electrical energy. The first such system is to be built in the Mojave Desert near Barstow, California, and is to be completed in 1980–1981. (Martin Marietta Corporation.)

Fundamental Problems: Population, Food, and Energy

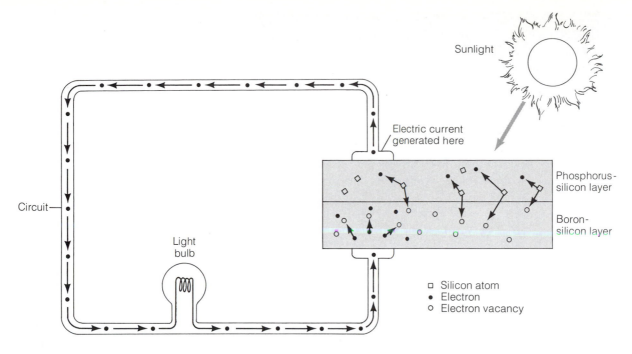

Sunlight

Electric current
generated here

Circuit —

Light
bulb

Phosphorus-
silicon layer

Boron-
silicon layer

□ Silicon atom
• Electron
○ Electron vacancy

Figure 19.9 A schematic drawing of a silicon solar cell. Sunlight frees electrons from silicon atoms, and the electrons move through the circuit and generate an electric current. (Federal Energy Administration.)

Quality estimates that by the year 2000, high-temperature power tower systems may produce up to 2 quads per year.

Solar collectors that produce temperatures between 100°C (212°F) and 315°C (600°F) (which is an intermediate temperature range) could be used in a wide variety of ways in industry. Such systems, which use less sophisticated tracking mirrors than the power tower system, are already available and should soon be utilized to produce steam in the chemical, textile, and food-processing industries. The Council on Environmental Quality estimates that these intermediate temperature systems may produce 2–5 quads per year by the year 2000.

An alternative to generating electrical power with solar-driven turbines is using photovoltaic cells, or, more simply, *solar cells*. These cells convert sunlight directly into electricity. Two very thin layers of specially treated metal crystals (silicon metal is commonly used) are sandwiched together to form a solar

cell. A schematic drawing of a silicon solar cell is presented in Figure 19.9. When sunlight strikes the junction between the two layers, direct current flows through wires connecting the layers. About forty solar cells must be linked together to produce the 12-volt potential of a car battery. Theoretically, silicon solar cells should be able to attain an efficiency of 25 percent, but in practice 14 percent efficiency is exhibited by commercially produced solar cells.

Usable quantities of electrical energy are produced when many small solar cells are connected together to form a solar electric-generating panel. An example is shown in Figure 19.10. Despite the simplicity of solar cells, they still are extremely expensive because of costly manufacturing processes. Thus, although a 6-by-6-meter (20-by-20-foot) panel of solar cells would produce enough electricity for an average home (assuming a maximum output of 5000 watts at midday and 13 percent operating efficiency), such a

Figure 19.10 Solar electricity-generating panels. (U.S. Department of Energy.)

panel would cost $60,000. Still, solar panels are durable and reliable, and have powered satellites since the 1950s. The goal of researchers is to mass-produce solar cells and reduce their production costs by a factor of twenty-five. If this goal can be achieved—and some promising developments can be cited—solar cells could be economically attractive by the mid-1980s. The Council on Environmental Quality estimates that photovoltaic solar cells may contribute 2–8 quads of this country's energy by the year 2000.

As with other solar techniques, solar panels are limited by sunshine variability due to cloudiness and nightfall. In response to this limitation, Peter Glaser, a solar energy researcher, has drawn up a scheme to launch a 65-square-kilometer (25-square-mile) solar panel into stationary, or geosynchronous, orbit 35,000 kilometers (22,000 miles) above the earth's surface. Such a system would receive about seven times more energy than an earthbound collector of equivalent size. The captured 10,000 megawatts of power would be beamed back to earth in the form of microwaves, which would be reconverted to electricity. An artist's concept of such a satellite panel is shown in Figure 19.11. The enormous size of such a panel, about ten-thousand times larger than any satellite launched so far, presents the greatest problem, but a vast array of other technical obstacles remain to be overcome. Perhaps in the next century such systems will provide us with electrical power.

We can summarize the state of solar energy technology to date by noting that though the means do

Figure 19.11 An artist's conception of a satellite solar panel in geosynchronous orbit that would beam the sun's energy back to earth in the form of microwaves, which would then be converted to electrical power. (NASA.)

exist for tapping significant amounts of clean, safe fuel—sunlight—the monetary costs of solar energy systems presently limit implementation. But as the costs of fossil fuels rise and the increased production of solar collectors lowers their costs, solar energy is expected to become competitive with fossil fuel sources. An added impetus for stepping up implementation is the fact that this source of power involves none of the hidden pollution costs that accompany the use of fossil and nuclear fuels.

Biofuels

As gaseous and liquid fuels become scarcer, we may renew our former reliance on *biofuels*—recoverable or harvestable plant or animal materials used to produce heat. Scientists today are looking at the possibility of growing plants solely for their heating value. In essence, power plants would burn biofuels and thereby convert solar energy (stored in food chains by way of photosynthesis) into steam for generating electrical power. A major advantage of harvesting solar energy via biofuels is that these materials can be used to generate power at any time, day or night, summer or winter. As we noted in the preceding sections, other solar methods are limited by the fact that collected energy must be either used immediately or stored in expensive auxiliary storage facilities. The limiting factor on biofuel use, availability, is discussed in Box 19.1.

Compared with solar panels, plants are very inefficient at collecting and storing the sun's energy. Efficiencies of 3 percent over a growing season are considered good, but the annual average efficiency is closer to 1 percent. Still, plants cover much of the earth's land surface, and nature can grow plants at less expense than that involved when we build solar panels. The raising of plants solely for their energy content is not likely to compete successfully in economic terms with raising agricultural crops, but large-scale tree farms (energy plantations) on nonagricultural land may be feasible sources of wood to fuel a power plant. One prominent energy expert, Dr. George Szego, calculates that 280–1600 square kilometers (110–630 square miles) of tree farms would be necessary to provide a sustained supply of wood to fuel a 1000-megawatt power plant. His cost estimates indicate that wood fuel costs would be competitive with coal. Foresters, however, are skeptical, since this scheme has yet to be tested for long-term reliability. Foresters are concerned that stresses such as insects, pests, and hostile weather would threaten sustained yields, thus rendering Dr. Szego's estimate too optimistic.

In addition to trees, other plants have been suggested as potential energy sources. Because of their luxuriant growth, water hyacinths and algae, including giant sea kelp, have been investigated as potential biofuels. The problem with water-grown biofuels, however, is that harvesting and drying the crops are prohibitively expensive. To circumvent these costs, scientists are examining the possibility of breaking down water plants anaerobically. For example, water hyacinths and giant sea kelp decay in the absence of oxygen to produce methane gas, which can be readily collected and burned. Animal manure is also under study as raw material for the anaerobic production of methane gas.

Other processes are available for producing fuels. For example, the pyrolysis of wood (the heating of wood in the absence of oxygen), used in World War II, may once again be reinstated as a means of producing gas, oil, and charcoal. Also, agricultural crops can be converted (fermented) to produce fuels such as alcohol. About 10 percent alcohol is blended in with gasoline to help extend gasoline supplies. This product, known as *gasohol*, is now being sold commercially at a few locations.

As we noted in Chapter 13, organic wastes can also be burned directly as fuel to generate electricity. About one-quarter of our organic wastes are readily available and easily collectible. Typically, the heating value of these wastes is about one-third that of coal. If all the collectible wastes were burned, they would furnish about 3 percent of the total amount of energy that was used in the nation in 1978. And the burning of organic wastes is economical. Because disposal costs are avoided, they can be entered on the positive side of the ledger. In some regions of the world, energy sources are so scarce that manure is the major source of fuel for cooking (Figure 19.12).

In summary, most biofuels, with the exception of organic wastes, are presently unable to compete economically with conventional fuels, and producing enough biofuels to provide significant amounts of energy would require large areas of land. But large tracts of land do exist where production of biofuels is possible, and these areas may be important energy sources in the future. We currently possess the technology for converting biofuels into electricity, but the process is so expensive that the prices of other fuels must rise before raising plants solely for fuel will be worthwhile. The Council on Environmental Quality's estimate is that by the year 2000, biofuels will contribute 3–5 quads of U.S. energy. In 1977, biofuels contributed about 1.3 quads.

Box 19.1

Firewood Stripping: The Other Energy Crisis

For the poor people of less developed countries, the energy crisis means something quite different from what it means to us. Some of us worry about whether we will have to pay more for gasoline, or if we may eventually have to switch from natural gas to coal for home heating. But roughly one-third of the world's people have a far more pressing worry: whether they can gather enough firewood to cook their meals.

In most less developed countries, wood is the main source of fuel for 90 percent or more of the population. As the populations of these countries increase, so does the demand for firewood—and in many areas this demand is growing faster than the local supplies. Firewood shortages are probably most severe in the Indian subcontinent, the semi-arid regions of Central Africa, and the Andean region of South America.

As firewood has become scarcer, it has also become more expensive. For example, in Naimey, the capital city of Niger, the average manual laborer must spend about 25 percent of his income to buy the imported firewood or charcoal he needs. The rise of firewood prices was prompted in part by the 1973 increase in the price of petroleum from the OPEC nations. This price increase also forced up the price of kerosene, the principal substitute fuel for firewood in poor countries. As more and more people found themselves unable to afford kerosene,

they turned to firewood—only to see its price rise even faster.

In many poor countries, firewood gatherers are stripping huge areas of their tree and brush cover. In Upper Volta, all the trees within a 40-kilometer (64-mile) radius of the city of Ouagadougou have been cleared and burned by the city's inhabitants. Such stripping has disastrous ecological effects. Soil erodes rapidly from the denuded land and silts up rivers and reservoirs. Soil erosion also reduces the productivity of agricultural land, thereby putting an additional burden on already hard-pressed people. And as firewood becomes scarcer, people often turn to dried animal dung for fuel. This practice further reduces agricultural production, since less animal manure is then available for use as fertilizer. Firewood stripping accelerates the process of desertification (see Box 5.2).

To date, little attention has been given to this other energy crisis. Most reforestation projects in the poorer countries have not been very successful, largely because grazing cattle, sheep, and goats get into replanted areas and eat the young trees. But the firewood crisis is beginning to force government leaders in these countries to rethink their citizens' relationship with the soil. A new balance in this relationship must be struck if the world's poor people are to have enough food and fuel.

Wind Power

Interest is currently being renewed in the ancient technology for utilizing *wind power*. For centuries, wind energy has been tapped by windmill blades to pump water, grind wheat, and generate electricity on a small scale. The first known windmills were built by the Persians as early as 250 B.C. And it was the

Figure 19.12 Animal dung to be used for cooking. (Agency for International Development.)

widespread use of windmills and sailing ships, also wind-powered, by the Dutch that allowed the Netherlands to become the most industrialized nation in the seventeenth century. Figure 19.13 shows some windmills in the Netherlands whose pumping function has been largely taken over by fuel-consuming pumps.

Today, scientists are using the techniques and materials of space-age technology to design and construct giant windmills for converting the wind's energy into electrical energy. Some of these windmills, such as that shown in Figure 19.14, are capable of generating 0.2 megawatts of electricity. If such machines were to sustain this level of power production continuously, they would provide enough power for a hundred average homes.

Theoretically, blades driven by the wind can convert 60 percent of the wind's energy into mechanical (motion) energy. Typically, however, wind-generating systems only extract about 35 percent of the wind's energy, and wind speeds must reach at least 19 kilometers (12 miles) per hour before most systems can operate. However, power produced by wind-driven generators is proportional to the cube of the wind speed, which means that for a given blade system, the power increases eightfold if the wind

Figure 19.13 Windmills in the Netherlands. (Photo KLM Aerocarto, the Netherlands.)

speed doubles. Hence, wind generators must be located at the windiest sites in a given region to take full advantage of the multiplying effect of added wind speed. Islands and hilltops make ideal sites.

Although scientists at NASA (National Aeronautics and Space Administration) believe that wind power could supply 5–10 percent of our energy needs by the year 2000, before such a goal could be reached serious technical and economic limitations would have to be overcome. Probably the most formidable technical problem stems from the extreme variability of wind itself. As indicated in Figure 19.15, the wind changes speed and direction from minute to minute and from season to season. Consequently, the electrical output of wind generators varies, and a wind power system must therefore contain a means of storing the energy generated during gusty periods for use during lulls. Usually, storage batteries such as car batteries are used for this purpose. Several thousands of dollars worth of batteries are required to provide a reasonable amount of storage in a wind power system for a typical

Figure 19.14 A 200-kilowatt wind generator at Clayton, New Mexico. This machine is currently being used by NASA to test the economics and performances of wind machines interconnected with conventional power plants. (U.S. Department of Energy.)

crease rapidly with the size of the generating plant. Thus, if the goal is to produce large amounts of wind-generated electricity, the use of many moderate-size windmills would be more economical than depending on a few larger ones. One wind power expert, Professor William Heronemus, has proposed that rows of windmills be anchored offshore to catch stiff sea breezes and produce electricity. The estimated costs of electric power produced by large grids of windmills offshore or on land are smaller than for conventional power plants. But electrical energy from such systems must be fed through an existing electrical distribution system to avoid high storage costs. Hence, the system is not as advantageous as we would like it to be.

Given these technical and economic limitations, wind has its greatest immediate potential in those regions where winds are consistent in direction and high in speed. In the United States, such regions include the western high plains, the Pacific northwest coast, the eastern Great Lakes, and the south coast of Texas. Low-powered wind systems have potential in small, isolated communities and on individual farms and ranches. The environmental impacts of wind systems are minimal: aesthetic concerns and the loss of some birds are thought to be the only such factors. In light of these considerations, it is ironic that in the 1930s and 1940s, in the interest of progress, the Rural Electrification Administration eliminated about fifty-thousand wind-powered electric-generating systems on midwestern farms.

The estimated wind energy potential in the United States excluding offshore areas is equivalent to 10–20 quads per year. The Council on Environmental Quality estimates that the equivalent of 4–8 quads of energy could be contributed by wind energy by the year 2000 if research activities and efforts to commercialize wind power systems are successful. Large quantities of metals for towers and storage systems and other resources would be necessary to support such an effort, but were a substantial market to develop, mass production would probably lead to significantly lower costs.

home. Roughly a 2-kilowatt generator is needed to meet the total electrical requirements, including heating, of a typical home. At this writing, total materials and construction costs for such a system—consisting of a tower, storage batteries, and the generator—would be about $5000.

Because wind speeds normally increase with altitude, windmills should be mounted on the tallest towers possible. However, the cost of towers also increases with height, and rapidly becomes prohibitively expensive. In addition, component costs in-

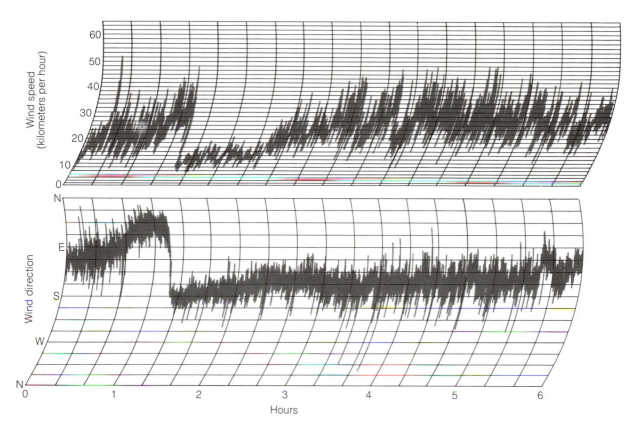

Figure 19.15 Variations in wind speed and direction from minute to minute over a period of six hours.

Hydroelectric Power

Another means of harnessing solar energy is the damming of rivers with *hydroelectric dams* and conversion of the energy of moving water into electrical energy. The average efficiency of hydropower generation is 75–80 percent, although in some newer installations, efficiencies are as high as 90 percent. Though we have been building hydroelectric dams for decades, we have tapped only 38 percent of the nation's hydroelectric potential. The worldwide figure of total potential tapped is 13 percent.

Why aren't we building more dams? One reason is that dams are expensive. Since dams must often be located far from the areas they serve, long transmission lines are needed, and these installations greatly increase both the cost of electrical transmission and the amounts of electricity lost. Also, many of our rivers are protected against dam construction by acts of Congress. These protective measures are necessary for a number of possible reasons, many having to do with the environmental and social costs associated with dam construction discussed in Chapters 8 and 15. For these reasons, few hydroelectric power facilities will be built. In fact, the Council on Environmental Quality estimates that new hydroelectric generation will come mostly from the addition of hydroelectric generating facilities to large and small dams existing. And this would increase the nation's energy supply by only 2.5 quads per year. The Council predicts that total energy generated by hydroelectric means in the year 2000 will be 4–6 quads, compared with the 3 quads presently being generated by this method.

Ocean Thermal Energy Conversion

Another solar energy conversion system currently being studied would exploit temperature differences between the cold bottom waters and warm surface waters of tropical oceans. Such temperature differences are known to be as high as 22°C (40°F). Floating electric power plants would be anchored in these regions. In these systems, called *ocean thermal energy conversion* systems, liquid and gaseous ammonia would be analogous to water and steam in a conventional turbine electric generator. Warm surface waters would be used to evaporate and pressurize ammonia, and the pressurized gas would be used to drive turbine generators. Cold waters lying as deep as 900 meters (3000 feet) below the plant could be brought up to condense the ammonia so the ammonia cycle can be continued. Electric power from such systems would be transmitted to shore via underwater cables or be used at sea to produce aluminum, ammonia, or other products requiring energy-intensive technology.

By the mid-1980s the Department of Energy plans to have a 100-megawatt ocean thermal energy conversion unit operating as a demonstration project to show the overall feasibility of this technology. Some anticipated problems involve the buildup of efficiency-reducing organisms such as barnacles on heat-exchanging components, and scientists have expressed concern that such systems could alter ocean current patterns and thus climate. But the advantage of these systems is that they could produce power continuously and thus be used to supply minimum daily power requirements. The long-range future of this option looks promising. The Council of Environmental Quality estimates that by the year 2000 these systems could be producing 1–3 quads of energy annually. Total potential capacity of ocean thermal conversion systems is estimated to be 15–40 quads.

The exact technologies that will be used to deliver future energy demands to Americans is not at all clear at this point in time. However, the Council on Environmental Quality's estimates for the year 2000 suggest that solar developments could make a substantial contribution if we aggressively pursue the development of solar-related technologies. The Council predicts that such technologies could furnish a total of 20–30 quads of our 80–120-quad projected demand for the year 2000. Projected totals for the year 2020 are considerably greater. Refer back to Table 19.1.

Tidal Power

Another type of hydroelectric power is *tidal energy*, derived from the ceaseless motion of ocean tides. The gravitational attraction among the earth, moon, and sun gives rise to a rhythmic oscillation of ocean water. Along coastlines, this pattern appears as a daily cycle of changing water levels, usually differing by 1–10 meters (3–30 feet) in height. Although tidal energy is reliable, only one major tidal-powered electricity-generating plant is now in operation in the world. This plant, situated at the mouth of La Rance River in France, has been operating since 1966 and has a capacity of 160 megawatts. Figure 19.16 provides a view of this plant.

The only sites considered feasible for tidal power plants are locales where the difference in sea level between low and high tides is anomalously large and where a relatively large bay or inlet exists to hold the in-coming tide. Because few such sites exist in the United States, tidal power will never be more than a minor local energy source. Recent proposals for the construction of tidal power plants at favorable sites (such as Passamaquoddy Bay, on the Maine coast)

Figure 19.16 The world's only tidal-powered electricity-generating plant. The plant spans the estuary at the mouth of La Rance River in France. (Electricité de France, Michel Briguad; French Engineering Bureau.)

have been scrapped because of the high capital cost of construction. The environmental effects of operating tidal power plants have not yet been fully evaluated.

Geothermal Power

Geothermal energy is the heat energy generated in the earth's interior. Most of this energy is released during the radioactive decay of certain elements deep within the crust and conducted to the surface. In certain localities, geologic processes concentrate geothermal energy to such a degree that exploiting it as

a power source is feasible. Such regions are sites of subsurface igneous activity where hot magma—measuring 1100°C (2000°F) or higher—comes into contact with groundwater. Where groundwater that contacts the hot magma is confined under high pressure, it becomes *superheated,* that is, it remains liquid even though it may be heated to several hundred degrees above the boiling point. A region in which this situation occurs is referred to as a *wet steam field.* In other cases, the groundwater is under less pressure, and when heated it turns directly into steam. Regions in which this situation occurs are called *dry steam fields.* In both types of steam fields, the heated groundwater is separated from the earth's

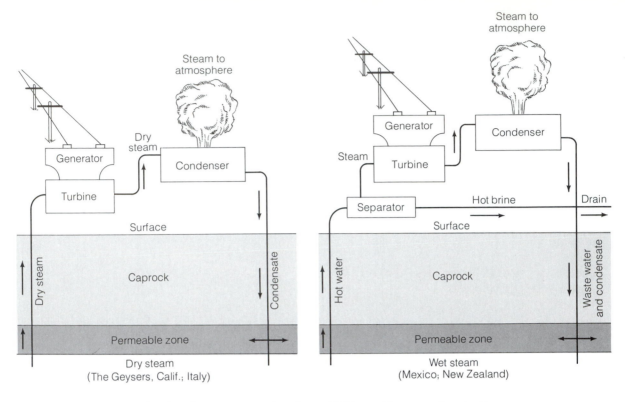

Figure 19.17 A schematic drawing of the two types of geothermal fields used for generating electric power. (After P. R. Ehrlich, A. H. Ehrlich, and J. P. Holdren, *Ecoscience: Population, Resources, Environment.* San Francisco: W. H. Freeman and Company, 1977.)

surface by an impermeable rock layer known as caprock.

Steam and superheated water are extracted through a well drilled through the caprock. In a dry steam field, the steam is used directly to drive a turbine, as shown in the left-hand drawing in Figure 19.17. But in a wet steam field, another step is required. Drilling releases the pressure, and, as a result, a portion of the superheated well water turns immediately to steam. The pressurized steam is then separated from the hot water and used to drive a turbine. In both systems depressurized steam from the turbine may then be released to the atmosphere. By condensing the spent steam, the output of the system is significantly enhanced. The warm water is then pumped into cooling towers where its heat dissipates.

One major problem with geothermal wells is the

disposal of waste water. In many instances, waste water contains at least trace amounts of sulfur dioxide, hydrogen sulfide, ammonia, and boron. Frequently, too, the waste water is extremely saline. For example, the yield of some wells in the Imperial Valley of California is only 20 percent steam; the rest is brine that can be up to thirty times more saline than ocean water. This saline water must be disposed of either at sea or by subsurface injection. At some sites, these dissolved salts may be recovered for commercial sale.

Another problem with geothermal energy is that it is a rapidly depletable resource. As steam or hot water is removed, the water table drops, since recharge by infiltrating surface waters is an extremely slow process. Also, heat is withdrawn perhaps hundreds of times more rapidly than it is renewed through water contact with magma. Thus, within ten years or so, the well dries up or the water turns cold, and a new well must be drilled elsewhere. With repeated drilling at many locations, however, a large geothermal field can be expected to last up to two or three hundred years.

Despite the limitations of geothermal energy, in certain areas a dry steam field operation can compete favorably with other sources of electrical power. Plant construction costs are generally two-thirds to three-fourths those of comparable fossil fuel power plants, and geothermal operations have the added advantage of requiring less maintenance.

The first dry steam field ever developed is located in Larderello, Italy. It has been in operation since 1904, and in 1974 produced 400 megawatts of power. But the largest known dry steam field is the Geysers facility, shown in Figure 19.18, located 145 kilometers (90 miles) north of San Francisco. Power production at the Geysers began in 1960, and has increased each year since then. In 1976, 500 megawatts of power were in service. An additional 100 megawatts of capacity will be added annually to the Geysers field. Total capacity is estimated to be 2000 megawatts, enough to meet one-half of the total power needs of the greater San Francisco area. Wet steam fields are in operation in New Zealand (where 145 megawatts were produced in 1975), Mexico (75 megawatts), and Japan (30 megawatts). Other plants exist in Iceland and the USSR.

With the growing interest in geothermal energy, and because known sources remain relatively untapped, the potential for growth is encouraging. Estimates made in 1974 indicate that 4000 megawatts of power may be generated by geothermal energy by 1985. By the year 2000, these figures could reach 60,000–100,000 megawatts. Estimates of potential production at undiscovered sites range from ten to two-hundred times this 60,000-megawatt value. But use conflicts will undoubtedly block extraction of much of the existing geothermal energy, since many steam fields are on federal lands. Some are in national parks, such as Yellowstone National Park, where drilling is prohibited. Should we impinge on these last vestiges of undisturbed land to help satiate our appetite for energy? Our decision could be influenced by the fact that even the most optimistic projections indicate that geothermal energy will account for less than 1 percent of our total electrical use in the year 2000.

Energy Conservation

We have seen in this chapter that before alternative energy sources become available to us to a significant degree, we will have to invest materials, money, time, and effort in developing these technologies. We will be at least into the 1990s before these options begin to make a difference in our energy supply. In the meantime, we can extend the lifetime of our fossil and nuclear fuels by implementing energy conservation strategies. In 1979, the President's Council on Environmental Quality assessed the potential effects of such conservation techniques, and determined that our national economy could continue to operate on 30–40 percent less energy if the efficiency of our fossil fuel utilization practices were improved. Thus, if we implement the technically and economically feasible conservation measures available to us, we should be able to fuel future growth by increasing efficiency rather than increasing future energy inputs. The economic and environmental benefits that we would derive from such efforts include reductions in the number of pollution-belching power plants and in imports of costly foreign oil. We turn now to a survey of the energy-saving measures that must permeate every segment of our society if we are to significantly reduce our total future energy demand.

Transportation Measures

In the 1930s, the automobile replaced railroad travel as the major means of personal transportation. The greater individual mobility provided by the automobile influenced the shape of our cities, determined the

Figure 19.18 Part of the geothermal electricity-generating plant at the Geysers, California, northeast of San Francisco. Pictured are two of the thirteen units that produce power for Pacific Gas and Electric Company. (Pacific Gas and Electric Company.)

average commuting distances between home and work and shopping centers, and molded, to a great extent, the patterns of our individual lives. Today, more than 134 million motor vehicles are registered in the United States. These vehicles travel 2.1 trillion kilometers (1.3 trillion miles) a year, and in the process they burn up more than 401 billion liters (106 billion gallons) of fuel—about 13 percent of the fuel used in the country. Because the automobile is such a voracious consumer of fuel, and because we have some personal control over the amount of fuel it consumes, automobile usage is a logical area in which to begin our conservation efforts.

One means of conserving fuel used in personal transportation is by reducing our dependency on individual personal vehicles and increasing our patronage of mass transit systems. A bus carrying fifty commuters uses considerably less fuel than fifty cars carrying fifty individuals. The more heavily mass transit systems are used, the more efficient they become. To attract more consumers to mass transit systems, local governments could designate special express lanes for buses on highways, provide comfortable and modern vehicles, improve night service,

Table 19.2 Energy required to move a passenger 1 kilometer by various modes of transportation.

Mode	Kilojoules per kilometer	Percent of transportation demand
Airplane	4700	9.3
Automobile	3500	88.4
Train	1700	0.7
Bus	1100	1.2
Walking	340	<0.1
Bicycle	200	<0.1

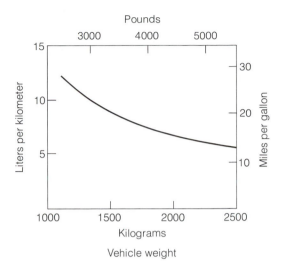

Figure 19.19 Fuel economy as a function of vehicle weight.

and ensure good security. Other alternatives to using automobiles, especially over short distances, are bicycling and walking. These are not only the most efficient modes of transportation, but also the healthiest. Table 19.2 shows the energy requirements for various modes of travel.

However, bicycling, walking, or using mass transit systems are not always practical, and nearly every aspect of the American way of life is dependent on convenient mobility. Thus, we must account for the fact that the private automobile, for the present at least, will remain in use. How can we cut down on the amount of fuel used by automobiles? One of the most important factors in the fuel economy of vehicles is weight (see Figure 19.19). For example, to attain a fuel consumption rate of 11 kilometers per liter (25 miles per gallon), a car must weigh no more than 1300 kilograms (2800 pounds), which is the size of a compact car. Small cars not only obtain better fuel economy than heavier cars, but they also require less material in their construction. Hence, less energy is used per car for mining, milling, and manufacturing. Automobile manufacturers are currently building smaller and lighter cars to meet federally mandated mileage standards.

Individuals can make more efficient use of their automobiles by adopting conservation measures as a matter of course. These measures are often based on simple logic. For example, of the 60 million people who commute to work, 51 million drive alone; only 9 million travel in car pools. Thus, if all car pools carried only two people, we could conceivably reduce the number of cars on the road during rush hour by 25 million. Car pools would also reduce congestion on highways, which in turn would prevent the loss in fuel economy experienced in stop-and-go traffic. And an even more effective means of reducing the amount of fuel used for commuting is van pooling. In van pooling, an employer makes a van solely available to about ten employees, who use it exclusively for commuting to and from work. Individuals can contribute to this effective conservation strategy by urging industries to implement van pools.

Proper car maintenance saves fuel and money too. Only by keeping carburetors and air filters clean and ignition systems well tuned can motorists obtain optimal energy efficiency from their car engines. Drivers can also save energy by refraining from using automobile air conditioners. Typically, these devices cut fuel economy by about 9 percent. Furthermore, using radial tires improves gasoline mileage by several percent. Driving habits, too, have a strong impact, not only on gas mileage, but on maintenance and safety as well. For instance, lurching starts and screeching stops not only reduce mileage but add to wear and tear on the vehicle. And driving at high speeds simultaneously reduces gas mileage and increases the risk of accidents. The 88-kilometer-per-hour (55-mile-per-hour) speed limit has not only conserved fuel, but has saved about one-hundred-thousand lives in the United States since it was imposed in 1973.

While most of the fuel used in vehicles is associated with transporting people (56 percent), freight hauling also accounts for a significant portion (36 percent) (the remainder is accounted for by military vehicles). Fuel conservation measures involving freight hauling focus chiefly on the mode of transport chosen. For example, moving freight by airplane requires three times more energy than moving it by train. Ships and barges are the means of freight hauling that consume the least fuel, and pipelines are the most fuel-conserving means of transporting gases and liquids. In fact, pipelines are now being utilized to transport coal in the form of a slurry (a mixture of finely ground coal and water), because this method is less energy and labor intensive than hauling coal overland. Transporting coal by pipelines is estimated to use 20 percent less energy than transporting by train. However, diverting water to these pipelines creates major environmental problems in the semiarid areas where much coal is located.

Estimates show that implementing cost-effective energy-saving measures could reduce energy consumption in the transportation sector of our economy by half the amount consumed in 1973. The chief means of achieving this reduction would be by increasing the fuel economy of automobiles by 150 percent from 1973 levels. Further savings of fuel would be achieved through conservation measures involving modes of transport other than the automobile. Overall, improvements in the efficiency of the transportation sector has the potential of reducing total energy demand by 12 percent from 1973 consumption rates, or a savings of 9 quads of energy per year.

Residential and Commercial Measures

If we make technical changes in the way we use energy in our homes and businesses, we could potentially reduce our energy demand in this economic sector by 40 percent from 1973 levels—a 15 percent saving in total energy demand. The technical improvements necessary to achieve these savings are summarized in Tables 19.3 and 19.4.

Reducing heat losses is the single most important means of conserving fuel in homes. Improving insulation, installing double- or triple-pane windows, and eliminating air leaks are all techniques for cutting down energy consumption. Most homes built prior to the 1960s are poorly insulated, but the cost associated with adding more insulation is generally recovered in four or five years through savings on fuel bills. Poorly insulated houses not only lose heat in the winter, but they gain heat in summer; thus, a greater demand is placed on heating and cooling systems. Even in many homes built since the 1960s, adding more insulation is worthwhile because of reasonable payback periods. Heat losses through cracks are easily reduced by caulking and weather stripping. Both procedures are inexpensive and quickly pay for themselves.

Homeowners can save energy in many other ways. For example, they can reduce water-heating costs by lowering hot-water heater temperature settings, covering the hot-water heater with an insulating blanket, or replacing a worn-out water heater with a new, better insulated model. And they can replace an

Table 19.3 Potential annual fuel-saving measures for the residential sector.

Conservation measure	Potential savings (in quads)
Reduce heat losses (insulation, etc.)	3.3
Reduce air conditioner load (insulation, etc.)	0.45
Cut water heating fuel requirements	1.05
Replace electric resistivity heating with heat pumps	0.60
Increase air conditioning efficiency	0.37
Increase refrigeration efficiency	0 37
Introduce total energy-efficient systems into 50 percent of multifamily units	0.30
Use microwave oven for 50 percent of cooking	0.22
Total	6.7

Source: President's Council on Environmental Quality, ''Good News About Energy,'' January 1979.

Table 19.4 Potential annual fuel-savings measures for the commercial sector.

Conservation measure	Potential savings (quads)
Reduce heating requirements	2.3
Reduce building lighting	1.0
Cut water heating requirements	0.3
Increase air conditioning efficiency	0.4
Increase refrigeration efficiency	0.2
Improve insulation, reduce ventilation rate and add heat recovery	0.2
Use total energy systems in 33 percent of all units	0.7
Use microwave oven for 50 percent of cooking	0.1
Total	5.2

Source: President's Council on Environmental Quality, ''Good News About Energy,'' January 1979.

electric resistance-heating system with heat pumps that transfer heat from outdoor air in winter or function in reverse as an air conditioner in summer. Builders and architects should be encouraged to incorporate into new homes such passive solar systems as overhangs, large south-facing windows, and movable, insulated window panels. In some cases, older homes can be partially retrofitted with some passive features, though, as we noted earlier, the cost of such efforts can be considerable.

The electrical energy used by appliances accounts for about 8 percent of the total energy used in the United States. Therefore, improving the operating efficiency of household appliances would significantly reduce overall electrical energy demand. The energy consumption of home appliances and lighting fixtures are listed in Table 19.5. Improving the efficiency of major energy-consuming appliances such as refrigerators would have the greatest effect in reducing household energy consumption. Presently, frost-free refrigerator-freezer combinations typically use 1800 kilowatt-hours of electrical energy per year. According to federal estimates, refrigerators with more insulation, greater motor efficiency, and improved cooling systems could reduce the annual energy consumption of these appliances by 45 percent. If such refrigerators were in use in all U.S. homes today, our electrical demand would be about 3 percent lower than it is now. Other appliances that could be redesigned to be more energy efficient include air conditioners (a 22 percent energy savings is possible) and television sets (government goals are to reduce energy consumption by 50 percent). The use of microwave ovens for cooking would further reduce energy requirements in the home.

Energy conservation measures available for the commercial sector are much the same as those for the residential sector. Lighting however, is of particular concern in businesses. In this regard, reductions in the amount of lighting used and improvements in efficiency could result in a significant reduction in energy consumption. The amount of light given off by an electric lamp per unit of energy input varies greatly with the lamp type. Fluorescent lights give off nearly four times the light of ordinary incandescent light bulbs. For outdoor lighting, sodium vapor lamps (yellow light) give off about three times the light of the popular mercury vapor (blue light) lamps usually used to light streets and parking areas. A final word regarding lighting applies to residences, businesses, and industries equally: to conserve energy, avoid all unnecessary lighting, including decorative lighting, and always turn off lights that are not in use.

Industrial Measures

Though we can take individual action to conserve fuel in our homes and our modes of transportation, opportunities to reduce energy consumption in commerce and industry are limited. Our most effective lever in this area is group action designed to encourage industries and commerce to save energy.

Recently, many industries have discovered that energy conservation pays dividends. Because of rising energy costs, investments in energy-conserving equipment are bringing greater returns than any other capital expenditures. Such capital investments include improvements in insulation for high-temperature processes and recovery of waste heat.

Waste heat recovery should be of particular interest to city planners. Many industrial processes produce more waste heat than the plant in which they occur could possibly use. For example, paper mills use about twelve times more energy in the paper-making process than they consume in their remaining operations. Thus, even if paper mills channeled their waste heat into other segments of their operation, they would still have large quantities of waste heat to vent into the air or to discharge as heated waste water. Therefore, a paper mill with a heat-recovery system would have heat to sell. In another example, petroleum-refining processes are estimated to produce twenty-eight times their own heat energy needs. It is clear from such examples that we would

Table 19.5 Energy consumption by home appliances and lighting.

Appliance	Annual energy consumption (in kilowatt-hours)	Annual cost of energy consumed*
Air conditioner	2000	$100.00
Electric blanket	150	7.50
Can opener	0.3	0.02
Clock	17	0.85
Clothes dryer	1200	60.00
Coffee maker	100	5.00
Dishwasher (with heater)	350	17.50
Fan (attic)	270	13.50
Fan (furnace)	480	24.00
Fluorescent light (3 fixtures)	260	13.00
Food freezer (16 cubic feet)	1200	60.00
Food mixer	10	0.50
Food waste disposer	30	1.50
Frying pan	240	12.00
Hair dryer	15	0.75
Hot plate (2 burner)	100	5.00
Iron (hand)	150	7.50
Light bulbs	1870	93.50
Radio (solid state)	20	1.00
Radio phonograph (solid state)	40	2.00
Range	1550	77.50
Refrigerator (frost-free, 12 cubic feet)	750	37.50
Sewing machine	10	0.50
Shaver	0.6	0.03
Television (black and white)	400	20.00
Television (color)	540	27.00
Toaster	40	2.00
Vacuum cleaner	45	2.25
Washer (automatic)	100	5.00
Totals	11,938 KWH	$596.90

*Cost of electricity = 5 cents per kilowatt-hour
Source: Citizen's Advisory Committee on Environmental Quality, Washington, D.C.

Box 19.2

Environment and Energy: An Optimistic View

We hear much today about the relative merits of the so-called "hard path" and the "soft path" energy technologies, as if they were mutually exclusive. I just don't accept that notion. In fact, in my view, not only are they not mutually exclusive, they are mutually supportive. And it is the environment, or at least our concern for the environment, that brings them together.

• • •

The so-called "hard-path" technologies include nuclear, the big coal-fired generating plants, the coal-to-gas plants, the coal-to-synfuel plants, and other large, high capital cost, and high production energy facilities. . . .

The "soft-path technologies" go under a number of names: "alternate," "renewables," "dispersed. . . ."

I like "dispersed," myself, because what we are really talking about is smaller, self-contained lower capacity, and more environmentally benign energy technologies—solar heating and cooling, photovoltaics, wind, geothermal, biomass, geopressured methane and nonconventional gas supplies. What is more important, they are in our near and mid-term future, not long term new supply technologies.

What I can't get through my head is why we have to have one or the other but not both the hard and soft. Nobody has been able to explain that to me.

I think the fallacy that most people have been tripping over is the idea that we have to have a single national solution, the so-called big fix, that we must do here in Texas exactly what we do in

New England, that there is only one way for the whole country to go. . . .

But I don't see why there has to be only a single national solution to our energy problems.

On the contrary, I can think of a lot of reasons why we ought to have many different energy solutions for many different places in our country, depending on their energy needs, their available energy resources, and the tastes, customs and traditions of the people who will be affected. And . . . in following that course, I believe we can continually upgrade the environmental posture of the country as a whole. . . .

It should be obvious that what I am talking about is a more local and regional approach to our energy problems. Let's take a few specific examples to see just how it would work.

Let's take as one example, a mythical little village in Southern Utah. They have about 10,000 people living there, and their energy needs probably can be served very well by a few small windmills plus some solar devices, and a geothermal plant. But they also have lots of coal around. Well, the historical way to serve this community . . . is to build a giant coal-fired plant—a hard-path approach—and string electrical wires to each of these communities, and the west coast.

But the cultural tradition in this section is that "small is beautiful." They don't . . . want big plants and big smoke stacks. They are farmers and ranchers and not interested in a "boom." They want the windmills, and the solar collectors, and the geothermal plant to serve their needs. Let them have it. Maybe they will let the coal be mined and shipped somewhere.

Now let's switch the scene to a heavily industrialized area in the Ohio Basin. They have lots of

---

Condensed from a speech delivered in 1978 by Robert D. Thorne, an assistant secretary of the U.S. Department of Energy.

factories, a big labor force and a very high energy demand. Those factories mean jobs and livelihoods for the people. To these people, small isn't necessarily beautiful.

. . .

This is not to say, however, that we can't put dispersed technologies in Ohio. As a matter of fact, we have a very large and very successful windmill demonstration in Plumbrook, Ohio, right now. . . . The more small communities using dispersed technologies in the environs of a large industrial center, the lower the environmental impact. The less demand put on the big central facility, consequently, the smaller that central facility need be.

. . .

Let's take one more example to see what would happen if we restricted ourselves to only one path, the soft path. What are we going to do about an area like Chicago with all of its industry and its huge energy demand?

. . .

First of all, you probably would have to cover every bit of open space and farm land in the area with solar farms and windmills and other such installations because it is a fact that soft path technologies usually take a lot of room for the amount of energy they produce. Secondly, you probably would have the area on the kind of energy rationing that would discourage any new industrial growth and make existing facilities very expensive.

. . .

In other words, the soft technologies do not make a very good natural fit with that area and the people's expectations.

. . .

The main point . . . is that if we look carefully enough, we can find good natural fits for most places in the country.

. . .

In the Pacific Northwest, they have lots of hydropower, moderate temperature geothermal, and lots of trees for gasification or burning, and water for nuclear plants. The climate is not very conducive to solar applications.

California is very high on solar, and has the climate for it and a lot of geothermal resources. Right now, they seem pretty opposed to nuclear and coal burning. Well, if that is their answer to their energy problems and they think it's the best answer for them, then why not?

Moving east, the western mountain states are rich in all kinds of energy resources—coal, oil, uranium, shale, solar, wind and, in some places, agriculture biomass. They are net energy exporters and we really don't have to worry about their energy demands.

. . .

Similarly the southwest is rich in oil, natural gases and sunshine. They export energy. . . .

The southeast has coal, vast resources of timber, and some oil and natural gas and agriculture biomass. They also have been reasonably receptive to nuclear power.

New England has been heavily dependent on foreign oil, but has a lot of potential from low-head hydro, from wind, a lot of timber, and they have been making the most extensive use of nuclear.

. . .

(Continued)

Many of the plains states have huge potential biomass resources, some coal, and plenty of wind and sunshine. In the non-industrialized areas where the energy demand is not too high, they have a wide variety of choices.

In Hawaii and Puerto Rico, they have sunshine, the trade winds for constantly turning windmills, some geothermal sites, ocean-thermal energy potential, and lots of biomass from their sugar cane.

• • •

Thus we are left with a relatively small section of the United States that is the most heavily industrialized, the Ohio Basin, Appalachia, the industrialized Mid-Atlantic areas. In reality, we will probably have to focus the hardpath technologies in these areas. Even here, though, there will be opportunities for soft-path technologies in individual communities, working hand-in-hand with the big centralized plants.

• • •

I think this approach would have an increasingly beneficial effect on the environment. By placing the environmentally benign technologies wherever they can meet the needs of the communities or regions, we can focus the big centralized plants more precisely in those places where they are absolutely needed. . . .

• • •

I have proposed the thesis that both the hard and soft path technologies can be utilized in a mutually supportive way that should be acceptable to most people in the country, and which should have a net effect of improving the environment. Energy development need not be a threat to the environment; it can also be used as a tool to improve it.

be wise to plan our industrial areas in the future so that waste heat from one industry could be used by neighboring industries, shopping centers, or other large facilities. Citizens can play an important role in encouraging this kind of thinking by calling such opportunities to the attention of planning agencies.

The improved management of energy-consuming equipment within industries can also reduce energy consumption. Such practices include adjusting temperatures for space heating and cooling, eliminating unnecessary lighting, shutting down machines not actually in use, and keeping boilers operating at peak efficiency. Other improvements include using only high-efficiency electric motors and replacing obsolete plants and equipment. These measures, plus other energy-saving technology too sophisticated to discuss here, could result in a 30 percent energy savings in the industrial sector. This reduction in turn would cut our total energy consumption by 14 percent, or about 10 quads.

In summary, energy-conservation measures alone could reduce our energy demand by 31 quads: 9 quads in the transportation sector, 7 quads in the residential sector, 5 quads in the commercial sector, and 10 quads in the industrial sector. These figures represent a potential 40 percent reduction from 1978 levels. However, we should remember that these savings will not come overnight. Only through incremental improvements over the next two or three decades will the dividends we have discussed be paid.

One clear conclusion we can draw from the statistics on energy use and potential conservation by sector is that the public cannot place the blame for our energy wastes solely on industry or commerce. Similarly, industrial leaders are in no position to blame the actions of citizens. The sooner we all recognize that all of us must conserve energy, the sooner our energy problems will be eased. And the good news is that our economy can still prosper, that environmental degradation can be reduced, and that it is possible for us to reduce our dependence on foreign fuels (see Box 19.2). Furthermore, energy conservation will make our fossil fuels last longer, buying us more time to evaluate and implement energy alternatives. These goals should be desirable enough to inspire each of us to make energy conservation a way of life.

Conclusions

Our image of the future with respect to energy is necessarily vague, but one thing can be stated with certainty: if we wish to maintain a reasonable quality of life, we must increase our conservation efforts and our reliance on alternative energy sources. But to which alternative should we turn? Geothermal, wind, and hydroelectric energy sources pose various combinations of technological, economic, and environmental difficulties, and even if these problems were resolved, all three options would hold the potential for meeting only local energy demands. And the multitude of hazards posed by nuclear breeder technology at this time virtually precludes its widespread use. Thus, in view of their almost limitless supply, plus the fact that their negative effects on the environment are minimal, solar radiation and nuclear fusion appear to be the most desirable long-term alternatives to conventional energy sources.

Will solar and nuclear fusion, then, at some time be our major energy sources? Many recent studies have analyzed our energy consumption patterns and attempted to project our future needs and sources. All such studies predict a gradual phasing out of our dependency on fossil fuels and a transition to a variety of alternate energy sources. But anticipating precisely which of these new technologies will be in use is a formidable task. As we have seen, many obstacles still exist to full-scale and immediate implementation of alternate energy sources. Overcoming these obstacles through research, development, and economic incentives will take time—perhaps many decades. Until these technologies are refined to the point at which we can rely on them, we will have to extend existing resources through conservation. We really have no other choice.

Summary Statements

A breeder reactor is a special type of nuclear reactor designed to fission more abundant uranium-238. Major obstacles in implementing this technology are the disposal of high-level radioactive wastes, and the securing of nuclear weapons materials (primarily plutonium-239) from terrorists.

The nuclear fusion of light elements yields energy. But the possibility that technology can be developed for sustained production of energy by nuclear fusion is still highly questionable.

Solar energy is abundant, but it is also diffuse and expensive to collect. Solar collectors are practical at this time in some geographical areas for space heating and water heating. Small test-scale solar power plants are currently under study.

Photovoltaic, or solar, cells convert the sun's energy directly into electricity. At this time, the cost of solar cells is prohibitive for most applications.

Growing plant materials (biofuels) as energy sources for power plants may be feasible in some regions. The environmental problems of such efforts, however, could be substantial. The organic fraction of garbage is also an alternate, readily available source of fuel for power plants.

The generation of electrical energy by means of wind power has been technically feasible for decades. However, wind-generated electricity requires a high initial investment and can only be produced at particularly well-suited sites.

Electricity is generated by hydroelectric dams through the damming of large rivers. Many negative effects are associated with such dams, including the flooding of large land areas, social and environmental disruption, and the loss of wildlife. Cost is another significant obstacle.

Ocean thermal energy conversion systems would exploit the differences between warm surface ocean waters and cold bottom waters to fuel floating electric power plants.

Tidal power can be utilized in only a few locations along coastlines where tidal oscillations are large. Also, capital costs for tidal power plants are prohibitive.

In some regions, hot rocks and hot subsurface water can be used to generate steam that can be employed to generate electricity. The western states hold the greatest potential for geothermal energy production.

Existing fossil fuel reserves can be extended through conservation. Energy conservation requires that we be constantly aware of how our energy-consuming activities can be performed more efficiently.

Since vehicles consume the highest percentage of the fuel used in the nation annually, it is essential that we alter our transportation practices to improve fuel economy. Proper maintenance of energy-consuming devices is another important step toward energy conservation. Keeping automobiles well tuned and insulating and weather stripping houses will help hold fuel consumption in these systems to a minimum. And householders purchasing new appliances should take energy efficiency into account in making their choices.

Although we are certain that oil and natural gas reserves will continue to diminish, the negative effects associated with both coal and nuclear fission motivate us to seek out alternative energy sources. Most such alternatives, however, will make only small contributions in meeting our total demands.

Solar energy and nuclear fusion appear to be the most desirable long-term alternative energy sources, but many obstacles to implementation of these technologies remain.

Questions and Projects

1. In your own words, write definitions for the terms italicized in this chapter. Compare your definitions with those in the text.

2. Distinguish between nuclear fission and nuclear fusion.

3. Compare the advantages and disadvantages of breeder fission reactors and conventional nuclear fission reactors for the generation of electric power.

4. Should we rely on nuclear fusion as a means of generating electrical power in the future?

5. Locate a solar heated home in your region and try to obtain some performance and cost data. Would you purchase such a home?

6. Using Figure 19.7, determine the amount of available solar radiation in your region. How does this amount of radiation available vary through the year?

7. Identify the advantages and disadvantages of harvesting the sun's energy using the following methods: solar collectors, solar cells, windmills, hydroelectric dams and biofuels. Overall, which option is the most practical on a large scale?

8. If the power plant in your area were to burn biofuels, what types of local vegetation would be used? Speculate on the environmental impact of such an operation.

9. What is the potential for using organic wastes (garbage, methane from landfills) in your community to generate or supplement conventional energy sources. If these sources are already being tapped, report on the operating details.

10. Comment on the potential of your locality for generating electricity through nuclear, hydro-, coal, solar, tidal, wind, and geothermal power. Which combination of these energy sources is the most likely to supply the future needs of your community?

11. Speculate on some of the environmental impacts of a tidal-powered generating facility.

12. What changes are you willing to make in your lifestyle to help conserve energy?

13. Contact your local energy utility or a heating and cooling contractor and ask about their energy-conserving strategies. What specific suggestions do they have for home energy conservation?

14. Develop and implement an energy conservation strategy for your home. Use electrical and fuel consumption figures to test the effectiveness of your plan.

15. Evaluate the merits of the following energy conservation strategies for the home: (a) allow bath water to cool after bathing instead of allowing warm water to go down the drain, thus permitting the heat from the water to warm the room; (b) turn down the thermostat and use an electric blanket at night; (c) discontinue the use of electric knives, garage door openers, and electric can openers.

16. Peak electrical demand in most communities occurs at about 4 P.M. each day. On an annual basis, peak demand occurs in summer. What are the advantages of reducing demand during peak periods? Suggest some practices that could be implemented in your community to reduce electrical energy demand during these periods.

17. What effect does the weight of an automobile have on fuel consumption?

18. How could public transportation be improved in your locality so that more people would use the system?

19. Has your campus reduced its energy consumption during recent years? If so, what conservation measures did it employ? If not, why?

20. Comment on the implications an extreme change in world climate would have on the energy supply.

Selected Readings

Boffey, P. M. 1975. "Energy: Plan to Use Peat as Fuel Stirs Concern in Minnesota," *Science 190*:1066 (December).

Calvin, M. 1979. "Petroleum Plantations for Fuel and Materials," *Bioscience 29*:533–538 (September). Reviews the potential of plants to supply certain desirable hydrocarbon fuels.

Council on Environmental Quality. 1978. *Solar Energy*. Washington, D.C.: U.S. Government Printing Office. Discussion of prospects for various solar technologies.

———. 1979. *The Good News About Energy*. Washington, D.C.: U.S. Government Printing Office. A survey of economically feasible strategies for reducing fossil fuel and electrical energy use. Presents projected energy savings.

Dorf, R. C. 1978. *Energy, Resources and Policy*. Reading, Massachusetts: Addison-Wesley. A textbook that examines the conventional and future energy technologies in light of resource availability and related policy decisions. Some sections are more technical than others.

Duffie, J. A., and W. A. Beckman. 1976. "Solar Heating and Cooling," *Science 191*:143–149 (January). A review of the status of the building architecture and economics of solar energy systems.

———. 1974. *Solar Energy Thermal Processes*. New York: John Wiley & Sons. A discussion of the scientific basis of solar collection systems.

Federal Energy Administration. 1976. *Buying Solar*. Washington, D.C.: U.S. Government Printing Office. An eighty-page pamphlet that provides information on variables people should consider before installing solar hardware in their homes.

Hammond, A. L. 1977. "Photosynthetic Solar Energy: Rediscovering Biomass Fuels," *Science 197*:745 (August). A review of recent developments in the use of biofuels as energy sources.

Hildebrandt, A. F., and L. L. Vant-Hall. 1977. "Power with Heliostats," *Science 197*:1139 (September). A discussion of sun-tracking mirrors for high-temperature solar applications.

Hirst, E. 1976. "Residential Energy Use Alternatives: 1976 to 2000," *Science 194:*1247–1252 (December). An evaluation of various energy conservation alternatives for the residential sector and their impact on residential energy demand.

————. 1977. *Transportation Energy Use and Conservation Potential.* Cambridge, Massachusetts: M.I.T. Press. A discussion of strategies for reducing energy consumption in the transportation sector of our economy.

Metz, W. D. 1977. "Solar Thermal Energy: Bringing the Pieces Together," *Science 197:*650 (August). A review of solar thermal projects.

————. 1977. "Ocean Thermal Energy: The Biggest Gamble in Solar Power," *Science 198:*178 (October). A look at proposed systems for utilizing temperature differences in ocean currents as an energy source.

Robinson, A. L. 1977. "Photovoltaics: The Semiconductor Revolution Comes to Solar Energy," *Science 197:*445 (July). A review of recent progress in converting solar energy directly into electricity.

Vendryes, G. A. 1977. "Superphénix: A Full-Scale Breeder Reactor," *Scientific American 236:*26–35 (March). An article on the first commercial-scale breeder reactor.

A freeway overpass in San Jose, California, that has been in place, unfinished, for ten years. Construction is just beginning again on the approach ramps. (Copyright © 1978 Barrie Rokeach.)

Chapter 20

Where Do We Go from Here?

Because the environmental problems we have identified in this book are global in scope, they are often difficult to conceive of in concrete terms. To counteract this difficulty, we have created a scenario in which these problems are expressed on a scale that is easier to comprehend. In this imaginary construction, the world's population is reduced proportionately to equal the population of a small town of 1000 people. In this town, based on world distributions, 50 inhabitants would be American and 950 non-American. The 50 Americans would earn about half the total income of the town, would own 15 times the number of possessions that the others could claim, and would use a highly disproportionate share of the electric power, fuel, steel, and general equipment available. Finally, the 50 Americans would produce nearly one-fifth of the town's food supply and eat about 70 percent more than their minimal needs required. Because many of the 950 non-Americans in the town would be hungry most of the time, they would bear ill feelings toward the 50 Americans, who would appear to them to be enormously rich and ridiculously overfed. Of these 950 non-Americans, 200 would suffer at some time from malaria, cholera,

typhus, or malnutrition. None of the 50 Americans would contract any of these diseases, and probably none would ever worry about contracting them.

This construction (based on figures developed by Dr. Henry Lepier) indicates that as Americans we experience a huge share of the world's advantages. However, often over the course of our history the process of achieving this affluent lifestyle has imposed unexpected costs upon us, and at present we are plagued with more unforeseen problems associated with our mode of life than ever before. Despite considerable cleanup efforts, pollution endangers our health, damages property, and degrades the natural environment. And, perhaps worse, pollution impairs many natural processes that are vital to our survival: the recycling of resources; the purification of the atmosphere, land, and waterways; the control of pests; and the regulation of cycles that maintain food production. Simultaneously, the resources on which we have depended for so long are dwindling away, and, as the need to import mineral and fuel resources grows, we are losing a measure of our national independence.

As we have pointed out often in this book, because

we are intimately related to our physical environment, our activities in one area have reverberations in many others. Thus, in our attempts to balance our need for imports by raising more food exports, we are degrading our rich agricultural lands. And now, with resource demands beginning to exceed supply in many areas, we are faced with scarcity-induced inflation, which reduces our ability—particularly among those of us with low incomes—to obtain the food, shelter, and transportation necessary for survival. Not surprisingly, as demands for housing, food, timber, petroleum, transportation, commercial goods, and recreation continue to grow, competing interests come into dramatic conflict over the use of our fixed supply of land—land that can satisfy at most only a few of the demands placed on it.

Throughout these developments, our population and our lifestyle expectations have continued—and appear likely to continue—to escalate. The conflicts that are inevitable as resource supplies diminish will sooner or later force us to choose among undesirable alternatives. And our choices will often be determined by availability rather than quality. Speaking simply, then, our environmental problems are basically matters of supply versus demand. In solving them, we must seek ways of attaining a balance without sacrificing the quality of our environment.

Supply Strategies

The one fact fundamental to any strategy we might develop is that we are living on a planet whose supply of resources is finite. Probably the most important lesson we learned from our journey to the moon and beyond was that the earth is our only home. We will not be establishing major colonies on other planets in the foreseeable future; nor will we be importing resources from other worlds. Hence, the resources on earth are all we have.

Some scientists, economists, and government offi-

cials believe that the concept of limited resources is misleading. They point out that the earth's crust and the oceans contain huge quantities of minerals, albeit in very dilute concentrations. They argue, too, that abundant energy sources are available to us: trillions of barrels of oil are locked up in oil shale, and virtually limitless quantities of fuel exist in the sea. These experts believe that, as economic conditions become more favorable and exploration techniques and mining technology are improved, copious supplies of many valuable resources will continue to be available far into the future. Let us examine the merits of this viewpoint.

It is true that there are vast, untapped reserves of minerals and energy sources that are presently too expensive to exploit. For very low-grade mineral deposits, recovery may never prove feasible unless inexpensive fuel becomes available, an unlikely prospect at best. In fact, all sources of energy are expected to grow increasingly more expensive. For example, as we have seen, even though solar energy is free, the cost of building a solar heat system is high. Granted, as the price of fossil fuel rises, solar energy is becoming more economically competitive. But the question remains as to who will be able to afford either alternative: fossil fuels or solar collection equipment. The minimum cost of even the cheapest fuel could conceivably outstrip the ability of those of us with low and average incomes to pay for it.

We have long depended on technology in our efforts to eliminate disease and improve our standard of living. Many Americans possess an almost blind faith that, like the cavalry, technology will come to our rescue. But the possibility that we can rely forever on new technologies to increase resource availability is questionable. Before we get our hopes too high, we should consider several limiting aspects of technological research and development.

To formulate solutions, we must first understand the problems. Most environmental problems, however, are vastly more complex than they initially appear. Hence, with respect to a particular problem, scientists and engineers may not possess sufficient understanding to focus their research efforts appropriately. For example, research on alternative energy sources is proceeding in a multitude of directions, basically because researchers do not yet know which of their efforts will be most fruitful. Furthermore, some technological problems are more difficult to solve than others, assuming that solutions do exist at all. Developing a commercially available nuclear fusion reactor, for instance, poses greater scientific challenges than did going to the moon.

Advances in technology usually require complex efforts involving considerable periods of time and many highly trained personnel. In addition, applying new technological principles often requires rare and exotic—and therefore expensive—materials. The example of nuclear fission power plants illustrates the huge investments involved in implementing technological advances. Although direct costs in a consumer's electrical bill are competitive with coal-fired generating costs, such a comparison is simplistic and misleading. Actually, for twenty years before the first nuclear power plant went into operation, consumers had been spending billions of dollars through federal support of research by governmental agencies, universities, and private industry. Federal funds have also underwritten costs of construction, fuel processing (including enrichment), health-hazard evaluation, and hazard insurance. Had taxpayers not subsidized these efforts, nuclear power would not be competitive with fossil fuel power today. In fact, without these funds, nuclear electricity-generating plants might still be on the drawing board.

Usually, the more complex the technology, the longer it takes to gear up to full-scale operation. The technology for power generation by nuclear fission reactors has been available for some twenty-five years, but nuclear power has not yet lived up to early expectations that it would become the major source of electrical power in the United States. Hence, even after modern technology brings a technique to the pilot or prototype stage, serious doubts remain as to whether full-scale implementation will occur in time to solve a supply problem.

Another consideration is that, as we have learned from experience, every new technological advance creates side-effects that detract substantially from the benefits it bestows. For example, any technique we might develop for using low-grade deposits of minerals is certain to involve the consumption of substantial amounts of fuel and other resources, such as water, and to generate large quantities of wastes. The existence of more waste products could, in turn, degrade other valuable resources—for example, by contaminating surface or groundwater, disrupting the landscape, or destroying wildlife habitats.

These economic and technological limitations, then, shed doubt on the belief that the earth still holds essentially unlimited stores of inexpensive resources. But by questioning this position we are not implying that resource exploration should be discontinued or technological development halted. Rather, we are suggesting that, as a society, we change the direction of our technological development in order to make better use of available resources. For example, in addition to increasing our oil supply by exploring for new deposits, we should also work to improve the technology for secondary and tertiary recovery of petroleum from producing oil fields. And, wherever possible, we should enhance resource supplies by improving recycling technology, particularly for the metals whose reserves are rapidly being depleted. We have seen that using scrap metals in place of virgin ores conserves energy, and that organic wastes can be converted into fuel or processed and fed to livestock. And we know that recycling solid wastes saves energy and reduces pollution. We need to intensify such efforts to conserve energy—for example, by increasing the energy efficiencies of appliances and machines. Where alternative technologies are available, we may be wiser in the long run to choose a more expensive technology than one that is more energy dependent. Hence, we are not proposing that our society scale down its dependence on technology in the future, but rather that it develop technology that is energy efficient, well suited to the real needs of society, and compatible with nature.

The latter consideration—designing technology that operates within ecosystem constraints—bestows the added benefit of reducing costs. For example, as we have noted, swamps, forests, and agricultural land can be used in some areas to provide tertiary treatment of wastewater effluents. By utilizing this natural cleansing technique, we could reduce our dependence on energy-intensive wastewater treatment technology, which is likely to become more expensive in the future. As a side-effect that might well prove to be of central importance, our use of natural systems in this way would add to the intrinsic value of these biomes, thus giving us more incentive to preserve them and the wildlife within them.

In summary, then, we seriously question the point of view that holds that inexpensive, abundant resources will be available in the future. There is no doubt that new supplies will be discovered and exploited, but at what costs? The prudent path is to make better use of our resources through reuse, recycling, and the recovery of the valuable materials in our wastes. Recycling, however, will not solve all our problems. Some resources, such as fossil fuels, cannot be recycled at all. And recycling cannot meet the demands of an expanding population, particularly if the per capita demand increases simultaneously. For this reason, many people believe that we must seek to solve our problems not only by increasing the supply of resources but also by reducing our consumption. We turn now to an exploration of the demand side of the problem.

Demand Strategies

Demand can be considered from two perspectives: demand by the total population and per capita demand. From the standpoint of population size, we have seen that the population growth rate has subsided greatly in the United States and that our population may become stationary within fifty to sixty years. This projection portends well. Because our population growth rate has declined so markedly, we have a better opportunity to cope with our problems of supply and demand.

However, per capita consumption in the United States is quite another matter. Despite increasing costs and complaints, the level of affluence of the average American continues to rise, and individual demands for consumer products grow unabated. Many American families now have air-conditioned homes equipped with at least two bathrooms, two television sets, and two cars. Many people own recreation vehicles such as power boats, campers, and motorcycles. Planned obsolesence has become an accepted way of life for American industry. Our society is infatuated with disposable items. On the whole, we are so affluent that we believe we can afford to use materials once or twice and then throw them away and buy more.

Trends in automobile use are a major index of the level of affluence in the United States, and they illustrate the conflicts that arise between affluence and conservation. For example, a 1977 survey of American households showed that, as a society, we are so affluent that large rises in gasoline prices will not faze most motorists. Thus, the idea that voluntary fuel conservation might significantly slow fuel use is optimistic at best. The fuel conservation features mandated by Congress and being incorporated in new lines by auto makers are intended to force conservation through improved design. Yet the full benefits of this fuel conservation program will not be realized until many more Americans change their consumptive habits. Better fuel economy is offset by our habit of buying cars—even small cars—equipped with energy-consuming extras such as air conditioning and power steering. Furthermore, we now drive our cars greater distances than we did before the Arab oil embargo of 1973.

In recent years, some people have indeed voluntarily reduced their level of fuel consumption. They are riding bicycles and small motorcycles for transportation, and placing greater emphasis on self-propelled sports such as jogging, backpacking, canoeing, and downhill and cross-country skiing. There is no doubt that such activities consume fewer resources and do less environmental damage than powerboats and campers. But, like those who require "extras" on small cars, many of these people want to buy the best that they can afford. For example, instead of buying three-speed bikes, more people purchase ten-speed bikes with components built from special metal alloys.

A handful of people have dramatically changed their lifestyles and consumer habits by leaving the comforts of home and taking up a pioneer existence in such remote locales as Alaska or northern Maine. Few of us, however, really have any desire to undertake the physical rigors and hardships of the "good old days." We appreciate the advantages that our society can provide in making our lives easier and, presumably, more enjoyable. Further, few people have the knowledge and skills required to grow most of their own food, to build and maintain their own houses, to sew and mend their own clothes, and simply to survive on their own. "Making a living" without modern conveniences would be extremely difficult for most of us.

What You Can Do

In the end, the responsibility for solving environmental problems rests with each individual. What can you do? First, you must decide if you want to do anything at all. Are the problems discussed in this book real to you? Do they affect you in any way? Are you concerned, and do you feel a responsibility to your fellow inhabitants of the earth? If you answer no to any one of these questions, you probably will not contribute to solving the problems we have discussed in this book. Choosing whether or not to act, then, is the first step.

Many Americans are concerned about the problems that plague the planet, but still believe that solutions—and unlimited resources—are readily available. If you take this position, then you too will probably do little to solve our problems. You will probably hold the opinion that, after all, the government and private industry will soon have these problems under control.

Many Americans, in contrast, are confused, pessimistic, and even fatalistic about prospects for the human race. They feel that it is useless to try to do anything, that the problems are so numerous and so complex, and the potential outcomes so threatening, that positive action to stave off catastrophe is futile. Having never lived through a prolonged economic depression, many of us have known only prosperity and good times. Hence, many young Americans lack the experience and the confidence to face potential changes. Even many of our political and religious leaders shy away from making critical decisions and providing vital leadership. Why should we be any different? However, as many of us realize from trying to solve our own personal problems, running away or ignoring critical issues seldom solves them. And refusing to act can lead to despair, while searching for solutions focuses our energies in a way that could pay off. It is in our own self-interest to make a realistic, if not optimistic, attempt at identifying and contributing to the solutions.

We can take heart that some efforts have actually paid off. Recall that such laws as the Clean Air Act, Clean Water Act, and Endangered Species Act are showing positive results. Emissions of some water pollutants and such air pollutants as sulfur dioxide and particulates are increasingly coming under control. Some endangered species are making a comeback. Several chlorinated hydrocarbon pesticides have been banned from general use in the United States. More energy-efficient cars are now being produced on American assembly lines. Population growth is beginning to slow down. Much is left to do as long as population and per capita consumption continue to rise, but we know that significant progress is possible in many areas.

Many people believe that as individuals we cannot do much about the important problems facing us today. But each of us can make important contributions. After all, the significant drop in birth rates in the United States in the 1970s was the result not of government action, but of the fact that individual couples decided to limit the size of their families. Each individual can have an impact on the problems in many ways—for instance, by changing his or her lifestyle and by making responsible decisions on the job. We can all decide to drive smaller cars that get better mileage and emit fewer pollutants, and we can participate in car pools to save gasoline and wear and tear on our autos. If we work for large companies, we can encourage them to purchase minibuses and make them available for car pools run by company employees. We can improve energy conservation in our homes by turning down thermostats, adding insulation, and using air conditioners only when they are really needed. Where space is available, we can grow our own gardens, thus gaining not only fresh vegetables but a sense of accomplishment and self-sufficiency as well. We can conserve water in a hundred different ways—for example, by fixing leaky faucets and taking showers rather than baths. And we can participate in recycling projects to conserve metals and reduce waste.

Another basic way to reduce consumption is to learn to make do with what we have. Just a few years ago, most people who traded in their old cars for new ones did so every two to three years. Now people are taking better care of their cars and keeping them longer. Few of us really need a new wardrobe each year. Changing styles in cars and fashions often lead us to buy more of what we really don't need. But the philosophy of "wear it out, make it do, use it up, and do without" is more in keeping with our resource situation and economic conditions than the idea of "when it goes out of style, throw it out."

Some of us may wish to communicate our concerns to our neighbors and acquaintances. If they are informed about our resource problems and the pos-sible approaches to solutions, many of them may be motivated to change their lifestyles, thereby adding to the social pressure for change. Since effective action on an individual level is often impossible, concerned citizens must form their own action groups or seek out the assistance of such major national organizations as the Audubon Society, National Wildlife Federation, and Sierra Club.

Many necessary changes would undoubtedly come about more quickly if the government took a greater leadership role. But government officials have to be pressured by citizens specifying their needs and expectations. Unfortunately, only a handful of the more than one-thousand registered lobbyist groups operating in our nation's capital have the interests of the average citizen at heart. The drafting and passage of responsible legislation, therefore, requires that we create strong grass-roots initiatives. To prevent powerful interests from misusing technology, the citizenry of a society such as ours must be well informed and active in the decision-making processes.

Students have the opportunity to contribute through their career choices. Some may choose to enter business administration, where they can help guide companies in their policy decisions regarding environmental quality and resource conservation. Some may be planning a career in the political arena, where reforms and changes in priorities are clearly needed. Others may enter research and development in order to contribute to technical solutions. Still others will become teachers, seizing the opportunity to expose minds of all ages to the value of a quality environment. However, no matter what career a student ultimately chooses, he or she can encourage conservation and efficiency in the workplace.

Above all, if each of us is to contribute to solving our critical environmental problems, we must keep informed. Each day the media bombard us with information. To sort it all out and resolve the contradictions, we may have to turn to public service organizations for help. In all situations, we must evaluate change in terms of basic ecological principles. Our society has a critical need for more watch-dogs alert to ill-considered decisions and for people skilled in interpreting and communicating the ramifications of technological developments.

Epilogue

If we do not delay and if we act wisely, we may have little to fear. By making major changes in our living patterns, we could substantially cut down our consumption of necessities and eliminate many luxuries altogether. We may be a little cooler in winter, but there will always be warm sweaters. We can still take vacations, though we may not be able to travel as far.

An interesting and significant sidelight of the Arab oil embargo was that Americans rediscovered each other. People tended to stay at home, spending more time with their families. They spent their leisure time doing such simple things as gardening, picnicking, hiking, and bicycling. Through these experiences, people acquired a better appreciation of nature and a simpler lifestyle. Many believed that this change in living patterns greatly enhanced the quality of their lives. Hence, they learned that a reduction in materialism need not mean a reduction in the quality of life. Too many of us have too often and for too long ignored and escaped from our surroundings—from our immediate families and from the communities in which we live. We now have the opportunity to renew family bonds and form closer ties with our neighbors.

We must make decisions. We must take action. What will you do to contribute to promoting ecological harmony on our small planet?

Selected Readings

Birch, C. 1975. *Confronting the Future.* Harmondsworth, England: Penguin. An optimistic look at the human predicament.

Brian, H., and J. D. Hallett. 1977. *Environment and Society: An Introductory Analysis.* Cambridge, Massachusetts: M.I.T. Press. A discussion of the politics and economics of environmental decision making from a social science perspective.

Brown, L. R. 1977. "Redefining National Security," *Worldwatch Paper 14.* Washington, D. C.: Worldwatch Institute. A discussion of our national security in the context of world resource supplies and allocations.

———. 1978. *The Twenty-Ninth Day: Accommodating Human Needs and Numbers to the Earth's Resources.* New York: W. W. Norton. An examination of world population problems and strategies for stabilizing world population.

Commoner, B. 1976. *The Poverty of Power.* New York: Bantam Books. An examination of the environmental and economic issues associated with the consumption of fossil fuel reserves and the changes we must make in our way of life.

Ehrlich, P. R., and A. H. Ehrlich. 1974. *The End of Affluence.* New York: Ballantine Books. A consideration of the crisis we will face in the next few decades and of how to cope with the changes that are inevitable.

Laszlo, E. 1978. *Goals for Mankind.* Bergenfield, New Jersey: New American Library. A summary of every nation's goals and the ways in which they must work together for the good of all humanity.

Peterson, R. W. 1979. "Impacts of Technology," *American Scientist* 67:28–31 (January-February). A cautious look at the technological future.

Ridker, R. G., and E. W. Cecelski. 1979. "Resources, Environment, and Population: The Nature of Future Limits," *Population Bulletin 34* (August). Washington, D.C.: Population Reference Bureau. Examines the balance between world supplies of resources and the demands presented by population and economic growth.

Schumacher, E. F. 1973. *Small Is Beautiful: Economics as If People Mattered.* New York: Harper & Row. A critical look at the economics of modern society. Many of the ideas are applicable to less developed countries.

Wildavsky, A. 1979. "No Risk Is the Highest Risk of All," *American Scientist 67:*32–37 (January–February). Warns against an overly cautious attitude toward new technologies.

Glossary

Abiotic Pertaining to nonliving components of the environment.

Abortion Extraction or expulsion of a fetus from the womb before the twentieth week of pregnancy.

Absolute Zero The theoretical temperature at which all molecular motion ceases (0° Kelvin, −273.15° Celsius, −459.6° Fahrenheit).

Acclimatization Adjustment of an organism to environmental change.

Acid Rain rain that becomes acidic after falling through and dissolving air pollutants, primarily sulfur dioxide.

Activated Sludge Treatment A sewage treatment process in which a portion of the decomposer bacteria present in the waste is recycled to the beginning of the process, thus providing the system with "acclimatized" organisms.

Active Solar Collectors Devices that concentrate energy from the sun.

Acute Exposure A single exposure to radiation or toxic substances that lasts seconds, minutes, or hours.

Adaptation A genetically controlled characteristic that enhances an organism's chances to survive and reproduce in the environment in which it resides.

Aerobic Decomposers Microorganisms that require oxygen to break down organic wastes into carbon dioxide and water.

Aerosols Solid and liquid particles suspended in the atmosphere.

Aestivation A condition similar to hibernation induced in some desert animals during periods of extreme heat and dryness.

Age Structure The distribution of age classes in a population.

A-horizon The upper layer of soil (topsoil), which contains humus and most plant roots. Considerable amounts of soluble soil constituents are leached from this zone.

Air Mass A volume of air covering thousands of square kilometers that is relatively uniform in temperature and water vapor content.

Air Pollution Episode A period in which air pollution concentrations reach levels that are hazardous to human health.

Air Pressure The weight of the atmosphere over a unit area of the earth's surface.

Air Stability The relative buoyancy of air parcels in the atmosphere.

Albedo The fraction of received radiation that is reflected by a surface.

Alpha Particle A particle emitted by certain radioactive materials; has the weakest penetrating power of the three major forms of radiation.

Amino Acids The basic building blocks of proteins.

Anaerobic Decomposers Microorganisms that, in the absence of oxygen, break down certain organic wastes into methane, water, and carbon dioxide.

Annual Growth Rate The crude birth rate minus the crude death rate.

Antagonism An interaction in which the total effect is less than the sum of the effects taken independently.

Anthracite Coal A hard coal that is relatively high in heat value and low in volatile matter (sulfur dioxide) and ash content.

Aquaculture The deliberate growing and harvesting of fish and shellfish, usually in some type of confinement.

Aquifer A highly permeable layer of rock or soil that holds or can transmit groundwater.

Area Landfill A natural topographic depression such as a canyon or valley that serves as a sanitary landfill.

Area Strip Mining A mining method used to extract near-surface rock and mineral deposits (usually coal) in relatively flat terrain. Heavy earth-moving equipment removes the overburden and target material from a series of trenches adjacent to each other.

Artesian Well A free-flowing well tapped into a pressurized aquifer.

Atmospheric Nitrogen Fixation A process in which the electrical energy in lightning changes nitrogen gas in the atmosphere into nitrate.

Auger Mining A mining method in which huge drills are used to extract coal from near-surface seams.

Bedrock Solid rock underlying soil and unconsolidated sediment.

Belt of Soil Moisture The top layer of soil from which plant roots obtain water.

Beta Particle A particle with moderate penetrating power that is emitted from certain radioactive materials.

B-horizon The soil layer beneath the A-horizon, consisting of weathered material and minerals leached from the A-horizon.

Biochemical Oxygen Demand (BOD) The amount of oxygen required by organisms to decompose the organic wastes in a given volume of water.

Biofuel Recoverable or harvestable plant or animal materials that can be used to produce heat.

Biological Control The regulation of a pest population through the use of natural predators, parasites, or disease-causing bacteria or viruses.

Biological Nitrogen Fixation The process by which blue-green algae and some bacteria convert atmospheric nitrogen gas into ammonia.

Biomass The total weight, or mass, of organisms in a given area.

Biome A major type of terrestrial ecosystem, such as the Arctic tundra or the Great Plains grasslands, that usually covers an extensive area.

Biotic Pertaining to the living components (organisms) of the environment.

Bituminous Coal A soft coal of moderate heat value, that is typically high in volatile material (sulfur dioxide) and moderate amounts of ash.

Breeder Reactor A type of nuclear reactor that produces slightly more fissionable material than it consumes.

British Thermal Unit (BTU) The quantity of heat needed to raise the temperature of 1 pound of water 1 degree Fahrenheit.

Calorie The quantity of heat needed to raise the temperature of 1 gram of water 1 degree Celsius.

Carcinogen A chemical substance or physical agent (such as radiation) capable of causing cancer.

Carnivore An animal that eats animals.

Carrying Capacity The largest population that a given area can sustain indefinitely.

Catalytic Converter An automobile exhaust control device that chemically changes hydrocarbons and carbon monoxide in the exhaust into carbon dioxide and water vapor.

Cell The basic structural unit of all organisms.

Chlorinated Hydrocarbons A class of chlorine containing chemicals, some of which are toxic, carcinogenic, and may bioaccumulate.

Chlorination The addition of chlorine to drinking water or treated sewage treatment plant effluent for the purpose of disinfection.

Chloroorganic Compounds Organic chemicals that contain chlorine as part of their molecular structure.

Chlorosis A yellowing of plant leaves due to chlorophyll loss.

C-horizon The soil layer beneath the B-horizon consisting of the broken or partially decomposed underlying bedrock.

Chronic Exposure Continuous or recurring exposure to radiation or toxic substances that lasts for days, months, or even years.

Cilia Tiny undulating hairs lining the respiratory tract that protect the respiratory system from inhaled particulates.

Clarification A purification process whereby suspended materials in water are allowed to settle out in quiescent basins. Chemicals are often added to enhance the process.

Climate The weather conditions of an area averaged over a period of time. Also, extremes in weather behavior.

Climatic Destabilization Climatic change that involves an increase in the frequency of extremes in weather behavior (record temperatures and precipitation, for example).

Climax Stage The end-point in an ecological succession series. The organisms present in the climax stage are referred to as climax organisms.

Cloud Seeding The introduction of tiny condensation nuclei, such as silver iodide, into clouds for the purpose of increasing rainfall.

Coalescence Process A process of precipitation formation that takes place in clouds whose temperature is above the freezing point of water.

Coal Gasification A process whereby coal is heated in the absence of oxygen and in the presence of steam to produce the combustible gases methane, carbon monoxide, and hydrogen.

Coal Liquefication A process whereby liquid fuels are obtained from coal by heating coal in the absence of oxygen.

Coevolution The process by which two interacting populations—for example, a parasite and a host—adapt to each other to their mutual benefit.

Cold Front The leading edge of an air mass that is displacing warmer air.

Coliform Bacteria A type of bacteria that resides in the human intestine whose presence in water is used to indicate whether water is potentially contaminated with disease organisms.

Combined Sewer A sewer system in which both runoff and sanitary wastes are transported by one large pipe to a sewage treatment plant.

Competition Interactions of organisms to secure a resource in short supply.

Competitive Exclusion Principle The generalization that two species having the same resource requirements cannot coexist indefinitely in the same area.

Composting The creation of a fertilizer or soil conditioner through the aerobic breakdown of cellulose waste, such as leaves, grass clippings, and newspaper.

Condensation The process whereby water molecules move from the vapor state to the liquid state, for example, in dew formation.

Condensation Nuclei Tiny solid or liquid particles that initiate deposition or condensation.

Conduction The transfer of heat from one object to another through direct contact.

Confusion Effect A flurry of activity produced by a group of prey that disorients a predator.

Consumers Organisms that use other organisms as a food source.

Continental Drift The extremely slow movement of the earths' crustal plates.

Contour Farming Plowing and planting across a slope rather than with it.

Contour Strip Mining A mining method used to extract near-surface rock and mineral deposits (usually coal) in hilly or mountainous terrain. Earth-moving equipment removes overburden and deposit from a series of benches cut along contours of elevation.

Convection Currents Movement of air resulting in a redistribution of heat.

Conversion Chemical or biochemical transformation of refuse into useful products such as fertilizer, fuels, and industrial chemicals.

Criteria Air Pollutants Those air pollutants for which the Environmental Protection Agency has established national standards for ambient air.

Critical Population Size The minimum number of individuals needed to ensure survival of a population.

Crude Birth Rate The number of births per one-thousand people in a population per year.

Crude Death Rate The number of deaths per one-thousand people in a population per year.

Cultural Eutrophication A hastening of the natural nutrient enrichment of surface waters by human activities such as agriculture and the discharge of nutrient-rich wastes.

Cyclone Collector A device that removes particulates from an industrial effluent air stream by inducing gravitational settling.

Deciduous Forest A forest composed of trees that lose their leaves during part of the year.

Decomposers Organisms such as bacteria, mushrooms, and maggots that feed on the remains of plants and animals.

Deep-Well Fluid Injection A method of disposing of liquid wastes by pumping them under pressure into subsurface cavities and pore spaces.

Delta A body of sediment deposited in an ocean or lake at the mouth of a river.

Demographic Transition A pattern of change in birth rates and death rates related to industrialization and common in developed countries in which a decline in the death rate is followed by a decline in the birth rate.

Denitrification The conversion of nitrates to nitrogen gas carried out by a specialized group of bacteria.

Density-Dependent Factors Interactions such as predation, parasitism, and competition whose influence on population size varies with the density of individuals in a habitat.

Deposition The process whereby water molecules in the vapor state directly form solid ice crystals, for example, in the formation of hoarfrost.

Desalination The process whereby the salt content of water is greatly reduced so that saline water can be used for human consumption or irrigation.

Desertification The degradation of terrestrial ecosystems through deforestation, overgrazing, and poor soil and irrigation management.

Detritus Freshly dead or partially decomposed remains of plants and animals.

Detritus Food Webs A food web based on decomposers feeding on detritus.

Discharge The volume of water flowing past a fixed point in a river, stream, or pipe per unit of time.

Dissolved Oxygen The oxygen dissolved in water. The amount is usually expressed in parts per million (ppm).

Distillation The process whereby a liquid is evaporated and recondensed as a purified liquid.

Doubling Time The length of time needed by a population to double in size.

Drainage Basin A geographical region drained by a river or stream.

Dredge Mining A process for mining streambed sands and gravel and placer deposits through the use of chain buckets and drag lines.

Dry Cooling Tower A mechanical device that cools hot water effluent by transferring heat to the air without relying on the evaporation of water.

Dry Steam Field A region in which the steam that is emitted from the earth's interior contains little condensed water.

Dust Dome A dome-shaped accumulation of particulates that often forms over urban-industrial complexes.

Dust Plume A dust dome elongated downwind from a city.

Earthquake The shaking of the ground caused by the sudden release of energy that has accumulated as stress in bedrock.

Ecological Efficiency The percentage of energy transferred from one trophic level to the next.

Ecological Equivalents Animals that exhibit similar

adaptations to comparable but geographically distinct biomes.

Ecological Succession A sequence of changes in which one type of ecosystem replaces another in a given area until an ecosystem is established that is best adapted to that environment.

Ecosystem A functional unit of the environment that includes all organisms and physical features within a given area.

Ecotone A transition zone between two or more distinct ecosystems.

Edge Effect The tendency toward increased species diversity in ecotones.

Electromagnetic Radiation Energy that can travel through a vacuum in the form of waves, for example, light.

Electromagnetic Spectrum The various forms of radiational energy arranged by wavelength and/or frequency.

Electrostatic Precipitator A device that removes particulates from an industrial effluent air stream by inducing an electric charge.

Endangered Species A species that is in immediate danger of extinction.

Energy Efficiency The fraction of energy that is channeled into a useful process.

Environmental Stress Changes in the environment that have adverse affects on the organisms involved.

Epiliminon The warm, relatively less dense top layer of water in a stratified lake.

Epiphytes Plants that grow on other plants, usually trees, but are not parasitic.

Erosion The breakdown of materials of the earth's crust by various physical and chemical processes and the transporting of the particles by wind, moving waer, or moving ice.

Essential Amino Acids Amino acids necessary to an organism's survival that cannot be synthesized by the organism, and hence they must be included in the diet.

Estuary A coastal ecosystem where fresh water and salt water meet.

Euphotic Zone The surface layer of a body of water to which photosynthesis is confined.

Eutrophication A natural process of nutrient enrichment whereby lakes gradually become more productive.

Eutrophic Lake A lake with a high rate of nutrient cycling and thus a high level of biological productivity.

Evaporation The process whereby liquid is transformed into the vapor state.

Evaporative Cooling Tower A mechanical device that relies on the evaporation of water for the cooling of heated water effluents.

Exponential (Geometric) Growth An increase by a constant percentage of the whole in a constant time period.

Extinction The process by which a species ceases to exist.

Fall Turnover A mixing process that occurs in autumn in a stratified lake whereby the surface water layer mixes with the bottom water layer.

Family Planning A voluntary program of fertility control used by a couple to achieve the family size of their choice.

Fault A fracture in a rock mass along which the rock has been displaced.

Floc Clumps of suspended materials such as bacteria that settle in water.

Flood Plain Land adjacent to a river channel that is covered by water when the river overflows its banks.

Fluorosis Poisoning resulting from the ingestion of fluorides.

Folding The bending of layers of bedrock by stresses within the earth's crust.

Food Chain A sequence of organisms, such as green plants, herbivores, and carnivores, through which energy and materials move within an ecosystem.

Food Chain Accumulation The process whereby certain chemicals become more highly concentrated in the bodies of organisms at the upper levels of a food chain.

Food Web A network of interconnected food chains.

Fossil Fuels The remains of ancient plant and animal life in the form of coal, oil, and natural gas.

Frequency The number of oscillations a wave makes in one second.

Fungicide A pesticide directed at killing fungi such as rust and molds.

Gamma Ray A highly energetic form of radiation emitted by certain radioactive materials that has a strong penetrating power.

Gasohol Fuel consisting of a gasoline-alcohol mixture.

Gene Pool The total amount of genetic information of all the individuals in a population.

Genetic Make-up The genetic information possessed by an individual.

Geothermal Energy Heat energy contained in the earth's interior.

Global-Scale Circulation The largest scale of atmospheric motion, which includes the wind systems that circle the globe.

Grazing Food Web A food web based on herbivores feeding on plants.

Greenhouse Effect The absorption and reradiation of terrestrial infrared radiation by atmospheric water vapor, carbon dioxide, and ozone.

Green Revolution A popular term for the dramatic increases in crop yields per hectare that sometimes accompany the introduction of new crop varieties and the improvement of soil, water, and pest management.

Ground Subsidence The sinking of the earth's surface that occurs when subsurface mineral deposits or fluids are withdrawn.

Groundwater Reservoir Extractable underground fresh water.

Growth Control Ordinance Regulations set by communities to limit their growth.

Half-Life The amount of time required for the radioactivity emanating from a particular radioactive substance to be reduced by one-half.

Heat of Fusion The quantity of heat required to change 1 gram of a substance from the solid state to the liquid state.

Heat of Vaporization The quantity of heat required to change 1 gram of a substance from the liquid state to the vapor state at the boiling point of the substance.

Herbicide A pesticide directed at killing weeds.

Herbivore An animal that eats plants.

Hibernation A state of dormancy exhibited by some animals during cold periods.

High-Level Radioactive Wastes Radioactive wastes that must be isolated to prevent radiation damage to living things.

Humus Organic materials that are slow to decay.

Hydraulic Mining A surface mining method that employs powerful jets of water to wash away overburden and reach rock and mineral deposits.

Hydrocarbon A class of chemical compounds, usually organic in origin, containing only carbon and hydrogen.

Hydroelectric Dam A structure on a river that converts the energy of falling water to electricity.

Hydrologic Budget Measurements of the relative amounts of water moving through the various subcycles of the hydrologic cycle.

Hydrologic Cycle The ceaseless flow of water from one reservoir to another.

Hygroscopic Having a special affinity or attraction for water molecules.

Hypolimnion The cold, relatively dense bottom layer of water in a stratified lake.

Ice-Crystal Process The process whereby snowflakes grow at the expense of supercooled water droplets in clouds.

Identified Resources Rock, mineral, or fuel deposits that have been assessed as to grade and quantity by actual field measurements.

Igneous Activity Flow of hot, molten rock.

Igneous Rock A rock formed by the crystallization and solidification of magma (or lava), either within the crust or on the earth's surface.

Infiltration Seepage of surface water into soil and rock layers.

Infrared Radiation Invisible radiation having wavelengths longer than the red end of the visible spectrum and felt as heat.

Inorganic Materials based on elements other than carbon, including water, oxygen, and minerals.

Inorganic Fertilizer Plant nutrients derived from mineral deposits (phosphate or potash) or from natural gas (ammonia).

Insecticide A chemical directed at killing insects.

Insight Learning The ability to respond correctly the first time to a new situation.

Intertidal Zone The zone on a beach that lies between high and low tide.

Integrated Pest Management A combination of pest control methods that is best suited for a particular pest and set of environmental and economic conditions.

Interspecific Competition Competition for a resource between populations of two or more species that adversely affects their well-being.

Intraspecific Competition Competition for a resource among members of the same species that adversely affects their well-being.

Intrauterine Device (IUD) A small metal or plastic device that is inserted into the uterus to prevent conception.

Isotopes Variations of the atomic mass within a particular element. Isotopes of a particular element have similar chemical properties but widely varying nuclear properties.

Juvenile Hormone A chemical whose relative concentration initiates the various stages of an insect's life cycle.

Landscape Evolution The gradual development of the landscape in response to weathering, erosion, and stresses acting from the earth's interior.

Landslide The gradual or abrupt downslope movement of rock, soil, mud, or a mixture of these materials.

Latent Heat Heat supplied or released during the phase changes of water.

Latent Heat Transfer The transport of heat energy from place to place as the result of changes in the phases of water.

Laterite A leached soil, high in iron and aluminum, that is found in many tropical areas.

Lava Hot, molten rock that is extruded from the crust onto the earth's surface through fissures in the ground or from volcanoes.

Law of Limiting Factors The principle that for each physical factor in the environment there is a minimum limit and a maximum limit beyond which no members of a given species can survive.

LD$_{50}$ The quantity of a substance that will kill 50 percent of a test population in a single dose.

Leachate Water that has passed through materials such as soil or garbage and that usually contains high concentrations of dissolved substances.

Leaching The dissolving, transporting, and redepositing of materials by water seeping downward.

Lignite A brownish coal of relatively low heating value and high ash content in which vegetative matter has been altered more than in peat but less than in bituminous coal.

Limiting Factor Any component of the environment that limits the ability of an organism to grow or reproduce.

Linear (Arithmetic) Growth An increase by a constant amount in a constant time period.

Liquefied Natural Gas Natural gas that has been liquefied by being cooled to $-162°C$ ($-259°F$) and thus reduced to $\frac{1}{600}$ its normal volume.

Longshore Current A component of water motion parallel to the shoreline, the result of coastal wave action that supplies sand to beaches.

Macrophages Specialized, free-living cells in the air sacs of the lung that engulf and digest foreign particles.

Magma Hot, molten rock material within the earth.

Marine Habitat A habitat associated with the oceans.

Meat Extender A substance, usually made of plant materials such as soybean meal, that is added to meat to increase its bulk.

Meat Substitute A substance, often spun soy protein, that is processed to simulate meat and sold as a source of protein to replace animal protein.

Meltdown The melting of a nuclear reactor's core through its container, caused by overheating.

Mesoscale Systems Weather systems that are relatively small and localized, for example, sea breezes, thunderstorms, and tornadoes.

Mesosphere The subdivision of the atmosphere situated between the stratosphere and an altitude of about 80 kilometers (50 miles).

Metamorphic Rock A rock formed when a preexisting rock is subjected to high temperatures, confining pressures, and chemically active fluids deep within the earth's crust.

Microscale Circulation The smallest scale of atmospheric motion; which includes circulation patterns within meters of the ground and vegetation.

Microwaves A form of elecromagnetic energy of relatively long wavelength that falls between infrared waves and radio waves in the electromagnetic spectrum.

Minamata Disease A disease of the central nervous system caused by mercury poisoning.

Mineral A solid characterized by an orderly internal arrangement of atoms, fixed chemical composition, and definite physical properties.

Minimum Tillage A farming procedure in which plowing, planting, fertilizer application, and weed control are combined in order that the number of necessary operations may be minimized.

Monoculture The cultivation of a single crop, such as corn or wheat, to the exclusion of other crops on the land.

Mortality The death rate.

Municipal Incineration The high-temperature burning of combustible solid waste and the melting of some noncombustibles in enclosed furnaces.

Mutagen A chemical substance or physical agent (such as radiation) capable of producing a change in an individual's genetic makeup.

Natality The rate of production of new individuals by birth, hatching, or germination.

National Forests Public lands set aside primarily to foster wise forest management and thereby insure sustained timber yield and watershed protection.

National Parks Lands regulated for the preservation of some natural feature as well as for the educational and recreational needs of the general public.

National Resource Lands Public lands managed by the Bureau of Land Management and used primarily for grazing and mining.

Natural Selection The differential survival and reproduction of individuals in nature leading to an increase in the frequency of some characteristics and a decline in the frequency of others.

Neritic Zone The region of the ocean that extends from the low tide line to the outer edge of the continental shelf.

Net Energy Efficiency The relative amount of energy available from a process or a resource after energy losses and outside energy inputs have been subtracted.

Neutron A fundamental particle found in the nuclei of atoms.

Nonpoint Sources Sources of pollutants in the landscape, for example, agricultural runoff.

Nuclear Chain Reaction A self-perpetuating sequence of fission reactions within the nuclei of certain atoms such as uranium-235.

Nuclear Fission The fragmenting of the nucleus of an element resulting in the release of energy, several neutrons, and the formation two new radioactive nuclei.

Nuclear Fusion A process by which the atomic nuclei of light elements are fused together to form heavier elements thereby releasing energy.

Nuclei The extremely small dense centers of atoms.

Ocean Thermal Energy Conversion A process that produces energy based on temperature differences in ocean waters.

Oceanic Zone The region of open ocean that lies beyond the continental shelf.

Oil Shale Shale deposits containing organic materials that can produce oil when the shale is heated to high temperatures.

Oligotrophic Lake A lake with a low rate of nutrient cycling and a low level of biological productivity.

Omnivore An animal that eats plants and other animals.

Onchorerciasis A waterborne disease, common in less developed countries, caused by parasitic worms spread by black flies, that can result in blindness.

Open Pit Mining A surface mining method that involves the removal of overburden from a large area so that rock and mineral deposits can be excavated to considerable depth.

Organic Materials produced by or derived from organisms or synthetic chemicals containing carbon.

Organic Fertilizer Plant nutrients derived from crop

and animal residues such as manure.

Organic Matter Materials usually of plant or animal origin.

Orographic Rainfall Rainfall triggered by topographical features such as mountain ranges.

Overfishing The harvesting of more fish than are produced, resulting in a decline and perhaps a depletion of a fish population.

Oxygen Sag Curve A graphic representation of the characteristic decrease and subsequent increase in dissolved oxygen levels in a river downstream from a discharge of organic wastes.

Ozone Shield A layer of ozone in the stratosphere whose formation filters out potentially lethal intensities of ultraviolet radiation. Thus, organisms on earth are shielded by ozone.

Parasitism An interaction in which one organism (the parasite) obtains energy and nutrients by living within or upon another organism (the host).

Parts Per Million (ppm) The unit measure of the concentration of a component substance; for example, a 1 ppm concentration of arsenic is 1 part of arsenic to 999,999 parts of other material.

Passive Solar Collectors Design features incorporated into a building to minimize energy utilization—for example, window placement minimizing solar capture in summer and maximizing capture in winter.

Permafrost A permanently frozen aggregate of ice and soil occurring in very cold regions.

Permeability The capability of soil or rock to transmit water or air.

Pesticide A chemical used to kill pests.

Photochemical Smog A noxious, hazy mixture of aerosols and gases formed by sunlight acting on oxides of nitrogen and hydrocarbons, common pollutants in urban-industrial air.

Photosynthesis The process by which green plants transform light energy into food energy.

Phytoplankton Free-floating, mostly microscopic aquatic plants.

The Pill An oral contraceptive consisting of a female hormone and a synthetic substance that is chemically similar to another female hormone.

Pioneer Stage The initial stage in ecological succession. The organisms present in the pioneer stage are referred to as pioneer organisms.

Placer Deposits Heavy minerals found in sand and gravel layers in streambeds or in coastal areas.

Point Sources Discernible conduits, such as pipes, ditches, channels, sewers, tunnels, or vessels, from which pollutants are discharged.

Pollutant A material or form of energy whose rate of transfer between two components of the environment is changed so that the well-being of organisms is adversely affected.

Pollution A change in the concentration of a material or form of energy, or the introduction of a material or form of energy, that adversely affects the well-being of organisms.

Polychlorinated Biphenyls (PCBs) A family of persistent industrial chemicals that accumulate in food chains and have adverse effects on humans and other animals.

Population A group of individuals of the same species occupying the same geographical region at the same time.

Population Control All factors that regulate the size of a population.

Population Momentum The tendency of a growing population to continue to grow for some time even if each couple, on the average, just replaces itself.

Porosity The volume of open space in a soil or rock layer.

Positive Crankcase Ventilation A device that reduces automobile hydrocarbon emissions by channeling crankcase blow-by gases back through the engine.

Precipitation Liquid or solid forms of water falling to earth from clouds.

Predation An interaction in which one organism (the predator) kills and eats another organism (the prey).

Predator Control Programs Organized efforts to control or eliminate populations of predators that are believed to kill livestock or game animals.

Primary Air Quality Standards Maximum exposure levels of criteria pollutants in ambient air set by the

Environmental Protection Agency and based on potential health effects on humans.

Primary Air Pollutants Substances introduced into the atmosphere that, in sufficient concentrations, pose serious hazards to environmental quality.

Primary Succession Ecological succession that begins in an area where soil is not present.

Primary Treatment The first stage in sewage treatment, which relies on screening and the settling out of insoluble materials.

Producers Organisms, mainly green plants, that manufacture their food by utilizing raw materials from air and soil.

Production The accumulation of energy at a trophic level by means of growth and reproduction.

Public lands Lands held in trust by the federal government and subject to multiple use regulations, for example, national forests, national parks, and wildlife refuges.

Pyrolysis A conversion technique whereby organic substances such as garbage, wood, or coal are broken down by intense heat applied in the absence of oxygen. Products include oils and combustible gases.

Quarry A small-scale surface mine or pit used for the recovery of rock.

Radiation The process by which energy flows from place to place as oscillating waves.

Radioactive A property of some substances whereby highly energetic, dangerous rays are emitted.

Radio Waves Relatively long-wave electromagnetic radiation transmitted by radio and television stations.

Rank of Coal A stage in the sequence of changes from peat to lignite to bituminous coal to anthracite coal.

Reclamation Those activities that foster natural succession on land disturbed by mining.

Recycling The recovery and reuse of resources.

Refuse-Derived Fuel (RDF) Fuel obtained as a product of solid-waste conversion processes.

Relative Humidity The percent of water vapor present in air compared with the maximum amount of water vapor in saturated air at a specified temperature.

Replacement Rate The population growth rate at which parents have only the number of children necessary to replace themselves.

Reserves That portion of a rock, mineral, or fuel deposit that can be extracted immediately both legally and economically.

Resource Recovery Facility A solid waste recycling plant where reclamation and conversion processes are combined.

Resources Rock and mineral deposits existing in concentrations sufficient to make extraction economically feasible.

Respiration The liberation of energy from food within an organism.

Rhythm Method A method of birth control based on abstention from sexual intercourse during a woman's fertile period.

Rock Cycle The transformation of a rock into another type of rock as a consequence of change in the rock's environment.

Rodenticide A chemical directed at killing rats, mice, and other rodent pests.

Salinity The amount of salt dissolved in water, usually expressed as parts of salt per thousand parts of water.

Sanitary Landfill A landfill consisting of layers of compacted solid waste sealed between layers of clean earth.

Sanitary Sewer A network of pipes that collects and carries domestic wastes to a sewage treatment plant.

Saturation The condition of air, soil, or water that is holding the maximum amount of a particular substance.

Schistosomiasis A debilitating waterborne disease, common in less developed countries, caused by parasitic worms that live in water, primarily irrigation canals.

Scrubber An industrial air pollution control device that removes soluble gases from effluents by spraying water through the effluent stream or by bubbling waste gases through water.

Secondary Air Pollutants Products of reactions among primary air pollutants.

Secondary Air Quality Standards Maximum exposure levels of criteria pollutants in ambient air set by the Environmental Protection Agency, designed primarily to minimize damage to property and crops.

Secondary Succession Ecological succession that begins on an area in which a stress such as agriculture or fire has removed the natural vegetation but in which the soil remains.

Secondary Treatment The removal of water pollutants from sewage through the growth and harvesting of bacterial cells.

Sediment Fragments of rock, minerals, or organic remains usually deposited by running water.

Sedimentary Rock A rock formed by the compaction and cementation of particles (sediments) of abiotic or biotic origin.

Seismic Gap Regions of quiescence within a seismically active belt.

Sensible Heat Transfer The transport of heat energy from place to place as a result of conduction and convection.

Separated Sewer System A sewer system employing two pipes, one to transport surface runoff water and the other, sanitary wastes.

Shielding Layer A shallow layer of air at ground level in which air is continually mixed by convective currents.

Shock Disease The collective term for the deviant behavior and physical stress that develop in individuals under conditions of extreme crowding.

Sigmoid Growth Curve An S-shaped curve illustrating leveling off after exponential population growth.

Slash-and-Burn Agriculture A farming method used in the tropics consisting of several steps: the clearing of overgrowth on a several-hectare plot, the burning of residue, the planting of crops for several years, and the abandoning of the site to allow the forest to reinvade it.

Sludge Suspended materials removed from water treatment processes and collected as a thick, pasty ooze.

Social Hierarchy The pattern of dominant-subordination relations among members of a population.

Soil Horizon A layer of soil distinguished from layers above and below by characteristic physical properties and a distinct chemical composition.

Soil Salination An accumulation of salts at the surface of a soil in an arid, poorly drained region.

Solar Cell A cell that can convert sunlight directly into electricity.

Solar Collector A device that collects the sun's energy.

Solution Mining The removal of deep deposits of soluble minerals through the pumping of water into an injection well to dissolve the minerals, and retrieval of solution via extraction wells.

Specific Heat Capacity The quantity of heat required to raise the temperature of a unit mass of a substance 1 degree Celsius.

Spring Turnover A mixing process that occurs in spring in a stratified lake whereby surface waters mix with bottom waters.

Stable Air Layer An air layer in which mixing of air is inhibited.

Stationary Population A population that is not changing in size.

Sterilization A procedure, usually surgical, that renders an individual incapable of reproduction. Exposure to radiation and certain chemicals also may produce sterility.

Stratosphere The subdivision of the atmosphere that lies between the troposphere and an altitude of about 50 kilometers (30 miles).

Stream Channelization The straightening of meandering streams for the purpose of transporting water more rapidly downstream.

Strip Cropping The planting technique that reduces soil erosion through the alternation of a crop of low soil-anchoring ability with a crop of high soil-stabilizing ability.

Sublimation The process whereby a solid such as ice is transformed directly into the vapor state without going through the intervening liquid state.

Subsurface Mining The extraction of deep mineral and fuel deposits through a system of drill holes, shafts, tunnels, or rooms blasted into the rock.

Supercooled Water Water that remains in the liquid state when cooled below its normal freezing point.

Superheated Water Water that remains in the liquid state when heated above its normal boiling point because of pressurization.

Surface Mining The extraction of near-surface deposits of rock, minerals, or fuels after the removal of overburden.

Suspended Materials Substances that do not settle out of fluids because of their small sizes.

Synergism An interaction in which the total effect is greater than the sum of the effects taken independently.

Synoptic-Scale Weather Circulation systems that influence the weather at the continental or oceanic scale, for example, air masses and high- and low-pressure centers.

Temperature A measure of the molecular activity of a substance; the faster the molecular motion, the higher is the temperature.

Temperature Inversion A temperature profile in the atmosphere characterized by an increase of temperature with altitude.

Teratogen A chemical substance or physical agent (such as radiation) capable of causing a birth defect.

Territorial Behavior Any activity whereby an individual actively defends an area against intruders of the same species.

Tertiary Treatment Any of several water treatment techniques used beyond conventional secondary treatment to improve water quality further.

Thermal Shock Interruption of the normal functioning of an organism because of a sudden change in the temperature of its environment.

Thermocline The transition zone in a stratified lake between the upper warm layer and the lower cold layer, characterized by a rapid temperature decline as depth increases.

Thermosphere The uppermost subdivision of the atmosphere, characterized by a continuous increase of temperature with height.

Threatened Species A species that is still abundant in parts of its range, but whose continued existence is in question because of a decline in its numbers.

Threshold Dosage The minimum harmful dosage.

Tidal Energy The energy associated with the tidal motion in the oceans, some of which can be recovered by hydroelectric generating stations located on oceanic tidal basins.

Tolerance Limits The limits on an environmental factor within which an organism can survive.

Toxicity The measure of the ability of a substance to induce discomfort, illness, or death.

Toxic Substances Substances that cause serious illness or death in a one-time dose or in low doses administered over a long time period.

Transpiration The loss of water vapor from plants through the leaves.

Trench Landfill A sanitary waste disposal technique whereby refuse is spread into a trench, compacted, and sealed each day with the dirt excavated from the trench.

Trial and Error Learning Learning that takes place in response to prior mistakes.

Trophic Level The feeding position occupied by a given organism in a food chain, measured by the number of steps removed from the producers.

Troposphere The lowest subdivision of the atmosphere, and the site of most weather events.

Tsunami A giant sea wave triggered by an earthquake.

Tundra A treeless, circumpolar biome characterized by a permanently frozen subsoil.

Ultraviolet Radiation Radiation more energetic than the violet end of the visible spectrum and capable of breaking some chemical bonds.

Undiscovered Resources Rock, mineral, or fuel deposits that could exist although the specific location, grade, and quantity are unknown.

Unstable Air Layer An air layer in which rising air parcels continue to rise, resulting in mixing within the layer.

Upwelling The upward movement of cold bottom water in the sea, which occurs when winds or currents displace the lighter surface water.

Urban Heat Island A region of relatively warm air centered over an urban-industrial area.

Vector An organism that transmits parasites from one host to another.

Warm Front The leading edge of an air mass that is displacing colder air.

Waste Reduction Resource conservation strategies that involve reducing the amount of per capita refuse generated.

Waterborne Pathogen A disease organism transmitted by water.

Water Table The upper surface of the groundwater reservoir.

Watershed The geographical region drained by a river or stream.

Wavelength The distance between successive crests or troughs of waves such as electromagnetic waves.

Weathering The chemical decomposition and mechanical disintegration of rock.

Wet Steam Field A region where steam is emitted from the earth's interior along with substantial quantities of hot water.

Wild and Scenic Rivers Rivers that have particular aesthetic or recreational value or are still in a natural, free-flowing state.

Wilderness Areas Portions of national forests, national parks and wildlife refuges where timbering, most commercial activities, motor vehicles and human-made structures are prohibited.

Wildlife Refuges Protected habitats of wildlife.

Wind Power The energy in the wind that can be channeled to perform work.

X-Rays Electromagnetic radiation of short wavelength capable of penetrating human tissue and breaking chemical bonds.

Zero Population Growth A state in which no population growth is occurring; synonymous with stationary population.

Zone of Aeration The layer of soil or rock in which pore spaces are partially filled with both air and water.

Zone of Saturation The layer of soil or rock in which pore spaces are completely filled with water.

Zooplankton Weakly swimming, mostly microscopic aquatic animals found near the water surface.

Appendix I

Conversion Factors

Units of Length

1 meter (m) = 100 centimeters (cm) = 1000 millimeters (mm) = 10^6 micrometers (μm)

1 meter = 39.37 inches (in.) = 3.281 feet (ft) = 1.094 yards (yd)

1 kilometer (km) = 1000 meter (m) = 0.6214 mile (mi)

Units of Area

1 square meter = 10^4 square centimeters = 10.76 square feet = 1.196 square yards

1 square kilometer = 10^6 square meters = 100 hectares = 247.1 acres = 0.3861 square mile

1 hectare = 2.471 acres = 1.076×10^6 square feet

Units of Volume

1 liter (l) = 1000 cubic centimeter = 33.81 fluid ounces = 1.057 U.S. quarts

1 cubic meter = 1000 cubic liters = 35.31 cubic feet = 264.2 U.S. gallons = 6.290 barrels

1 cubic kilometer = 10^9 cubic meters = 0.2399 cubic miles = 8.106×10^5 acre-feet

Units of Mass

1 gram (g) = 1000 milligrams (mg)

1 kilogram (kg) = 1000 grams = 2.205 pounds (lbs)

1 metric ton (MT) = 1000 kg = 2205 lbs = 1.102 short tons

Units of Energy

1 kilocalorie (kcal) = 1000 calorie (cal)

1 joule (J) = 1 watt-second = 0.2390

1 kilojoule = 10^3 joules = 0.9484 BTUs (British thermal units) = 737.6 foot-pounds

1 kilowatt-hour (KWH) = 3600 kilojoules = 3413 BTUs = 860.4 kilocalories

1 quad = 10^{15} BTUs = 1.054×10^{15} kilojoules = energy in 172×10^6 barrels of oil

Units of Power

1 kilowatt (kw) = 1 kilojoule/second = 1.341 horsepower

1 langley/minute = 0.6974 kilowatt/square meter = 3.687 BTU/square foot/minute

1 megawatt = 10^3 kilowatts = 10^6 watts

Miscellaneous Conversions

Units of Flow

1 cubic foot/second (cfs) = 448.9 gallon/minute = 8.931×10^5 cubic meter/year = 724.0 acre-feet/year

Units of Temperature

degrees Celsius (°C) = degrees Kelvin (°K) −273.16

degrees Celsius = (5/9) ×(degrees Fahrenheit °F − 32)

The Geologic Time Scale

Era	When period began (millions of years ago)	Period	Animal life	Plant life	Major geologic events
Cenozoic	2	Quaternary	Rise of civilizations	Increase in number of herbs and grasses	Ice Age
	65	Tertiary	Appearance of first man dominance on land of mammals, birds, and insects	Dominance of land by flowering plants	
Mesozoic	135	Cretaceous		Dominance of land by conifers; first flowering plants appear	Building of the Rocky Mountains
	180	Jurassic	Age of dinosaurs		
	225	Triassic	First birds		
Paleozoic	275	Permian	Expansion of reptiles		Building of the Appalachian Mountains
	350	Carboniferous	Age of amphibians	Formation of great coal swamps	
	413	Devonian	Age of fishes		
	430	Silurian	Invasion of land by invertebrates	Invasion of land by primitive plants	
	500	Ordovician	Appearance of first vertebrates (fish)	Abundant marine algae	
	570	Cambrian	Abundant marine invertebrates	Appearance of primitive marine algae	
Precambrian		—	Primitive marine life		

Appendix III

Expressing Numbers as Powers of Ten

Scientific notation—that is, expressions of numbers as powers of 10—is a means of writing very large or very small numbers without using the long string of zeros that follows or precedes them in conventional notation. The number 4,000,000, for example, has six zeros; writing them out every time the number is used could become awkward. But since 4,000,000 is obtained by multiplying 4 by 10 six times $(4 \times 10 \times 10 \times 10 \times 10 \times 10 \times 10)$, in scientific notation this number can be written as 4×10^6. On the other hand, a smaller number, such as 0.00004, is obtained by dividing 4 by 10 five times:

$$\left(\frac{4}{10 \times 10 \times 10 \times 10 \times 10}\right).$$

In scientific notation, this number is written as 4×10^{-5}. A part of the system of scientific notation is shown below:

$$
\begin{aligned}
1,000,000 &= 10 \times 10 \times 10 \times 10 \times 10 \times 10 = 10^6 \\
100,000 &= 10 \times 10 \times 10 \times 10 \times 10 = 10^5 \\
10,000 &= 10 \times 10 \times 10 \times 10 = 10^4 \\
1,000 &= 10 \times 10 \times 10 = 10^3 \\
100 &= 10 \times 10 = 10^2 \\
10 &= 10 = 10^1 \\
1 &= 1 = 10^0 \\
0.1 &= 1/10 = 10^{-1} \\
0.01 &= 1/100 = 10^{-2} \\
0.001 &= 1/1,000 = 10^{-3} \\
0.0001 &= 1/10,000 = 10^{-4} \\
0.00001 &= 1/100,000 = 10^{-5}
\end{aligned}
$$

Index

Alabama, 428
Alaska, 121, 423, 439
  earthquake in, 456
  oil and natural gas, 543, 546, 549,
    551, 554
Alaska pipeline, 21, 107, 110, 113
Albedo, 262
Alcan gas pipeline, 543
Alcohol, as a fuel, 332, 594
Aldrin, 194, 428, 498
Alfalfa, 497, 501, 514, 516
Algae. *See also* Blue-green algae; Kelp
  adaptations of, 136, 139
  in eutrophication, 134, 229
  in food chains, 13, 137, 141
  nutrients and, 44–46, 132, 191
  population dynamics of, 80
  in water supplies, 223
Algeria, 118
Alien species, 84–85, 417–418, 428
Allen, J., 196
Alligator, 410, 428
Alpha particles, 557
Altitudinal gradients, in biomes, 127
Alum, in water treatment, 229, 233
Aluminum, 347, 600
  recycling of, 395–396
Amazon Basin, 126, 502–504
American Can Company, 399
American Industrial Health Council
  (AIHC), 327
American River, 454
Ames, Iowa, 398
Amino acids, 521
Ammonia, 199, 514, 600, 602
  nitrogen cycle and, 31–33, 36, 189,
    191
Amoco Cadiz, 199–201
Anaerobic decay, 188–189, 315, 316,
  355, 383, 388–389, 393
Angina pectoris, 292–293
Animal wastes (manure), 58, 514, 516,
  520, 594, 595
Animals. *See also by type*
  air pollution effects on, 65–66, 295
  in biomes. *See by type*
  in food chains, 7–23
  husbandry of, 20, 519–520, 522
  in nutrient cycles, 25–36
  oil effects on, 201–202
  pesticide effects on, 496–497
  as test organisms, 50, 196–198, 409
Annual growth rate, 473, 476
Antagonism, 49
Anthracite, 356, 375, 550
Antibiotics, 409, 471
Appalachia, 361, 363, 551
Appalachian Trail, 444
Aquaculture, 205, 524
Aquatic ecosystems, 130–144. *See also
  by type*
Aqueducts, 152–153
Aquifers, 166–167, 168
Arab oil embargo, 444, 538, 541, 543

Aransas National Wildlife Refuge, 425
Arboretums, 428
Arctic ecosystems, 98, 107–110, 113
Argentina, 34
Arithmetic growth, 471
Armadillo, 409
Arroyos, 169
Arsenic, 194
Artesian well, 166
Asbestos, 51, 294, 324, 326
Asbestosis, 294
Ash, fly, 551, 572–573
Asia. *See also countries by name*
  biomes of, 124
  food consumption in, 19
  population growth of, 478, 480, 487
Atmosphere. *See also* Air
  circulation of, 269
  composition and structure, 257–265
  cycles and, 25, 26, 30–35
  dynamism of, 261–274
  pollution of, 286–318
Atomic Energy Act, 226
Attwater's Prairie Chicken, 61
Auburn Dam, 452, 454
Audubon Society, 624
Australia
  agriculture in, 504
  biomes of, 124
  loss of wildlife, 417
  population growth in, 476
  rabbits in, 85–86
Austria, 476
Automobiles. *See* Motor vehicles

Bacteria. *See also* Aerobic decay; An-
  aerobic decay; Decomposers
  decomposition by, 8, 9, 10, 13
  in nutrient cycles, 25, 28, 31, 134,
    516
Bald Eagle, 416, 421
Bangladesh, 477, 480, 486, 487
Barges, 200–237
Bedrock, 350, 352, 353
Beef, 16, 432
  consumption of, 13
  energy subsidy to, 20–21, 519–520,
    522
Belgium, 476
Benzene, 315, 326
Benzo(a)pyrene, 301, 315
Beryllium, 324
Beta particle, 557
Bethlehem Steel Corporation, 339
Bicarbonate, 183
Big Thompson Canyon, Colorado, 451
Biochemical oxygen demand, 188, 191,
  229, 232, 336
Bioconversion, 332, 396–397
Biofuels, 593–595, 610–612
Biological control, 81, 498–499, 500
Biological nitrogen fixation, 31, 516
Biomes, 107. *See also by type*

Birds. *See also by name*
  DDT in, 194, 416, 428
  as endangered species, 415, 417, 420
  in food chains, 141, 144
  pollution effect on, 201
  population dynamics of, 78, 85, 87,
    90–91
Birth control methods, 481–485. *See
  also by type*
Birth defects, 47, 51
Birth rate
  family planning and, 480–481, 487
  infant mortality and, 480
  in population dynamics, 78, 79, 473,
    475, 477, 478, 480, 486, 490
Bituminous coal, 355–356, 550
Blue-green algae, 31, 134, 189, 516
Boreal forest. *See* Northern coniferous
  forest
Borlaug, N., 527
Bormann, F., 62
Boron, 602
Boston, 460
Bottle laws, 395
Boulder, Colorado, 450
Brazil, 126. *See also* Amazon Basin
Breakwater, 447
Breeder reactor, 579–582
British Thermal Unit (BTU), 272
Brookhaven National Laboratory, 49, 53, 54
Broom sedge grass, 93–94
Brown, L., 527
Brown pelican, 416, 428
Bubonic plague, 469–470
Bureau of Land Management (BLM),
  443, 454
Bureau of Reclamation, 454
Butterfly, 67
Byssinosis, 294

Cacti, 117, 120, 420, 426
Cadmium, 51, 194, 301
Calcium, in water, 61, 183, 208, 233
Calcium carbonate, 107. *See also* Limestone
Calcium hydroxide, 229. *See also* Lime
Calhoun, J. B., 90
California. *See also specific cities*
  agriculture in, 81, 410, 510, 512
  air pollution in, 323
  biomes of, 121, 123
  geothermal resources, 602–603
  land subsidence in, 241
  solar energy and, 611
  solid waste management, 17, 387
  water pollution in, 141, 191–192, 210
  water supply in, 151, 244, 245–246
Calorie, 271–272
Campbell Soup Company, 395
Canada, 24, 53, 423, 514
Cancer
  causative factors, 51
  deaths from, 47
  and the pill, 482–483
  radiation and, 555–557

DDT, 62, 81
  in food chains, 192, 194, 196, 236,
    416
  malaria and, 251
  regulation of, 236-237, 428, 498
Death rate
  and abortion, 483
  in Americans, 47
  from cancer, 47
  fuel cycle and, 568-573
  and population dynamics, 78, 79,
    473, 477, 478, 481, 482, 483, 484
  risk of, 569
Decomposers, 8, 10, 11, 13, 55, 131,
  134, 187-189, 226-229, 502, 594.
  See also Anaerobic decay
Decomposition. See Aerobic decay; An-
  aerobic decay
Deep-well injection, 391-392
Deer, 78, 82
Deforestation, 19, 60-61, 118, 126, 432,
  503-504
Delta, 138-139, 447
Demography. See Birth rate; Death rate;
  Population
Demographic transition, 476-477
Denitrification, 33
Density dependent competition, 84,
  87-88, 89, 92-93
Deposition, 156-157
Desalination, 176, 245
Desertification, 117, 118-119, 496
Deserts, 117-121, 124. See also by name
Detergents, 36
Detritus, 132, 136, 137, 141
Detritus food webs, 11, 13, 137
Detroit River, 219
Deuterium, 582
Developed countries. See also countries
    by name
  migration and, 477-478, 479
  population growth in, 476-477
Development, economic, population
    growth and, 476-477, 478, 480
Diaphragm, 483
Dieldrin, 194, 428, 498
Diesel engines, 536, 549, 567
Dioxin, 290-291, 384
Discharge, of rivers, 168-170
Diseases, 47, 50-51, 77. See also by
    name
  from air pollution, 50-51, 288-290,
    292-295, 570
  deficiency, 520-521
  and population growth, 469-471
  from radiation exposure, 50-51,
    555-556
  from trace metals, 193, 195
  waterborne, 182, 186-187, 210, 231,
    250-251
Disinfection. See Chlorination
Dismal Swamp, 136, 355
Dissolved oxygen, 130-131, 132, 136,
  187-188, 191, 233

and temperature, 131
Distillation, 156
Diversity, 107, 111, 112, 125-126, 128,
  133-134, 188, 190
  radiation and, 51-53
  stability and, 57-60, 410
Dodo bird, 417
Dolomite, 25, 183, 208
Donora, Pennsylvania, 287, 312-318
Doubling time, 473
Down's syndrome, 556
Drainage basin, 167-169, 182
Dredge spoils, 188, 207, 392-394
Dredging, 138, 392, 393
Drought, 116, 118, 151, 155, 242, 248,
  495-496, 510-511
Drugs, 409
Dry ice, 244
Dubos, René, 411, 413
Duluth, Minnesota, 385-386
Dust bowl, 116, 118, 207
Dust dome, 300
Dust plume, 303
Dutch elm disease, 85
Dysentery, 182, 186-187, 210

Earth
  internal structure, 347
  landscape evolution, 347-350
Earthquakes, 454-457, 458, 477, 571
  prediction of, 456
  risk of, 458
Ecological efficiency, 3-23, 522
Ecological equivalents, 124-125
Ecosystems. See also by specific type
  definition of, 7
  stability of, 23-24, 57-58, 60
Ecotones, 127-128, 136-138
  and climatic change, 128, 281, 495-
    496, 510
Ecuador, 478
Edge effect, 128
Education and population, 480
Effluents. See Air pollution; Water pol-
    lution
Egypt, 99, 152, 250
Eisenhower, Dwight, 448
Electric power
  from biofuels, 593-595
  from geothermal sources, 601-603
  generation of, 536, 542, 549, 550,
    559-567, 567-568, 579-582, 589-
    593, 596-598, 599-603, 610
  historical uses of, 534
  from ocean thermal conversion, 600
  from tides, 600-601
  water use for, 202-206
  from wind, 595-598
Electromagnetic radiation, 261, 264-265
Electromagnetic spectrum, 262, 264-265
Electrostatic precipitation, 305, 328
Emissions standards. See Air pollution;
    Water pollution
Emphysema, 294

Endangered species, 61, 405-432. See
    also by name
  management of, 422-423, 426-428
Endangered Species Act, 428-429, 431
Energy. See also specific forms
  agriculture and, 515, 516-520
  alternate sources, 579-603
  conservation of, 538, 542, 603-613
  consumption and resources of, 538,
    540-541, 541-554, 567
  efficiency of use, 568
  historical use of, 533-538
  in mining, 367, 370
  policy for, 538, 567-573
  recovery of, 206
  recycling and, 371
  subsidies, 20, 137-138, 518-520
  thermodynamics, laws of, 14-15
  uses of, 539-541
England, 65-66, 481, 482
Entropy, 15
Environmental Protection Agency
  on air quality, 324-325, 550, 570
  on pesticides, 194, 501
  on water quality, 186, 217, 219, 223,
    224-225, 232, 233
Epidemics, 469-470, 477
Epilimnion, 131-132
Epiphytes, 94, 122, 126
Eric the Red, 281
Erie Canal, 218
Erosion. See Sediment; Soil erosion
Estuaries, 136-138, 143, 207, 447, 526
Euphotic zone, 131, 137, 140, 141
Europe. See also countries by name
  biomes of, 124
  fertilizer use in, 515
  population of, 469, 476
Eutrophication, 133-134, 189-192, 229,
  242, 455. See also Cultural
    eutrophication
Evaporation, 156-157, 162, 205, 245
Everglades National Park, 136, 405,
  410, 415
Evolution. See Natural selection
Exponential growth, 471-472
Extinction, 97, 405, 410, 411-413, 417-
  418

Fall turnover, 132
Family planning, 480-481, 487
Famine, 469, 471, 477, 488
  potato, 23
  in Sahel, 495-496
Faults, 348-349, 350, 456
  San Andreas, 350, 352
Federal Energy Administration, 567
Federal Flood Disaster Protection Act
  of 1973, 454
Federal Insurance Administration, 450
Federal Land Policy and Management
  Act of 1976, 443
Feed lots, 188
Ferric chloride, 229

Fertility, human, 474, 480
Fertilizers, 197, 514–516, 567. *See also*
    Animal wastes
    and energy consumption, 515
    high-yield crop varieties, 508
    inorganic and organic compared, 514
    in less developed countries, 514–516
    nutrient cycles and, 191
    from sewage sludge, 228–229
    supply of, 514, 595
    use of, 58, 60, 515
Fetus, human, 556
Filters, 328, 329
Fire, 57, 63, 98, 115, 123, 126, 158, 415
Fire retardant chemicals. *See* PBBs;
    PCBs
Firewood, 63, 595
Fish, 132, 134, 136, 144, 187, 188,
    203–205, 207, 242. *See also by type*
    pollution and, 196–199, 220–221,
        236–237, 238, 526
    production of, 137, 140, 141, 189,
        193, 195, 196, 201, 524–526
Fish kills, by acid rains, 304–305
Fission. *See* Reactors, nuclear
Floc, 227, 233
Fleming, A., 471
Flood, 450–454, 455
    causes of, 451–452
    flash, 450, 452, 453
    hundred year, 450
    structural control of, 452–454, 455
Floodplains, 170–171
    habitation of, 450
    management of, 452, 454
Floodway, 454
Florida, 205, 237, 361, 454–455
Fluoride, 199, 295, 298
Fluorosis, 295
Folding, of rock, 347, 348, 350
Food and Agriculture Organization of
    the United Nations (FAO), 249,
    505, 506, 512, 514, 520
Food and Drug Administration, 193,
    198, 482–485
Food chains, 10–24, 57–58, 134, 410
    accumulation in, 192–199, 228, 393,
        416, 558–559
Food supply. *See* Agriculture; Crops;
    and by name
Ford, Gerald, 374
Forest, 206. *See also* Deforestation; *and
    by type*
    biofuels from, 594
    logging, 122, 442, 443, 452, 460
    tree farming, 57, 417
Fossil fuels. *See also* Coal; Energy; Oil;
    Natural gas
    agriculture and, 516–520
    carbon dioxide and, 36
    generation of, 25, 354–356
    types of, 354–356, 541–554
Four Corners power plant, 323
Fox River, Wisconsin, 218

Fracturing. *See also* Faults
    of rock, 347, 348
France, 200, 566, 582, 600
Franconia Notch State Park, New
    Hampshire, 460
Freeways, 448–449
Freons. *See* Halocarbons
Frequency, 264
Fresh water. *See* Water
Fruits, nuts, and vegetables, 504, 518–
    520
Fuel. *See* Energy; Fossil fuels; Nuclear
    energy
Fuel economy, 395, 605–606
Fungi, 8, 24, 85, 121, 187, 409, 471,
    502
Fungicides, 24, 498
Fusion. *See* Reactors, nuclear

Galapagos Islands, 417
Galveston, Texas, 335
Gamma radiation, 265, 557
Gandhi, Indira, 481
Gases. *See by name*
Gasohol, 332, 594
Gasoline. *See also* Fuel economy
    unleaded, 317
Gene pool, 408–410, 428, 504
General Electric Company, 198
Generation time, 79–80
Genetic variability, 24, 409, 428, 504
Geologic hazards, 450–458
    earthquakes, 454–457, 458
    floods, 450–454, 455
    landslides, 123, 457–458
    permafrost, 113
Geologic time, 347, 348, 643
Geometric growth, 471–472
Georgia, 428
Geothermal electricity generation, 601–
    603, 611
Geysers, The, 602
Ghana, 476, 496
Glaciation, 174–175, 347, 411–413
Glacier National Park, 422–423
Glaciers, 155, 162, 174–175
Glaser, P., 592
Glass, 396, 589
Glucose, 10, 11, 17
Gold, 199, 358, 360
Grains, *See crops by name*
Granite, 183, 350, 564
Grasslands, 114–117, 124, 416
Gravitational settling, 313, 328
Gray's Lake National Wildlife Refuge,
    425
Grazing food webs, 11, 13, 141
Great Auk, 418
Great Lakes, 172, 218, 229, 447. *See
    also by specific name*
Greece, 155, 409
Green Revolution, 508, 509, 578, 526
Greenhouse effect, 266–267, 305–306
Greenland, 281

Grizzly bear, 421–423, 430–431
Ground subsidence
    by groundwater withdrawal, 240–241
    by mining, 364–365
    by permafrost melting, 113
Groundwater, 155, 162–167, 170
    contamination of, 168, 192, 208–212,
        288, 392
    geothermal energy and, 601–603
    minerals deposited by, 354, 356
    recharge of, 162, 239–242
    water supply from, 232–233, 235
Growth control ordinance, 450
Guatemala, 478

Haber Process, 36, 514
Habitat island, 418
Haddock, 525
Half-life, 557–559
Halocarbons, 231, 299–300
    ban on, 299, 306
Hawaii, 24, 85–86, 160, 348, 418, 428,
    612
Heat of fusion, 272, 273
Heat island, 300, 302
Heat pump, 206, 608
Heat of vaporization, 272, 273
Heath hen, 418
Helium, 582
Hepatitis, infectious, 182, 186, 187, 210
Heptachlor, 194, 498
Herbicides, 60–61, 416
Herbivores, 7–23 *passim. See also by
    name*
Herring, 141, 525
Hessian fly, 500
Hibernation, 64
High pressure system, 259, 269
    and air pollution potential, 308, 310,
        312
Highway Act of 1973, 449
Honeybees, 496–497
Hooker Chemical Company, 384
Houston, 335, 479
Hubbard Brook Forest, 60–61, 62
Hubbert, M. King, 554
Hudson River, 198
Human wastes, 188
Humidity, relative, 158, 205
Humus, 35, 55, 106, 125, 502
Hurricane Agnes, 452
Hydrocarbons, 301, 315, 324, 329, 331,
    334, 335, 336. *See also*
    Chlorinated hydrocarbons
Hydrogen, 543
Hydrogen sulfide, 189, 293, 316, 388,
    602
Hydrologic budget, 160–162
Hydrologic cycle, 150–177
Hydropower, 242, 250, 536, 599–601,
    611
Hygroscopic nuclei, 158
Hypolimnion, 131–132, 134, 188, 204–
    205

Macrophages, 290, 292
Magma, 348, 350, 354
Magnesium, 61, 183, 208, 233, 347, 348
Magnesium chloride, 158
Magnetite, 353
Maine, 90, 395
Malaria, 84, 251, 471
Malaysia, 480
Mali, 495
Malnutrition, 488, 520, 521, 522
Manganese, 209
Manganese nodules, 370-371
Manure. *See* Animal wastes
Marble, 352
Marine ecosystems, 139-144
Marine Protection, Research, and Sanctuaries Act of 1972, 226, 393
Marshes, 11, 134-136, 166, 207, 234-235, 416
Martin, P. S., 412
Mass transit. *See* Transportation
Massachusetts, 387, 418
Meat. *See specific types*
Meat extender, 521
Meat substitute, 521
Mercaptans, 316
Mercury, 51, 193, 324
Mercury chloride, 195
Mercury, methyl, 192, 195
Mesosphere, 261
Metals, toxic, 51, 62, 193-194, 228-229, 231
Methane, 189, 228, 315, 356, 543, 594, 610
  from landfills, 388-389
Methemoglobinemia, 210
Mexico
  biomes of, 118
  food production in, 508
  geothermal, 603
  groundwater in, 209, 241
  population of, 468, 473, 476, 478, 486
  water pollution in, 209, 244
Miami, 479
Michigan, 196-197, 219, 232, 395, 405, 415
Microorganisms. *See* Algae; Bacteria
Microwaves, 265, 592
Middle East, 504. *See also countries by name*
Midnight dumpers, 385
Migration, 78, 477-478, 479, 496
Milwaukee, Wisconsin, 217, 228, 399
Mimicry, 67
Mindanao, Philippines, 456
Mine tailings. *See* Waste
Mineral Leasing Act of 1920, 346
Minerals, 353. *See also* Reserves; Resources
  conservation of, 370-374, 376
  critical, 366-367
  exploration for, 367-369
  mining of, 358-365
  from seabed, 370-371

shortage of, 366-367
  U.S. imports, 366, 368
  U.S. production versus consumption, 366-367
  U.S. reserves of, 366-367
Minimata disease, 193, 195
Minimum tillage, 512, 514, 520
Mining, 358-365, 551
  and agriculture, 374
  auger, 361
  dredge, 358, 360
  and energy, 367, 534
  hydraulic, 358, 360
  impact of, 211, 358-365, 452
  legislation, 345, 346, 347, 374
  of low-grade deposits, 366-367
  open pit, 344, 358
  on public lands, 369, 444, 446
  quarry, 358, 359
  reclamation, 374-375
  solution, 363
  subsurface, 363-365
  strip, 358-359, 361, 362
  surface, 358-363
  waste from, 353-354, 363, 382, 383
Mining Law of 1872, 345
Minnesota, 82, 192, 224, 385-386, 426
Mississippi, 428
Mississippi River, 198, 534
Mississippi sandhill crane, 97
Missouri River, 218
Moment, G., 430
Monarch butterfly, 67
Mongoose, 499
Monoculture, 23-24, 58, 60, 63, 501
Monsanto Chemical Company, 199, 236
Monsoon rains, 307, 495, 496
Montana, 422-423, 549, 552-553
Morocco, 514
Mortality. *See* Death rate
Mosquito, 84, 85, 86, 250-251, 431
Motor vehicles
  and air pollution, 313, 314
  emissions control devices, 330-333
  emissions standards, 324
  emissions trends, 336
  energy use by, 534, 603-606
Mountain building, 348, 349
Mountain View, California, 389
Muir, J., 411
Multiple Use-Sustained Yield Act of 1960, 442
Muskrats, 98
Mutagen, 47, 51, 556
Mutation, 24
Myxomatosis, 84-85

Natality. *See* Birth rate
National Academy of Sciences, 501, 507
National Aeronautics and Space Administration, 597
National Cancer Institute, 51
National Center for Resource Recovery, 398

National Commission on Materials Policy, 347
National Conservation Commission, 440
National Energy Act, 538
National Environmental Policy Act of 1969, 2
National Forest Management Act of 1976, 442
National forests, 440, 442-443
National Institute of Occupational Safety and Health, 51, 325-326
National Park Service, 444
National parks, 444. *See also by name*
National Recreation Trails, 444
National Research Council, 223, 233, 375
National resource lands, 443-444
National System of Interstate and Defense Highways, 448
National Technical Advisory Committee, 203-204
National Trails System, 444
National Wild and Scenic Rivers System, 444, 446
National Wilderness Preservation System, 444
National Wildlife Federation, 624
National wildlife refuges, 426, 444. *See also by name*
Natural gas. *See also* Pipelines
  consumption and resources of, 541-543, 610-611
  extraction of, 363
  origin of, 356
  transport of, 536
  uses of, 514, 534, 536, 539, 567-568
Natural selection, 64-66, 98, 497
Nebraska, 240, 375
Necrosis, 296
Neritic zone, 139-141
Netherlands, The, 596
Neutralization of acids, 328, 394
Neutrons, 555, 580
Nevada, 117, 240
New Hampshire, 60-61, 460
New Jersey, 164-165
New Mexico, 118, 245, 323
New Orleans, 223, 450
New York City, 181, 479, 533, 534
  air pollution in, 334-335
  waste disposal in, 385, 386
New York State, 49, 192, 198, 241, 244, 384, 565, 571
New Zealand, 417, 476, 603
Ngana, 89
Niagara Falls, New York, 384
Nickel, 51, 62, 194
Niger, 495, 595
Nigeria, 496
Nile River, 139, 152
Nitrate, 31, 33, 34, 210
  in water, 33, 36, 60-61, 189, 191-192
Nitric acid, 304, 316
Nitric oxide, 315-316